SOURCE BOOK ON ELECTRON BEAM AND LASER WELDING

Other Source Books in This Series:

Brazing and Brazing Technology
Gear Design, Technology and Performance
Copper and Copper Alloys
Materials for Elevated-Temperature Applications
Maraging Steels
Powder Metallurgy
Wear Control Technology
Selection and Fabrication of Aluminum Alloys
Industrial Alloy and Engineering Data
Materials Selection (Volumes I and II)
Nitriding
Ductile Iron
Forming of Steel Sheet
Stainless Steels
Heat Treating (Volumes I and II)
Cold Forming
Failure Analysis

Metals Science Source Book Series:

The Metallurgical Evolution of Stainless Steels
Hydrogen Damage

SOURCE BOOK ON ELECTRON BEAM AND LASER WELDING

A comprehensive collection of outstanding articles from the periodical and reference literature

Compiled by
Consulting Editor

Melvin M. Schwartz

Manager, Manufacturing Technology
Manufacturing Planning/Engineering
Sikorsky Aircraft
Division of United Technologies

American Society for Metals
Metals Park, Ohio 44073

ISBN: 0-87170-104-9

PRINTED IN THE UNITED STATES OF AMERICA

Contributors to This Source Book

M. J. ADAMS
The Welding Institute

YOSHIAKI ARATA
Osaka University, Japan

CONRAD M. BANAS
United Technologies Research Center

F. L. BECKER
Battelle
Pacific Northwest Laboratories

F. J. BECKET
Electron Beam Processes Ltd.

LEE BENSON
Holobeam, Inc.

VICTOR E. BOCCELLI
Re-Entry and Environmental Systems Div., General Electric Co.

N. F. BRATKOVICH
Allison Div., General Motors Corp.

EDWARD M. BREINAN
United Technologies Research Center

K. W. BROWN
The Welding Institute

HUGH CASEY
University of California
Los Alamos Scientific Laboratory

J. N. CLARK
CEGB Marchwood Engineering Laboratory

S. R. DOCTOR
Battelle
Pacific Northwest Laboratories

P. DUMONTE
Sciaky-France

H. P. ELLISON
Grumman Aerospace Corp.

SIMON L. ENGEL
GTE Sylvania Inc.

J. H. FINK
Westinghouse Research Laboratories

FRED FLICK
University of California
Los Alamos Scientific Laboratory

W. R. FREEMAN
Lycoming Div., AVCO Corp.

LEO GOLIN
Chrysler Corp.

M. A. GREENFIELD
Air Force Wright Aeronautical Laboratories

SHŌJI HARADA
Osaka University, Japan

Y. HAYAKAWA
Mitsubishi Heavy Industries, Ltd.

K. HIGUCHI
Mitsubishi Heavy Industries, Ltd.

ROBERT R. IRVING
Iron Age

TED B. JEFFERSON
Welding Design & Fabrication

R. L. JONES
Central Dockyard Laboratory
Portsmouth, U.K.

H. N. KRISHNASWAMY
Re-Entry and Environmental Systems Div., General Electric Co.

R. J. LANYI
Industrial Equipment Div.
Westinghouse Electric Corp.

JOHN LEE
Chrysler Corp.

HENRY LEFER
Welding Design & Fabrication

J. F. LOWRY
Westinghouse Research Laboratories

W. P. McGREGOR
Solar Div.
International Harvester Co.

W. L. McINTIRE
Allison Div., General Motors Corp.

J. G. MACIORA
Grumman Aerospace Corp.

H. MARQUEZ
Solar Div.
International Harvester Co.

FUKUHISA MATSUDA
Osaka University, Japan

J. H. P. C. MEGAW
UKAEA Culham Laboratory

E. A. METZBOWER
Material Science and Technology Div.
Naval Research Laboratory

D. W. MOON
Material Science and Technology Div.
Naval Research Laboratory

DENNIS MULVANEY
Holobeam, Inc.

HIROYOSHI NAGAI
Kawasaki Heavy Industries Ltd.

NOTE: Affiliations given were applicable at date of contribution.

T. NAKAMURA
Dengensha Manufacturing Co.

K. NODA
Mitsubishi Heavy Industries, Ltd.

JOSEPH NUSSENBAUM
Holobeam, Inc.

MARIO L. OCHIEANO
Lockheed-California Co.

M. OHSUMI
Mitsubishi Heavy Industries, Ltd.

CARLOS PAEZ
Grumman Aerospace Corp.

ROBERT E. PURDY
Allison Div.
General Motors Corp.

HENRY J. RIDDER
Magnaflux Corp.

A. J. RODGERS
The Welding Institute

J. D. RUSSELL
The Welding Institute

A. SANDERSON
The Welding Institute

G. SAYEGH
Sciaky-France

GEORGE SCHAFFER
American Machinist

DUANE J. SCHMATZ
Ford Motor Co.

B. W. SCHUMACHER
Westinghouse Research Laboratories

JOHN R. SHEWELL
Lasertool Consultants

S. M. SILVERSTEIN
Lycoming Div., AVCO Corp.

WILLIAM H. SPAULDING
AiResearch Manufacturing Co.

R. J. STEARN
The Welding Institute

STANLEY J. STEFANSKI
Sundstrand Aviation Corp.

V. STRAUTMAN
Lycoming Div., AVCO Corp.

THOMAS TAGLARINE
Grumman Aerospace Corp.

A. N. TAYLOR
The Welding Institute

KIYOHIDE TERAI
Kawasaki Heavy Industries Ltd.

VICTOR VIGIL
University of California
Los Alamos Scientific Laboratory

RICHARD G. WILCOX
Wallace Barnes Steel

R. A. WILLGOSS
CEGB Marchwood Engineering Laboratory

ROBERT H. WITT
Grumman Aerospace Corp.

WALTON R. YERGER
Wallace Barnes Steel

MICHAEL YESSICK
Ford Motor Co.

CONTENTS

PREFACE

Ralph Waldo Emerson, the eminent essayist, transcendentalist, and poet, who never dabbled even casually in the sciences, is responsible for a statement — a truism, in fact — that implies much more than a man of his parochial interests probably intended. "Invention," he wrote, "breeds invention." It was a line more becoming the notebook of an Edison or a Cyrus McCormick. Yet there remained unsaid an even more discerning corollary — namely, that as breeders, some inventions are far more prolific than others. Electron beam and laser technologies are among the prolific breeders, and it was inevitable that both would find application in joining processes.

The wide interest in electron beam and laser welding for manufacturing applications does not rest merely on the curiosity accorded many new processes. It persists and grows because of a very basic interest in high productivity *combined* with high quality. With EBW, for example, deep-penetration welds can be obtained with a simple, square butt preparation; the fusion zone is only 5 to 10% of that produced by arc welding; heat input is low but power density is high, so that degradation of properties is minimal; deformation and distortion are extremely low; and the process is suitable for joining virtually all metals in similar or dissimilar combinations. Laser welding, although generally not as versatile, has its own impressive list of advantages.

SOURCE BOOK ON ELECTRON BEAM AND LASER WELDING offers a broad selection of articles covering all important aspects of both welding processes. The book is divided into two parts, containing a total of nine principal sections. Its scope and contents are briefly reviewed in the following summary.

Part A: Electron Beam Welding

Processes and Equipment. A series of four articles in this section provides both a general and a specialized discussion of EBW processes and equipment. The range is of particular value to the engineer or production manager whose familiarity with EBW is slight, covering high, intermediate, and low beam power equipment and partial vacuum and nonvacuum processing, both of which are suitable for high production requirements. A mobile vacuum chamber is also considered.

Design. An article by F. J. Becket explains how EBW offers the designer a new concept in the fabrication and joining of metals and gives specific guidance on butt and lap joints, the fabrication of angle, tee, channel, box, and "H" sections and flange and shaft joints. Insight into the use of EBW for aircraft applications is provided by S. J. Stefanski in the second article in the section.

Materials Welded. As stated previously, the EBW process is suitable for joining virtually all metals in similar or dissimilar combinations. The nine articles in this section, therefore, are merely indicative of the versatility of the process in joining both common and relatively uncommon metals and alloys. Much attention is given to cleaning prior to welding and the avoidance of weld porosity.

Applications. Predictably, much of the literature dealing with EBW applications concentrates to some extent on feasibility studies: many industries, including the automotive, aerospace, nuclear power, and oil pipeline industries, are intent upon mastering the techniques of a new process and benefiting from its many advantages. Many of these areas of interest are covered in the eight articles featured in this section.

Quality and Process Control. The EBW process is generally selected for precision assemblies requiring high-quality welds. As emphasized in the article by Wilcox and Yerger, a variety of factors affect quality, including raw material inspection and preparation, welding techniques and equipment, personnel, postweld thermal and mechanical processes, and preliminary and final weld inspections. Acting in concert, these factors determine weld precision, weld soundness, and mechanical and metallurgical properties. Other quality and process control techniques, including in-process monitoring and the use of a microcomputer-based data acquisition system, are also discussed in this section.

Part B: Laser Welding

Introduction. Over the years, the U.S. Air Force Materials Laboratory has commissioned state-of-the-art reviews of the technical literature pertaining to significant innovations in materials technology. One such review of laser welding covers the developments in laser systems from their inception in 1960; it is the first article in this section and encompasses all the highlights of laser technology, with particular emphasis on welding, in an extensive and noteworthy report. In contrast, the second article in the section is a terse but meritorious treatment of a few typical case histories in the industrial application of laser welding.

Design and Tooling. The opening article, by J. R. Shewell, after explaining the specific advantages of laser welding, provides details on gas shielding, joint preparation and fit-up, beam size and absorption, welding temperatures, weld penetration and welding speed, and the selection of welding parameters. Most of these processing details are elaborated upon in the two articles that follow, the concluding article concentrating on the role of lasers in spotwelding.

Applications. The applications considered in this section, although limited in number, cover both deep- and shallow-penetration welding — categories into which all applications inevitably fall. Applications range from power plant welding at one extreme to the hermetic sealing of flatpack packages for microcircuits at the other.

Properties of Welds. Although improved mechanical and metallurgical properties have been obtained in many significant applications, the assumption of improvement in virtually all applications with all metals cannot be generalized, as is demonstrated in the review of mechanical properties and fracture toughnesses of laser welds of high-strength alloys by Metzbower and Moon of the Naval Research Laboratory. In the final article in the book, Breinan and Banas discuss the interesting phenomenon of fusion-zone purification during welding with high-power CO_2 lasers — a very significant metallurgical advantage.

For his expertise and technical judgment in selecting the articles in this book, the American Society for Metals extends its most sincere appreciation to Mr. Melvin M. Schwartz, Consulting Editor, who has contributed previously to this series. His intimate knowledge of joining processes has again proved invaluable. Most grateful acknowledgment is extended to the many authors whose work appears in this book and to their publishers.

PAUL M. UNTERWEISER
Staff Editor
Assistant Director of Reference Publications
American Society for Metals

WILLIAM H. CUBBERLY
Director of Reference Publications
American Society for Metals

Cover photos, which are pantographic patterns that illustrate the superlative degree of precise beam control obtainable with electron beam welding units, are courtesy of J. E. Jenkins, Senior Metallurgist, Research and Development, Leybold-Heraus Vacuum Systems, Inc.

PART A:
Electron Beam Welding

The application of electron beam welding Part 1 – Equipment

A general review based on an IIW paper, by **Kiyohide Terai*** and **Hiroyoshi Nagai***, from a Japanese company for which The Welding Institute built an EB welder

High productivity in the welding of steels is attained by the use of high speed welding processes, typically in conjunction with high welding currents. In general, this approach can be used only with low strength steels unless special precautions are taken to avoid reductions in the strength or toughness of the weld or heat affected zone (HAZ). Increases in the service performance of structures, however, necessitate the use of higher strength steels with specific compositional or heat treatment requirements which, in turn, may limit the heat input used in the welding operation.

In the majority of cases, a reduction in joint properties cannot be avoided under high speed welding conditions, and hence a reduction in productivity or efficiency occurs. For instance, when welding a high strength steel such as HY130, with a yield strength of 1000 N/mm^2, the permissible heat input is low and the number of welding passes therefore increases.

Electron beam (EB) welding, which is recognised as an excellent welding process for high performance materials, offers the following advantages:

1 Deep penetration welds can be obtained with a simple square butt preparation, and the fusion zone is only 5-10% of that produced by arc welding. This feature becomes more noticeable as the plate thickness increases up to the present limits of one pass EB welding (Fig. 1).

2 Even at the limiting plate thickness, the heat input is low but the power density is high (Table 1) so that degradation of properties, even in thick high strength structural materials, is minimal in comparison with arc welding (Fig. 2). With HY130 steel, the heat input for welding should be limited to 1.78 kJ/mm. For TIG welding of 30mm thick HY130 steel this means that 48 passes are required compared with one pass with EB welding at the same heat input. Another important advantage is that the low heat input substantially reduces the risk of weld decay in welded austenitic stainless steels, thereby eliminating the need for solution treatment after welding.

3 Deformation caused by EB welding, such as shrinkage or angular distortion is very low, typically less than 10% of that caused by arc welding (Table 2). As a result, some of the normal arc welding considerations regarding the location and sequencing of weld beads to avoid distortion are rendered unnecessary.

4 In contrast with arc welding, EB welding is suitable for joining almost all metals, whether in similar or dissimilar combinations. Even in difficult combinations, the use of an insert can frequently solve most problems, including degradation of weld joint properties. This technique has been proved with respect to the fracture strength of HY130 steel and other high strength steels for applications at low temperature and also to the long term rupture strength of austenitic stainless steels for use at high temperatures in nuclear pressure vessels. Properties have been obtained equivalent to those of the base material or at least better than those produced by the conventionally accepted TIG process. Comparisons of the efficiency of EB and submerged arc, MIG and TIG processes are shown in Fig. 3 and 4.

Despite its advantages, EB welding has been limited in its application mainly to small items which can be placed within an evacuated work chamber. This had been the case since the first applications of EB welding to the canning of nuclear reactor fuel elements in 1957, but in recent years, more serious consideration has been given to the application of EB welding to large structures. Three basic approaches have resulted:

1-Local chamber EB welding: With this system, the EB gun is moved over the workpiece and only the small volume between the gun and the workpiece is maintained under vacuum. The main problem is to provide a vacuum seal, although work by the authors has overcome some of these problems in shielding butt joints. This approach overcomes problems of size limitation associated with large fabrications.

2-Full chamber EB welding: Full chamber EB welding, with the use of ancillary chambers to accommodate the shape of the workpiece is an extension of the conventional EB technique which it is claimed, with only small amounts of further development could offer a practical solution.

3-Non-vacuum EB welding: Here, the electron beam, which is produced under high vacuum, is led through a nozzle with a gradual reduction in the vacuum down to atmospheric pressure. Although overcoming problems of size of workpiece, the technique is limited currently to a maximum of 23mm for both working distance and penetration. This is because the largest commercial non-vacuum type of EB system available at present has a maximum accelerating voltage of 200kV and an output of 25kW. However, a new system developed by Westinghouse with a capacity of 175kV, 56kW is claimed to be capable of welding 50mm thick steel out of vacuum, with a working distance of 18mm.

Of the three techniques outlined above, the non-vacuum type is still some way from extensive commercial application because of the limited working distance and penetration depth. In the following sections, details are given of the various commercial approaches to local and full chamber electron beam welding, with some applications.

**Dr Eng Terai is Manager and Dr Eng Nagai is Research Engineer, Welding Research Department, Kawasaki Heavy Industries Ltd, Japan*

Table 1 Power density of heat sources

Welding heat source		Power density, W/mm^2
Gas flame	($O_2+C_2H_2$)	10
Jet flame	(O_2+H_2)	30
Argon arc	(200A)	150
Electron beam	Pulse	$\geqslant 10^5$—10^6
	Continuous	10^4—10^7
Laser beam	Pulse	10^5—10^{11}
	Continuous	10^3—10^7

Table 2 Heat input and transverse shrinkage in various welding processes

Welding method	Plate thickness, mm	Heat input, kJ/mm	Transverse shrinkage, mm
EBW	12	0.2—0.5	0.067—0.07
Manual welding	12	3.0—3.2	0.8—0.9
Submerged arc welding	12	5.6—6.0	0.75—0.8
TIG	12	1.2—1.3	0.38—0.42

Table 3 Classification of local chamber EB welding equipment

Items		Local chamber A	B	C	D
Shielding	Type	Sealing materials*	Sandwich type double chamber	Fastener sealing	Stringer bead ground for sealing
	Unit	Single use		Modular use	—
EB gun	Type	Movable in the atmosphere	Movable in the vacuum	Periscope	Covered by double mobile chamber
	Power	Usually low voltage 12—60 kW	Low voltage 15—30 kW	Low to middle voltage 60 kW	Usually low voltage ≤ 30 kW
	Pressure, torr	$< 5 \times 10^{-2}$	$< 2 \times 10^{-4}$	$< 5 \times 10^{-2}$	$< 5 \times 10^{-2}$
Pumping	System	Single unit: rotary pump mechanical booster	Twin units: rotary pumps mechanical boosters diffusion pumps	Single unit rotary pump mechanical booster	
Positioner and jig	Facility	Not necessary	Not necessary	Manipulator and carriages	Not necessary
	Welding position	Flat, horizontal, vertical	Vertical	Flat, horizontal, vertical	
Pressure vessel	Shape	Flat plate, cylindrical only	Special shape only	Cylindrical, spherical, only	Cylindrical only
	Penetration (steel)	30—120mm	70mm	120mm	70mm

* roll-on, roll-off metal tapes or single belt material

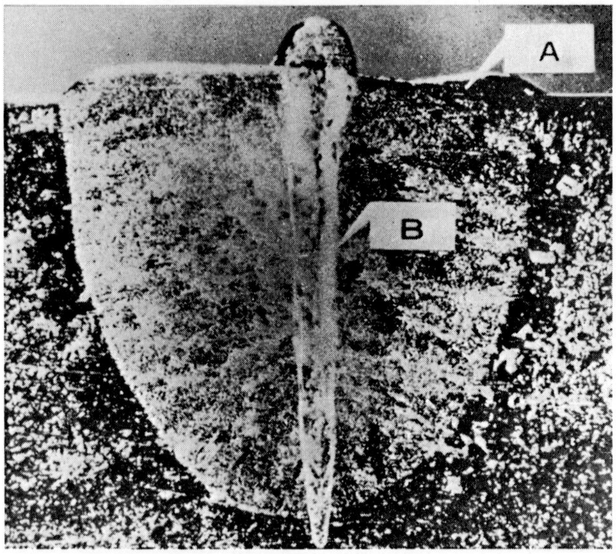

1 *Comparison of penetration and size of fusion zone between conventional TIG welding (A) and EB welding (B).*

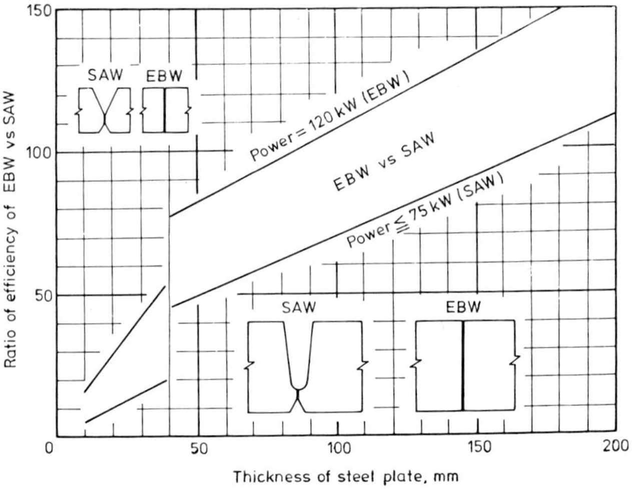

3 *Comparison of efficiency between EB welding and submerged arc welding.*

2 *Heat input requirement and welding sequence in various welding processes for thick steel plate. Numbers ( ) indicate runs per weld.*

Steel	Thickness of plate (mm)	Max. heat input kJ/mm	Welding process: EBW	TIG	Manual arc welding	MIG	Submerged arc welding	ESW
Mild steel	——	No requirement	200 mm (1)	(280)	(180)	(80)	(60)	(1)
HT-60 *NS	t > 16	10	100 mm (1)	(140)	(80)	(32)	(25)	——
HY 90 †MIL-STD	t<12·6	1·8	35 mm (1)	(50)	(30)	(25)	——	——
	t>12·6	2·2						
HY 130 †MIL-STD	15·8~22·1	1·6	30 mm (1)	(48)	——	——	——	——
	22.1~34.8	1·8						
	34.8~	2·0						

*NS: Japanese defence agency standard
†MIL-STD: American military standard

4 *Comparison of efficiency in various welding processes.*

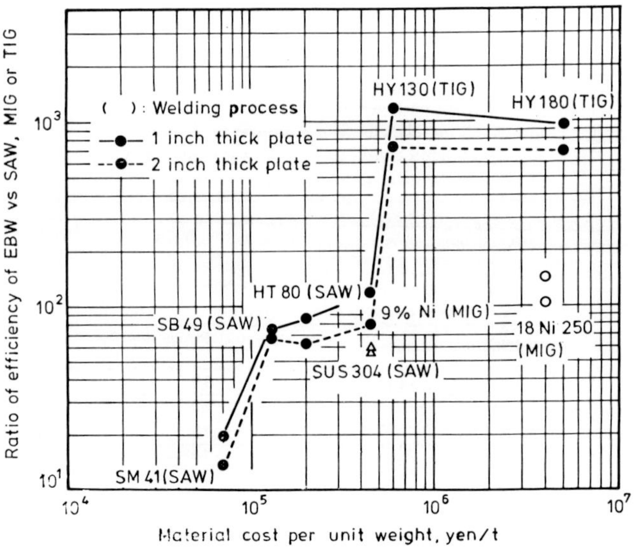

(a)

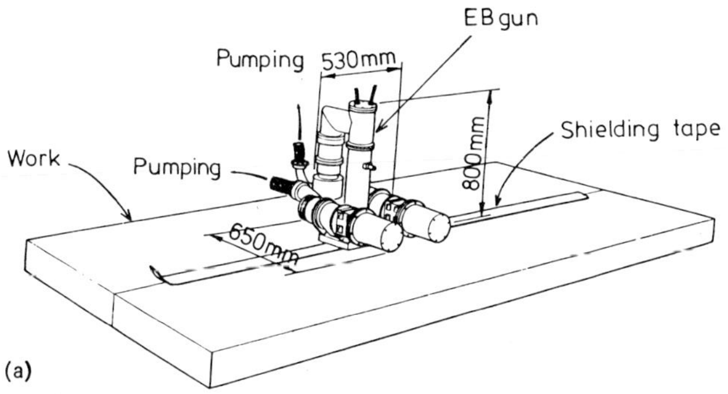

(b)

5 *Type A local chamber EB welding (a) The mechanism (b) Practical application.*

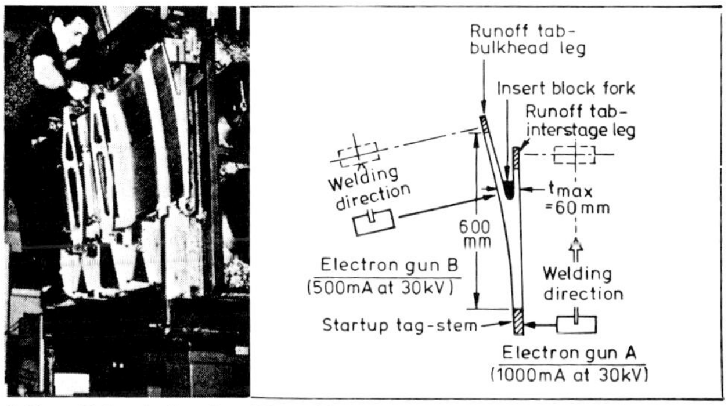

6 *Type B local chamber EB welding equipment.*

7 *Another example of type B local chamber EB welding equipment.*

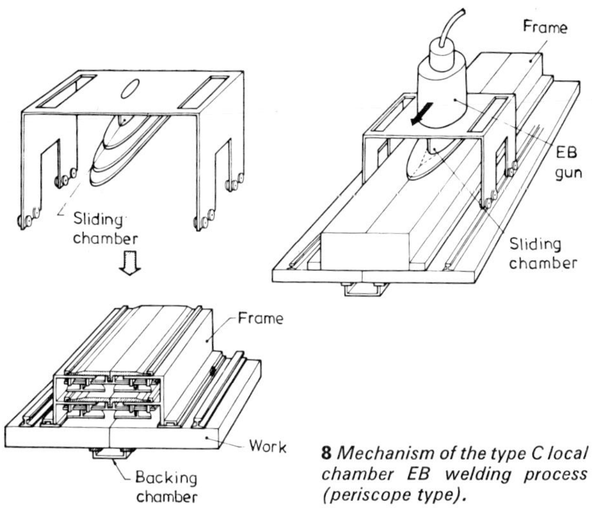

8 *Mechanism of the type C local chamber EB welding process (periscope type).*

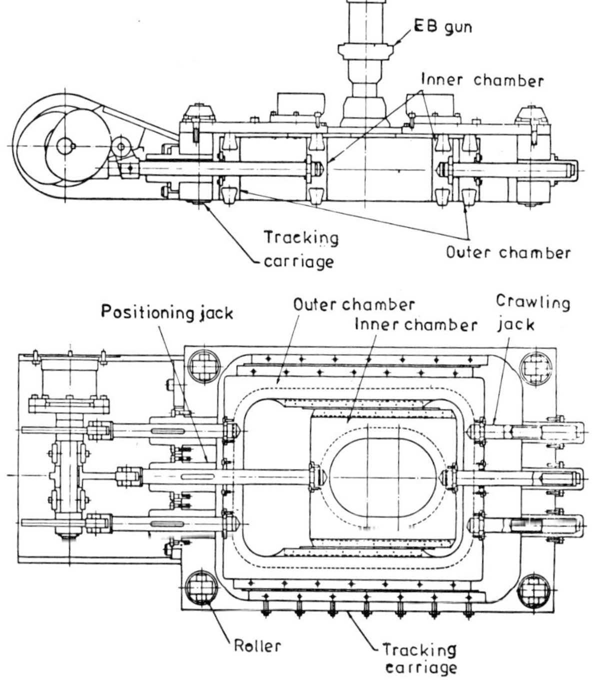

9 *Mechanism of the type D local chamber EB welding process.*

10 *Welding of a cylindrical structure by type D local chamber EB welding equipment.*

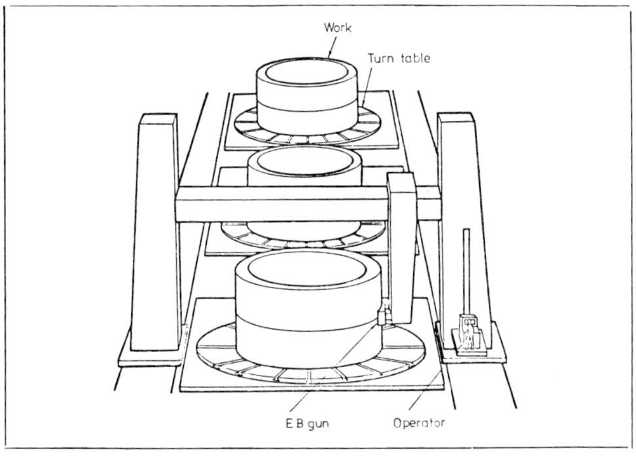

Table 4 Classification of full chamber EB welding equipment

Items		Full chamber A	B	C	D
Chamber	Type	Large size	Work itself	Large size	Docking type
	Unit	Single use	Single use	Single use	Modular use
EB gun	Type	Movable	Movable	Stationary	Corking type
	Initial position	Free	Free	Roughly adjustable	Minutely adjustable
	Power	Low voltage 30 kW	Middle voltage 100 kW	Usually high voltage 75—100kW	Usually high voltage 100—120kW
Pumping	Pressure (torr)	$< 2\times10^{-4}$	$<2\times10^{-4}$	$< 5\times10^{-2}$	$<5\times10^{-2}$
	System	Multiple units: rotary pumps mechanical boosters diffusion pumps	Single unit: rotary pump mechanical booster diffusion pumps	Multiple units: rotary pumps mechanical booster	Multiple units: rotary pumps mechanical boosters (corking type)
Positioner	Facility	Large positioner and welding jig	Welding jig	Large positioner	
	Welding position	Horizontal, flat	Horizontal, flat	Horizontal, flat, vertical	
Pressure vessel	Shape	Cylindrical, spherical or any other shape	Cylindrical	Cylindrical, spherical or any other shape	
	Max. penetration (steel)	70mm	200mm	250mm	350mm

Local chamber EB welding equipment

Table 3 gives details of four different approaches to local chamber equipment. Type A, developed by the Nuclear Agency of France and Clover Co, has a slit in the welding frame above the weld joint which is covered by a metal tape or silicone rubber (Fig. 5). Modifications carried out by Kawasaki, however, have improved the stability and application of the equipment. The EB gun is generally of the low voltage type (60kV) with an output of about 60kW and the unit is restricted to flat plates or cylindrical structures.

Type B equipment (Table 3) was developed specifically by Sciaky Co of USA for welding the ring of a complex Y branched member which formed part of the liquid oxygen fuel tank for the Saturn rocket. The equipment (Fig. 6) consists of two EB guns of 30kV, 15kW and 30kV, 30kW respectively, located one each side of the ring. The vacuum shielding was applied from each side by a split chamber around the ring. Both guns are numerically controlled to take account of changes in plate thickness. One of the problems with this equipment is in maintaining the vacuum despite any interference of the ring with the chamber closure seal. This particular equipment is suitable only for the specific application, but Fig. 7 shows an alternative system, still of type B, developed by Kawasaki. The basic component is a vertical type A unit incorporating two 45kW EB guns. Details of the application of this equipment to the welding of a spherical tank for an LNG carrier are given later.

Kawasaki are also working on another approach, (type C in Table 3), whereby a small vacuum chamber incorporating the EB gun approaches the workpiece through the welding frame or jig, rather like a submarine periscope. The welding frame also incorporates a double shielding device (Fig. 8) with a backing chamber. This system provides for higher accuracy and will accommodate both flat and curved components.

The final system, (type D in Table 3,) comprises an EB gun mounted on a tracking carriage with a double chamber (Fig. 9). The vacuum is higher in the inner chamber and each chamber may be moved readily. Reducing the size of the chamber, particularly the inner one, increases the range of applications, since even quite a large curvature of the plate surface then appears flat to the seal (Fig. 10).

Left **11** *Type A full chamber EB welding equipment. 13m×11m.*

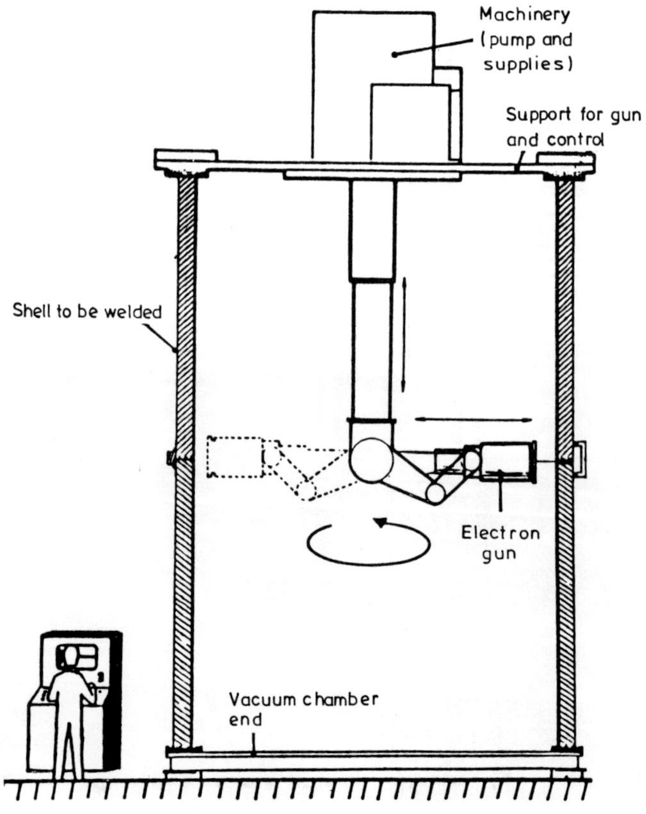

Right **12** *Mechanism of type B full chamber EB welding equipment.*

Above **13** *Type C full chamber EB welding equipment built by The Welding Institute*

Right and below **14** *Type D full chamber EB welding equipment: (a) independent units; (b) units in pairs; (c) units in line.*

(a)

(b)

(c)

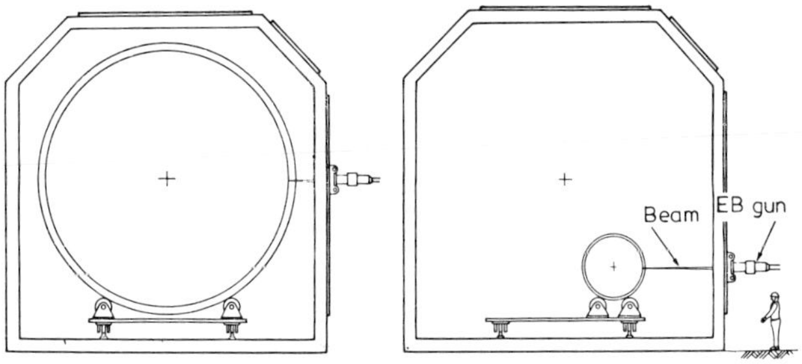

16 *Adjustment of an EB gun to accommodate the size of welded structure.*

Full chamber EB equipment

This can also be divided into four types (Table 4). Type A, developed jointly by Sciaky Co of France and General Electric Co (Fig. 11), comprises a 13m high by 11m diameter sphere into which the welding jigs are fitted, with a 30kV, 30kW EB gun. A wide variety of shapes and sizes of structure can be welded providing that a high vacuum is used in conjunction with short gun to workpiece separation.

The Type B equipment comes from Sciaky of France, and the workpiece forms its own vacuum chamber (Fig. 12) with the EB gun rotating inside it. Despite the higher power of the EB gun—100kV, 100kW—the unit is suitable only for short gun-to-workpiece distances in cylindrical objects.

Figure 13 shows a Type C unit developed by The Welding Institute in the UK, and now in use by the Kawasaki group. It comprises a 75-100kW gun mounted on a large vacuum chamber and produces a beam allowing longer working distances under conditions of low vacuum (1×10^{-1} to 1×10^{-2} torr). This gun has been designed to overcome the problems of gases or vaporised material from the structure being welded causing a discharge within the electron gun, a feature typical of equipment of over 50kW rating. Gun ports are provided in the horizontal, 45° and vertical positions with ready location of the gun at each port.

The fourth type of full chamber, although new in its concept, is essentially a derivative of the type C system. By suitable location of several chambers the production line can operate as independent units (Fig. 14a) or as pairs of units (Fig. 14b) or as one long line (Fig. 14c). Figure 15 shows the automatic method of corking the electron gun, and methods have also been developed for positioning the gun for easy adjustment relative to the workpiece. The 180kV, 120kW gun is sufficiently powerful to allow large variations in the gun to workpiece distance (Fig. 16), and would enable a large range of structures to be accommodated using only one EB gun.

15 *Mechanism of the automatic 'corking' operation of an EB gun.*

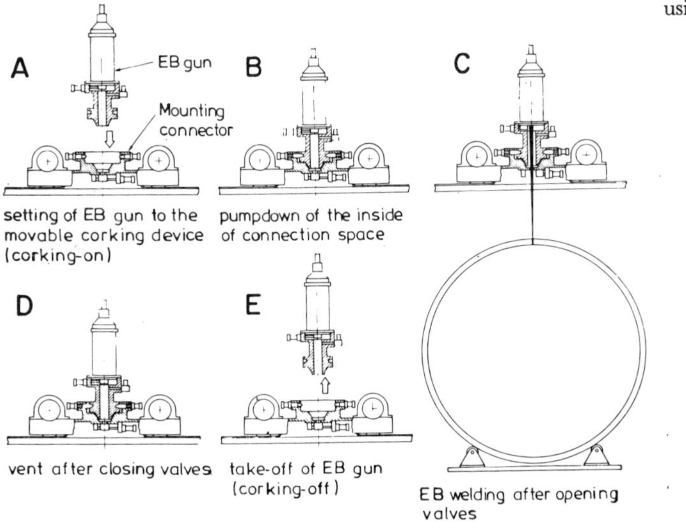

A major advance in high-power electron-beam welding in air

J. F. Lowry, J. H. Fink, and B. W. Schumacher

Westinghouse Research Laboratories, Pittsburgh, Pennsylvania 15235
(Received 18 June 1975)

A major extension of the capabilities of nonvacuum electron-beam welding follows from the development of a machine with higher beam power. Tests with a new atmospheric electron gun operating at 60 kW have shown that the advantages of nonvacuum electron-beam welding must be reevaluated. Gas heating produced by the beam itself becomes very pronounced, so that electron scattering is reduced causing the high-power density of the beam to be retained over larger working distances. Single-pass butt welds with a depth-to-width ratio of 4:1 can be made in 3.8-cm-thick steel at a speed of 0.77 cm/sec, while the ultimate welding depth exceeds 5 cm. Medium thick material, e.g., 1.3-cm steel, can be welded at a 5-cm work distance, with a depth-to-width ratio of 3:1. The large work distance permits access to more complex structures and interior corners such at T sections, which can now be fabricated from a plate, 1.3 cm thick, at a speed of 2–3 cm/sec, making a full through weld from one side. Also, seam welds of two 3-mm-thick hot-rolled steel sheets can be produced at speeds of up to 15 cm/sec. High-power machines need not be significantly larger or costlier than low-power guns, and the present 60 kW does not represent any technical limit. Welding efficiency improves with higher power; when welding steel of 2 cm thickness or more the energy efficiency of the process at 60 kW is better than 50%, while at 40 kW it is merely 30%. In addition, the high power permits greater welding speed. These developments translate directly into improved cost justification for electron-beam welding and a broad expansion of its possible applications.

PACS numbers: 81.30.Q, 06.60.V, 79.20.K, 89.20.

I. INTRODUCTION

Electron-beam (EB) welding in the atmosphere began historically with very-low-power beams.[1] Yet even a beam of 12—24 kW, which is representative of today's commercial machines, must be considered low power, as we shall see. A certain threshold value of the beam power density is also of significance, namely, that value at which the melting or vaporization process becomes instantaneous and where no heat is lost by conduction. Then melting proceeds with maximum efficiency and will reach a maximum rate above which, in spite of higher-power input, only vaporization is obtained. A theoretical study[2] showed that beam powers in excess of 50 kW may be needed for this type of efficient, "quasi-adiabatic" welding. But no atmospheric-type electron guns with beam powers above 36 kW had been built before to study this kind of high-power process.

In air, the beam power density is degraded by scattering of the electrons by gas molecules; this has led to the general conclusion that EB welding in air requires a close proximity of workpiece and electron gun muzzle, because if the *power density* is reduced, the rapid melting conditions and the desirable deep narrow weld zones (similar to in-vacuum EB welding) cannot be maintained. Yet, even with a beam of only 9 kW conventional looking welds can be obtained at 3-cm work distance and at high speed,[3] since power as such is not lost at the larger standoff distance.

With beams of high power *and* high power density several aspects of the process change. A beam of 40 kW or more has a power density at the gun of more than 2 MW/cm^2, which heats the gas in the beam path to temperatures exceeding 1000 °C. This reduces the gas density and consequently the electron scattering, permitting the original high beam power density to be retained further than might be expected; the ultimate range is also increased. An increase in voltage, even as small as from 150 to 175 kV, reduces the scattering cross section, thereby introducing a similar beneficial effect, which is quite noticeable. Recently, the range and power density increase as a function of beam power was investigated[4,5] and it was found that at a given distance from the gun the power density could be 50 times greater than one would expect without taking into account the gas heating effect. As a result of the higher power density, the adiabatic threshold at the workpiece can be exceeded even at large work distances. Of course, the same argument is valid for a beam passing through the metal vapor in the region of the weld.

Thus, at higher beam powers the electron beam in air can make narrow and deep welds with profiles approaching those of welds made in vacuum. Typical high-power working distances of 2—5 cm will permit access to most ordinary workpieces, while at larger standoff, 7.5 cm and more, shallow welds can be made at very high speeds. With the exception of in-vacuum EB welding, high-power EB welding in air is the most efficient welding method that has been devised.

In the following, we will discuss some specific examples of high-power welding processes without directing our considerations to particular commercial applications. However, these data will form a useful basis for planning the future applications of this new tool. It should also be recognized that the experimental results included as examples in this text were directed at demonstrating process potential (speed, penetration, standoff capability, weld profiles) and do not represent the highest-quality welds which might be attained with the process.

II. THE ELECTRON GUN

The 60-kW 175-kV experimental machine used for these tests is shown in Fig. 1. It closely resembles the 24/36-kW guns previously described,[6] in that it uses the same size beam transfer column to bring the beam from the vacuum to the atmosphere, the same cathode

FIG. 1. Laboratory prototype of 60-kW 175-kV electron gun for delivering a beam into the atmosphere. (1) power supply (cable-connected to the electron gun), (2) high vacuum chamber for the electron gun cathode, (3) transfer column for bringing the beam to the atmosphere, (4) magnetic lens, and (5) beam exit aperture.

cartridge, and a high-voltage insulator and vacuum chamber which are only slightly larger to accommodate the higher voltages. A suitable compact inexpensive gas-insulated power supply of the desired higher output rating was developed, having an ac voltage ripple of approximately 3% regardless of load, while the dc level is controlled to ± 2% accuracy at a given current loading. Cathode and beam focusing optics were developed for higher currents at higher voltages, along with auxiliary power supplies of suitable ratings.

Although this may seem unexpected, it turns out that the beam power density of this new gun geometry is nearly proportional to the total beam power. While the cathode diameter must be increased by a factor of $\sqrt{2}$ to double the current, this mainly alters the angular divergence of the beam but does not change the diameter of the focal spot.[7] For a beam in air this small change in angular divergence is completely obscured by the scattering.

The focus setting on the machine will, of course, affect the beam diameter and the welding results. In all of the following tests, the focus was set for optimum beam transmission efficiency. With high-power operation the beam transmission losses are less than 5%; cathode power requirements are less than 1% of the total. As a result the general efficiency of converting electrical power to electron-beam power exceeds 90%. This, together with the superior welding efficiency, makes the EB processes particularly interesting at a time of energy shortage.

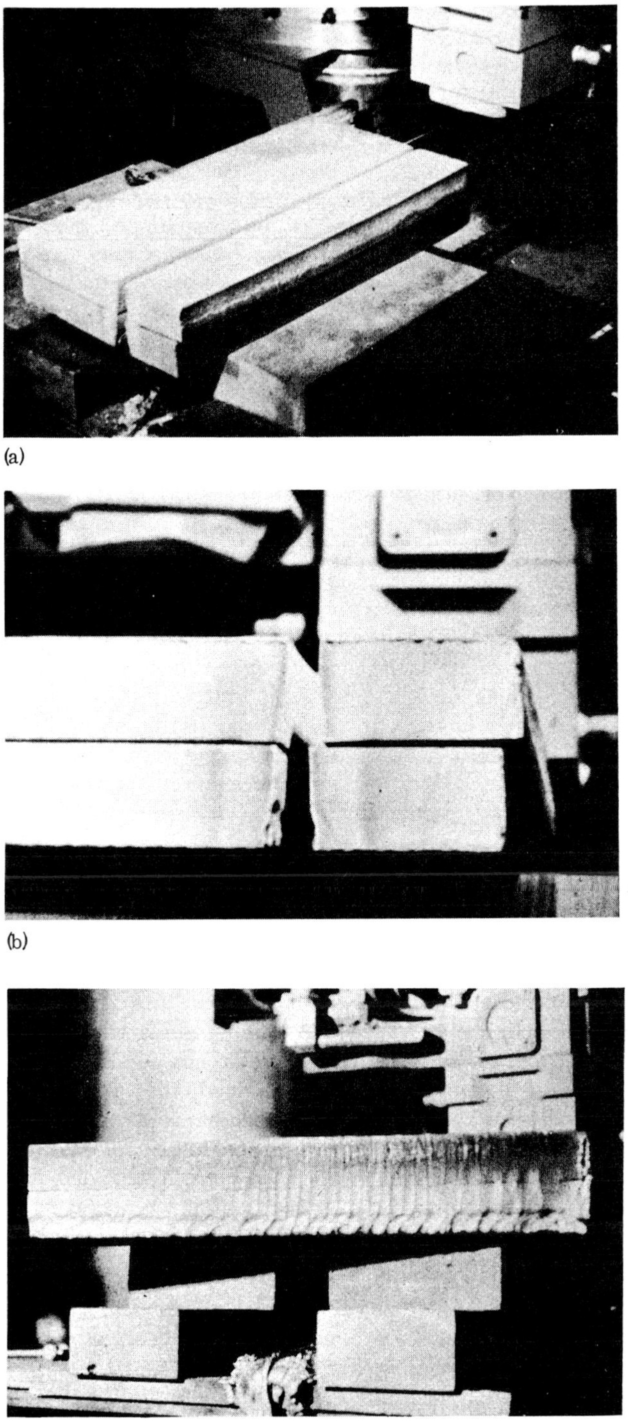

(a)

(b)

(c)

FIG. 2. Electron-beam cut through 5 cm of steel: beam 51 kW, 165 kV; speed 0.66 cm/sec; work distance 1.2 cm. (a) Plates have moved out from under the gun. (b) Profile of the cut. (c) Appearance of the cut faces.

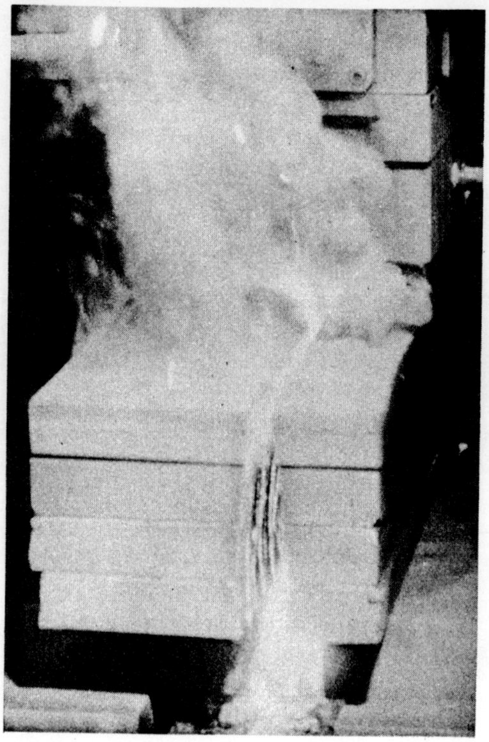

FIG. 3. Electron-beam cut through 10 cm of steel: beam 51 kW, 165 kV; speed 0.14 cm/sec; work distance 1.2 cm.

The auxiliary equipment for the higher-power gun is roughly the same as for the lower-power version; hence, capital costs do not increase in proportion to power, but at a much lower rate. Moreover, as shown in the following, the welding speed capability goes up more than proportionally with the beam power. Thus, the high-power machines greatly reduce EB welding costs, while providing excellent energy efficiency.

It should also be mentioned that the 60-kW rating of the present experimental machine does not represent any theoretical or practical limit. Further power increases can bring further advantages.

As for the appearance of the beam in air see Refs. 3–5.

III. MELTING, CUTTING, AND WELDING AT DIFFERENT POWERS

It obviously depends upon the beam power and power density whether a particular metal target will begin to melt when hit by a beam moving across its surface. If the power density is very low, the beam energy can be carried away by heat conduction so that no melting will occur. If the power density exceeds a critical minimum level, then instantaneous melting will occur (in fact, it cannot be prevented) and only a negligible amount of the energy will be lost by heat conduction. At an even higher threshold of power density instantaneous evaporation will occur and nothing can be done to prevent it.[2] At intermediate power densities some intermediate condition between these extremes may be encountered.

In principle, one can cut and weld with an electron beam when the power input is sufficient to melt the metal. But it is clear that the process will look drastically different when the beam has just enough power to cause melting and when it is capable of instantaneous melting. Since this is a function of welding speed and beam power density, and since the power density of a beam in air (and in metal vapor) depends upon the distance it has traveled, the depth or thickness to which the beam can cut or weld clearly depends upon speed, beam power density, and "standoff" or work distance; having only a low-power beam, rather than high power, we cannot expect to make the same cut or the same weld simply by doing it slower. We have elaborated this point so extensively because it is the second key factor for understanding what follows. (Gas heating and scatter reduction by a high-power beam itself are the other factors.)

The electron beam will produce a melt cut only if the molten metal can flow away. For plates more than 1 cm thick, gravity will cause the liquid metal to drain out of the slot, despite the surface tension which tends to hold it in place; the latter is the case with thin workpieces, e.g., plates of 0.5 cm thickness or less. Therefore, the cutting of thick plates is much easier. Figure 2 shows a cut made through two steel plates, each 2.5 cm thick, while Fig. 3 shows a cut in progress going through 10.2 cm of steel, and Fig. 4 shows the profile of a cut through 12 cm of aluminum. In each case, the width of the cut increases with depth as a result of electron scattering which widens the beam. We have not tried to optimize the cutting process or to evaluate its potential. Cutting speeds, as observed, are given in the captions of Figs. 2–4 and also in Fig. 12.

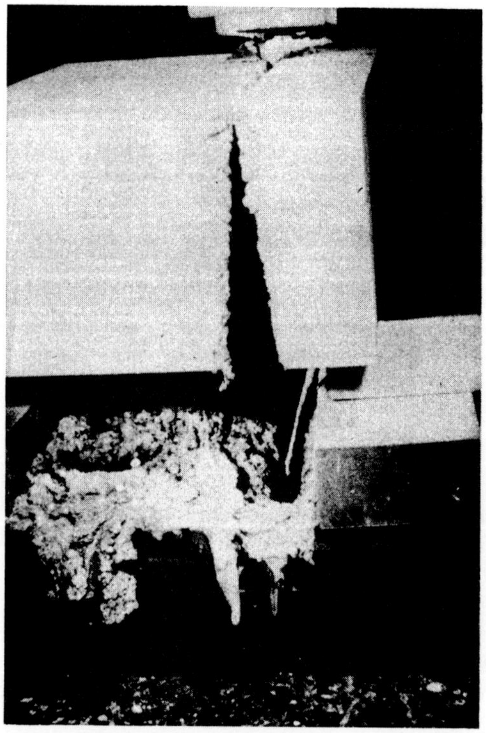

FIG. 4. Completed electron-beam cut through 12 cm of aluminum: beam 51 kW, 165 kV; speed 0.1 cm/sec; work distance 1.2 cm.

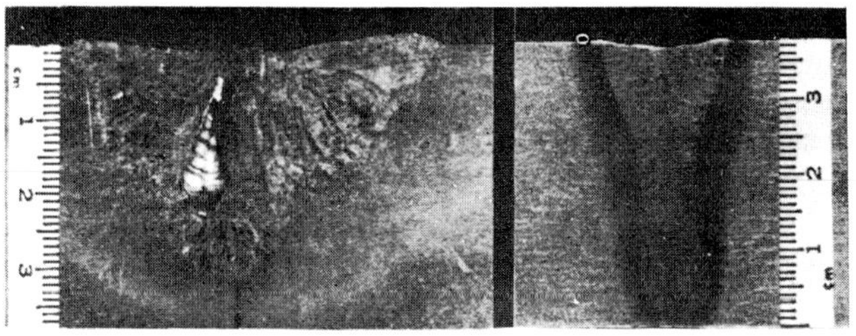

FIG. 5. Types of welds obtainable in air, with different electron-beam power levels. The excessive porosity in the left sample could have been reduced, probably, by using shield gas. A 2.5× power increase yields a 10× speed increase and produces a much better weld. Both samples 1020 hot-rolled steel. Left: 20-kW 150-kV 1.2-cm work distance, speed 0.07 cm/sec. Right: 51-kW 165-kV 1.42-cm work distance, speed 0.77 cm/sec.

To produce a thick-section weld the molten metal must be prevented from flowing out of the weld zone. One must start the weld sufficiently far inside the edge of the workpiece or a "starter block", and one must either avoid full penetration or use a cooled backing plate, as is well known. For the penetration, the available beam power will set a limit. This is not immediately obvious; the process can best be understood by looking at it as follows. Any beam impinging on a semiinfinite target in a stationary mode will ultimately produce a semispherical melt pool whose radius or depth is determined by beam power, time, and heat conduction. If the beam then starts traveling very slowly over the surface, a semicylindrical melt zone will be left behind, which is twice as wide on top as it is deep, i.e., with a depth-to-width ratio of 0.5:1. As the beam accelerates, traveling faster and faster across the surface, one would expect the radius of the semicylindrical melt zone to get smaller, but nothing else to happen. This is true, as long as heat conduction is the dominant factor governing the spread of the energy, i.e., as long as the beam power is low. But, as already mentioned, a high-power density electron beam can produce instantaneous melting over the area where it impinges. Moving along the surface it can even generate a rather deep melt hole or "keyhole". We must come back to this point later. The result is that the fast-moving high-power beam will generate a deep and narrow melt zone, which is the ideal shape for a weld. The difference between a low-power weld and a high-power weld of about the same depth is shown in Fig. 5.

The practical limit to the thickness of the workpiece which can be welded with a given beam power is reached with a weld shape somewhere between the two cases shown in Fig. 5. The depth-to-width ratio may reach a value of 2:1. This situation exists in the test welds shown in Fig. 6; these can be considered as lap welds of two 2.5-cm plates or simply as beads on plate. In this case, the workpiece started moving first and the beam was turned on when it fell well inside the edge of the plates. Figure 7 shows a 5-cm weld being made. In Fig. 6, the melt zones go through both plates, while the molten steel was prevented from flowing out by a water-cooled-copper backing plate. But due to warpage of the steel plates (which were not clamped to the backing plate) the liquid metal ran out at the rear half of the weld on the left, and a very interesting shape was exposed. The surface we see is obviously the liquid/solid interface at the moment the molten metal ran out. The narrow cut towards the rear shows the actual width of the electron beam itself. The freezing zone toward the finished part (front) of the weld looks like a bow wave; in fact, the surface of most EB welds shows a pattern of frozen bow waves from which the direction in which the beam was moving can be determined. The crater between the finished weld and the cut resembles the end crater which is often found when the beam is turned off. The metal which later fills this crater forms the familiar triangular top part or nail head of the weld. The metal in this triangle cannot have been melted by stray electrons surrounding the beam, because a similar zone would also have been melted on both sides of any cut. As shown in Fig. 6, this is not the case. This

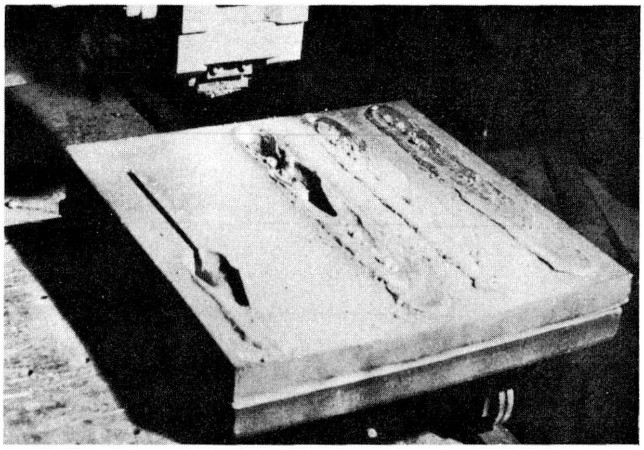

FIG. 6. An illustration of welding and cutting depending on the retention of the molten metal. Lap weld of two steel plates each 2.5 cm thick; in the left weld the copper backing plate did not prevent run out of the molten metal and a cut resulted in the rear part of the run; the crater shows the solid/liquid interface at the moment of the run out; the width of the cut equals the width of the beam. Beam power 50 kW, speed 0.42 cm/sec, 1.4-cm work distance.

FIG. 7. Lap weld of two 2.5-cm-thick steel plates in progress. The beam was started inside the front edge of the plates. One can see the spread of the heat on the vertical face; a copper plate is used as backup.

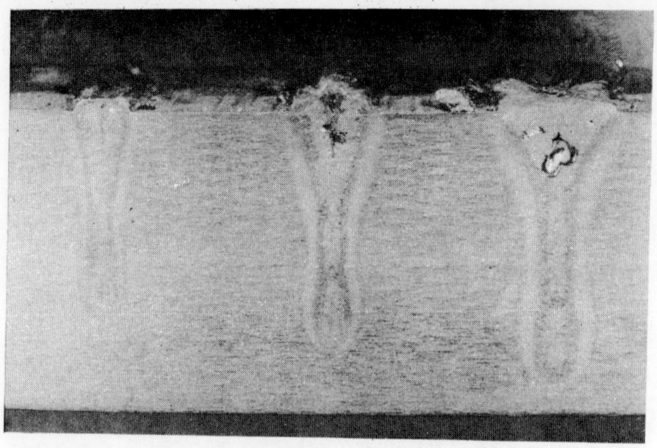

FIG. 8. Depth vs weld speed in 1020 hot-rolled steel; beam power 50 kW at 165 kV; work distance 1 cm; speed 1.5/1.1/0.76 cm/sec. Plate is 3.8 cm thick.

triangular melt zone must be the result of heat conduction, which occurs to a much greater extent with slow welding. There is some very hot liquid metal near the top of the melt zone; the heat stored there can readily account for the further melting. The crater in Fig. 6 intersects the groove of the cut at a right angle; it seems correct to assume that the melt front, where the beam has hit the metal, was further ahead in the groove at the moment of the run out.

It remains to be explained how the electron beam produces the deep melt hole or key hole in the first place. As long as metal is flowing back into the beam path, vaporization can take place, thereby removing some metal; this vapor oxidizes, forming the "smoke" which is often seen. Meanwhile the vapor pressure will help to keep the cavity open. However, it is impossible that later in the welding process all the metal passes through the vapor phase. It seems reasonable to assume that the liquid in the melt front under the beam is near the boiling point where surface tension would be zero; whereas in the nearby freezing zone the surface tension has its maximum value, which is very high for all metals. Hence the hot liquid is pulled into the freezing zone by the force of surface tension. Whether there is still another mechanism at work, we do not know. In fact, as can be seen from Fig. 22, the fluid dynamics of the welding process is not fully understood, even in this much simpler case.

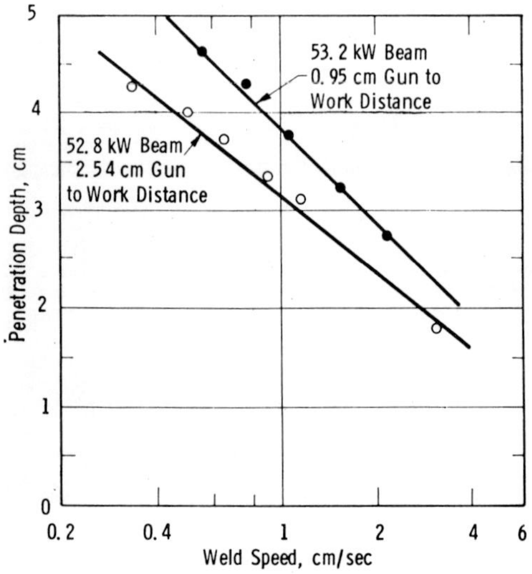

FIG. 9. Depth as function of speed and work distance for 1020 hot-rolled steel.

From a practical point of view it is of interest to know the shape and depth of the melt zone as a function of beam power, working distance, and welding speed. Figures 5, 8, and 9 show variations of depth with speed and beam power for mild steel. To ascertain these relationships for stainless steel a series of bead-on-plate runs were made and the cross sections of the melt zones obtained are shown in Figs. 10–12. The point in Fig. 12 indicating that a cut has occurred lies at a speed where the extrapolated depth curve has not yet reached the thickness of the plate (3.3 cm). This is predominantly a heat conduction effect as the heat sink is inadequate when the melt approaches the rear face of the workpiece.

At smaller work distances, deeper welds are obtained, indicating the importance of the beam power density. (The total beam power does not change with distance.) A more interesting question concerns the influence of the total beam power. Figure 13 shows the results of tests at the constant standoff distance of 2.54 cm, with beam powers of 56 and 44 kW, respectively. The 44-kW tests were made at two different beam voltages and, surprisingly, the depth values obtained at the lower beam voltage of 145 kV are slightly greater than the ones obtained at 175 kV. However, not much

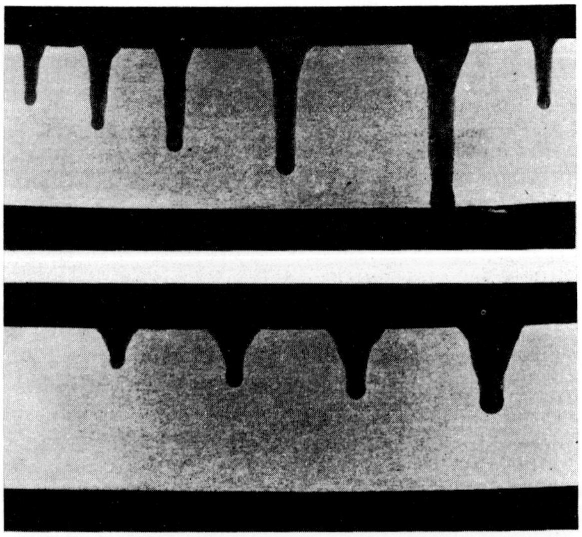

FIG. 10. Appearance of the melt trace for different weld speeds and working distances but identical beam power of 56 kW at 175 kV; type 304 stainless steel; 3.18 cm thick. Top: Work distance 2.54 cm; speed (left to right): 8.7/5.2/3.8/2.5/1.8/8.0 cm/sec. Bottom: Work distance 5.1 cm; speed (left to right): 8.3/5.1/3.9/2.5 cm/sec. While there are signs of faint root or centerline cracks in some of the welds, the through weld (top, second from right) is completely sound, and so were all the other through welds which we saw.

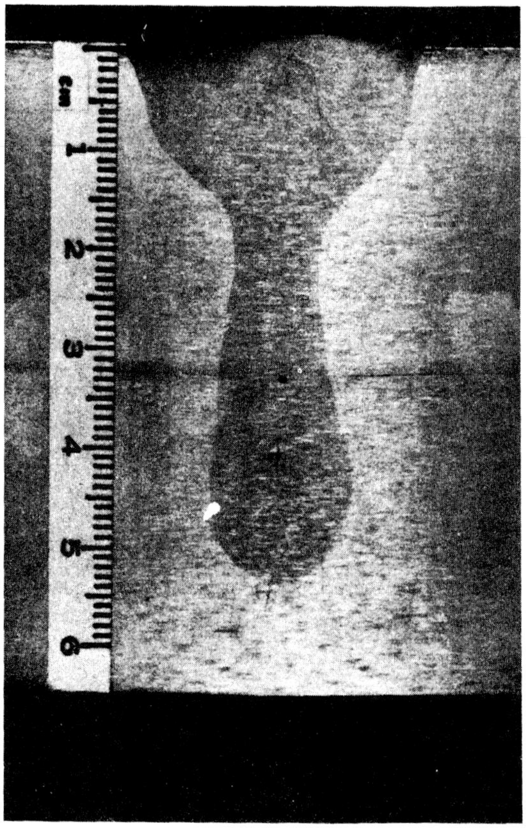

FIG. 11. Cross section of slow-speed deep-penetration lap weld in type 304 stainless steel. Beam power: 52 kW at 165 kV; work distance: 1.6 cm; speed: 0.25 cm/sec.

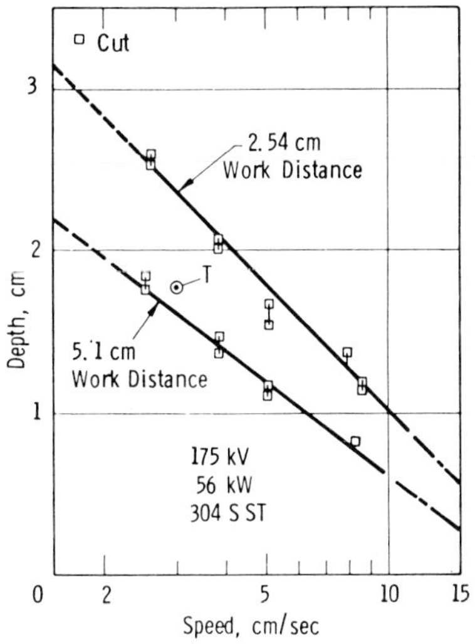

FIG. 12. Depth vs speed at constant beam power, for stainless steel. The points show measured depth values at different cross sections of the same bead. The point T denotes the welding speed and median-line depth of the T-section welds shown in Figs. 17 and 20, where the work distance was 5 cm.

significance should be attached to this, because we cannot fully and freely choose the power density (focus) for any given voltage and any total power in the present gun, and we do have evidence that the lower voltage beams are better focused with a higher peak power density on the workpiece. Measuring the actual power density in any welding test is extremely difficult. When comparing any welding results, here and from other tests, it should be recognized that the actual power density is, unfortunately, a largely unknown parameter.

The influence of the total power on the depth-vs-speed curve, as shown in Fig. 13(a), deserves a particularly detailed discussion because it reveals some of the more subtle facts of EB welding. The power ratio of the two beams for which the curves of Fig. 13(a) were obtained is 1.27; yet the ratio of the speeds at any given weld depth is greater. For a 2.5-cm depth the speed ratio is 2.0, but it drops to 1.8 for a 2-cm depth, and to 1.6 for a 1.25-cm depth. If the nature of the process did not change, the speed ratio would remain the same as the power ratio of the two beams. But under existing conditions the melting efficiency also changes with welding speed. This has been calculated and the re-

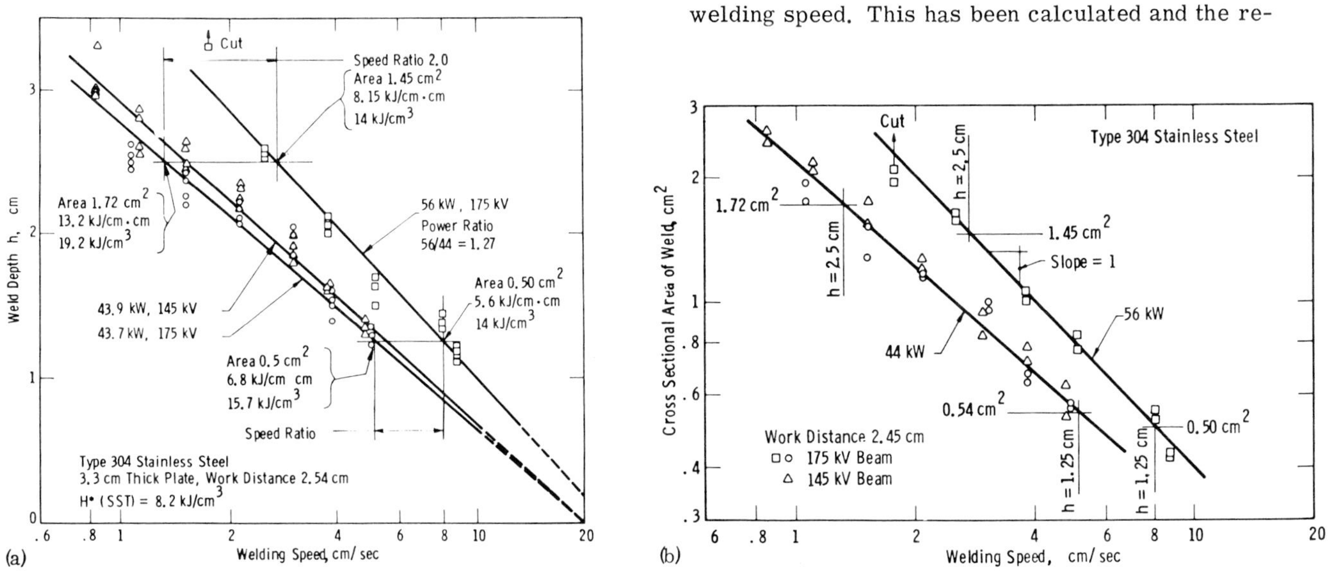

FIG. 13. (a) Weld depth vs welding speed for different total beam power values. Energy input per unit weld length, and energy per unit volume of welded metal are shown for 2.5- and 1.25 cm depth. For further explanation see the main text. (b) Cross-sectional area of the welds of Fig. 13(a) as determined by planimetry of the photographs of two sections spaced 1.5 cm apart.

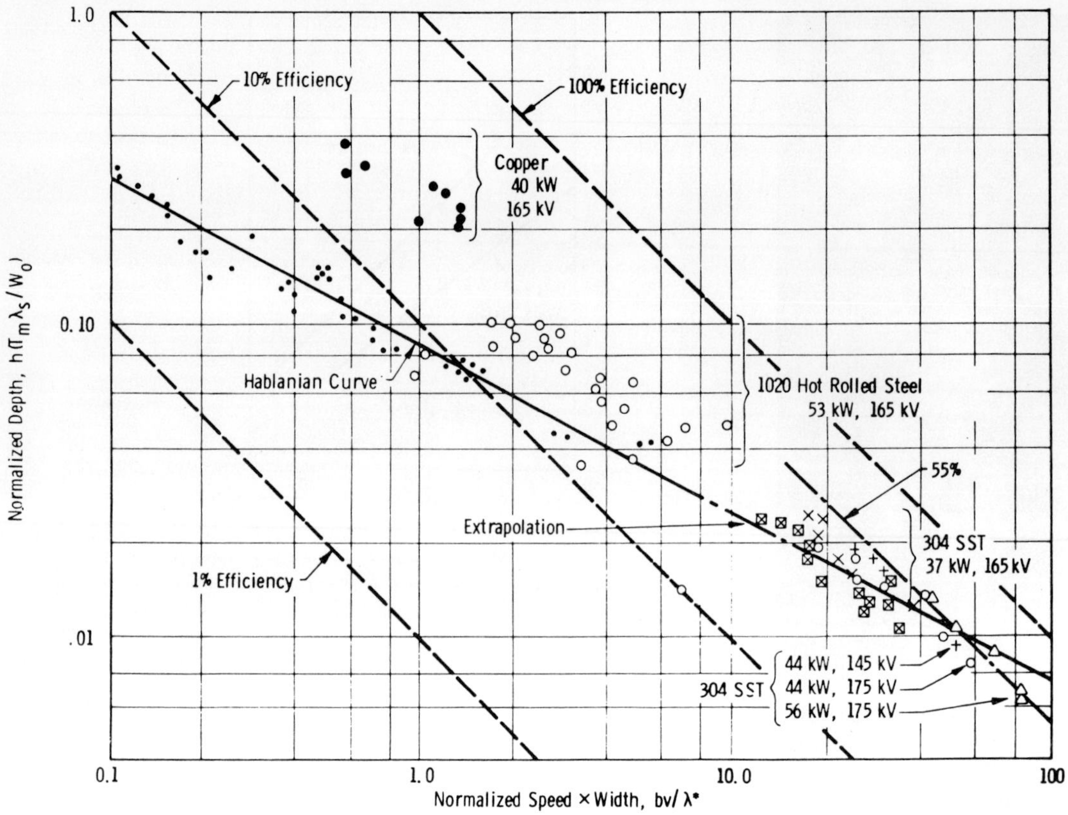

FIG. 14. Representation of EB welding results in a Hablanian plot (see discussion of figure in text).

sults are shown in Fig. 13(a). Dividing the beam power by the product of depth and speed gives the energy expanded per unit length and depth of the weld; this value is customarily given in joules per centimeter per centimeter rather than joules per square centimeter. If we measure the area of the cross section of the melt zone we can calculate the energy required to melt 1 cm^3 of metal at different welding speeds. The theoretical minimum for heating and melting 1 cm^3 of stainless steel is approximately $H^* = 8.2$ kJ/cm^3 so that any additional energy must have gone into heat conduction losses (and some small additional losses due to vaporization). From the values in Fig. 13(a), it is clear that the energy expended per unit length and depth of the weld is inversely proportional to the welding speed. But it is also lower for the higher-power beam. This difference between the beams is even more pronounced when we look at the energy expended per unit volume of melted metal. For the 56-kW beam we find about $1.7H^*$ is needed, whereas for the 44-kW beam it is $2.3H^*$. The cross section of the welds is also slightly larger at 44 kW. A value of $1.7H^*$ means the efficiency is 59%.

To determine the weld cross sections, which is so important in this analysis, two cuts about 1.5 cm apart were photographed, enlarged, prints made, and the areas determined by planimetry. Figure 13(b) shows the cross-sectional areas as a function of welding speed in log-log coordinates. A straight line with a slope of unity is obtained for the 56-kW beam, which means by doubling the speed the cross section is reduced to one-half its former value; this is what one would expect if a constant fraction of the energy input is used for melting. For the 44-kW beams, with either 175 or 145 kV, another straight line is obtained but with a definitely smaller slope of 0.9, which is another indication of the onset or increasing importance of certain energy-loss mechanisms.

Summarizing these observations we can say: at higher beam powers the welding process is noticeably faster as one would expect, but also more efficient, which is not self-evident; less heat goes into the workpiece, and the shape of the melt zones is more favorable. A similar observation has been reported for lower-power beams when welding 0.6-cm aluminum at small standoff distances[8]; only now we have attained power levels where the same efficiency is achieved for a weld thickness of 3 cm. It also follows from these observations that a power level of even 44 kW is not yet "high" in the sense of giving good efficiency when welding 3-cm-thick steel with the power density as it exists at the workpiece at larger working distances.

A more refined analysis of the EB welding process, with comparisons between different materials, requires consideration of a great number of variables such as thermal diffusivity, melting temperature, power, weld width, etc. Hablanian[9] has shown a simple way to present such data by means of a set of dimensionless parameters. Fink has shown[10] that the welding efficiency can also be displayed on the same graph (Fig. 14). Although extrapolation from low-energy welding results seemed to predict reduced efficiency for deeper welds, as shown by Hablanian's curve in Fig. 14, actual high-power welding test results are in the 20—40% region of

FIG. 15. Fabrication of a T section by high-speed EB welding. Beam 56 kW at 175 kV; cold-rolled mild steel 1.3 cm thick; speed 3 cm/sec; work distance 5 cm.

efficiency. The earlier expectations were too pessimistic!

In the plot of Fig. 14, h is the depth of the weld, W_0 is the beam power, hence h/W_0 represents the depth achieved per unit beam power; the factor $T_m\lambda_s$ accounts for the materials properties [λ_s is the thermal conductivity; $T_m = T_M + H_F/c_s$, the corrected melting temperature (T_M is the melting point, H_F the heat of fusion, c_s the specific heat)]. Thus, the vertical axis represents essentially depth per unit beam power (dimensionless because of the materials factor).

The horizontal axis represents welding speed v, multiplied by weld width assumed equal to beam width b (i.e., the surface area swept over by the beam per unit time), normalized by division with the thermal diffusivity λ_s^* of the material.

The original empirical curve of Hablanian fits the low-power welding data obtained earlier; the small dots are experimental points for various materials and welding parameters (all in-vacuum 4.8-kW maximum beam power) collected from the literature by Hablanian.[9] The high-power data from our tests are shown by larger symbols.

The lines for welding efficiency were introduced by Fink.[10] The high-power data show a higher efficiency than the original Hablanian curve; yet this curve, if extrapolated would lead to over 100% efficiency, which is impossible. The data points for 56 kW and stainless steel fall on a constant efficiency line [as to be expected from Fig. 13(b)] rather than on the extrapolated Hablanian curve. Since the 44-kW data did not show constant efficiency [see Fig. 13(b)], they appear here in the space between a constant efficiency line and the Hablanian curve. For further comments see Ref. 10.

With a beam power of 50–60 kW at 160–175 kV, butt welds between 5-cm-thick steel plates or lap welds between two 2.5-cm-thick plates can be made in one pass, with good depth-to-width ratios, and with low heat input. While the properties of these welds have not been evaluated except for the inspection of numerous cross sections, it is to be expected that they are as satisfactory as those of other thick EB welds made in vacuum[11,12] as well as previous lower-power welds in air.[13] With respect to one of the specific problems encountered with EB welds, namely, the presence of root cracks, we have observed that they never seem to occur on full penetration welds where sufficient metal comes in contact with a cooled backing plate. Perhaps the difference is that in this case the metal freezes from the root up and not from the top down; besides, it can shrink three dimensionally.

The greater standoff which can be tolerated with beams of higher power and voltage has additional significance, namely, one gains greater accessibility to the workpiece. The muzzle of the electron gun with its beam exit apertures has a minimum size since the nozzles must be fairly massive, water cooled, and exchangeable. Thus, the width of the front end of the gun, measured from the beam axis, is at least 1.5 cm on one side. With a 5-cm distance the machine can therefore approach a flat workpiece surface at an angle of 20°. At this angle we can weld the inside corner of a T joint with "legs" of any length.

The larger standoff also leaves room for rugged exchangeable guard pieces between the workpiece and the beam exit nozzles to protect the nozzles from backsplashing metal and assure long nozzle life even at high welding power.

The metal flow is obviously greatly influenced by the position of the workpiece. When the finished weld moves along a downward slope, the electron beam staying on the high side, the melt flows away from the beam. On the other hand, if the workpiece moves upward, liquid metal flows back toward the electron beam. We have observed that even if the deviation from horizontal of the workpiece motion is only 5°–10°, the "downhill weld" always shows an exaggerated "crown", whereas the "uphill weld" shows a rather flat upper surface. But, we found *no* difference in the depth of these welds when measured from the original workpiece surface, as we always do.

FIG. 16. Welding speed was too low, thus a cut resulted; it shows the width of the beam at the workpiece.

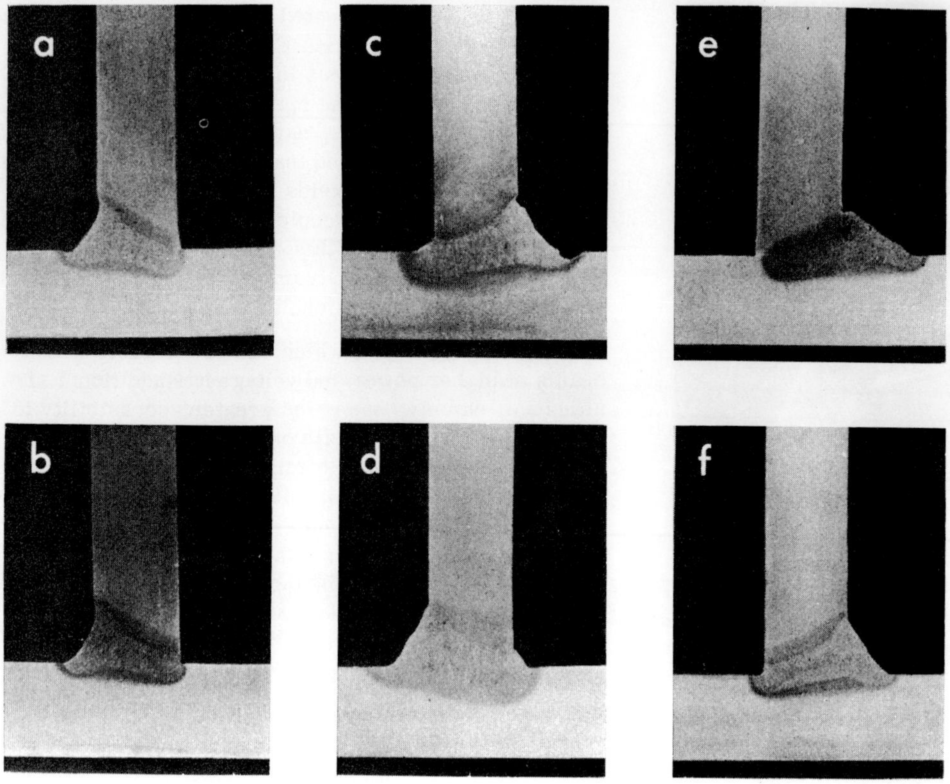

FIG. 17. Fabrication of T sections from 1.3-cm-thick cold-rolled steel plate by high-power high-speed EB welding. Assurance of proper corner fillets by the addition of filler metal and a chamfered chill bar. Beam power 56 kW at 175 kV; work distance (beam exit nozzle to inner corner of T) 5.1 cm. Welding speed and other data as follows: (a) 3.6 cm/sec, no filler, undercut, not fully through, beam aimed too low. (b) 3.75 cm/sec, no filler, undercut, through, chill bar not chamfered. (c) 2.12 cm/sec, some, but not enough filler, heavily chamfered chill bar. (d) 2.12 cm/sec, square filler rod, no undercut, heavily chamfered chill bar. (e) 3.0 cm/sec, 0.5-cm square filler rod, beam aimed too low, not through. (f) 2.96 cm/sec, 0.24-cm square filler, chill bar with small chamfer, no undercut, through, acceptable weld. Of the numerous cross sections which were made none showed any porosity or any cracks. See Fig. 12 for a comparison of welding depth.

IV. HIGH-SPEED WELDING OF MEDIUM-THICKNESS WORKPIECES

Another advantage besides that of welding very thick sections is gained with high-power beams: It is possible to weld medium-thick stock, e.g., mild steel of 1.2–1.5 cm thickness, with higher speed than ever before. For test purposes we used a T section, which is a very basic shape from which many more complex units can be composed. Flat pieces of cold-rolled steel 1.3 cm thick were held by C clamps in a T configuration with surfaces and edges as rolled but wire brushed and painted with deoxaluminite. The foot plate of the T made a 16° angle with the vertical beam (see Figs. 15 and 16). The beam was aimed at a point 1 mm beyond the inner edge, away from the foot plate. A standoff distance of 5 cm was used. Figure 15 shows the process in action. The results are shown in Figs. 16 and 17. The first run was too slow and produced a partial cut, shown in Fig. 16, which gives a vivid impression of the direction and width of the electron beam at the workpiece level.

Since a through weld is essential for the strength of the T piece, the run out of molten metal had to be stopped by a cooled copper bar in the rear corner of the T. This bar was chamfered to various degrees.

Figure 17(a) shows a weld not penetrating fully to the chill bar, because it was aimed too low and the speed was slightly too high. The corners at the top side show undercut, as one would expect with the liquid metal forming a smooth fillet in what was originally a sharp corner. Figure 17(b) shows a weld with an unchamfered chill bar; there is no fillet in the rear corner and there is again some undercut at the top or crown side of the weld. Figure 17(c) shows what is obtained at lower speeds and with a heavily chamfered chill bar. The undercut is enhanced because the metal formed another fillet in the rear corner.

To eliminate undercut, one obviously must add metal, which we did by placing a metal rod in the corner prior to welding and allowing it to melt during the welding run. The result can be seen in Figs. 17(d) and 17(f). The electron beam has to be properly aimed; if it is too close to the foot plate the shape shown in Fig. 17(e) results. The ideal shape is shown in Fig. 17(f). Figure 18 gives a side view of such a T section, showing start and finish of the weld.

It seems that one should also try to weld this kind of T section in a dual run, making one smaller weld on each side. The welding speed would more than double when welding half as deep, as we know from the depth-vs-speed curves. The result of two of many such tests is shown in Fig. 19. Oddly enough, it was never possible to avoid root cracks in the *second* of the two welds,

FIG. 18. View of EB welded T section showing start and finish of the weld with remainder of filler rod.

whereas the first one never showed any cracks. It is an open question whether a pair of such welds if made with two beams simultaneously might be perfectly sound.

With a weld as in Fig. 17(f) shrinkage pulled the center bar of the T out of alignment by 1.5°. However, it was possible to compensate for this effect by machining the edge of the center bar at an angle of 88.5° to the face. This would be no problem in large scale production.

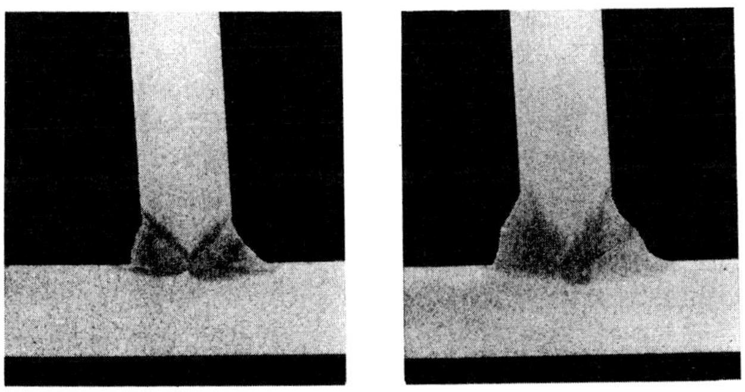

FIG. 19. Unsuccessful *consecutive* dual T-section welds. The *second* weld always showed centerline cracks, but never the first one. If these welds were made simultaneously, with two beams, cracking might be avoided and a speed advantage gained.

FIG. 20. T-section weld of type 304 stainless steel, made with a beam power of 56 kW at 5-cm work distance. The welding speed is 2.9 cm/sec.

As the photographs of Fig. 17 show there is a slight curvature produced in the footplate of the T section due to the weld shrinkage. Should it be disturbing, then one would compensate for it by rolling the foot plate with an original curvature or bend in the opposite direction. A double-sided weld, as in Fig. 19, would reduce all thrinkages considerably.

With stainless steel the same type of weld can be

FIG. 21. High-speed seam welds joining two pieces of hot-rolled steel, each 3 mm thick. The piece in the rear had offset edges (see bottom corner) which did not interfere with the production of a good weld. The piece in the middle has a perfect smooth crown. The piece in front was welded with too narrow a beam, and too fast; only a partial crown was produced. Beam power for all welds 56 kW at 175 kV; beam vertically downward. The edges of the plates were wire-brushed and painted with Deoxaluminate (Special Chem. Corp., Ossining, N.Y.). Work distance (rear to front): 6.3 cm, 6.3 cm, 5.7 cm; welding speed (rear to front): 13.8 cm/sec, 13.8 cm/sec, 16.9 cm/sec.

(a) (b) (c) (d) (e)

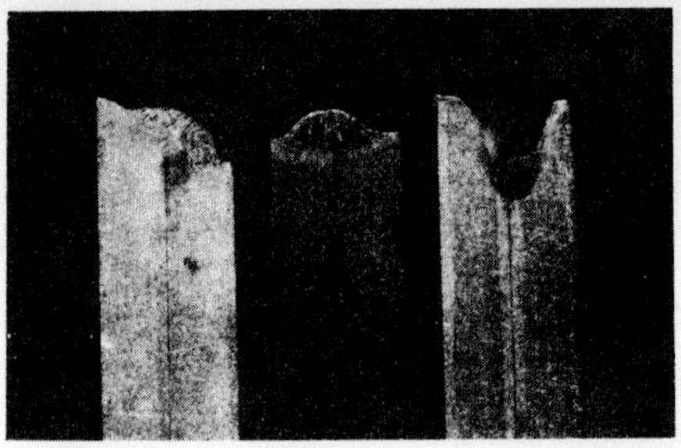

(f) (g) (h)

FIG. 22. Cross sections of seam welds made with a beam power of 56 kW (at 175 kV). Speed (in cm/sec) and work distances (in cm) were as follows: (a) 14.8 cm/sec, 6.6 cm; (b) 14.8 cm/sec, 7.0 cm; (c) 16.0 cm/sec, 7.0 cm; (d) 16.9 cm/sec, 7.6 cm; (e) 13.8 cm/sec, 6.4 cm; (f) 16.9 cm/sec, 6.0 cm; (g) 16.9 cm/sec, 5.7 cm; (h) 25.4 cm/sec, 2.5 cm. At the lowest speed and work distance (a) the most metal is melted; at (c) speed is too high to melt the full face; yet at larger work distance such as in (d) speed may even be higher and a good looking but shallow crown is obtained. At smaller work distances (f)—(h), the crown does not cover the full width of the sheets. The deep weld in (h) is very irregular and, at least in this material, could not be controlled closely enough to be useful. For (e) the two pieces were offset approximately 3 mm.

obtained, as one would expect. Figure 20 shows an example. The welding speed and energy input per unit length is practically the same as for mild steel, because we are already in the speed range where the lower heat conduction of the stainless steel is no longer very important and the specific heat of melting is about the same for both types of steel.

V. RAPID SEAM WELDS

Another application of high-power beams, which is most attractive from an economics point of view, is high-speed seam welding of rather thin material, e.g., 3 mm thick or less. With 36-kW beam power this process was very successful,[6] but with higher power it can be applied to heavier material. The welds shown in Fig. 21 are such welds. With a 56-kW beam at 6—7.5-cm standoff distance the practical range of welding speeds is 12—17 cm/sec (7.2—10 m/min). Obviously less metal is melted at the higher speeds [Fig. 22(a)—22(d)]; the process can tolerate considerable offset of the edges, as Fig. 21 also shows.

In this welding mode, with large standoff and, therefore, a rather wide beam at the workpiece, a wide trace of metal is melted. Although higher welding speeds can be obtained at closer distances, the metal flow processes may no longer permit making a good-looking sound weld. What happens can be seen in Figs. 22(f)—22(h) which shows weld cross sections obtained at smaller work distances. The deep weld of Fig. 22(h) is remarkable; melting occurred so fast that the outside of the 3-mm-thick sheets show no heat effect. The rope or bead of metal seen in Figs. 22(f) and 22(g) could not be made to flow fully across the two plates, although the whole width of these plates was clean, free of oxides, and, in fact, melted down. There are yet many interesting facets of this process to be explored and explained.

These high-speed welds are not particularly brittle, as Fig. 23 shows, where a pair of these seam-welded plates were gradually pried apart.

VI. CONCLUSIONS

The development of an electron-beam gun with higher-power capabilities and the associated high voltage and auxiliary power supplies now permits the generation of beams carrying 60 kW into the atmosphere. At this power level the electron beam in air can make narrow and deep welds with profiles approaching those of welds made in vacuum. Heating of the gas in the beam path by the beam itself causes a significant decrease in gas density and, consequently, a marked reduction in beam scattering. Thus the high-power density characteristic of the focused electron beam is retained to greater working distances, permitting access to a much greater variety of workpiece configurations.

Clearly, the higher beam power permits increased welding speed for a given depth of penetration. And because of the decrease in heat-conduction losses at high welding speeds, the welding process becomes increasingly efficient at higher beam powers, approaching 55% at 60-kW power.

Since 60 kW does not represent either a theoretical or practical limit to the electron beam power that can be delivered into the atmosphere, further increases in the already very good efficiency of the welding process may be expected at even higher beam power levels. The excellent conversion efficiency of electrical power into EB power will be a decisive advantage with respect to energy consumption and cost.

Thus nonvacuum EB welding should become quite competitive with vacuum EB welding with regard to weld

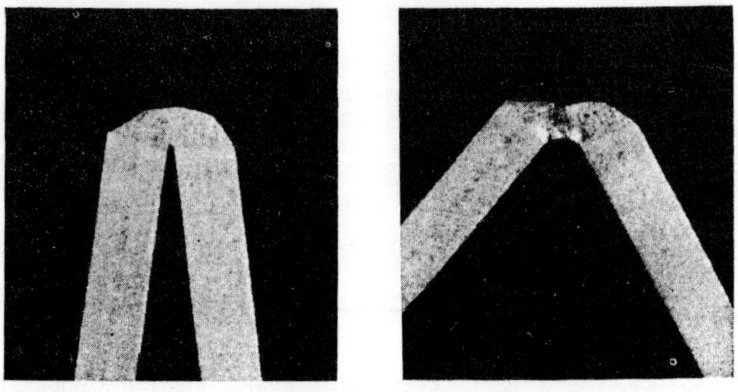

FIG. 23. A portion of the plates shown also in Fig. 22(b) is gradually pried apart; only after an angle of about 60° was reached did cracking begin in the inner corner.

profiles, weld depth versus speed, weld quality, and power efficiency.

APPENDIX: SOME COMMENTS ON THE HISTORY OF ELECTRON-BEAM TECHNOLOGY

In the course of the years a number of "histories" of EB technology, and EB welding in particular, have been published, which are frequently quoted. Unfortunately they are far from being complete and correct. There is in particular the article by Bunshah[14] who does not even list the contributions of the Nobel prize winners Hertz and Lenard, although this "history" goes back as far as 1705. The book by Meleka[15] erroneously, yet curiously at great length, tries to give the impression that Steigerwald had more to do with the atmospheric electron gun developments by Schopper and Schumacher at the University of Stuttgart, than to supply, on request, one of the new types of cathodes he had just described in the open literature. As a matter of fact, he had nothing to do with the design of the gun and beam-transfer system to air, nor with the suggestion to use the beam for welding or thermal blasting of materials.

In view of these less than accurate histories, we thought the reader may want to know the following facts: The first non vacuum electron gun in which a *focused* electron beam was transported to the atmosphere was designed and described in detail by Schumacher[16]; an earlier version was mentioned in 1950 by Schopper and Schumacher.[17] (This gun did not use thin foil windows of the Hertz/Lenard type.) Thin metal sheets were cut and fine wires welded in the 50-W beam of this machine. Schopper and Schumacher considered, and mentioned in patents, lectures, and publications the possible uses of these atmospheric electron beams for localized and controlled melting, cutting, welding and thermal blasting of materials.[18] We believe the last paper of Ref. 18 to be the first mention of EB welding in public print. It predates the first use of EB welding in vacuum. Revised design procedures for the beam-transfer system, with explicit mention of welding in the atmosphere, were published in 1961.[19]

Only thereafter, in October 1961, did Alloyd Electronics Corp., Cambridge, Mass., obtain a contract from AFSC/WPAFB for Development of Non-vacuum Electron Beam Welding.[20] For this work an electron gun of the Schopper/Schumacher type was bought from Heraeus GmbH, who, since 1954, have sold a commercial model with 4-kW beam power, but have not studied welding. In 1964, Meier[21] of Hamilton Standard Div., United Aircraft Corp., described nonvacuum welding with a 7-kW beam; in 1966 he showed 12-kW welds.[22] Dietrich showed welds made with the Heraeus gun at the Third International Congress on Vacuum Science and Technology, Stuttgart, 1965.[23]

In the mean time Schumacher developed the EB fluorescence and scatter probe for supersonic gas flow analysis, and an atmospheric microprobe for thickness gaging.

Also in 1965 a nonvacuum EB welder with a small lightweight gun head was described by Lempert, Lowry, Seaman, and Williams.[24] This became the first EB welder with a mobile gun head and stationary workpiece, from which the present Westinghouse machines evolved. As for the cathodes, the bolt cathode of Bas was used and further developed to yield higher currents, because it possesses much superior electron optical characteristics compared with any directly heated filament cathode.

[1] B. W. Schumacher, Optik (Stuttgart) **10**, 116 (1953); Ann. Phys. **13**, 404 (1953); *Transactions of the* 1961 *Vacuum Symposium and Second International Congress, Washington, D.C.* (Pergamon, New York, 1962), p. 1192–1200; E. Schopper and B.W. Schumacher, U.S. Patent 2,899,556; Stuttgarter Nachrichten **119**, 14 (1953). The latter reference is, as far as we know, the first time that EB welding, in air or vacuum, has ever been mentioned in print.

[2] B.W. Schumacher, Optik (Stuttgart) **39**, 558 (1974).

[3] B.W. Schumacher, Welding J. **52**, 312 (1973).

[4] R.C. Smith and B.W. Schumacher, Nucl. Instrum. Methods **118**, 73 (1974).

[5] J.F. Lowry and B.W. Schumacher, Nucl. Instrum. Methods (to be published).

[6] R.J. Lanyi, B.W. Schumacher, and J.M. Wells, Westinghouse Eng. **34**, 98 (1974).

[7] B.W. Schumacher and J.H. Fink, IEEE Trans. Nucl. Sci. **NS-20**, 403 (1973); and J.H. Fink and B.W. Schumacher, Optik (Stuttgart) **39**, 543 (1974).

[8] B.W. Schumacher and F.D. Seaman, in *Tenth Symposium on Electron, Ion, and Laser Beam Technology* (San Francisco Press, San Francisco, 1969), p. 29-42.

[9] M.H. Hablanian, in *Proceedings from the Fifth Symposium on Electron-Beam Technology*, 1963, edited by J.R. Moreley (Alloyd Electronics Corp., Cambridge, Mass., 1963), p. 262.

[10] J.H. Fink, Welding J., Res. Suppl. 1375 (1975).

[11] J.G. Bjeletich, in Air Force Materials Laboratory Technical Report No. AFML-TR-73-197, 1973 (unpublished).

[12] P.J. Konkol, P.M. Smith, C.F. Willebrand, and L.P. Connor, Welding J. **50**, 765 (1971).

[13] R.M. Evans, Metals and Ceramics Information Center Report No. MCIC-73-13, 1973 (Battelle, Columbus, 1973), p. 11.

[14] R.F. Bunshah, in *Introduction to Electron Beam Technology*, edited by R. Bakish (Wiley, New York, 1962).

[15] A.H. Meleka, *Electron-Beam Welding* (McGraw-Hill, New York, 1971).

[16] B.W. Schumacher, Optik (Stuttgart) **10**, 116 (1953).

[17] E. Schopper and B.W. Schumacher, Phys. Verh. **6**, 73 (1950); Z. Naturfch. A **6**, 700 (1951).

[18] German Patent 971610 (1952); Swiss Patent 332,987; French Patent 1,090,183; British Patents 77,426/427/428; U.S. Patent 2,899,556; E. Schopper and B.W. Schumacher, Stuttgarter Nachrichten **119**, 14 (1953).

[19] *Proceedings of the Second International Congress and the Eighth Symposium on Vacuum Science and Technology, Washington, D.C.* (Pergamon, New York, 1961), p. 1192.

[20] ASD Technical Report No. 63-7-926, 1963 (unpublished).

[21] J.W. Meier, in *First International Conference on Electron and Ion Beams in Science and Technology*, edited by R. Bakish (Wiley, New York, 1965).

[22] J.W. Meier, Metall. Soc. Conf. **51**, 505 (1966).

[23] W. Dietrich, in *Proceedings of the Third International Congress on Vacuum Science and Technology, Stuttgart*, 1965 (Pergamon, New York, 1966), p. 595.

[24] J. Lempert, J. Lowry, F. Seaman, and C. Williams, in *Proceedings of the Electron and Laser Beam Symposium, Penn State University*, 1965, edited by A. El-Kareh (San Francisco Press, San Francisco, 1965), p. 393.

Nonvacuum electron beam welding comes of age

Exciting progress has been made in welding out of a vacuum. The improvements show up in faster welding, better welds, and lower costs.

Electron Beam welding can produce welds of high quality at speeds five to ten times faster than Mig or submerged arc welding and do it at costs that are competitive in operation and in equipment. EB can also be automated to further reduce costs.

The EB welding equipment for high volume production may be of the partial vacuum type or the nonvacuum. Partial vacuum equipment requires pumpdown time before welding to reach operating vacuum level. The time may range from several seconds to as long as 30 minutes. This requirement generally limits high production partial vacuum EB welding to parts of less than 1 cubic foot, which require pumpdowns of 3 to 7 seconds, or where sliding vacuum seals can be employed. Nonvacuum EB welding requires no vacuum chamber for the work. The work can be any size and stays at atmospheric pressure during the welding cycle. The job requires no preweld pumpdown, and some of the handling equipment is simplified. A nonvacuum EB welder generates free electrons at a specific rate (beam current), accelerates the electrons to high speed (beam voltage), and limits the beam diameter at the work (spot size); the welder does all these tasks with precision control.

How a nonvacuum welder works

The 36-kw. welder recently developed by Westinghouse, generates free electrons by heating a small-diameter tungsten rod hot enough for electrons to boil off the rod at the required rate. Electron bombardment from two auxiliary filaments next to the tungsten rod heat the rod to its emission temperature. The main emitter and the auxiliary filaments are integral parts of the same, replaceable, plug-in assembly. The plug-in cathode lasts better than 40 hours at 36-kw. and it can be rebuilt. An operator can change cathode assemblies in 30 minutes.

In operation, the cathode assembly is maintained at a high negative potential (-160 kv.) relative to a ground potential anode plate positioned 1.2 inches from the cathode. The electrons from the filament, after being accelerated and shaped by the electric field, pass through a hole in the anode plate which aims them towards the work. On the way to the work, the beam diverges due to coulomb repulsion and the angular aperture of the gun. A magnetic lens along the path refocuses the beam.

To keep the cathode assembly from oxidizing and from shorting to the surrounding ground-potential surfaces, the cathode assembly must be operated in a vacuum environment of 10^{-4} Torr. A semivacuum is also needed along the beam transmission path to limit the beam spread caused by collisions between electrons and air molecules. The required vacuum levels are maintained by a series of vacuum pumps which progressively evacuate several small chambers that form the beam transfer column. The chambers are separated by small orifices through which the electrons must pass until they reach atmospheric pressure at the end of the column.

Over-pressurizing the end of the beam exit nozzle to slightly above atmospheric pressure keeps the metal vapor from being ingested into the transfer column. The diameter of the beam, directly below the exit orifice, is primarily determined by the molecular weight of the gas used to over-pressurize. Helium, (molecular weight: 4) makes for considerably less beam spread than does air (molecular weight: 29). Therefore helium makes for greater penetration or faster welding at a given distance from the work. With air as the effluent gas, the distance between the exit nozzle and the work must be about 1.2 inches. Using helium you can go to 1.8 inches. All electron beams release X-rays when fast-moving electrons collide with molecules and equipment must be shielded.

The machine welds with the beam vertical (down), horizontal, or at any angle in between. The welding head connects to remote mechanical vac-

PRINCIPAL PARTS OF A 36-KV. NVEB WELDER

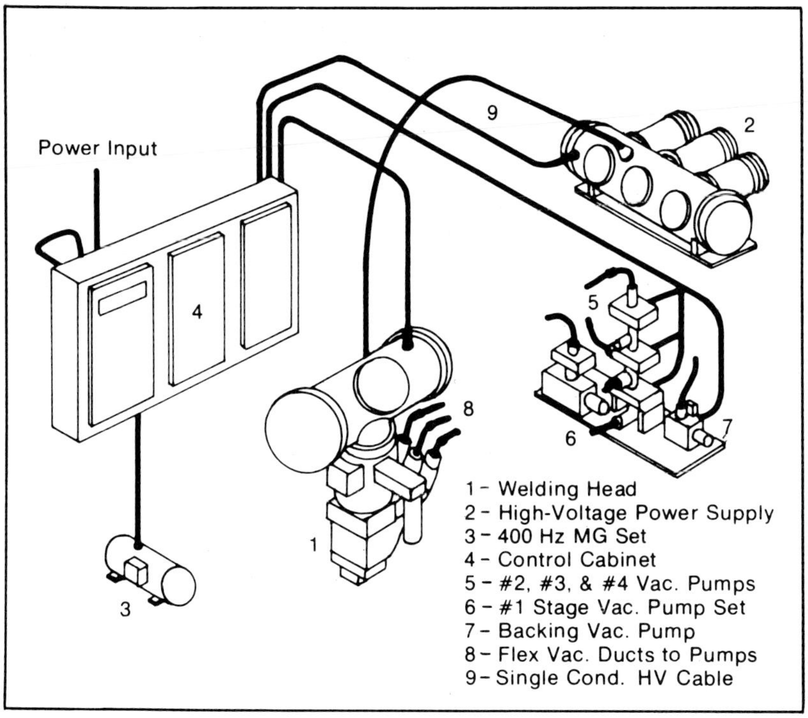

WELDING HEAD

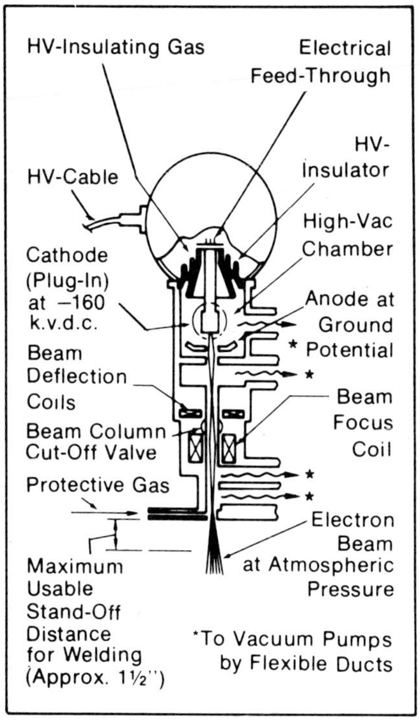

uum pumps, and to a remote high-voltage power supply. The EB welder can be in a fixed-position moved at high speeds during welding.

The 36-kw. high-voltage power supply, which generates the 160-kv. accelerating potential for the electrons, has three 12-kw. elements. A 400-hz. motor-generator provides regulated input power to the supply and also isolates the system from 60-hz. powerline fluctuations. Instead of oil, sulfur hexafluoride (SF_6) gas serves as high-voltage power supply and in the cathode power supply section of the welding head. This insulation simplifies maintenance.

The control cabinet contains setup controls, instrumentation, high-voltage and beam-current regulators, and general sequencing control. The controls are designed to ensure correct settings before welding starts and to require only slight attention from the operator.

Process capability

Using air as the effluent gas, a 36-kw. welder welds anything in steel from a 0.1-inch buttweld at 350 i.p.m. to a 1.0-inch buttweld at 25 i.p.m. With helium as the effluent, penetration can be increased by factors of from 1.3 at the low end of the speed range to 2.3 at the high end.

A wide range of materials can be welded — 300 and 400 series stainless steels, low to medium carbon steels, aluminum-killed and rimmed steels, aluminum alloys, titanium and Inconel. Weld quality depends on such variables as material, required penetration, joint design, welding speed, fitup, and surface preparation. High quality welds may require gas shielding to prevent oxidation and porosity. To compensate for poor fitup, a cold- or hot-wire feed may be needed.

Operating costs

Excluding special jobs requiring filler metal or shielding gas, the main costs of a nonvacuum EB welder are electrical power, electron emitters, beam exit nozzles, and the effluent gas.

Nonvacuum EB welders are efficient devices delivering 85 to 95 percent of the generated beam energy to the work. If ancillary equipment, such as vacuum pumps, are used, the overall efficiency drops 55 to 60 percent. At 2 cents per kw. hour, the cost is $1.30 an hour for electrical power.

Cathode assemblies are guaranteed for 40 hours at the 36-kw. power level. Factory-rebuilding costs $140, bringing cathode costs to $3.50 an hour.

So the total operating cost for a 36-kw. welder using air as the effluent gas, would be $6.40 an hour. Helium boosts the figure to $16.40 an hour. Sixty-five percent of the helium used for over-pressurization enters the first-stage vacuum pump, and can be recycled by using a recovery system. This recovery reduces helium consumption to 35 percent of the nonrecovered value, and decreases operating cost to about $10 an hour.

Production adaptability

Because the NVEB process can weld so fast and for such low unit operating costs it cries for fully automated work handling systems to take advantage of its potential for high-volume. Handling equipment ranges from simply moving the work to and from the welding position, to precise positioning of the work at the beam. Sometimes the same equipment does both jobs, like the setup for welding the longitudinal seam of tubing continuously. Other jobs may need a computer-controlled positioner. Westinghouse build a prototype CNC (computer numerical controlled) system for welding auto frame components. Two 36-kw. welders mount with beam axes horizontal on two independent four-axis CNC positioners. The two welders face each other and simultaneously contour weld the two butt seams of a complex box section (stationary during welding) at 280 i.p.m. Dual synchronous shuttles, each holding two 36-inch-long parts, transfer the parts to and from the weld position. They go through radiation gates at each end of a booth which encloses the positioners and welders. When one shuttle is in weld position, the second is outside the radiation booth for automatic unloading and manual reloading. Total machine cycle time is 22 seconds for welding two parts, each 72 inches long. Gross production: 325 parts an hour. With two operators to manually load parts, the line can boast of an operating efficiency of 80 percent. Using air effluent and no filler wire, the operating cost is 11 cents — much better than 26 cents estimated for manual welding with filler wire.

Another example of highly automated NVEBW equipment is a three-axis CNC system for welding edge seams of emission control devices. It has combined welder and part motion. The machine rotates the part around its centerline — A, moves the horizontally mounted EB head in and out — B, and rotates the head about its beam — C. Only A and B need precise control.

The CNC synchronizes all motions so that the electron beam impinges on the work at a 90° angle and moves at a constant speed relative to the work. While the welding head moves rapidly to stay perpendicular to the wrok, the weld puddle is steady and produces consistently high-quality weld. One CNC can control several welders at once.

Radiation tunnels eliminate the need for radiation gates in transferring parts. Tunnels have three low-speed indexing transfers having a U-shaped floor arrangement. The transfers deliver and remove parts from a four-station dial index table. The table moves the parts to and from the weld station. The time required for nonwelding motions is less than 3 seconds of the 14-second total machine cycle time. A programmable controller sequences the machine allowing the CNC unit to be dedicated to the motion control of the three servo axes.

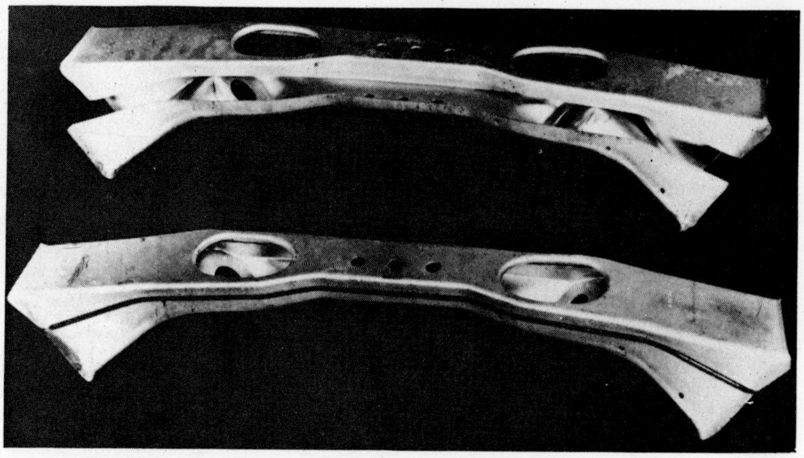

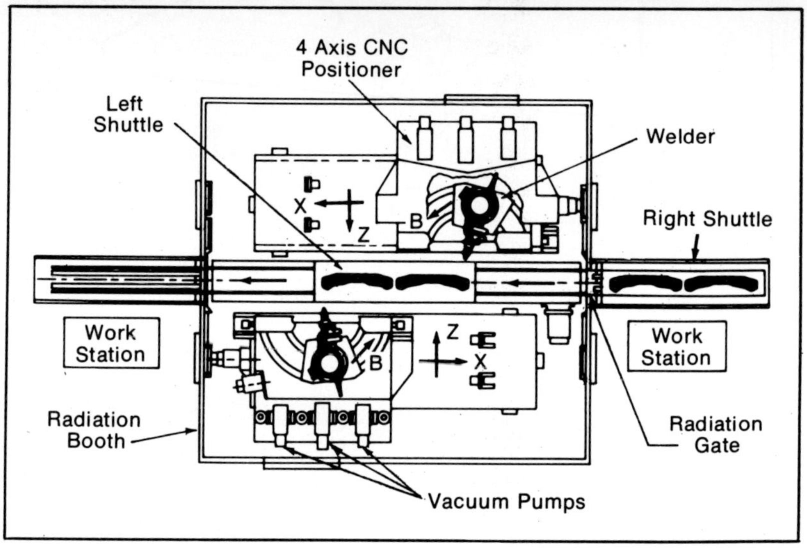

Computer numerical controlled (CNC) setup for NVEB welding auto frame crossmembers. Two 36-kw. welders mount on four-axis CNC positioners. The welders contour weld at once two butt seams of the complex box section (shown) at 280 i.p.m. Synchronized shuttles hold two crossmembers each, transfer them to and from welding position. Radiation gates are at each end of a booth that encloses positioners and welders. When one shuttle is in weld position, the second is outside the booth for automatic unloading and manual reloading. Cycle for two 72-inch parts: 22 seconds; in an hour, 325 parts. Operating efficiency with two workers: 80 percent.

The present and the future

Approximately 14 non-vacuum EB welding systems are in production use. Another 12 are being installed for welding automative parts. Ten more have been installed in welding development laboratories. Reason: quality welds, high speeds, and economy. For low-volume jobs, one of the more common welding methods may be cheaper. However, for high-volume applications requiring welds of high quality the capability and versatility of NVEBW may justify the higher initial investment. ■

Based on a paper presented by R. J. Lanyi of the Industrial Equipment Div. of Westinghouse Electric Corp. at the Westinghouse Machine Tool Forum.

Principles and applications of an electron beam welding machine with mobile local vacuum chamber

"SUCTION CUP OR VENTOUSE* MACHINE"

G. Sayegh**, P. Dumonte**, T. Nakamura***

I. Introduction

When large workpieces are to be electron beam welded, in shipbuilding and oil industry for example, the necessity of creating vacuum in a chamber containing the components to be welded represents a serious handicap and even one of the limiting factors of the process. To obviate the problems arising from the huge dimensions of the vacuum chamber, Sciaky have developed in their laboratories in Vitry-France a special machine featured by the so-called "Ventouse" type mobile vacuum chamber. This machine welds bulky parts without having to introduce them in a vacuum chamber.

Below are described the operating principle and the technical specifications of this machine, some laboratory experimental results and several potential industrial applications of this equipment.

1. Operating principle

Machine operating principle merely consists in creating vacuum only around the area to be welded and having an electron beam produced by a gun integral with a mobile carriage, strike the works.

Sealing at the lower part of the works can be achieved by
- continuous arc-weld with limited penetration,
- hermetically fixed metal support or
- counter-chamber pumped to 10^{-2} Torr approximately.

The "ventouse" is fixed at the upper part of the works (see fig. 1 next page). This Ventouse is composed of a chamber (A) with seals located at its lower part to ensure tightness against workpiece face. A table integral with this chamber (B) has a longitudinal "Vee" (C) notch whose axis coincides with the plane of joint. This notch is fitted with a special silicon seal armoured with a metal braid. The vacuum in the chamber when pumped from one of the ends is 10^{-2} torr. A carriage supporting an electron gun (D) slides on this table over the whole length of the notch.

The most important problem consists in having the electron beam generated by the gun strike the works through the notch of the table in which is housed the silicone seal. This is achieved by a specific system composed of a series of rollers (E) designed to lift the "Vee" seal ahead of the carriage and to move it laterally away from the notch, thereby leaving space for the electron beam

* ventouse is the name is french.

** G. Sayegh and P. Dumonte are with Sciaky-France

*** T. Nakamura is with Dengensha Manufacturing Co. (Japan)

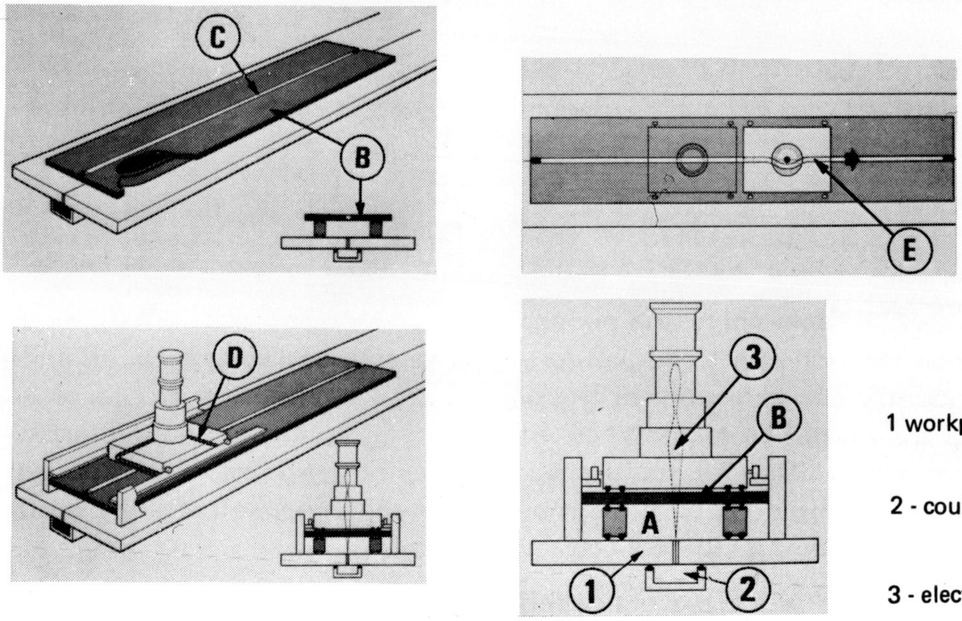

1 workpieces

2 - counter chamber

3 - electron beam

Fig.1 - operating principle of the ventouse machine

to flow.Then, the seal is replaced in the notch at the rear of the carriage. A local vacuum chamber connecting the gun with the works is thus created and as the carriage is fed, it moves the local vacuum chamber accordingly along the joint, and the beam can perform the weld bead.

2. Machine technical specifications

A - electron gun

B - moving carriage

C - driving motor

D - control panel

E - T.V. screen

Fig.2 - general view of the ventouse

The machine shown in fig. 2 has a linear weld capacity of 2.50 m. The overall weight of the equipment is 2,000 daN approximately but could be reduced if special alloys were used. The approximate weight of the mobile unit (carriage + gun) is 350 daN.

left : bottom view of the chamber

right : upper view of the table and endless screw

Fig.3 - some details

The table is made of non-magnetic stainless steel (fig. 3) to eliminate the influence of unwanted magnetic fields on the beam. Carriage lower section is equipped with two seals designed to ensure carriage tightness in the chamber. These seals are mobile on the upper surface of the table and this is why the english term "sliding seal machine" is sometimes applied to this type of equipment. In addition, the carriage is fitted with rollers which travel on rolling tracks thanks to a suitable guide system. Carriage speed is adjustable from 0.05 to 1 metre per minute by motor-driven endless screw but this speed can be increased if necessary using other motors and gears.

The "Vee" seal housed in the table notch is an essential component of the system. Indeed, due to its important role, this seal must withstand significant tensile and shearing stresses and it is of utmost importance that the seal be accurately guided during carriage travel so as to avoid jamming with the subsequent damages. The mechanical properties and the chemical composition of the seal and the metal armour have been determined after long tests.

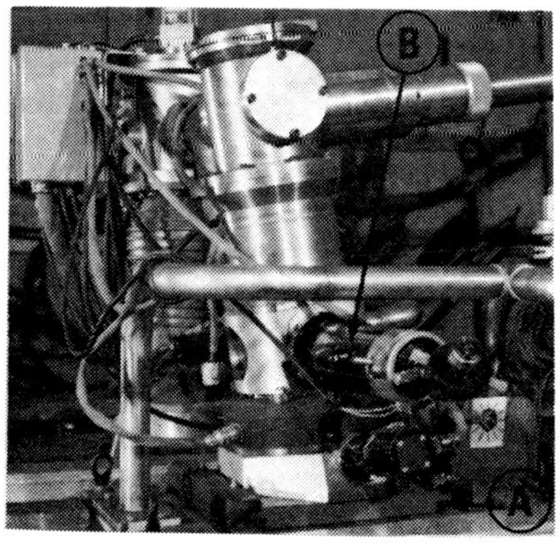

Fig.4 - lateral displacement of the gun

A - sliding table

B - viewing optics

The gun is fixed on a table which moves laterally (at right angles with weld direction) through a suitable drive system (figure 4). The sideways motion ($\pm$ 5 mm) permits accurate positioning of the beam on the joint. It is possible to associate a seam tracking system (mechanical sensor or electronic seam tracker) in order to detect joint position and make the necessary corrections to gun sideways motion control motor so as to accurately align the beam on the joint.

It can be noted that gun axis is slightly inclined with respect to the vertical plane in order to reduce metal vapour emission towards the gun. Further to previous studies conducted by Sciaky on the influence of the pressure in the work chamber on power distribution within the electron beam (1), we have selected a pressure of 30 - 40 microns as maximum vacuum level in the chamber. From our laboratory tests it can be concluded that in an electron beam accelerated by 45 - 60 kV, the power distribution is slightly affected when the vacuum level does not exceed 30 - 40 microns. The actual vacuum in the chamber is in the order of 20 microns and increases to 25 - 30 microns during the welding operation with a 12 kW beam.

The various components of the gun (electrodes, magnetic coils...) have been optimized by computer programme (2) . A 30 kW beam accelerated by 60 kV giving a spot diameter below 0.6 mm at a gun-to-work distance of 350 mm was thus obtained. The gun is supplied by a flexible cable of adequate length to ensure carriage free motion. The deflecting coils are supplied by a suitably rated generator which ensures beam vibration on the joint and power distribution regulation during welding.

Joint viewing is achieved by using optics integral with the carriage or TV camera. Therefore, beam alignment on the joint is obtained by first moving the carriage without the beam, next the weld can be effected with the desired beam power. A filler wire feed system can be added when deep penetration is required, i.e. when the gap is about 0.5 to 1 mm.

Fig.5 - vertical position of the machine

Generally speaking, the welds are effected in the flat position with a vertical gun but the equipment can be arranged vertically for cornise welds (figure 5) - an interesting feature when thick gauges (above 40 mm) are to be joined. The Ventouse will then be maintained on the workpieces by the atmospheric pressure while the gun moves either horizontally (from left to right or vice-versa) or vertically (from bottom to top). In the second case, solidification of the first melted layers represents an horizontal support for the molten bath, thereby reducing the influence of gravity on bath flow.

3. Typical laboratory applications

Sciaky recently conducted butt-welding tests with 15 to 40 mm thick steel sheets (14 M4 - 16 M5 grades) for shipbuilding industry and similar tests with 5 to 20 mmthick AG5 aluminium for the oil industry.

Reference :

(1) P. Dumonte, G. Sayegh, J.P. Vignau -VI Kolloquium Strahltechnik - Braunschweig, mars 73

(2) G. Sayegh, P. Dumonte - 3rd Seminar on EB applications, Stratford/Avon mars 74

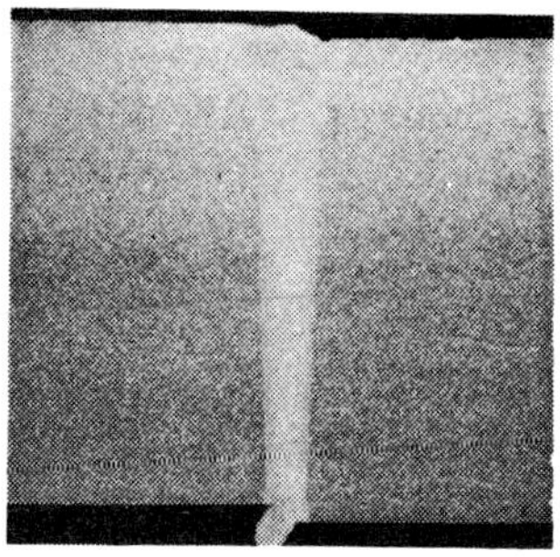

Fig.6 - weld in steel sheets of 25 mm thick

Fig. 6 shows a 2.5 m long weld on a 25 mm thick plate. A consistent, steady bead was obtained and the macrograph of the transverse section reveals a parallel faced joint free of defects. The welding parameters were : accelerating voltage : 55 kV, current : 220 mA , speed : 0.9 metre per minute.

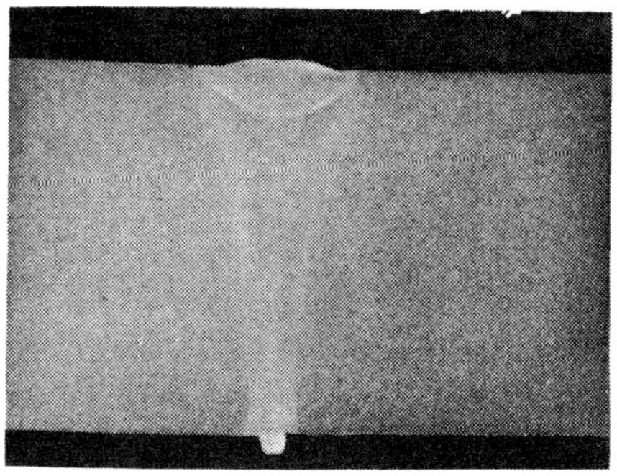

Fig.7a -

Fig. 7a is a macrograph of a 15 mm plate. In this example, a finishing pass was performed with a low power beam which permitted to correct a small bead undercutting.

Specimens from the welded plates were subjected to mechanical tensile and bending tests. Despite criticisms concerning the value of such tests in characterizing E.B. welds properties, it is interesting to examine the results. The tensile tests showed that rupture always occurs outside of the welded area and that the average breaking stress is 54 hbars. In addition, bending tests at both faces revealed no defect in the joint.

We have welded 5 to 20 mm thick aluminium plates.

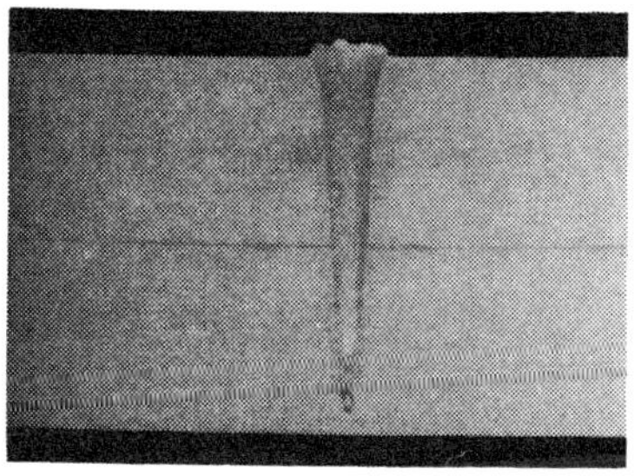

Fig.7b -

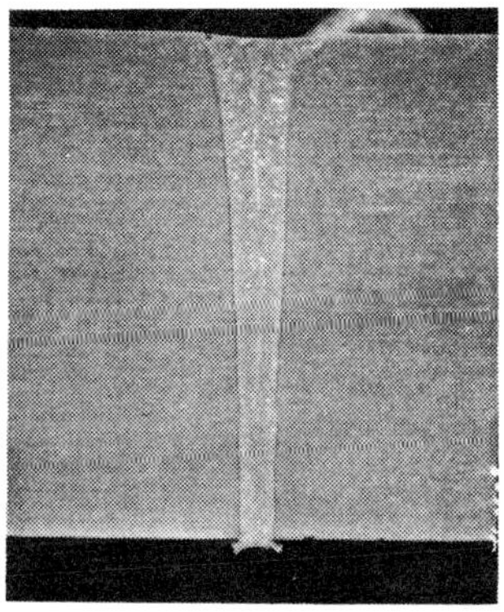

Fig.7c -

Fig. 7b shows a macrograph in 5 mm sheets. A backing support requested by the client can be noted. Fig. 7c is a cross-section in a 20 mm thick weld without backing support and the consistency of the molten bath is an interesting factor. The welding parameters were : Accelerating voltage : 48 kV - Current : 180 mA - Speed : 1 m/min.

Shrinkage after welding was measured on a base of 50 cm length. It was noted that in 5 mm thick sheets, shrinkage in a direction perpendicular to the joint is 0.02 mm while it was so small in a direction parallel to the seam that it could not even be measured.

4. Advantages and potential applications of the "ventouse"

The most interesting feature of the Ventouse is its suitability to make assemblies with the advantages of E.B. welding (negligible distorsions, deep penetration...) without having to introduce the works in vacuum chambers.

The equipment described in this paper is obviously of small capacity but the initial results have shown that the operating principle is reliable enough to permit extrapolation in two directions :

- towards the manufacture of 10 to 12 m long linear Ventouses which may find applications in Naval Engineering. It has been estimated that the cost of such an equipment can be amortized in one year with the sole savings on consumables when it is used for welding 30 mm thick plates. This equipment can produce a 10 m long weld in less than 30 minutes (figure 8)

- towards the manufacture of curved Ventouses suitable for spherical or cylindrical reservoirs with small curvature radius.

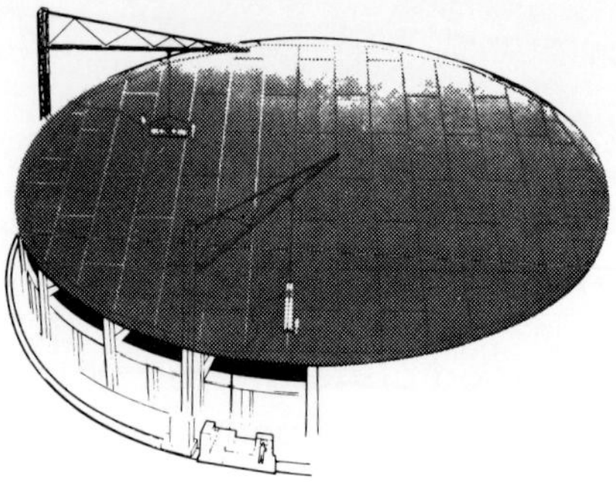

Fig.9 - welding of shields for nuclear reactors

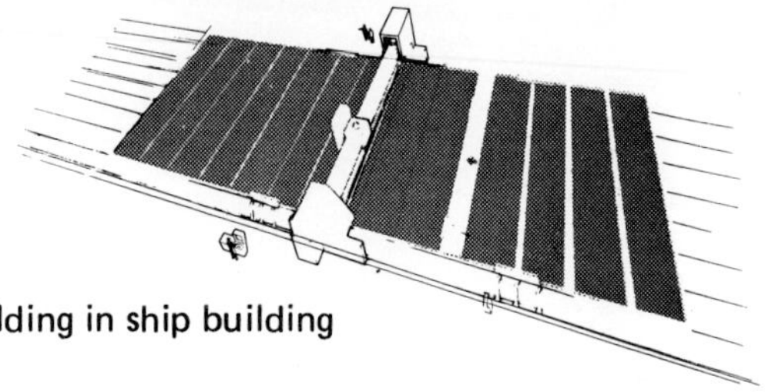

Fig.8 - welding in ship building

Another example of application for this equipment as described in this paper is the welding of shields for nuclear reactors. These shields are stainless steel circular plates with diameters of 15 to 20 m and thicknesses exceeding 100 mm. They are manufactured from 0.6 x 2 m butt-welded straight plates. A Ventouse equipment could possibly be designed for this butt-welding operation (figure 9).

Conclusion

We have described the operating principle and the technical specifications of an E.B. welder capable of joining large workpieces without making use of vacuum chambers.

From the first results, the conclusion is that this principle is reliable and sound ; developments aiming at greater lengths and curved surfaces can therefore be contemplated.

This process appears to follow the right way to industrial applications and future prospects are promising.

DESIGNING FOR ELECTRON-BEAM WELDING

F. J. BECKET*

SYNOPSIS: Electron-beam welding offers a new design concept for the fabrication of complex assemblies from simple machined parts. Full penetration welds can be made with minimal distortion and without a filler rod so that in many instances welding is carried out on finish machined components as a last stage of assembly. Applications ranging from the aero-space, electronics, automobile, food processing, nuclear, chemical and general engineering fields are included in the work now handled by a few specialist EBW jobbing shops, offering a complete sub-contract service for this advanced technology welding process.

ELECTRON-BEAM welding offers the designer a new concept in the fabrication and joining of metals. To exploit fully the process, however, an understanding of the techniques used is needed together with an acceptance of a few basic rules relating to the design and preparation of weld joints.

Among the many advantages to be gained from this metal fusion process the more obvious are minimal distortion, the elimination of oxidation effects, and the ability to fully penetrate thick plate sections without the use of a filler rod. The metals and alloys that can be joined include a wide range of low and medium carbon steels, stainless steels, nickel-base alloys, copper-base alloys, aluminium, and refractory metals.

In conventional welding practice, it is often necessary to prepare parts for welding in a semi-machined condition with the final machining carried out as a post-weld operation. In many applications, E.B. welding allows the post-weld machining stage to be eliminated due to the drastically reduced distortion allowances which need to be made. When comparing the economics of other metal joining techniques (e.g. conventional welding and brazing) it is important that cost savings resulting from any associated machining operations should be taken into account.

*General Manager, Electron Beam Processes Ltd., Byfleet, Surrey.

Basic Principles

Electrons, accelerated to a high velocity and focused on to a target, give up their kinetic energy in the form of heat. This is the essential mechanism of an electron-beam welding gun. The method of generating the electron beam is usually the basic triode system common to the TV tube and comprises a heated filament (cathode), a bias cup and anode as shown circled in Fig. 1. The accelerated beam is focused by the electrostatic lens formed by this arrangement but not to a degree satisfactory for welding. Therefore, provision is made for fine focusing the beam by means of a separate electro-magnetic lens system. An additional system of coils allows the beam to be deflected permanently by applying a steady direct current or to produce an oscillating or circular motion by means of an alternating current which may be of variable frequency up to 20,000 Hz.

For thermionic emission to occur, the electron gun must be operated in a vacuum of the order of 10^{-4} torr, and the vacuum environment also serves to insulate the high-voltage anode from the rest of the gun system. High vacuum is also essential if the filament, usually of tungsten wire or strip, is to operate at the optimum emitting temperature for a reasonable time between changes.

In the welding chamber, however, the order of vacuum can be relaxed to the range 10^{-1}—10^{-2} torr without suffering a serious loss of performance in welding due to beam scatter. The trend in modern electron-beam welding machine design has been towards separately evacuated systems in which the gun or column is pumped to a high vacuum by a fractionating oil diffusion pump and rotary pump; the welding chamber is partially evacuated by a high capacity Rootes pump backed by a matching rotary pump. Such an arrangement significantly improves the overall performance of a machine by reducing pump-down time and thereby the floor-to-floor time for completing a welding cycle.

An electron has an exceedingly small mass ($9{\cdot}1 \times 10^{-28}$g) but it can be accelerated to a velocity of about half the speed of light over a short working distance by using a voltage in the region of 150 kV. The higher the accelerating voltage,

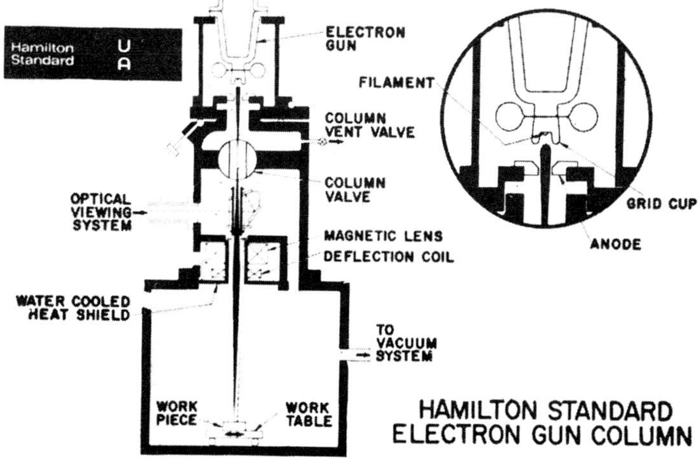

Fig. 1. Schematic diagram of E.B. welding machine and basic triode gun. (Courtesy: Hawker Siddeley Dynamics Ltd.)

Table 1.

Material or Material Combination	Example	Weldability
Low Carbon steels	EN.3	Good
Stainless steels	EN.58B	Good
Copper and Copper-base alloys	Cu-Ni; Al-Br; Ph-Br; Be-Cu	Good
Nickel-base alloys	Nimonic 75; Inconel; Nispan C	Good
Aluminium alloys—low copper grades	NS.4, 5 & 6. HS.30	Good
Refractory metals	Ti, Zr, Mo, Ta, Nb	Good
Medium and high carbon steels Nickel-chromium alloys Nickel-chromium-Molybdenum alloys; Cobalt-base alloys High strength Aluminium alloys	Higher Nimonics, Hastelloy, Stellite, Hiduminium 48, H.E.30 (extruded section)	More difficult—requires careful preparation and weld development.
Tungsten and Tungsten Carbide Cast Iron Tool and die steels High speed steels		Very difficult—usually only welded to a dissimilar metal, e.g. stainless steel.
Brass, cadmium and similar metals having a high vapour pressure		Impossible. Volatile constituents boil off under vacuum.
Dissimilar metals Successful welds have been achieved on many dissimilar metal combinations	Mild steel, Stainless steel, Nickel-base alloys, Copper-base alloys to Copper base alloys, Nickel base alloys, Stainless steel, Mild steel e.g. EN.58B to Phos. Bronze Copper to EN.3 Tungsten to Tantalum	Weld development required on certain combinations

the greater the kinetic energy imparted to the electron. The larger the numbers of electrons, i.e., beam current, the greater the energy produced. For welding applications E.B. guns of from 150 W up to 75 kW are used. In practical terms, machines now in use cover the full range of section thickness from foil up to 8 in. (203 mm) weld penetration.

With regard to accelerating voltage, a fairly arbitrary division has been made by the equipment manufacturers between so-called high-voltage and low-voltage guns. High-voltage systems are broadly speaking those in the accelerating voltage range of 60—150 kV and for low-voltage equipment 15—30 kV guns are used.

In an electron bombardment process, X-rays are generated and their penetration increases with the accelerating voltage of the beam. The work chamber must, therefore, be designed to shield the user from harmful radiation and on low-voltage machines this is fairly easily achieved by using $\frac{1}{2}$ in. (12·7 mm) thick stainless steel for the vacuum vessel. Sighting windows would be constructed of lead glass to complete the shielding. Simply using a heavy plate section is inadequate for shielding a high-voltage machine and normal practice in this instance is to provide a cladding of lead sheet over the chamber walls.

For a more comprehensive study of the process and its applications, the reader is referred to a Welding Institute publication on electron-beam welding[1].

Advantages of the Process

These can be summarized as follows:

1. Low Heat Input. Deep penetration welds having a very narrow fusion zone can be achieved (Fig. 2). Typically, the heat input in an E.B. weld can be as little as one-fortieth of an arc weld giving similar penetration. Thus, distortion is reduced—often to a negligible level—allowing finish machined components to be welded as a last stage of manufacture. This feature is illustrated by the method of manufacturing the components for a liquid density meter shown in Fig. 3.

2. Freedom from Oxidation, Because welding takes place in a vacuum there is no scaling or surface oxidation effects. While the vacuum environment is beneficial to most metals, it is of particular significance in welding the refractory metals, e.g., tungsten, tantalum, molybdenum, zirconium niobium, and titanium—all of which are susceptible to oxidation attack during welding.

Fig. 2. Typical E.B. weld illustrating the narrow fusion zone and deep penetration characteristics.

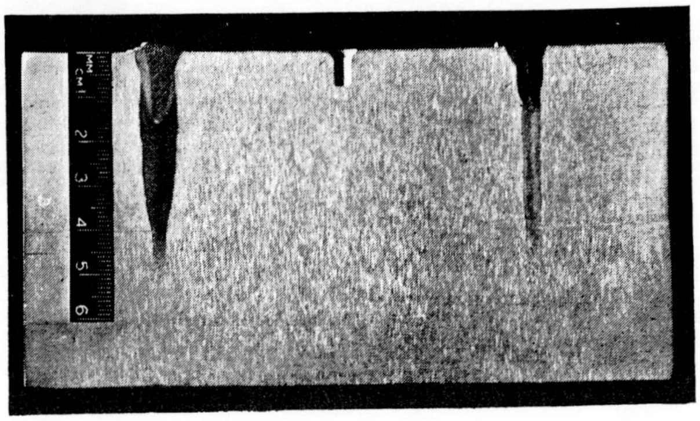

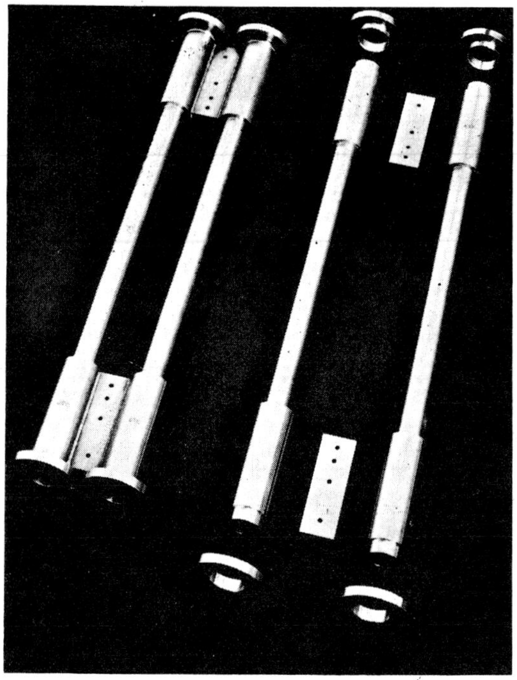

Fig. 3. Component parts for a liquid density measuring system shown as supplied and as a welded assembly. Note the difficulties of carrying out any further machining operations after assembly. E.B. welding allows this part to be produced from finish machined stainless steel flanges and tubes, and Nispan C bridge pieces. (Courtesy: The Solartron Electronic Group Ltd.)

Fig. 4. Butt weld preparations.

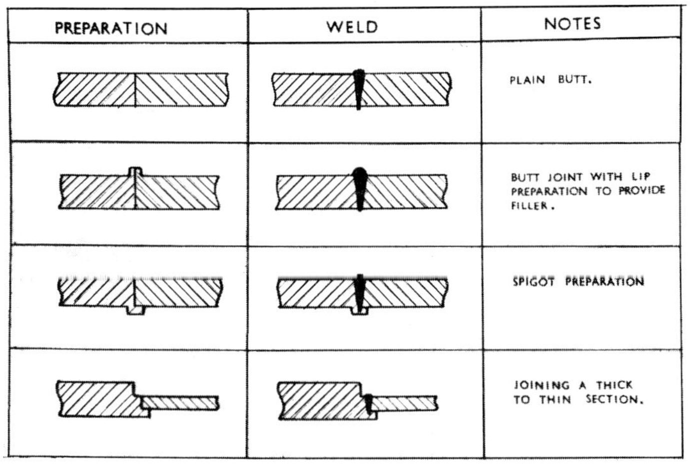

Fig. 5. Lap joint preparations.

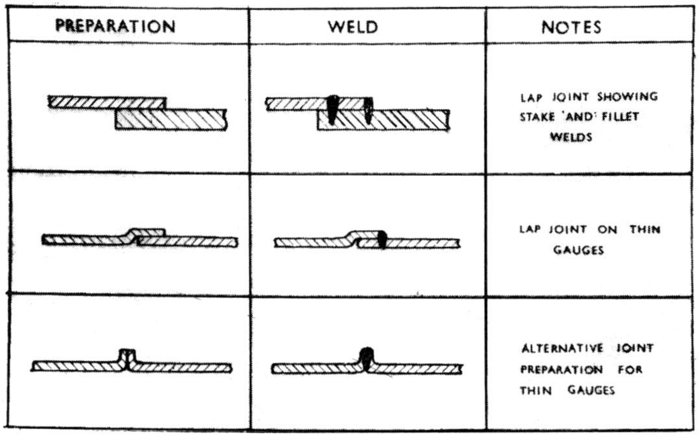

3. *Good Repeatability and Quality Control.* Electron-beam welding is a "machine process" in which the control of beam power, focus, and welding speed are regulated by dial settings. Hence, good reproducibility is achieved with less reliance on the manual dexterity of the welding operator than with conventional welding techniques.

4. *Design Opportunities.* Precision fabrications can be made from simple, cheaply machined parts. Dissimilar metals can be joined. E.B. welding is an autogenous process, no filler materials are usually employed. Welding can be made in inaccessible places; if the joint can be positioned in optical line of sight with the beam it can be welded.

Materials that can be E.B. Welded

Generally speaking, materials that can be joined by conventional welding methods can also be E.B. welded and some additional materials and dissimilar metals can be also joined by E.B. welding. Table 1 lists the range of materials that have been investigated and successfully welded.

Joint Design

E.B. welding being an autogenous process, it follows that a different approach is required to weld preparation than that applied in more conventional processes. The following notes set out the basic types of joints commonly used and show by schematic and practical illustrations some typical applications.

Butt joints. Taking the simplest and most common weld joint, that of butt welding two plate sections as an example illustrating the basic difference in joint preparation; for a conventional arc weld, a Vee preparation would be used and the weld built up by successive passes using filler material. For E.B. welding, however, the mating edges of the plate sections are machined off square so as to present a solid metal interface to the beam which fuses the joint in a single pass.

From this example, it can be seen that while the E.B. weld joint preparation is the simpler, a somewhat higher standard of machining is called for to ensure that the mating surfaces are in close abutment. For a plain butt joint on a $\frac{1}{2}$ in. (12·7 mm) plate a maximum gap of 0·005 in. (0·12 mm) could be allowed but this would be reduced for thinner sections.

Fig. 4 shows some variations on plain butt joints, on both similar and dissimilar section thicknesses. The lip and spigoted preparations shown are devices used to counteract undercutting and drop-through of weld metal.

Lap Joints. Used mainly in joining thin gauge metals or a thin to a thick section employing a stake or nail weld. Typical preparations are shown in Fig. 5 and the instrument flexure, Fig. 6, is a good example of simulating spot welds with the electron beam. In this application, Nispan C link pieces are welded to the EN58B stainless steel limbs of the flexure which is used in a liquid flow measurement system, the fatigue life of the component was doubled by changing from spot welding to E.B. welding due to the elimination of oxidation in the weld zone.

Fabricating Angle, Tee, Channel, Box and H Sections

Having established that relatively distortion-free welds can be made on machined components, it is possible to consider the process as a method for fabricating various shapes in which the resulting section is built up from a number of easily machined parts.

Fig. 7 illustrates some basic joint preparations used for butt welding angle and Tee sections and these can also be applied to channel, box and H sections.

The HE.30 aluminium alloy frame shown in Fig. 8 is a good example of precision fabrication by E.B. welding. The

Fig. 6. Instrument flexure illustrating the joining of thin to thick sections, in this instance 0·010 in. (0·25 mm) Nispan C links E.B. spot welded to 0·060 in. (1·5 mm) stainless steel limbs. (Courtesy: Foxboro Yoxall Ltd.)

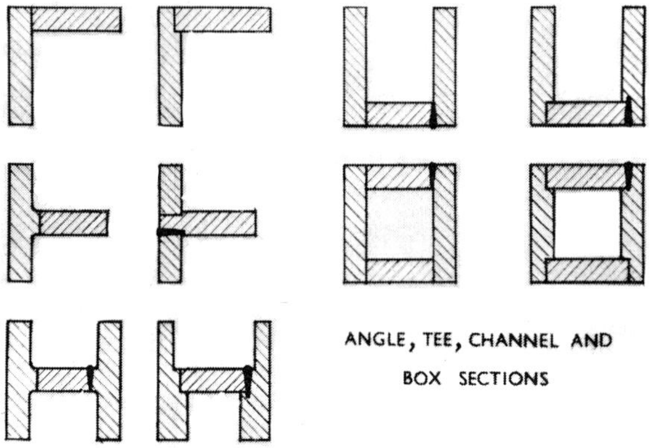

Fig. 7. Angle and Tee joint preparations.

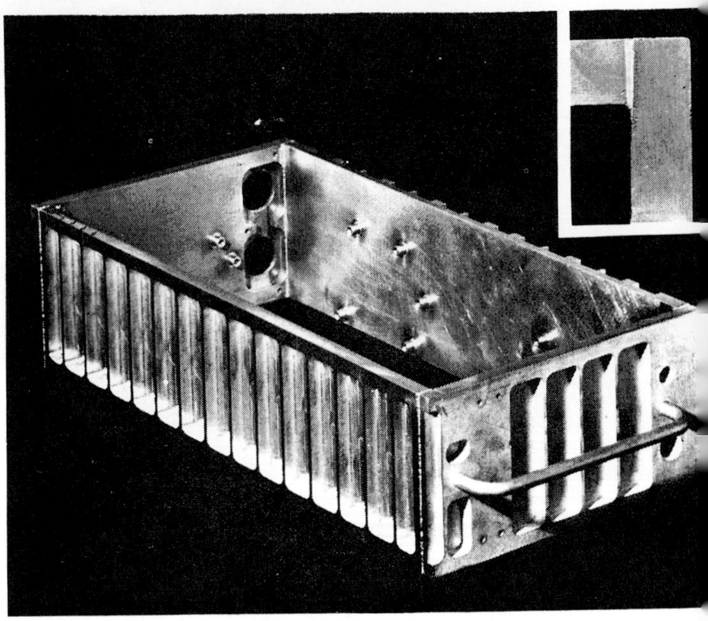

Fig. 8. Aluminium alloy frame illustrating precision fabrication by E.B. welding. The assembly is built up from four fully machined HE.30 panels, butt welded on each corner joint. Note the narrow and almost uniform width of weld fusion zone shown in the smaller picture.

four flat plates are fully machined prior to assembly into a welding fixture which doubles as an acceptance jig to ensure that the piece parts are within dimensional tolerance when assembled for welding.

Flange and Shaft Joints

Flanges and gears can be joined to shafts and tubes by means of either planetary or circumferential joints as illustrated in Fig. 9. The planetary weld preparation normally requires an interference fit between the mating parts, usually of the order of 0·001 in. (0·025 mm) per inch (25·4 mm) diameter for shafts up to about 10 in. (254 mm) diameter. Failure to make allowance for weld shrinkage in this type of joint can result in the material being left in a stressed condition and, therefore, subject to cracking occurring during welding, in subsequent heat-treatment, or in service.

The alternative plain or spigoted butt preparation does not present the same problem because the component parts are now more free to contract during welding.

The cover picture of this issue of this journal shows a number of pinion gears being unloaded from a HSD model 433 Dynaweld machine. The gears are in EN36 and in the fully machined and hardened condition. The EN3B flange

Fig. 9. Flange and shaft joint preparations.

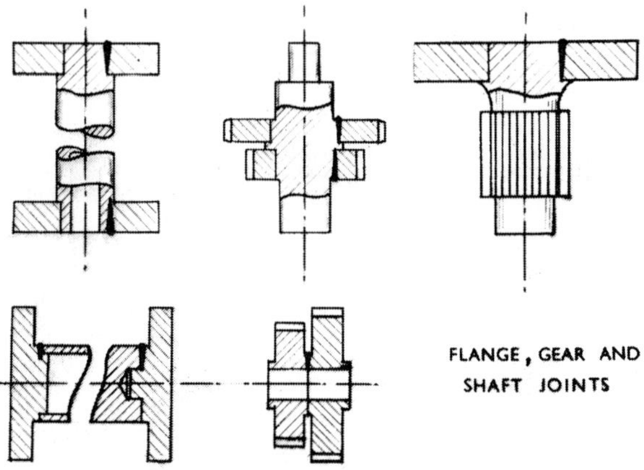

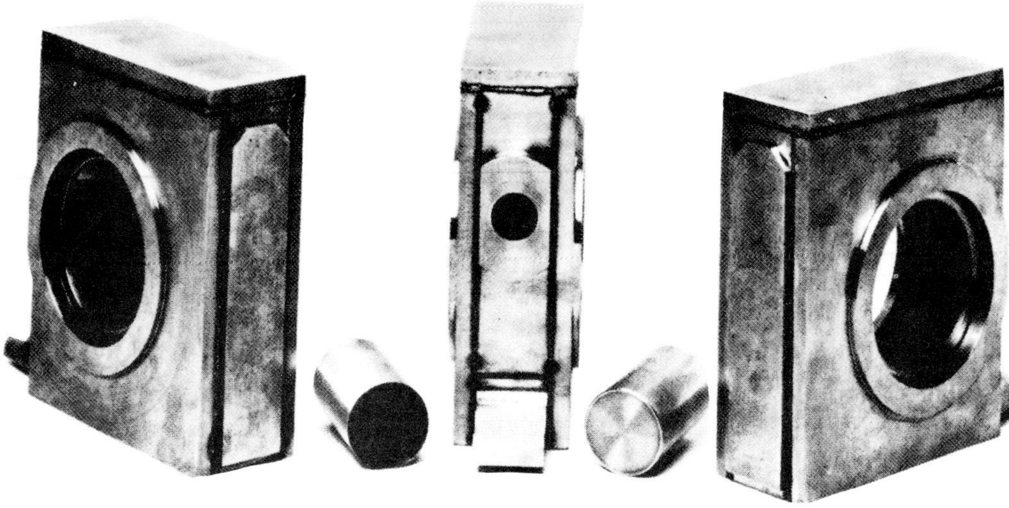

Fig. 10. Fabricated "box" assembly in EN58B stainless steel, illustrating the concept of joining simple machined piece parts to form a complex assembly. Plain butt welds are made on all joints. By contrast, the thin gauge cylindrical capsules are made from stock size stainless tube with an end disc welded in.

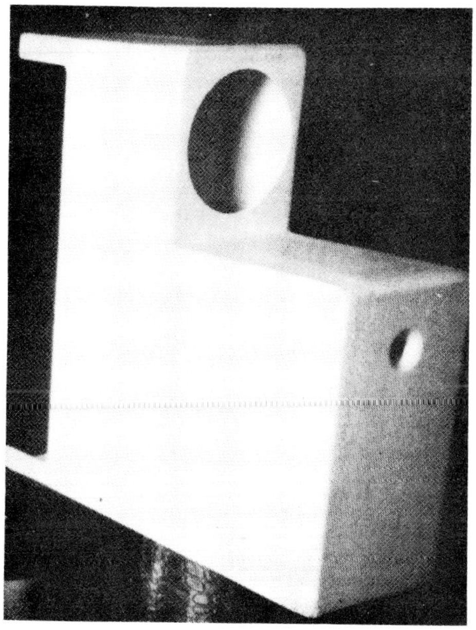

Fig. 11. Precision fabrication in thin gauge aluminium. An HS.30 cover box formed in three sections from 0·036 in. (0·9 mm) sheet and butt welded without distortion. The dimensional accuracy achieved allows the cover to fit snugly on to a machined housing.

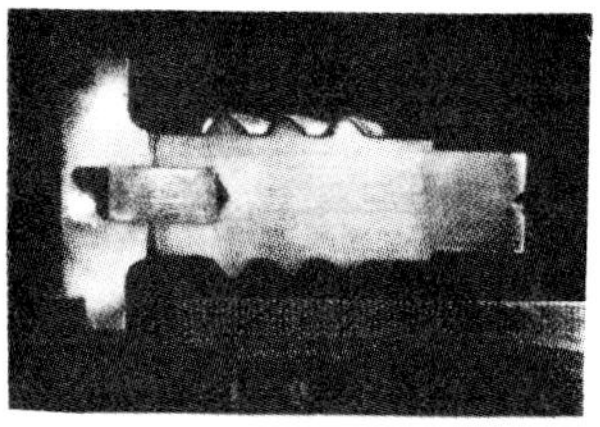

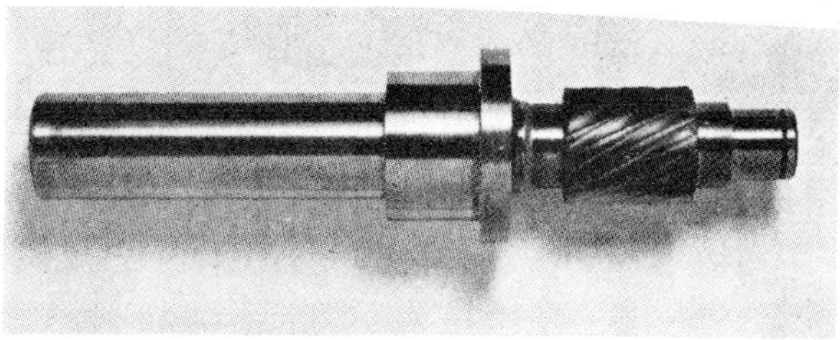

Fig. 12. E.B. welding of composite gear/shafts. An EN34 gear and FV520 shaft joined by an EN58B dowel forming a locating spigot at the welding stage. The gear is in a hardened and ground condition and a final machining allowance is left on the shaft. (Courtesy: Weir Pumps Ltd.)

plates are in the rough turned state when welded. A spigoted butt joint preparation is used. The motorized table on which the gears are rotated during welding can be set to orbit on the diameter of the weld joint and allows multi-loading of identical components. Special tooling of this type greatly improves the utilization of a general purpose welding machine by reducing the floor-to-floor time for a repetitive operation.

Electron-Beam Welding as a Design Concept

The process gives every encouragement for the designer to think in terms of fabricating rather than casting, forging or "hewing from the solid". It is usually substantially cheaper to machine a number of simple sections rather than a single but complex shape. The stainless steel box fabrication shown in Fig. 10 illustrates the point; practically all the machining operations being completed before welding and the finished assembly being a very difficult section to machine, particularly on the internal surfaces.

Other typical applications bring out the low distortion aspects of E.B. welding and show how a creative design approach provides unusual solutions to metal fabrication problems. The cover for an electronics "package" shown in Fig. 11 is a good example of the possibilities for using E.B. welding as a precision fabrication method for thin gauge sheet metalwork. In this instance, 0·036 in. thick (0·9 mm) HS.30 aluminium alloy is formed to make a closure to dimensional tolerances which allow it to fit snugly on to the machined casting carrying the electronics package.

Fig. 12 illustrates the use of dissimilar materials in fabricating a gear and shaft assembly for a small pump. The gear in EN34 is finish machined and hardened before being E.B. welded to the partially machined FV520 shaft. An EN58B dowel pin provides a locating spigot and simplifies the machining preparation for welding. Two reamed holes and a ground dowel are marginally cheaper than a machined spigot.

Composite shafts are also employed in chemical engineering where a corrosive resistant alloy shaft is joined to a wear-resistant driving gear, for instance.

A particularly fine example of precision fabrication is shown in Fig. 13. A prism rod linkage for an aircraft sextant is constructed from two precision ground ball bearings which are E.B. welded to stainless steel stems joined to a centre tube. The spherical accuracy of the balls is virtually

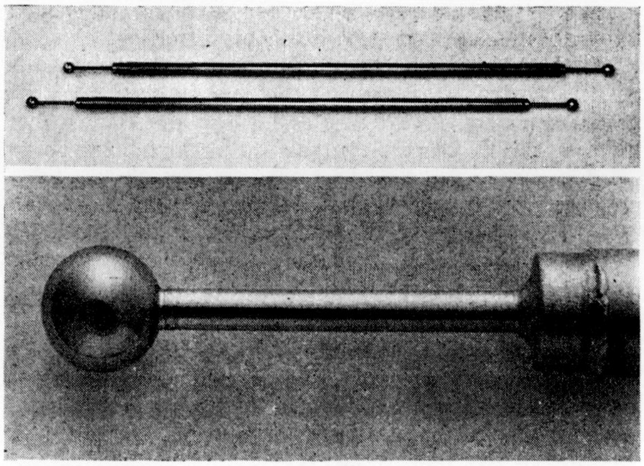

Fig. 13. Prism rod linkage for an aircraft sextant construction from two precision ground ball bearings joined by E.B. welding to stems and centre tube. The sensitivity of the mechanism requires that the spherical accuracy of the bearings remain virtually undisturbed by the welding operation. The lower part of the illustration shows a ×8 magnification of the weld. (By courtesy of Kelvin-Hughes—Division of Smiths Industries Ltd.)

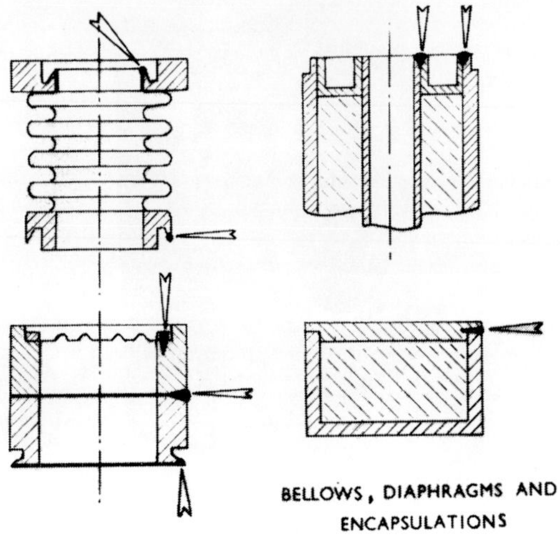

Fig. 14. Weld preparations for bellows, diaphragms and encapsulations.

unaffected by the welding operation and by good fixturing matched pairs of rods are held to within a "few thou" on ball centres.

The weld preparations used for joining thin gauge bellows and diaphragms are illustrated in Fig. 14. The extremely low heat input from an E.B. weld also makes it possible to hermetically seal chemical substances in containers and to encapsulate potted electronic components.

Economics

Factors which determine the cost of E.B. welding are the following:

The size, complexity and quantity of components to be welded.

Welding machine time; including loading and unloading of parts, evacuation and actual weld cycle.

The quality standard required of the weld and the level of inspection, non-destructive testing, etc.

The capital and running costs of the machine; utilization of the equipment, e.g., operated on a single-, double-, or three-shift system.

Because of these variables it is impracticable to relate costs to "the weld price per inch run" as is sometimes possible with conventional welding. The capital cost of the equipment plays a major part in determining a "machine hour rate" for operating the types of general purpose machines most suited to E.B. weld jobbing shops. The precise figure used in costing is, however, influenced by the class of work processed. One can give two examples illustrating extreme cases. For instance, a salvage operation to repair a single expensively machined casting could be priced at a machine hour rate of £15 with an open limit on final cost due to the high degree of risk and technical expertise involved in ensuring a successful result. Within this rate, metallurgical and N.D.T. back-up facilities would be provided as required. Conversely, a commercial quality volume production job assuring continuous machine utilization on a double shift basis over several months could be priced at £8—£10 an hour.

These figures apply to the medium-sized high voltage machines with welding guns of 6—10 kW. Low voltage welders are appreciably cheaper in capital cost and offer economic advantages in reducing operating costs for certain types of welding—generally covering the lighter gauge components quite adequately.

Ideally, for offering a comprehensive sub-contract welding service a full range of machines are needed from large chamber high voltage equipment to small chamber low voltage welders.

Using the Process

The following notes are intended to guide the prospective user of the process as to the likely progression of a typical E.B. weld job from the design stage onwards.

At Design Stage. Consult E.B. welding engineer to establish technical and economic viability. If possible, submit drawing of component giving details of material specification and condition at time of welding (e.g., rough machined, finish machined, unhardened, hardened, except in weld zone, etc.). Agree weld preparation and limits and fit of parts. (The welder can mark up drawings to show any design changes recommended).

Weld Development. Depending on the component, material specification and quality of weld required, it may be necessary to prepare test-pieces simulating the actual part. These will be used (a) for weld parameters to be established, and (b) for the customer's approval of the technique before proceeding to the production stage. The welder may need several test-pieces for parameter development and these will usually be scrapped off through sectioning. The end result is always better for having sufficient test-pieces during the development stage.

Tooling. Any special tooling (e.g., mandrels, tie bars or other welding fixtures) can be designed and manufactured by the welding engineer to enable economic batches of parts to be processed in the machine, thus giving optimum unit cost.

Pre-Production Stage. A pilot batch of parts welded using the special tooling, from sets of piece parts supplied. This allows both the tooling and the welding parameters to be proved on the actual component.

Production Stage. Full production can now be commenced. The E.B. welding shop should be responsible for: Checking incoming piece parts against manufacturing drawing(s). Degreasing and demagnetizing parts. Assembly of parts for welding. Welding to established parameters. Post-weld inspection and any non-destructive testing (X-ray, penetrant or ultrasonics) called for. Preparation of release documents, etc., in the case of A.I.D., C.A.A. or similarly controlled inspection procedures.

Summary

Electron-beam welding is a versatile and extremely controllable electro-heat process which can be applied equally well to the most delicate of components or to a deep penetration weld on a section thickness more likely to be found in heavy engineering.

From the relatively limited number of examples described, applications ranging from the aero-space to food processing industries are covered and few modern processes enjoy such a wide field of potential use as E.B. welding.

The importance of approaching the subject with an open mind cannot be over emphasized. E.B. welding is not a magic word for solving impossible metal joining problems, neither is it simply an alternative to conventional welding processes. Used sensibly, however, it offers the creative designer a new concept in precision metal fabrication whereby simply shaped piece parts are machined and welded as a final assembly stage, with negligible distortion.

Outside the aero-space and nuclear fields, economic factors have largely controlled the rate of growth of the process in general engineering, and it is in this context that the true value of the services provided by the jobbing shops can be assessed. In offering welding machine capacity, toolroom, NDT and design consultancy facilities to all sectors of industry, a significant contribution has been made to the advancement of an important metal fusion process.

Acknowledgements

The author wishes to thank the companies who have kindly given permission for the publication of the various illustrations used. Thanks are also due to the Directors of the Southborough Group of Companies for their permission to publish this paper.

Reference

1. Meleka, A. H., Electron Beam Welding: Principles and Practice. Published by McGraw-Hill for The Welding Institute.

ELECTRON BEAM WELD JOINT DESIGN

Stanley J. Stefanski
Sundstrand Aviation Corporation
Rockford, Illinois

This paper emphasizes the importance of joint design, covering the self-fixturing aspect, part fabrication, surface and material cleanliness, economics of joint design, and inspection of the joint. There are two basic circular welds, the radial and circumferential, plus another type of weld which is the special design or the angular weld. This discussion covers the advantages and disadvantages of each.

Success in the field of electron beam welding depends tremendously upon the correct joint design. To take a part from an engineer and weld it successfully on a one-shot proposition is nice. It is much more efficient and dependable to educate the engineers and design people with respect to correct design prior to actual welding of the joint. Sundstrand took about a year to educate its people in the correct technique. When low-hardenability materials are used the joint design can be based more upon the economics of part fabrication and the self-fixturing joint designs. When materials of high hardenability such as tool steel or AISI 52100 are used, proper joint design can greatly reduce cracking, which is the major problem in welding these materials.

Joint design is also influenced by the type of inspection procedure to be used. It is a sound idea to review the joint design with a nondestructive test expert. One can weld many things that can never be checked, especially if, as in our situation, these one-shot propositions are infrequent. Most of our output is large production; but some of our parts are long-term parts, in fact, one specific part ran into millions of units. This is all done automatically using a 52100 material and this, again, is where we need to know whether we are having problems such as cracks, porosity, and/or insufficient penetration. The parts are relatively inexpensive up to the weld operation but after the weld operation, major processing costs build up tremendously.

There are two basic circular welds and two variations of each. These are the radial and circumferential lap and butt joint welds. The lap weld often is the easiest to weld but sometimes requires extensive fixturing. As a rule, these types of welds do not lend themselves to ultrasonic or X-ray inspection.

Figures 1 and 2 are examples of a part that can be welded either as a radial butt weld, such as in Figure 1, or as a radial lap weld, as in Figure 2. Figures 3 and 4 are examples of a part which could be welded either as a circumferential butt weld (Figure 3) or a circumferential lap weld (Figure 4). Figures 5 and 6 are examples of a part which could be welded as an angular butt weld or as a lap weld. As stated before, these are designs of special need that may warrant some consideration. Electron beam welding lends itself well to making angular welds better than any other commercially available process.

It may be noted that in Figures 1 through 6 there is not a step at the entrance of the weld joint, and this is as it should be. If a butt joint of different material thicknesses is required, the step should not be at the beam entrance point. Figures 7 and 8 illustrate this point. Here there is no mismatch; the parts are fairly flush together from the beam entrance standpoint. If there is any mismatch or step it should be from the underside of the weld where, when you overpenetrate, you are not going to hurt it. The closeness of a step, which is labeled A in Figure 9, to the entrance to the weld joint should be limited to roughly .010 to .020 in., or B over A, whichever is greater. In Figure 7, if the material has a high hardenability, this value should be increased to .040 to .060, or B over 2, whichever is greater. We prefer not to have any protrusions in the way at the beam entrance point. The beam becomes defocussed as you go farther from the focal point, so that, in the case of protrusions, you do not get good penetration.

The radial butt weld is most commonly used in a circular weld joint because it is self-fixturing. This self-fixturing ability allows the part to be cleaned, assembled, stored, and then welded without costly tooling. The ability to store parts prior to welding can be of great economic value for high-production items. This is especially true if the welding operation is somewhat slower than assembly time. Self-fixturing of a part generally reduces machine load and unload time. Smaller self-fixturing parts in many cases can be magazined or hopper-fit for maximum production and reduced downtime. Another important advantage of the radial butt weld is that, in most cases, it lends itself better to ultrasonic inspection than any other joint configuration. In many present EB applications the joint design is dictated solely by the ability to perform nondestructive testing of the weld. This very important aspect of weld joint inspection is overlooked on critical highly stressed or high-hardenability parts. The primary disadvantage of the radial butt weld from an EB standpoint is that it is difficult

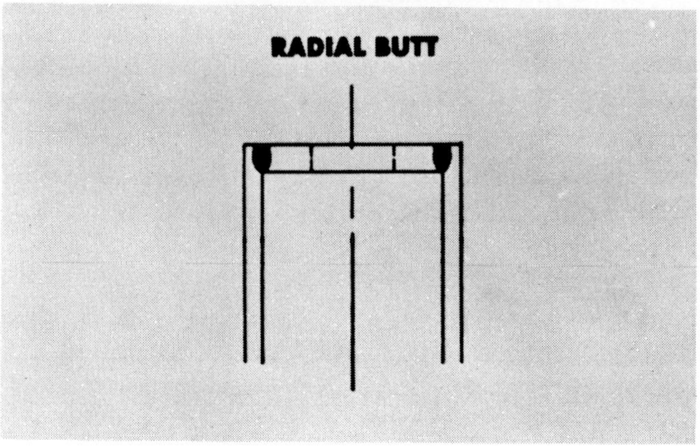

Figure 1

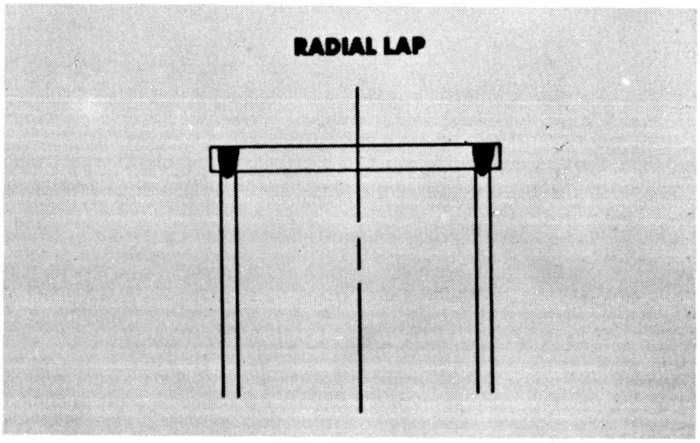

Figure 2

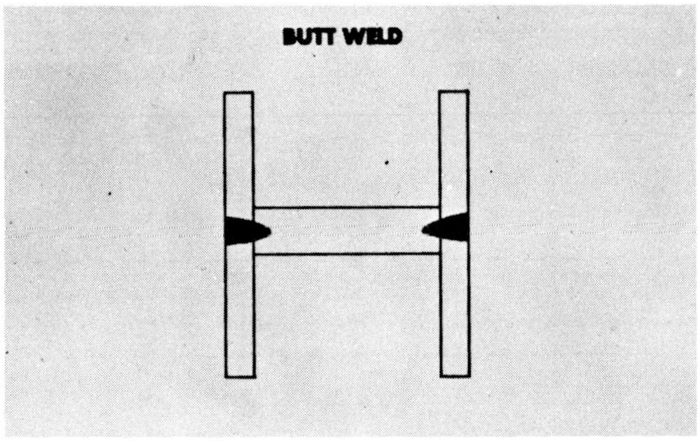

Figure 3

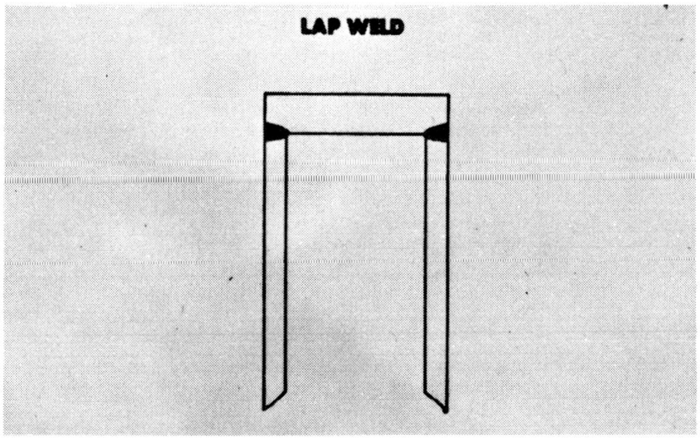

Figure 4

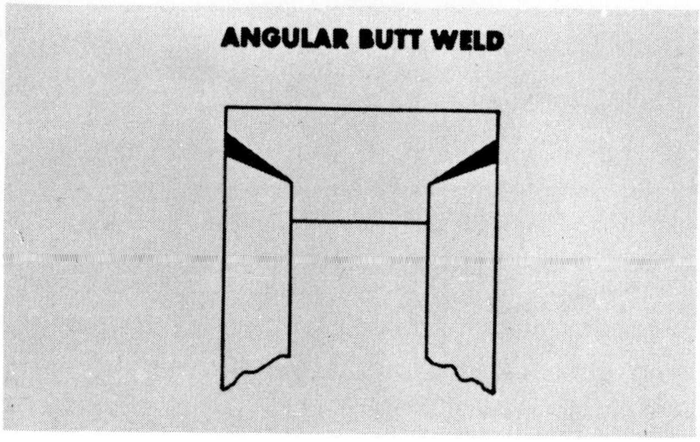

Figure 5

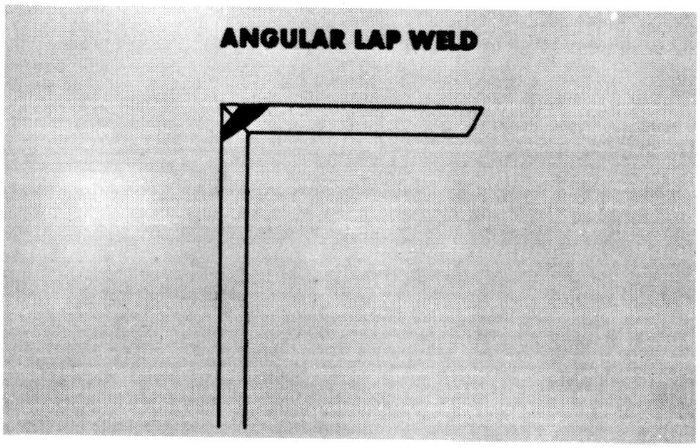

Figure 6

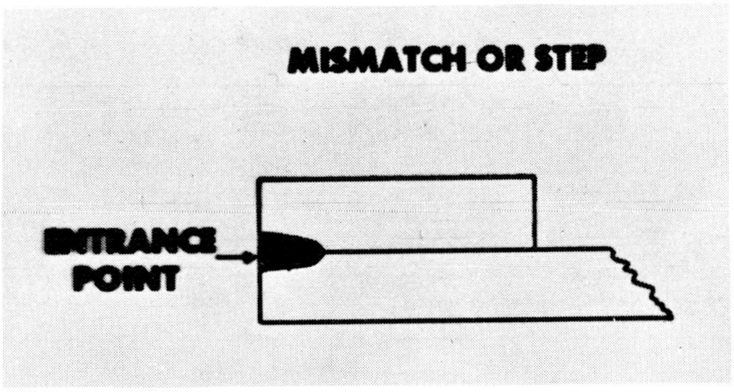

Figure 7

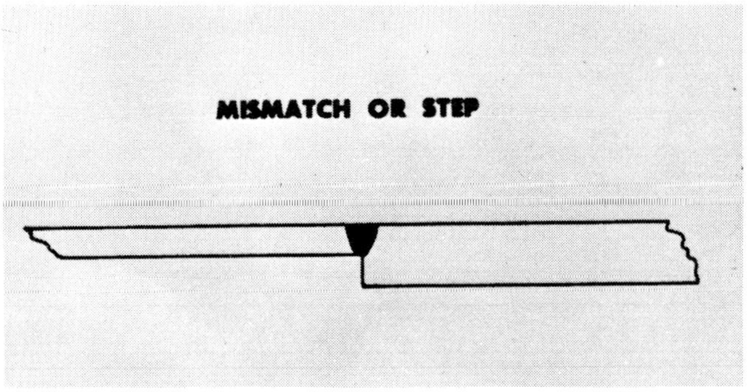

Figure 8

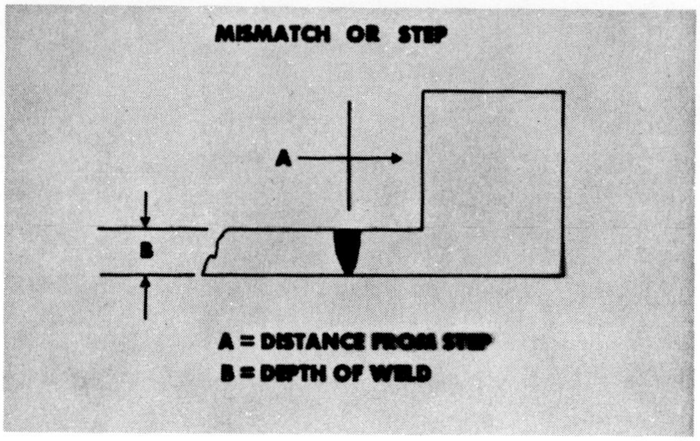

Figure 9

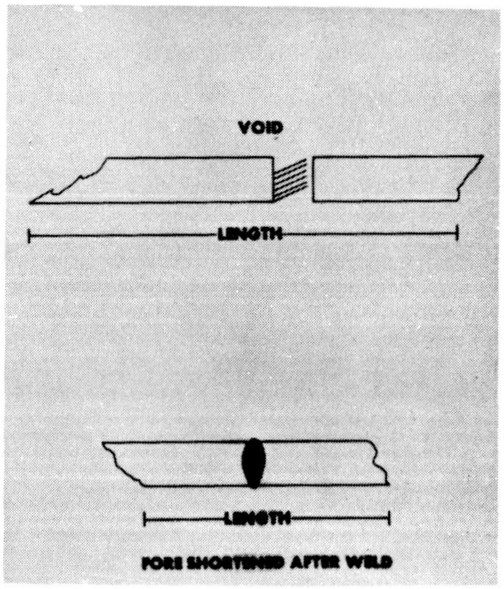

Figure 10

to obtain crack-free welds. This is especially true of high-hardenability materials such as 52100. From an engineering standpoint, 52100 is a good material. It holds up and wears very well. It has been said that we are probably the most experienced and successful welders of 52100, in fact, 92 percent of our work is in 52100.

To return to crack-free welds, distortion is another disadvantage when you do not want to draw the hardness from the surrounding areas. The mechanics of this can be explained as follows. Prior to welding, parts with a high interference ratio have some voids. After the welds have been completed, these voids have been filled with metal. This metal can only be supplied by drawing the two parts to one another, which places the main bar of the weld in tension. It took quite awhile for us to explain to the engineers that when they specify a micro-surface finish number on two mating parts, that the smoother the finish the less voids you have on the surfaces of contact, or the area to be welded, because the micro-finish rating indicates the variation of the height and the depth on the surface of the part. You can see that the distance from peak to the valley is what is actually a void. Moreover, when you push a part with an interference fit, you still have some voids. Material is taken from the base metal, as in Figure 10, in forming the upper and lower weld bead. This shrinkage phenomenon causes distortion. We like to specify parts that have no more than .0005 to .001 maximum clearance, since we do a lot of high-production work. This is a good round figure to work with and, from a self-fixturing standpoint, these parts do hold up quite well. Figure 11 shows a part after welding, highly stressed and distorted. If we consider what this part would look like when half welded, as in Figure 12, the OD wall would be in stress but, due to shrinkage, the plug would pull up and away from the opposite side. To overcome this condition, the part must be tack-welded in several places around the joint prior to welding. This is a problem common to many self-fixturing parts.

Another problem with the butt weld is the effect of the stress ratios near the weld joint. In Figure 13, heat is applied in equal amounts to the inner and outer parts at the weld joint. The outer member would become hotter and expand away from the inner part. This could be a shroud on some part where you want to add some stock to it. Here you have different mass ratios. The result is that the outer member becomes egg shaped during the last half or last third of the weld and pulls away so that we no longer have a joint of proper fitup. The problem may be aggravated if the expansion ratio of the outer material is greater than that of the inner material. This problem can be minimized by the following

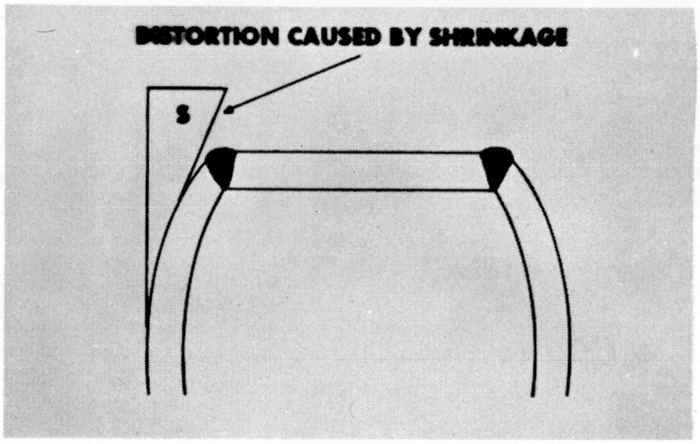

Figure 11

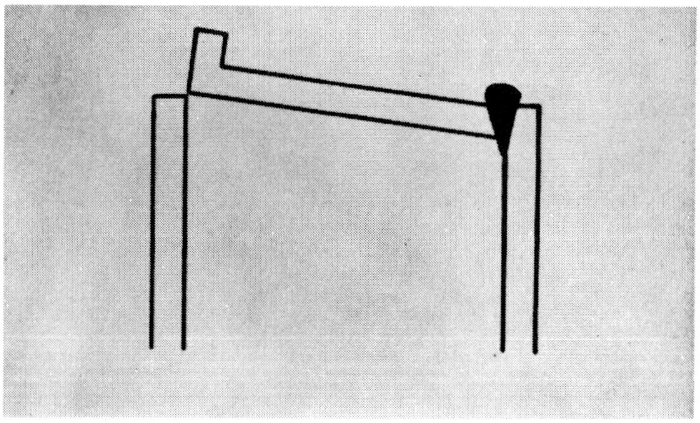

Figure 12

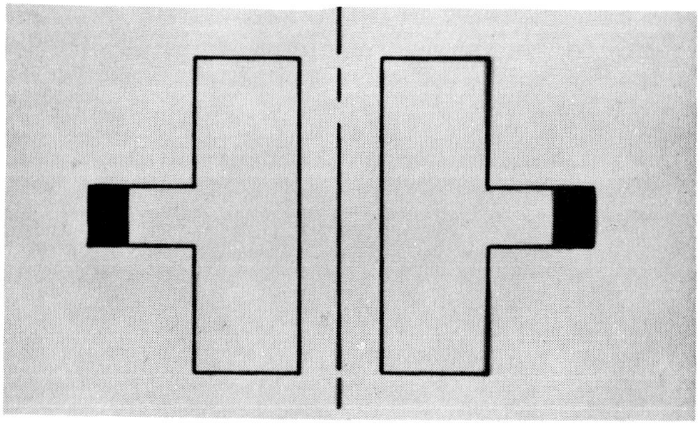

Figure 13

approaches. First, if the part configuration cannot be changed, a copper or aluminum chill can be fitted to the OD or the size of the outer member. The more intimate the contact between the part and chill, the less chance of separation during welding. The purpose of this chill is to increase the thermal mass of the outer member. A second step would be to increase the mass of the outer member and machine it to size after welding. The time required to machine the outside can be an economic disadvantage, so that if this is a high-production-type part several inexpensive chills would be the best solution. Another economic disadvantage is that close tolerances are required on each part to be welded; this is especially true in a case of self-fixturing parts. It may be wise to redesign the weld joint configuration and build a few close-tolerance fixtures instead of putting more money into individual piece parts.

Since circumferential welding does not induce high stresses, distortion, or shrinkage, it is suitable to certain design configurations such as in Figures 3 and 4. Figure 3 shows a plug in a bore or, say, a hole. Penetration is at a minimum and so are stresses. Penetration is through the OD wall, just to tack the plug in place, involving very little stress setup. Figure 4 is an example of a cap welded to a tube of thin-wall cross section. Again, very little penetration is required and the plug and the cap represented in these two figures do not represent the close-tolerance dimensions or fit required in the radial weld. A drawback of this type of weld design is fixturing. Since there is not as much penetration as in a radial weld, the circumferential weld does not have much strength. All the strength is in the thickness of the weld joint cross section and inspection accessibility to verify weld results is very restricted (as in Figure 3).

DISCUSSION

Question: In the parts you are welding of 52100, what post-heat treatment do you find necessary to take care of retained austenite?
Answer: Well, we have just made a breakthrough in the last part of September and in early October, 1966. Prior to this time the thinking was that in order to get away from this condition you had to preheat the part, weld it, and then post-heat treat it. We have eliminated this costly time because we are actually using the beam to do this. This is a step in the right direction in eliminating this entirely. Welding 52100, after 1-1/2 years of pulling out our hair, now seems to be child's play.

Question: How hard is the 52100 when it is welded? Answer: Most of our 52100 is around R_c 25-28 when we weld it. We have some parts now, especially on finished gears, where the finished gear has to end at a sharp corner, and is very difficult to machine it any other way than as a two-piece item. The hardenability of this gear, or the hardness of this gear, on one job we have now is 58-64 R_c. They want this gear welded on a little 1/32-in. flange but they don't want you to take any of the hardness of the gear away or draw it down in hardness. The technique we have developed seems to be paying off right now.

Question: Do you use a radial pipe joint on 52100? Answer: Yes we do. We cover the full spectrum as far as joint configurations on most of our parts.

Question: How did you eliminate the preheat? Answer: We were able to eliminate the preheat. I don't know if I would be willing to divulge that right now. As I say, it took 1-1/2 years to do it and we are still not sure it is going to work out, but it has worked in the last 1-1/2 months that we have been using this technique. When I left the shop we were running 250,000 parts on which we are using this technique. Those 250,000 parts are quite an investment just on the say-so that we can lick the time that it took to preheat these parts.

Question: Why do you feel that it is necessary to angle the beam? Answer: If you want to come straight down the side of this shoulder, if you would call it that, the closest you would like to get is somewhere around .060; more times you get around .030. We don't like it that close -- we angle the beam slightly and we come in and do get the necessary penetration.

Question: You said you have eliminated the preheat in the 52100 outside the chamber? Answer: We are talking about preheat with the beam.

Question: What is your technique for lowering the tendency for cracking in radial lap joints? Answer: You just have to have a smooth fitup and intimate contact. Say that you are putting a plug in a thin-walled pipe or anything like that, the closer the fit and the smoother the microsurface, the less cracking you have. We find that the smoother the fit and the closer the fit, without actually being a press fit, the better the things are.

Question: What are the surface finish requirements? Answer: On some of these parts we were running around 125 microinches, which is standard in our shop on a lot of production items, but now we are getting down to around 25 to 30 microinches' finish, which is very nice. On some of our parts, such as cylindrical objects that have a plug welded on the end, we are going through what they call a rollerizing operation. We have a $60 apparatus which, attached to the idle station on a Brown and Sharpe machine, does quite a nice job of sizing your part plus adding a good surface finish to it.

Question: How thick are the weld joint sections? Answer: The weld joint thickness varies from .089 to .400 in 52100.

Question: In welding 1/4-in. wall, what type of inspection do you use? Answer: We do 100% inspection, since we don't want to let bad parts get out of the department and into the stream of final manufacture. We inspect these parts 100% using Sperry ultrasonic equipment. When first setting up the job, we may check one part out of ten just to see what kind of penetration we do have, then in the next step we will check all ten and find out what the percentage is of good to bad. After we get running we will take these parts down and use the Sperry C-scan. Then we will group these parts in a very neat arrangement and will C-scan right through the side wall to check the penetration.

Materials Welded

Electron-beam welding of structural steels

by **J. D. Russell, A. J. Rodgers** and **R. J. Stearn,** The Welding Institute

A study of the EB welding of 22 mm thick pressure-vessel steels is described in this abbreviated version of Welding Institute Report P54/72

The electron-beam welding of a range of 22 mm thick steels, typical of those used in the pressure vessel industry, has been studied. The aims were to examine their susceptibility to defects and to assess the joint properties of welds produced by welding procedures developed to control defects.

Most of the steels gave rise to porosity and, to a lesser extent, weld-metal and heat-affected zone (HAZ) cracking. These defects can be controlled by the use of the correct welding procedure. Joint properties were assessed according to BS 1515 and found to be acceptable. COD tests indicated that the fracture toughness of the weld-metal after stress-relief heat treatment was at least as good as that of the parent metal.

It is concluded that weldability problems are not insurmountable when applying electron-beam welding to pressure- vessel steels.

The possibility of using electron-beam welding in the fabrication of large thick wall structures has been speculated upon since deep welding was discovered in the late 1950's. There are, however, two main technical problem areas in applying this technique in heavy fabrication:

a) engineering problems, associated with vacuum and joint fit-up requirements;
b) material weldability aspects.

The problems of operating in a vacuum envrionment, demands of good joint fit-up and high capital cost of the equipment, have probably been the main reasons why the applications of this apparently attractive, single-pass welding procedure has not found readier acceptance in place of the conventional multi-pass welding procedures in, for example, the pressure vessel industry. The problems of providing a vacuum environment, whether by totally enclosing chambers or local enclosures, together with precision work or gun handling, require expensive development work in order to provide a solution. Some aspects of the feasibility of applying electron beam welding to large structures have already been examined at The Welding Institute (1), although this survey was concerned with a specific case involving vessels of constant diameter and wall thickness constructed from very high quality material. Where general structural steels and vessels with varying wall thicknesses and diameters are involved, it is less easy to find a general solution. Therefore, before embarking on the expensive development work necessary in the general context of heavy structures, it is advisable to establish the technical feasibility of producing satisfactory welds in the materials of interest. Having carried out this preliminary work to define equipment and process requirements, it would then be logical to examine the economics and practicability of carrying out these requirements in a typical manufacturing environment.

The ex-Ministry of Technology committee set up to review the pressure vessel industry, considered that such a programme would be a worthwhile contribution to their survey on welding. This report deals specifically with the investigation of the weldability problems likely to arise in applying electron-beam welding to pressure vessel steels of about 25 mm thickness.

EQUIPMENT AND MATERIALS

Materials

Experience in electron-beam welding a variety of steels has shown that the occurrence of defects, mainly porosity and cracking, is associated with material impurities such as gas content, sulphur and phosphorus. Since electron-beam welding is essentially an autogenous process involving square-edge joints, there is little scope for the addition of beneficial filler material in comparison with arc welding.

The materials selected for this programme were intended to cover the potential best and worst examples of the types in current use in the pressure vessel industry. This selection was made from the BS 1501 range covering both high and low temperature applications as follows:

BS 1501 — 151 — 28 A, carbon steel (semi-killed)
BS 1501 — 161 — 28 A, carbon steel (silicon-killed)
BS 1501 — 213 — 28 A, carbon: manganese steel (semi-killed, niobium treated)
BS 1501 — 224 — 28 A, carbon: manganese steel (silicon-killed, aluminium treated)
BS 1501 — 271 manganese, chromium, molybdenum, vanadium steel
BS 1501 — 622 — 2¼% chromium, 1% molybdenum steel.

The 6 kW 150 kV welding machine used for this work is capable of penetrating 50 mm in steel, but in order to allow some flexibility in choice of process parameters such as travel speed and beam focus it was decided to limit the tests to nominally 25 mm thick materials. Thus all the materials were rolled to nominally 25 mm with the exception of materials M8 and M9 which were 25 mm thick slices cut from a 150 mm thick billet. After removal of rust and scale from the plate surface the materials were all reduced to a standard 22 mm thickness.

Equipment

The welding equipment used for this investigation was a standard 150 kV, 6 kW Hamilton Standard machine with the addition of a high-frequency beam spinner/oscillator and a beam slope-up and slope-down device designed and built by The Welding Institute. The Institute's beam probe device was also used where necessary to define the focus position accurately and to examine the beam energy distribution.

EXPERIMENTAL WORK

The experimental programme has two main aims:

a) to determine the susceptibility of the materials to weld defect formation i.e. porosity and cracking, and to examine the effects of process parameters in controlling these defects;
b) following the successful development of welding procedures, to establish the basic weld properties in accordance with BS 1515 together with an assessment of fracture toughness as measured by COD.

Porosity and cracking tests
The applications envisaged for the electron-beam welding mainly involve full penetration square-edge butt joints and, therefore, the work was confined to examining the types of defect which occur in simple butt joints in 22 mm thick material, welded in a single pass. However, in order to avoid the possibility at this stage of inconsistencies due to joint fit-up and surface contamination, it was decided to use melt run tests on clean degreased plate instead of butt joints.

It had been established that, with the exception of the two aspects mentioned above, melt run tests stimulate a butt weld with sufficient accuracy for the present purposes.

Weld-metal porosity is known to be the main problem in electron-beam welding steels with high gas content and inclusion levels. Cracking is also a possibility in autogenous welds in steels of this type. Two basic tests were therefore employed to investigate these aspects from the point of view of comparing the relative behaviour of the various ypes of steel examined.

Porosity. Simple melt run tests accross plates 100 mm wide x 150 mm long were made in the 100 mm direction transverse to the rolling direction and 10 mm apart. Preliminary tests were carried out to examine the effects of focus position over the range of travel speeds possible (i.e. up to 380 mm/min) with a 6 kW beam. Having established the optimum focus condition to give paralell-sided welds, a series of experiments was then carried out with focus and travel speeds covering the range 100-380 mm/min with power adjusted to give a positive underbead for each speed.

A similar series of tests was carried out over a range of travel speeds to examine the effect of circular beam deflection at 2 kHz with amplitudes in the range 0.2-1.0 mm. This technique has previously been found to be effective in eliminating porosity in steels. In addition to the basic beam circle generation, the effects of a transverse beam oscillation and elliptical beam deflection patterns were also investigated.

All the test runs were assessed by radiography normal to the plane of the weld as illustrated in figure 1(a). Transverse sections were also taken, and polished and etched to examine the weld profile.

Cracking. In order to achieve the maximum level of consistent restraint with small test pieces, circular patch type melt runs were made, as shown in figure 1(b), to supplement the crack susceptibility information from the straight line runs in the porosity tests.

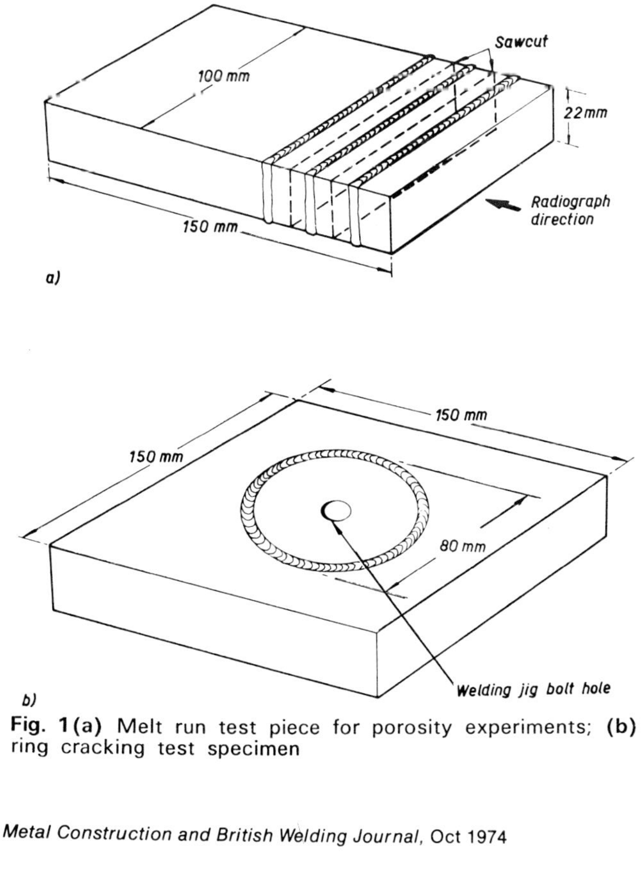

Fig. 1(a) Melt run test piece for porosity experiments; **(b)** ring cracking test specimen

Melt runs were used in preference to butt welds, in order to avoid the influence of fit-up and to simplify test piece machining. Two test welds were made for each material and travel speed. After grinding off the top and bottom beads, the welds were radiographed and examined ultrasonically. Sections were subsequently taken from both the overlap region and its diametrical opposite and polished for metallographic examination.

Joint properties
From the results of the two previous test series, welding conditions were selected which would produce defect-free butt joints in the materials chosen. Slight modifications were necessary to these conditions in order to achieve a satisfactory weld bead shape, because of the influences of the joint compared with a melt run. Test plates of finished size 300 mm x 300 mm were produced with the weld transverse to the rolling direction across the centre of the plate. Two test plates were prepared in three selected materials (one plate to be tested as-welded and the other after post-weld heat treatment), BS 1501-161, BS 1501-271 and BS 1501-622. Before the test plates were cut for test piece manufacture, they were radiographed and sections examined for defects.

Samples were cut from these plates in order to carry out the mechanical tests as required by BS 1515 i.e. transverse weld tensile, side bend and V-notch Charpy. COD tests were also carried out using square ligament specimens with 18 mm below the saw cut notch.

The tensile and bend tests were carried out at room temperature whereas the Charpy and COD tests covered the temperature range —196° to +50°C.

Metallographic examination
Sections were examined for weld defects and to compare weld profiles at all stages during the development of welding procedures.

Representative samples of the parent materials were also examined and the types and numbers of inclusions compared in order to correlate material cleanliness, nominal analysis and susceptibility to defect formation. Hardness traverses were made across typical transverse sections of each material covering fusion and HAZs in both as welded and heat treated conditions where applicable.

EXPERIMENTAL RESULTS
Porosity tests
Typical radiographs from these tests are reproduced in figure 2(a) which shows runs in a particular material with a normal beam, and in figure 2(b) radiographs of runs with beam-spinning at 2 kHz with 1.0 mm amplitude in BS 1501-161. The fusion zones tend to be wider at the lower travel speeds (as can be seen in figure 3) which shows transverse sections of the runs radiographed in figure 2) and hence for more accurate assessments of percentage porosity it would be necessary to modify the two-dimensional assessment based on the radiograph by the fusion-zone width. Nevertheless it can be seen that it is possible to eliminate porosity completely by the use of the optimum combination of travel speed and beam deflection.

Of the process parameters examined, travel speed, beam focus position, beam power and beam deflection, all were found to have an effect on the occurrence of porosity although not all factors are equally important. Also although low travel speeds in general are beneficial, there appeared to be an optimum speed at about 200-250 mm/min. Beam power and focus position are the main parameters which affect weld profile, and it was found that parallel-sided welds were much less prone to porosity than tapered welds. By far the worst weld profile is that with a wide bottom, narrow top as shown in figure 4.

In general, the use of circle deflection with an amplitude of 1.0 mm gave porosity-free welds. The majority of the pores were of the type shown in figure 2, i.e. elongated with only occasional rounded pores as are typically found in arc weld deposits.

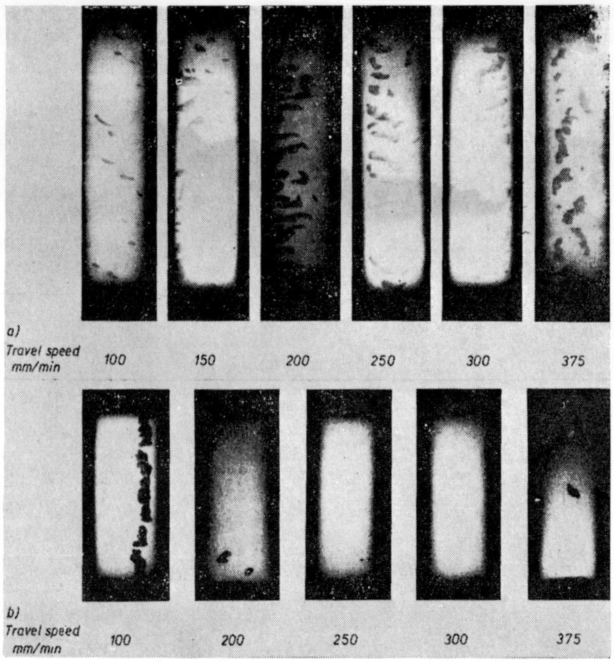

Fig. 2 Radiographs of melt runs in **(a)** BS 1501-224 material with normal beam at various travel speeds, **(b)** BS 1501-161 with beam spinning at 2 kHz, 1.0 mm amplitude, over range of travel speeds.

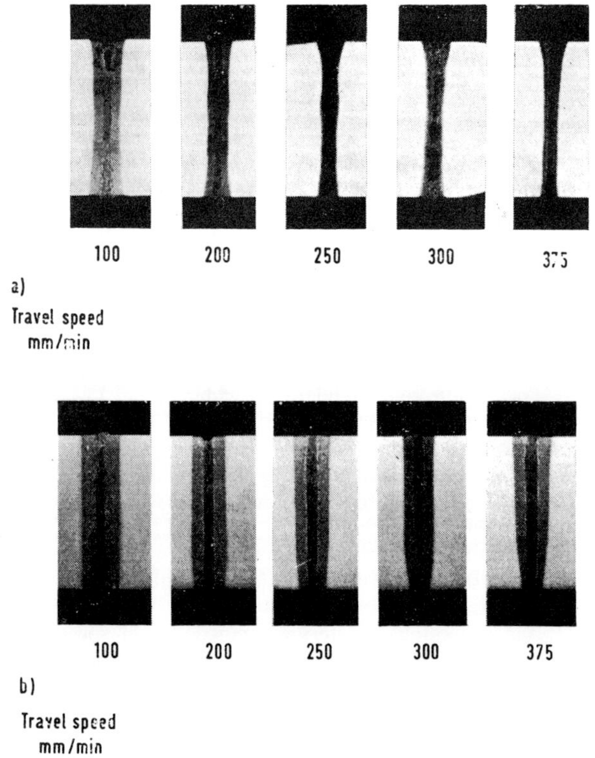

Fig. 3 Transverse sections of melt runs in BS 1501-224 over range of travel speeds. **(a)** Normal beam. **(b)** Beam spinning at 2 kHz, 1.0 mm deflection. × 1.5

Cracking tests

Both radiography and ultrasonic examination failed to reveal cracks in any of the materials, indicating freedom from major cracking. Some of the materials, however, exhibited small cracks on the polished sections. The worst material from the cracking point of view was BS 1501-161 (both casts), in which both HAZ tearing and solidification cracking were observed at 250 mm/min travel speed and above. Figure 5 shows the worst example but, as can be seen, even here the cracks are very small. In order to establish the length of these cracks along the weld length, thin sections were radiographed as for the porosity tests in figure 1(a). This showed that the cracks were never longer than 2.5 mm. Cracking at the lower travel speed (200 mm/min) was confined to small HAZ tears never longer than 0.1 mm.

Transverse solidification cracks similar to that in figure 5 were also found in material BS 1501-213, whilst HAZ tearing was also found to a small degree in BS 1501-271, in both cases at 300 mm/min travel speed and above. There was no significant difference in the extent of cracking between the ring cracking tests and the straight melt runs used for porosity tests made at the same travel speed. It should be noted that cracking which could possibly be significant in service occurred with travel speeds of 250 mm/min and above in both casts of BS 1501-161, and 300 mm/min and above for BS-1501-213 and BS 1501-271.

Joint properties

All the tensile tests failed in the parent material in both the as-welded and post-weld heat-treated states. A typical post-weld heat-treated specimen after tensile test is shown in figure 6, together with side bend test pieces in both as-welded and heat-treated states. It can be seen that in all cases the weld area has yielded less than the parent metal. The side-bend tests in the three materials passed 180° bend

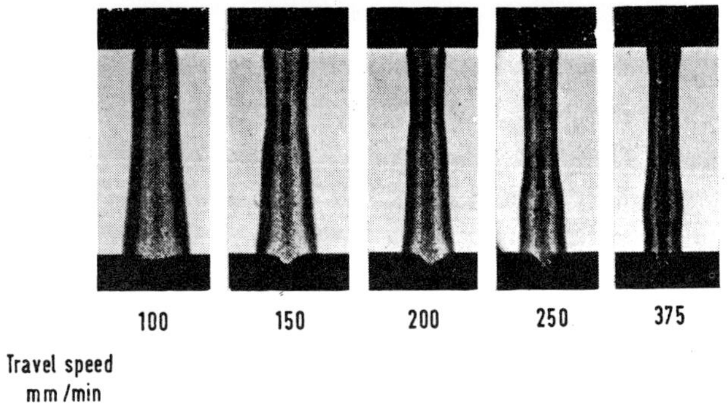

Fig. 4 Transverse weld sections over range of travel speeds with beam focussed to give wider bottom than top

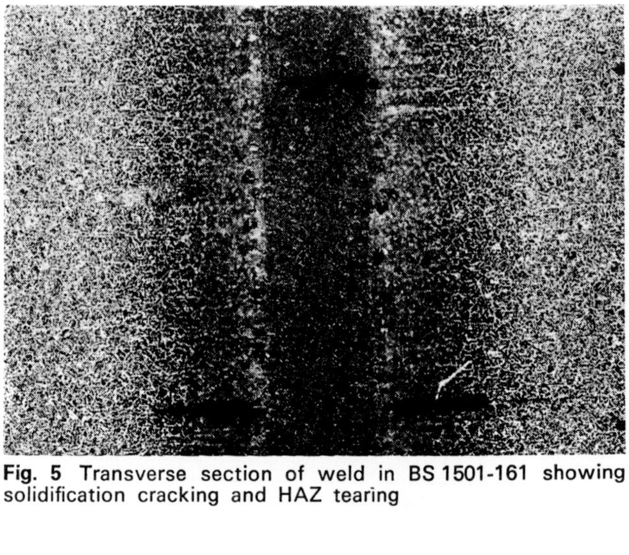

Fig. 5 Transverse section of weld in BS 1501-161 showing solidification cracking and HAZ tearing

successfully, the difference between the heat-treated and as-welded samples being only in the peaking of the weld region on the crown of the bend. A typical hardness traverse is shown in figure 7 for BS 1501-271, for both as-welded material and after heat treatment for one hour at 650°C.

Fig. 6 **Above:** transverse plate tensile specimen showing failure away from weld in post-weld heat-treated BS 1501-622. **Below:** side bend tests in BS 1501-161 as-welded and after post-weld heat treatment

In order to explore the possibility of softening the weld area in BS 1501-271 further than 320 VPN reported above, samples were heat treated for two hours at 650°C. These treatments resulted in average weld-metal hardnesses of 290 VPN and 270 VPN respectively.

V-notch Charpy values with the notch in the fusion zone are shown in figure 8 and COD values over the same temperature range in figure 9. The results are for as-welded and heat-treated welds for both toughness tests. Saw-cut notches were used for the square ligament COD tests.

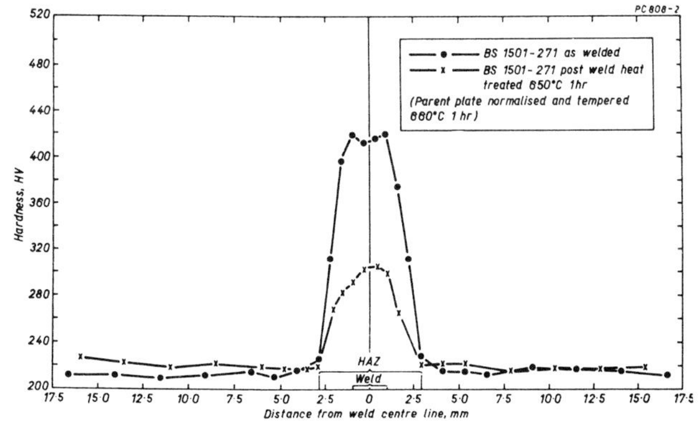

Fig. 7 Hardness traverse on BS 1501-271 on tranverse weld section

Metallographic examination

As reported above, cracking was revealed only by examination of polished sections. Two types of cracking were identified, solidification cracking parallel with the plate surface and HAZ tearing associated with inclusion lines also parallel with the plate surface.

The weld-metal structures were all fine-grained, and varied from upper bainite with ferrite or acicular ferrite to martensite depending on the degree of alloying in the steel. In the HAZ immediately adjacent to the fusion boundary, there was, in every case, a very narrow band of coarse grained transformed parent material — as can be seen in figure 5.

DISCUSSION

Occurrence of defects

It is possible to interpret the results of the porosity and cracking tests in two ways: first as a means of assessing the defects likely to arise when electron-beam welding these types of materials, and secondly for comparing the relative weldability of the various grades in order to suggest possible composition changes in the worst materials to effect an improvement. All the materials with one exception showed some tendency to form porosity of the elongated form typical of electron-beam welds. The exception was a $2\frac{1}{4}$ Cr 1 Mo steel of foreign manufacture which, it is believed, may have been vacuum-degassed. However, it was shown that with all the materials it was possible to produce porosity-free welds, provided that the welding procedure was developed to promote vacuum degassing during welding. Materials with high oxygen content tend to be more prone to porosity. Hydrogen and nitrogen levels appear to have no significant effect, at least over the ranges examined, particularly in the case of nitrogen for which the best steels have highest nitrogen levels. Also, high inclusion contents do not necessarily mean that a steel will be prone to porosity. For example, the best material has an intermediate inclusion level; but the important point here may well be that the amount of silicate is very low.

The materials tended to fall into distinct categories of tendency to form porosity. Thus, after the high quality Cr : Mo steel came the three BS 150-271 silicon-killed steels, all of which had relatively low oxygen and inclusion levels. Only slightly worse were the next group, the silicon-killed aluminium-treated type, which had slightly higher oxygen and inclusion levels. The materials with the greatest tendency to porosity were the semi-killed steels, namely, BS 1501-151 and 213 which, although they had relatively low inclusion levels in comparison with the fully silicon-killed steels, probably still retained some FeO. Almost indistinguishable from these steels was the fully killed BS 1501-161, of which both casts were very 'dirty' and had the highest inclusion levels of any of the steels.

In general it can be concluded that the BS 1501-622 and BS 1501-271 steels can be welded with relative ease; that is a wide range of process parameters can be used before porosity becomes a problem. The silicon-killed aluminium treated steels can be welded fairly easily, provided degasing welding procedures are used (this is advisable for any material because of the risk of joint contamination). The remaining steels examined can be welded with sufficient care in the selection of operating parameters, the individual problems depending on the adequacy of the degassing operation during steel-making and the subsequent removal of deoxidation products from the melt.

Weld-metal cracking was not found to be a serious problem as it occurred only in a few steels and the cracks were always very small. This result is not surprising in view of the relatively low sulphur and phosphorus levels in these steels. It is possible that cracking could be more

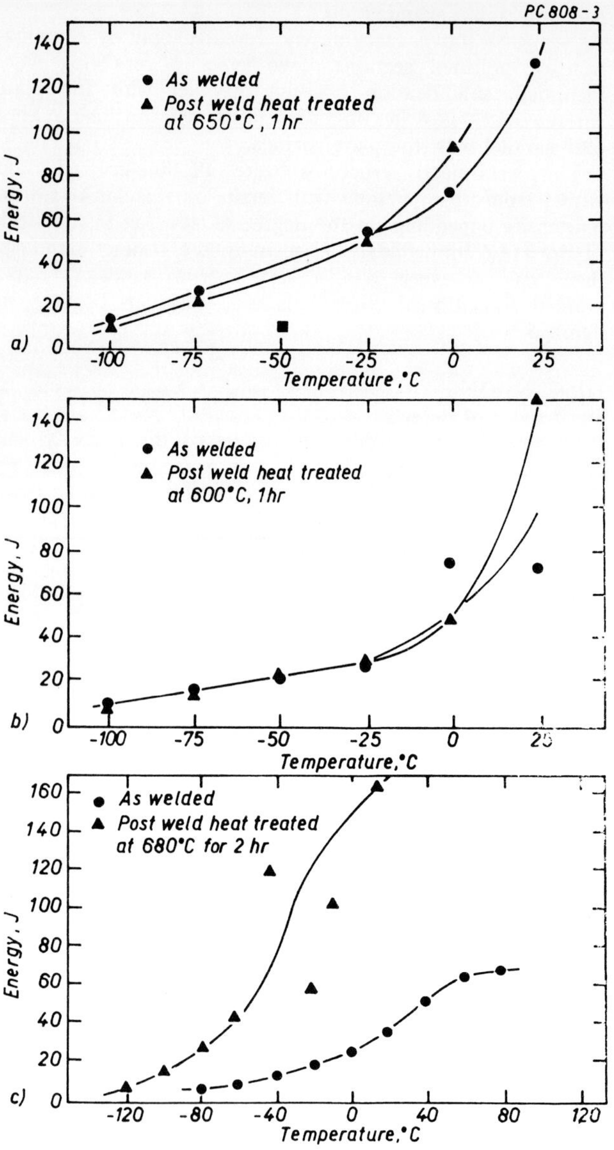

Fig. 8 Charpy V notch results for welds in (a) BS 1501-271, (b) BS 1501-161, (c) BS 1501-622

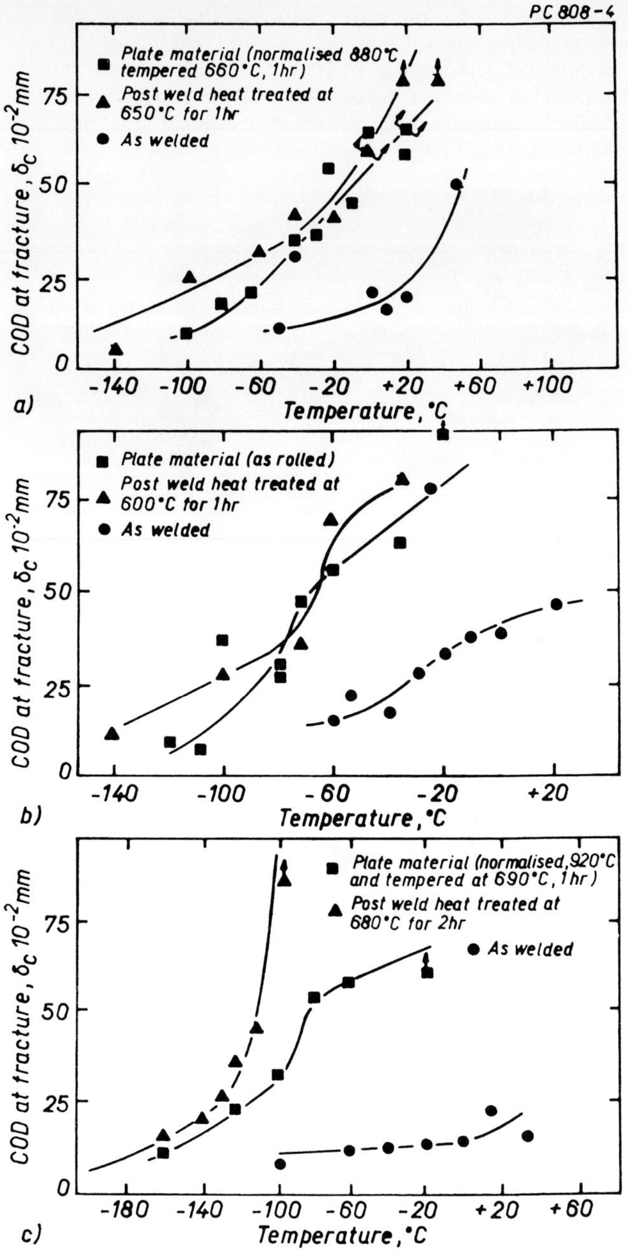

severe if the upper limits allowed by BS 1501 specifications were reached. The steels in which cracking did occur had the highest sulphur and phosphorus levels, the worst being BS 1501-161. Similarly, BS 1501-213 with fairly high sulphur and phosphorus content also showed a tendency to solidification cracking. Heat-affected zone tearing occurred mainly in the dirtiest steels, both casts of BS 1501-161, and to a small degree in BS 1501-271 which, although it did not have a particularly high total inclusion level, had a high percentage of sulphide inclusions. Cracking, therefore, does not appear to be a significant problem in electron-beam welding these steels. It was found (for the test pieces and materials examined) that at speeds of 200 mm/min and below, significant cracking did not occur in any material. Also it may be said that the cracking which was observed at higher welding speeds is unlikely to be significant in practice, as may be inferred from the fracture toughness assessment discussed below. It must be borne in mind, however, that the degree of restraint imposed by these test pieces was not very high and it is possible that with large weldments and greater material thicknesses cracking could be a greater problem.

Fig. 9 COD values for weld-metal in (a) BS 1501-271, (b) BS 1501-161, (c) BS 1501-622

Mechanical tests

The tensile tests gave satisfactory results, all specimens failing in the parent material. This is as would be expected from the hardness values, since in nearly all cases the weld area was harder than the parent material. Converting the hardness values for the weld-metal into approximate yield strengths suggests that even after post-weld heat treatment the weld overmatches the parent material by some 50% in the case of BS 1501-271 and 40% for BS 1501-161. Only with BS 1501-622 does post-weld heat treatment result in comparable weld and parent metal properties.

The attempts to soften the BS 1501-271 weld further by additional treatment time at 650°C showed that some improvement could be made. Softening was greater with the 685°C for 1 hour treatment, but unfortunately this also applied to the parent material since the latter had been previously tempered at 660°C. The standard treatment (1 hour at 650°C), however, improved the weld-metal toughness substantially, in spite of the high weld-metal hardness, as can be seen from the COD values given in figure 9(a). The beneficial effect of post-weld heat treatment also applies to the other two materials tested [figures 9(b) and (c)]. The Charpy values also show an improvement after heat treatment for BS 1501-622, but unlike the COD data this test does not discriminate between the as-welded and heat-treated conditions for the other two materials.

The smallest defect tolerance for all materials is for unstress relieved nozzle welds in BS 1501-271. It is unlikely that any of the materials could be used satisfactorily in the nozzle region without stress relief, with the possible exception of BS 1501-161 for which the maximum permissible defect size is 15 mm.

In the stress-relieved condition, all weld-metals would be safe to use in all joints. The defect tolerances for stress-relieved weld-metal in the critical nozzle region vary from 10 mm for BS 1501-271 to 48 mm for BS 1501-161; these are substantially larger than any actual defects found in this investigation. In the less critical membrane regions, the critical flow sizes cover the range 36-152 mm for the same materials. For all the materials it can be said that stress-relieved weld-metal is at least as good as parent material from the toughness point of view, and in the case of BS 1501-622 it might even be suggested that pressure vessels should be made out of all electron-beam weld-metal!

Potential of EBW for structural steels

The types of defect examined have been those concerned with material properties, whereas there are other defects such as a missed joint which are concerned with process control. These latter would need to be considered in an overall assessment of the chance of having significant but undetected defects in an electron-beam welded structure. So far as defects in the present materials are concerned, it has been shown that porosity can be controlled to acceptable levels by suitable process development and that weld cracking (at least with the eleven casts examined) is not a serious problem. Even if welding cracking were a more troublesome problem in other materials, or with highly restrained joints, it would be possible by process control, to ensure that such cracking was always parallel with the plate surface. Therefore, the defects could not be longer than the weld width, and hence by its nature would be both of short length and intermittent along the weld, and thus such defects might be acceptable. The most dangerous defect in electron-beam welding is probably the missed joint, but this ought to be detectable by surface NDT procedures or ultrasonics.

The materials examined in this report have all been steels associated with the pressure vessel industry, and the defect tolerance calculations were based on pressure vessel practice. These steels, however, are also probably typical of many structural steels from the points of view of gas content, impurities and inclusions. Therefore, the overall results should be relevant to other industries and in such applications as valves, rotors, etc.

General conclusions

1. The eleven casts of steel examined can be all satisfactorily joined at 22 mm thickness by EB welding, provided that weld pool degassing welding procedures are adopted. Cracking when present was small and parallel with the plate surface, and was not observed when travel speeds were reduced to 200 mm/min and below.
2. However, in order to improve the weldability of pressure vessel steels from the EB welding point of view, it would be advisable to restrict the levels of oxygen, sulphur and phosphorus below the present levels.
3. The joint properties assessed according to BS 1515 gave satisfactory results in all cases, and COD values for stress-relieved weld-metals were at least as good as the parent material.

Recommendations

The results of the preliminary tests reported show that satisfactory EB welds can be made in structural steels and it is, therefore, suggested that the following work would be a logical consequence in the case of pressure vessel welding.

1. Survey areas of potential application in heavy engineering and assess the technical and economic aspects of applying electron-beam welding in these areas.
2. Carry out further weldability trials on a wider range of materials and plate thicknesses, and assess joint properties in both weld-metal and HAZ.
3. On satisfactory completion of the above, carrying out larger scale tests and produce model vessel for testing.
4. Investigate joint fit-up tolerances and means of providing the required fit-up in production for a number of different joint designs, e.g. main seams, nozzle connection.
5. Investigate methods of providing required vacuum environment, e.g. complete vacuum chamber or local and sliding seals.
6. Correlate findings of the suggested programme and make final economic assessment.

Many of these aspects are common to the application of EB welding in the general context of heavy section components such as rotors but without significant vacuum engineering problems.

Acknowledgements

We gratefully acknowledge the co-operation and assistance of Messrs G. R. Egan and A. C. Steel.

References

1 Russell, J. D., Brown, K. W. and Adams, M. J.: 'A feasibility study of electron beam welding large thick-walled pressure vessels'. The Welding Institute Report P47/70.
2 Egan, G. R.: 'An analysis of defects in welds using fracture mechanics — a survey of the literature'. The Welding Institute Report E/41/71.
3 Harrison, J. D., Burdekin, F. M. and Young, J. G.: 'A proposed acceptance standard for welding defects based on suitability for service'. Second Conference on Significance of Defects in Welds, London, May 1968.

Electron-beam weldability of three aluminium alloys

By A. Sanderson, A.I.M., A.M.C.S.T., M.Weld.I., A. N. Taylor
and **R. J. Stearn,** The Welding Institute.

The electron-beam weldability of three alloys—8mm (5/16 in) H15 (Al;4.5%Cu), 6.35mm (1/4 in) H30 (Al; 1%Mg; 1%Si) and 5.35mm (0.21 in) Hiduminium 48 (Al; 4%Zn; 2.5% Mg)—studied as part of a research programme for the Underwater Weapon Research Establishment. Both high (150 kV) and low (30)kV) voltage electron-beam welding have been used for comparison. All the alloys required cleaning prior to welding to avoid porosity formation.

Electron beam welding can offer many advantages over other welding processes although each material has to be considered carefully. Some generalisations can be made about certain groups of similar alloys.

Initial trials on the H15 and H30 alloys showed that they were prone to solidification cracking and for all the alloys, cleansing prior to welding minimised porosity formation. The alloys were pre-heat-treated to meet practical considerations of fabrication. Thus, heat input, weld widths and the ability to recover properties after welding were thought to be important. Post-weld ageing was used to optimise heat-treatment temperature and times for the H15 and H30 alloys and the Hiduminium 48 alloys were given four post-weld heat treatments to compare results.

EQUIPMENT

The equipment used was a full-vacuum modified, single-lens 30 kV 9 kW Brad Thompson Industries (BTI) machine [1,2] and a 6 kW, 150 kV Hamilton Standard (HS) machine. Maximum travel speeds available were 1 m/m (40 in/min.) and 9 m/m (360 in/min.) respectively. Of the two different guns available—a Rogowski (R40) and a Steigerwald (S32)—the S32 was capable of producing narrower fusion zones but would not operate without arcing on aluminium even though several techniques were tried. On several of the materials, filler-wires were necessary to eliminate weld-bead notching and a special feed unit was devised to dispense spooled or short runs of wire (figure 1).

Beam oscillation was used on the HS machine to improve bead profile with a deflection frequency of up to 2 kHz. Generally the amplitude was less than 0.5 mm.

TECHNIQUES AND RESULTS

H30 and H15 alloys

Preliminary welds made in these materials invariably contained porosity and solidification cracking. Two high-voltage melt runs are shown in figure 2 for the H15 alloy—in this case a double thickness joint was attempted which resulted in cracking at the backing/plate interface. Investigating the four main variables—plate preparation, welding speed, backing thickness and penetration depth—which seemed to affect these defects, it was found that of the four plate preparations used, mechanical scraping or dry machining was effective. Chemical pickling was better than wire brushing for removing the surface oxide but appeared to leave chemical deposits. Solidification cracking, aggravated by porosity formation, depended mainly on welding speed. In all cases where 1.6 to 8 mm thick backing material was used, cracking occurred at the backing/plate interface; however, cracking and porosity were reduced if the welds penetrated fully.

At welding speeds of 0.8 to 2.5 m/min (30 to 100 in/min),

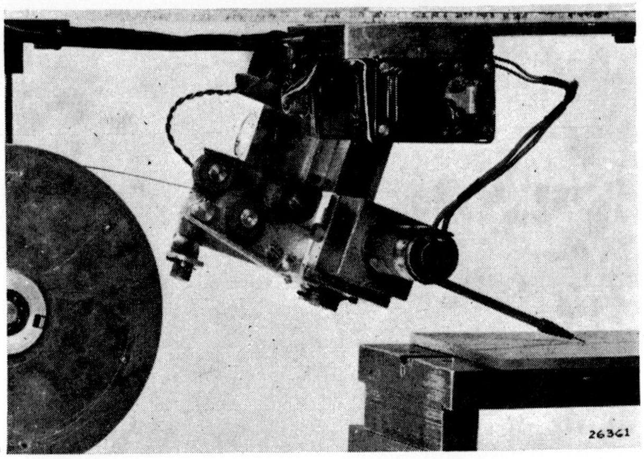

Fig. 1 Wire-feed unit used in the high-voltage electron-beam welding machine, showing wire-straightening rollers, feed motor and guide tube.

the weld metal was free from solidification cracking but increasing the speed to 3.8 m/min produced some cracks. Unfortunately it was found difficult to control weld-bead shape below about 4 m/min; there was a tendency to produce notches at the plate surfaces. The situation was improved by beam oscillation at 2 kHZ with an amplitude of 0.4 mm (0.016in) across the joint, and slight defocussing.

However, filler-wire (see figure 1) or pre-placed filler shim had to be used to eliminate notching completely. Figure 3 shows an example of a joint made by this technique; further smoothing of the weld beads was achieved using a low-current defocussed beam (cosmetic pass).

In general, low-voltage welds tended to be wider and more tapered in section than corresponding high-voltage welds;

Fig. 2 High-voltage melt runs in wire-brushed and degreased 8 mm (5/16 in) H15 plate with matching backing showing gross porosity and cracks at backing/plate interface.

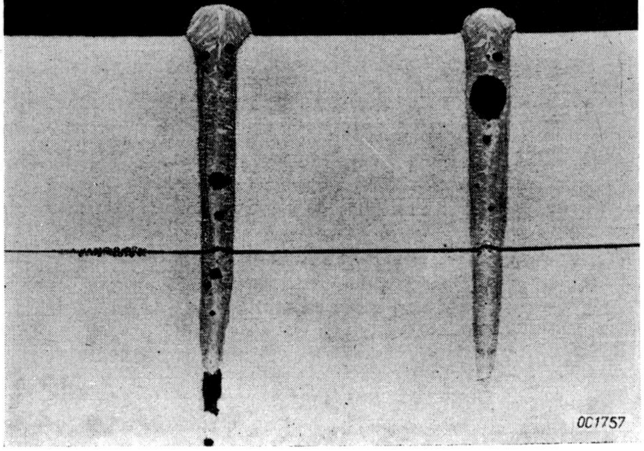

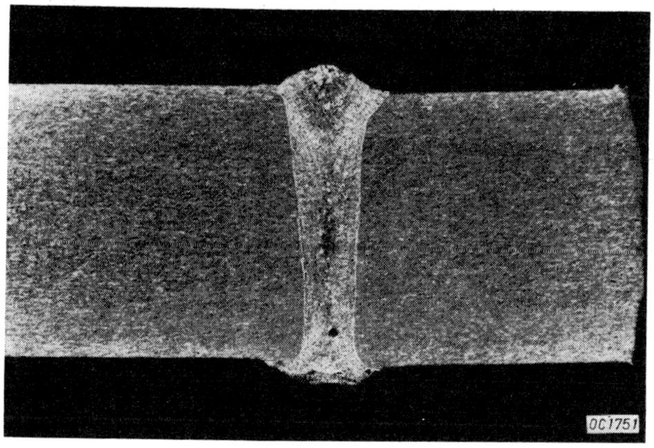

Fig. 3 High-voltage weld in H30 alloy made using shim technique. Shows proud weld beads (X10). Welding conditions* are given below.

Machine	Accel. voltage, kV	Beam current, mA	Travel speed	
			m/min	(in/min)
HS (R40)	140	26.0	2.29	(90)

thus there was more difficulty in obtaining proud weld beads.

An example of a microhardness survey, carried out across a transverse section of a weld for different post-weld ageing treatments, is given in figure 4. For a high-voltage H30 alloy weld, the fusion zone width is approximately 1.5mm, but the HAZ extends to some 3 times this value. During ageing the FZB (fusion zone boundary) recovers some 30 points in 20 hours at 171°C; further ageing, however, causes loss of hardness in weld and parent plate. As-welded, there is an expected overaged region, found some 2 mm from the weld centre-line, which does not recover completely. Adjacent to the FZB there is evidence of a hardening effect.

The H15 alloy showed a similar trend although recovery was less effective in the overaged region and the FZB. Low-voltage welds also gave similar results to the high-voltage H15 and H30 work but the FZB and HAZ were considerably wider and ageing was slightly less effective. On the basis of microhardness surveys, ageing treatments of 20 hours for the H15 alloy and 10 hours for the H30 at 171°C were selected. Tensile results for 19 mm (0.75 in) wide specimens are given in Table 1.

19 mm (0.75 in) wide specimens were subjected to a longitudinal bend test. Both the low-voltage and high-voltage H30 post-weld-aged specimens withstood 180° bends with no signs of failure, but the high-voltage and low-voltage welds failed at bend angles of 25 and 30°.

Hiduminium 48

Prior to welding, the 5.35 mm (0.21 in) plate material was solution-treated at 460°C for one hour, water-quenched and aged for 18 hours at 120°C. The high-voltage machine was used for this alloy and suitable welding conditions were established on mechanically-scraped plate using a systematic optimisation procedure in which focus position, beam current, beam oscillation and travel speed were considered as important variables. The accelerating voltage and working distance were fixed at 140 kV and 100 mm (4 in). An optimum melt run gave proud top and bottom beads with only slight intermittent undercutting (maximum depth less than 0.2 mm) and transverse radiographs of a 4 mm longitudinal slice containing the weld revealed only a small number of pores.

*Beam-oscillation and focus-position data have been omitted for brevity. Full details are given in the original report[8].

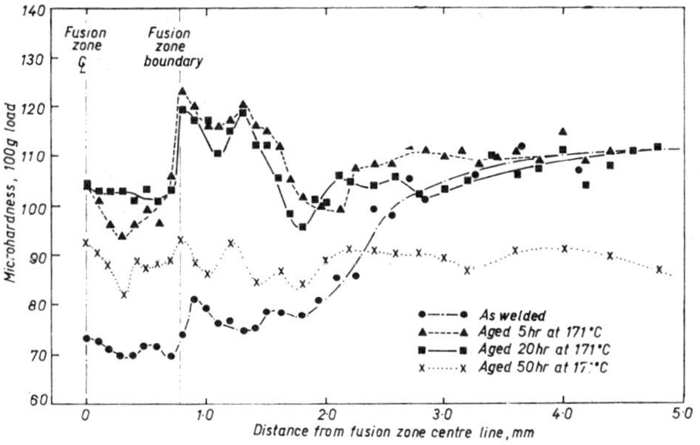

Fig. 4 Microhardness plots for high-voltage melt run in H30 alloy for 'as-welded' condition and three subsequent ageing heat treatments (indentations positioned on plate centre-line).

Unfortunately, the optimum conditions established on solid plate were not reproduced with butt welds. There was a tendency for weld-bead notches to form when a joint-line was included in the fusion zone. Reducing the maximum gap to 0.01 mm (0.0005 in.) and the use of stepped joints and optimisation of conditions did not improve the under-cutting problem substantially; it was concluded that some filler technique was essential and a wire-feed system was used. A suitable welding condition which produced proud top and bottom beads was found using 1 mm (0.04 in) Hiduminium 68 wire. A cosmetic pass (figure 5) smoothed the top bead. There were no cracks and negligible porosity in micro sections examined.

A large oscillation amplitude has been employed to obtain suitable wire melt-off rates. For the cosmetic pass, the beam was conduction-limited by oscillation and the use of a defocussed medium-voltage (60 kV) beam. Joints for mechanical testing were prepared as for the H15 and H30 alloys.

Four batches of specimens were treated in the following way.

1. Welded: naturally aged for 28 days.
2. Welded: naturally aged for 6 days; aged 40 h at 120°C.
3. Welded: naturally aged for 6 days; aged 16 h at 100°C plus 8 h at 150°C.

Fig. 5 Micrograph of Hiduminium 48 alloy fusion zone showing proud weld beads—top bead smoothed by application of cosmetic pass. Welding conditions are given below.:

Machine	Accel. voltage kV	Beam current, mA	Travel speed	
			m/min	(in/min)
HS (R40) butt weld	140	25	2.3	(90)
HS (R40) cosmetic pass	60	10	0.76	(30)

4. Welded: naturally aged for 6 days; aged 16 h at 100°C plus 2 h at 150°C plus 2 h at 180°C.

The first treatment was to simulate the practical 'as welded' state. Six days were allowed to elapse before subsequent treatment of the other three specimen batches to reduce differences in the amount of natural ageing after welding. The ageing treatments chosen for these batches were selected as follows[4].

1. Simple single-stage treatment; but possibility of embrittlement.
2. Double treatment to give a good combination of tensile and toughness properties.
3. Three-stage treatment to give a maximum chance of

Table 1 High–voltage and low–voltage electron–beam weld tensile test results for H15 and H30 alloys

Material	Material conditions	Weld type	0.1% proof stress N/mm²	0.1% proof stress (tonf/in²)	UTS N/mm²	UTS (tonf/in²)	Elongation %	Location of failure
H15	Parent plate (7)							
	W.P. condition	—	324	(21.0)	417	(27.0)	8	—
	W. condition	—	232	(15.0)	386	(25.0)	12	—
	As welded	H.V.	195	(12.6)	361	(23.4)	1.0	Fusion zone boundary
		H.V.	239	(15.5)	366	(23.7)	1.0	
		H.V.	229	(14.8)	411	(26.6)	1.5	
	Aged for 20 hr at 171°C	H.V.	181	(11.7)	415	(26.9)	2.5	Fusion zone boundary
		H.V.	199	(12.9)	414	(26.8)	3.0	
		H.V.	178	(11.5)	405	(26.2)	2.5	
	As welded	L.V.	327	(21.2)	327	(21.2)	1.3	Fusion zone boundary
		L.V.	344	(22.3)	344	(22.3)	1.3	
		L.V.	338	(21.9)	366	(23.7)	1.0	
H15	Aged for 20 hr at 171°C	L.V.	193	(12.5)	402	(26.0)	1.0	Fusion zone boundary
		L.V.	196	(12.7)	371	(24.0)	0.5	
		L.V.	199	(12.9)	423	(27.4)	1.0	
H30	Parent plate (7)							
	W.P. condition	—	232	(15.0)	293	(19.0)	8	—
	W. condition	—	108	(7.0)	201	(13.0)	15	—
	As welded	H.V.	—		261	(16.9)	4.5	Across weld metal
		H.V.	208	(13.5)	270	(17.5)	3.0	Across weld metal
		H.V.	158	(10.2)	269	(17.4)	3.75	Across weld metal
	Aged for 10 hr at 171°C	H.V.	236	(15.3)	292	(18.9)	6.25	Across weld metal
		H.V.	205	(13.3)	290	(18.8)	7.0	Across weld metal
		H.V.	266	(17.2)	289	(18.7)	11.0	Parent plate
H30	As welded	L.V.	221	(14.3)	250	(16.2)	4.0	Fusion zone boundary
		L.V.	229	(14.8)	242	(15.7)	3.5	Fusion zone boundary
		L.V.	221	(14.3)	255	(16.5)	4.8	45° shear failure across fusion zone
	Aged for 10 hr at 171°C	L.V.	168	(10.9)	310	(20.1)	8	FZB and 45° shear in HAZ
		L.V.	198	(12.8)	307	(19.9)	8	45° shear failure across fusion zone
		L.V.	151	(9.8)	307	(19.9)	8	FZB and 45° shear in HAZ

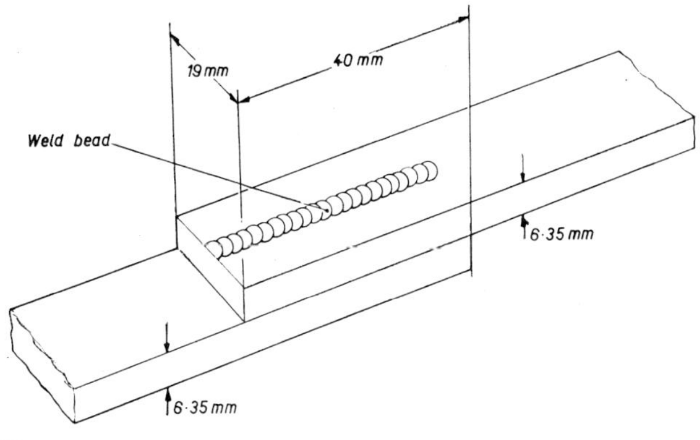

Fig. 6a Shear test specimen showing location of stake weld.

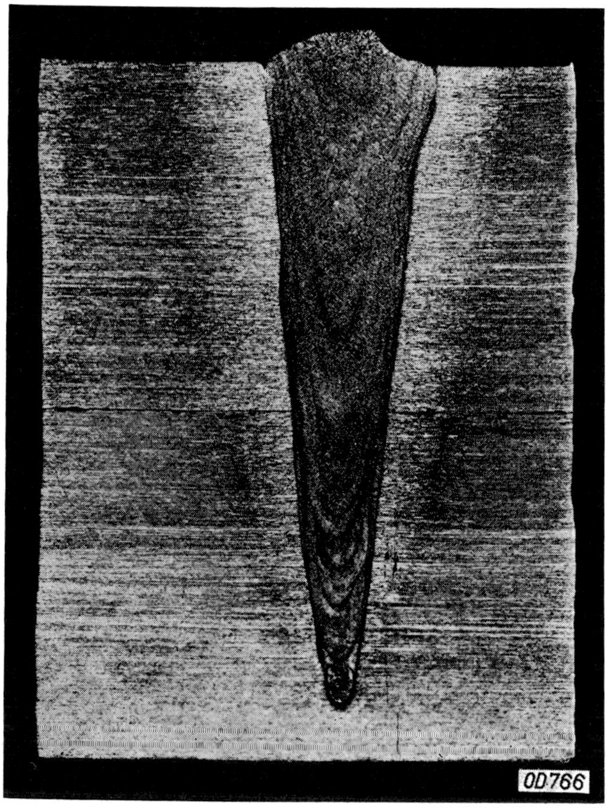

Fig. 6b Micrograph of stake weld in Hiduminium 48 alloy. Note freedom from cracks at backing/plate interface but some root defects. Welding conditions are given below:

Machine	Accel. voltage kV	Beam current, mA	Travel speed m/min	Travel speed (in/min)
HS (R40)	140	18	1.0	(40)

avoiding embrittlement.

In addition, shear test specimens (figure 6a) were prepared using a high-current defocused beam to obtain substantial finger penetration widths at the plate/backing interface (figure 6b). The macrosection shows some cracking and porosity in the root of the weld but the joint would probably be acceptable for many low-stress applications.

The result of the tensile and shear tests are given in figure 7 (maximum and minimum results are given as 'I' shaped markers; the average of three values is also shown). The face-bend test carried out on a 19 mm (0.75 in) radius roller shows that none of the ageing treatments, including natural ageing for 28 days, caused embrittlement. All specimens withstood a 180° bend without failure.

The double artificial ageing heat treatment gave the highest UTS and proof stress results for the transverse tensile test and also the highest shear strength for the stake welds (figure 7). All of the tensile specimens failed in the weld area although there were some differences in the fracture type from batch to batch. The fracture surfaces showed some evidence of a small amount of porosity in the weld but this would probably have a negligible effect on the mechanical properties. Elongation was measured on 50.8 mm (2 in) and 12.7 mm (0.5 in) to locate the position of maximum deformation. These results are also shown in figure 7.

DISCUSSION

Control of weld bead shape was found to be a common problem on all of the three alloys studied. The use of high welding speeds was not practicable because of cracking difficulties, especially on the H15 and H30 alloys. This could be minimised by using beam oscillation but weld bead notching could not be entirely eliminated. Moreover conditions, when established, were found to be critical and there was sufficient difference between melt run results and welds to cause a recurrence of notching. Slight changes in machine performance were sufficient to give the same trouble.

Mechanical properties

As welded, the H15 UTS results for the high-voltage (H.V.) and low-voltage (L.V.) welds were similar to parent plate material in the W condition, the elongation was, however, very low and the L.V. 0.1% proof stress results were anomalously high. After ageing, the H.V. welds recovered some ductility but the L.V. results were poor. The UTS approached the parent plate WP-condition values in both cases but the proof stress value remained low. As welded, the H30 0.1% PS and UTS results were better than the WP condition but the ductility was again reduced. After ageing, the properties, in the case of the H.V. welds, attained parent plate values but the

Fig. 7 Tensile and shear test results for Hiduminium 48 alloy for four post-weld ageing treatments.

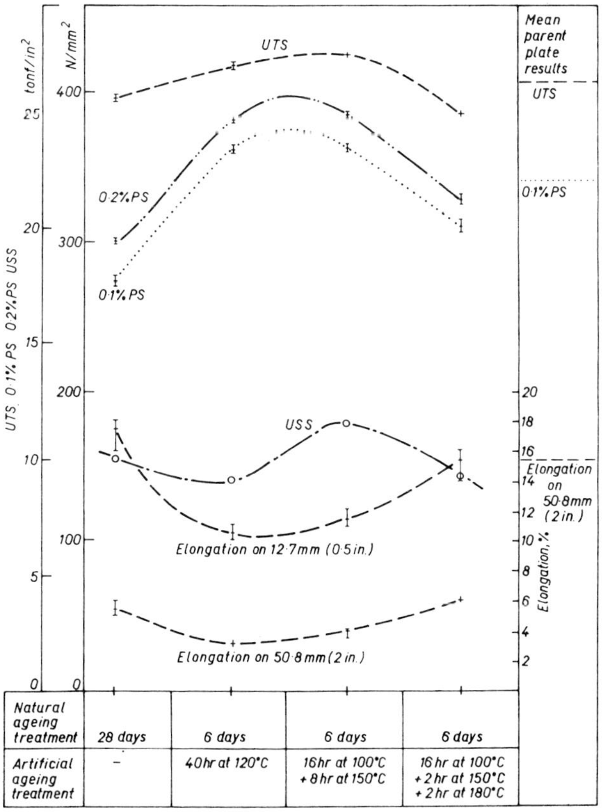

L.V. proof stress results were low. The recovery of ductility for this alloy was reflected in the bend test results. Aluminium-magnesium-silicon (H30) alloys are known to be prone to cracking when a conventional fusion welding process such as TIG is used; normally thicknesses in excess of 3 mm cannot be welded successfully unless an aluminium silicon filler is used. Unfortunately, this addition may reduce joint strength by as much as 40% and although previous attempts have been made to weld this type of alloy using EBW, cracking difficulties have been experienced[5,6]. The H15 type of alloy is not considered to be weldable by the TIG process unless a filler such as Al:Cu:Si is added.

It can be concluded that EBW is possible for H15 and H30 alloys in thicknesses up to at least 8 mm, without the need to add any weakening filler. However, cracking does occur if speeds in excess of 4 m/min. are used, and porosity caused by inadequate cleaning appears to promote cracking.

Porosity formation is critically dependent upon the cleaning technique used. Aluminium alloys are usually covered by an invisible layer of hydrated oxide; it is this which causes porosity formation. Wire brushing followed by degreasing may be sufficient for other non-ferrous alloys but on aluminium and its alloys this merely scuffs the oxide into the surface. Pickling is a superior technique but it is easy to leave chemical deposits on the surface which can cause porosity[6].

By far the best technique is to 'dry machine' the material by milling, turning or by mechanically scraping it. Welding should then be performed as soon as possible to avoid oxidation and moisture absorption. If storage is necessary then this should be done in a heated, dry environment[6]. The Hiduminium 48 alloy was superior in all respects to the H15 and H30 materials. The longitudinal bend test results indicated that it is not susceptible to HAZ embrittlement with any of the post-weld ageing treatments applied. This was confirmed by the elongation data obtained. From figure 7, it is clear that the great majority of deformation occurred in the weld zone, because the elongation values on 12.7 mm plate are several times greater than those measured on a 50.8 mm gauge length.

HAZ embrittlement is encountered in MIG welding, particularly if the welds are naturally aged, and in some circumstances this may even occur when artificial ageing is used[4]. In general, TIG welds are not prone to embrittlement. Bartle[4] has suggested that HAZ embrittlement in this type of alloy is a result of solid solution concentration of the alloying and impurity elements at, and in, close proximity to the grain boundaries. This concentration, which forms during welding as a consequence of liquefaction constitutes a highly alloyed material. Such material, when naturally aged, is notch sensitive. Artificial ageing depletes the grain boundary concentration by the formation of relatively large precipitates in the vicinity of the grain boundary, thus effectively eliminating or reducing the quantity of potentially brittle material.

Our work suggests that electron-beam welding does not appear to cause HAZ embrittlement even when the welds are naturally aged. This indicates that the grain boundary concentration of alloying elements has not reached a critical level. It is likely that electron-beam welding is superior to MIG welding in this respect due to differences in the heating and cooling rates within the HAZ, since the formation and extent of grain boundary liquation depends on the diffusion rate of solute atoms. In TIG welding the heating rate in the HAZ is generally slow enough to allow diffusion of solute atoms away from the grain boundary and may even prevent any liquation taking place. On the other hand, in MIG welding, the heating rate is considerably faster and there is probably insufficient time for diffusion of solute atoms away from the grain boundaries; liquation occurs and since the solute atoms are more soluble in liquid than in solid, there is further diffusion into the liquated region. The net result in this case is a high concentration of solute atoms in the vicinity of the grain boundary which can later lead to embrittlement during natural ageing.

In electron-beam welding, it is suggested that because the heating and cooling rates are greater than for MIG welding, there is insufficient time for the diffusion of solute atoms into the liquated regions. Consequently the solute atom grain boundary concentrations are insufficient to cause embrittlement during natural ageing. It is evident from figure 7 that the double artificial-ageing heat treatment gives higher proof stress and UTS values than the original parent plate and also the highest ultimate shear stress for the stake welds. In the 'as-welded' state, the 0.1% proof stress and UTS attained was 80% and 90% of the parent plate values respectively. Elongation, however, measured on 50.8 mm (2 in) was only 5 6%, although the great majority of deformation occurred in the weld area. The reduction in elongation for the single and double artificial-ageing treatments corresponds with the increase in proof and ultimate tensile strength and is probably not an indication of any embrittlement.

The shear specimens demonstrate that stake welding could be used for certain applications; in comparison, the H15 and H30 alloys could not be welded in this way.

Hiduminium 48 alloy is not only a stronger alternative to H30, but even naturally aged welds give better mechanical properties than fully heat-treated un-welded H30 alloy.

CONCLUSIONS

H15, H30 and Hiduminium 48 alloys can be electron beam welded in thicknesses at least up to 8 mm without the addition of filler material to modify the chemical composition of the weld. Welding speeds less than 2.5 m/min. (100 in/min) must be used for the H15 and H30 alloys to avoid cracking. The surfaces to be welded and the plate in the vicinity of the weld must be cleaned to avoid porosity formation. Wire brushing tends to scuff oxide on to the surface. Chemical pickling can cause contamination. Dry machining (e.g. milling or turning) or mechanical scraping should preferably be used. Welding should be carried out immediately after cleaning wherever possible; if storage is necessary this should be done in a dry, heated environment. It is difficult to produce satisfactory bead profiles (proud weld beads) without adding a matching filler material. This may be added as wire or shim or by incorporation of extra material into the joint design as integral filler. Either high or low voltage welding may be used. The latter may give wider welds but this does not, on the whole, affect the mechanical properties although distortion is usually greater. *H15 alloy:* a post-weld ageing treatment of 20 hours at 171°C gives UTS values comparable with fully aged parent plate material; 0.1% proof strength and elongation are, however, low. *H30 alloy:* a post-weld ageing treatment of 10 hours at 171°C gives UTS and elongation values comparable with fully aged parent plate material; the 0.1% proof strength, however, may be low. In this alloy, the as-welded properties may be sufficient for some applications. The H15 alloy is brittle in the post-weld artificially-aged condition; the H30 alloy is not. *Hiduminium 48 alloy* has very good mechanical properties even in the post-weld naturally aged condition and no HAZ embrittlement problems were encountered—even after post-weld natural ageing. The highest mechanical properties were obtained using a double post-weld ageing treatment (16 hours at 100°C plus 8 hours at 150°C). Shear test results suggested that a stake weld might give adequate performance in the Hiduminium 48 alloy. The H15 and H30 alloys cannot be satisfactorily welded in this manner.

ACKNOWLEDGEMENTS

We would like to thank AUWRE for permission to publish this article. A large proportion of the experimental work reported here was carried out by A. N. Taylor. We deeply regret his recent death.

APPENDIX 1

Pickling treatment for aluminium alloys.

1. Degrease.
2. Soak for 1 min. in solution of 10% NaOH and 0.3% NaS at 80°C.
3. Immerse in water and scrub to remove loose deposits.
4. Immerse in 10% nitric acid for 5 min.
5. Wash in methanol and dry.

Although this treatment gave better results than wire brushing followed by degreasing, there is a danger that deposits may be left on the plate surface from the chemical solutions[7].

References

1. Sanderson, A. and Adams, M. J. "Beam characteristics of a 30 kV electron gun", The Welding Institute report P27/68.
2. Adams, M. J. and Sanderson, A. "Single lens focus and penetration characteristics of a low voltage electron beam system", The Welding Institute report P28/68.
3. The properties of aluminium and its alloys. Information Bulletin 2, *Aluminium Federation Publication*, March 1966.
4. Bartle, P. M.: "Heat affected zone embrittlement in Al:Zn:Mg alloys". Select conference on weldable Al-Zn-Mg alloys. Welding Institute 23-25 September, 1969, Abington.
5. Adams, C. M. *EB Welding Symposium,* Editor: Edward and Funk, November 7-8, 1967, pp. 31-34.
6. Armstrong, R. E. as reference 5, page 88.
7. "Aluminium prep. methods can cause contamination". *Welding Engineer,* Part 1, October 1968, p. 50, Part 2, November, 1968, p. 50.
8. Sanderson, A., Taylor, A. N. and Stearn, R. J.: An investigation into the welding of five non-ferrous alloys. *Welding International* vol. 2, No. 2, 1972.

Electron Beam Welding of High Strength Aluminum Alloy

M. Ohsumi*, K. Higuchi*, Y. Hayakawa*, K. Noda*

1. Introduction

It has been a general consideration that super super dulalmin, the representative of which is 7075, can not be successfully welded with conventional fusion welding method. This is due to the fact that it has a severe inclination of weld cracking and extreme detraction in weld strength.

We have been investigating the application of EB welding process to this sort of materials. We are reporting the penetrasion characteristics and the mechanical properties of 7075.

Considerable differences in welding penetration phenomenon were found between alloys depending on the sort and amount of alloying element, when the penetrations do not reach to the bottom of the plate thickness (partial penetration). In particular, the penetration of 7075 which contains Zn and Mg of higher vapourization pressure as pricipal alloying elements is much deeper than 2xxx aluminum alloys whose main alloying element is Cu. This is probably due to the elevated pressure in molten pool capillary caused by the vaporization of Zn and Mg during welding, and in this case, weld defects such as spikes and cold shuts were apt to be formed at the root of the weld bead.

On the other hand, from the view point of mechanical properties, the properly EB welded 7075 exibits tensile strength comparable to the endurance limit of base metal, but the fracture mode seems to be rather brittle with small elongation. The fracture toughness of the weld, however, is superior to that of base metal. This is a remarkable fact, but incomplete fusion is suspected on some fractured surfaces, which are considered to be cold shuts.

2. Penetration Characteristics of Aluminum Alloys.

2.1 Materials and Procedure of Experiment

The chemical compositions of aluminum alloys used in this experiment are listed in Table 1. Test pieces shown in Fig. 1 were successively set in line and were bead-on-plate welded in one chamber. Welding schedule of Table 2 was applied.

2.2 Results and Discussion of Experiment

Penetration geometries of aluminum alloys listed in Table1are shown in photo. 1, from which it is evident that there are considerable differenes among materials. As for bead appearance, 7075 and 5083 are irregular and rugged as if eruption has occured, 2024 and 2014 are fairly smooth and wide, and most beautiful, uniform and widest are 2021 and 2219. As for the depth of penetration, 7075 and 5083 are especially narrow and deep, approximately twice those of other materials such as 2014, 2024, 2021, 2219, and penetration becomes shallower and rounder in the order of 2024, 2014, 2021, 2219. Penetration depth and bead appearance are conversely related. Longitudinal cross-section of weld beads were taken to observe the ripple of penetration,

* Mitsubishi Heavy Ind., Ltd., Nagoya Aircraft Works, First Engineering Department, Fabrication Research Section 10, Oye-cho, Minato-ku, Nagoya, 456, Japan

Table 1 Chemical composition of materials used (%)

Material	Cu	Si	Fe	Mn	Mg	Zn	Ti	Sn	V	Cd	Zr	Cr
2014	3.9-5.0	0.5-1.2	<1.0	0.4-1.2	0.2-1.2	<0.25	<0.15	-	-	-	-	-
2021	5.8-6.8	<0.2	<0.3	0.2-0.4	<0.02	<0.1	0.02-0.1	0.03-0.08	0.05-0.15	0.02-0.2	0.1-0.25	-
2024	3.8-4.9	<0.5	<0.5	0.3-0.9	1.2-1.8	<0.25	-	-	-	-	-	<0.1
2219	5.8-6.8	<0.2	<0.3	0.2-0.4	<0.02	<0.1	0.2-0.1	-	0.05-0.15	-	0.1-0.25	-
5083	< 0.1	<0.4	<0.4	0.3-1.0	3.8-4.8	<0.1	<0.2	-	-	-	-	<0.5
7075	1.2-2.0	< 0.4	<0.5	< 0.3	2.1-2.9	5.1-6.1	< 0.2	-	-	-	-	0.18-0.35

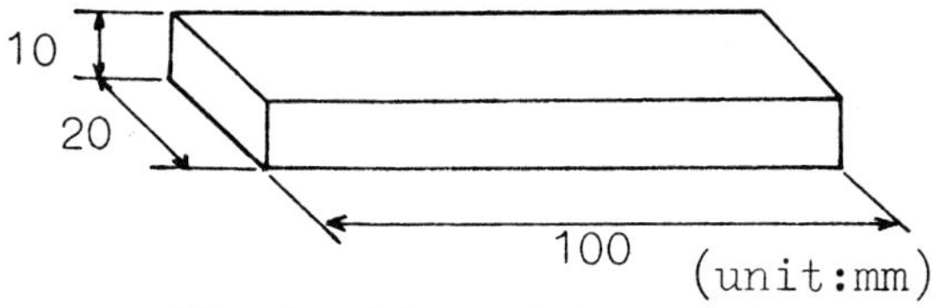

Fig.1 Size of test piece

Table 2 Welding schedule

Parameter	
Accelerating voltage	150 KV
Beam current	7 mA
Welding speed	1.5 m/min
Work distance	150 mm
Focus	sharp(ab=1 at small current)
Atmosphere	$1x10^{-4}$ Torr

which has made clear that the same influence of chemical composition as on spikes also exists. Alloys 7075 and 5083 were accompanied with root porosities and cold shuts at almost all of the spike areas.

In the case of fully penetrated welding, 7075 still exibits, not conspicuously, above phenomena, obtaining narrower penetration than 2219 etc. (Photo. 2).

Closely related to the remarkably deep penetration of 7075 and 5083 are probably the evaporation of Zn and Mg during welding which are the principal alloying elements of these materials.

A vapor pressure vs. temperature curves for individual alloying elements are shown in Fig. 2. In Fig. 2, solid lines are for pure metals, while dotted lines are for alloy corrected by Raoult's law according to the following equation;

$$P_{AS} = \gamma_A N_A P_A \qquad (1)$$

where

P_{AS} = the vapor pressure of alloying element
γ_A = activity coefficient assuming 1 ($\gamma_A = 1$)
N_A = mole fraction of alloying element
P_A = vapor pressure of alloying element as pure metal

$$N_A = \frac{Wt\%_A/M_A}{Wt\%_A/M_A + Wt\%_B/M_B}$$

A: alloying element

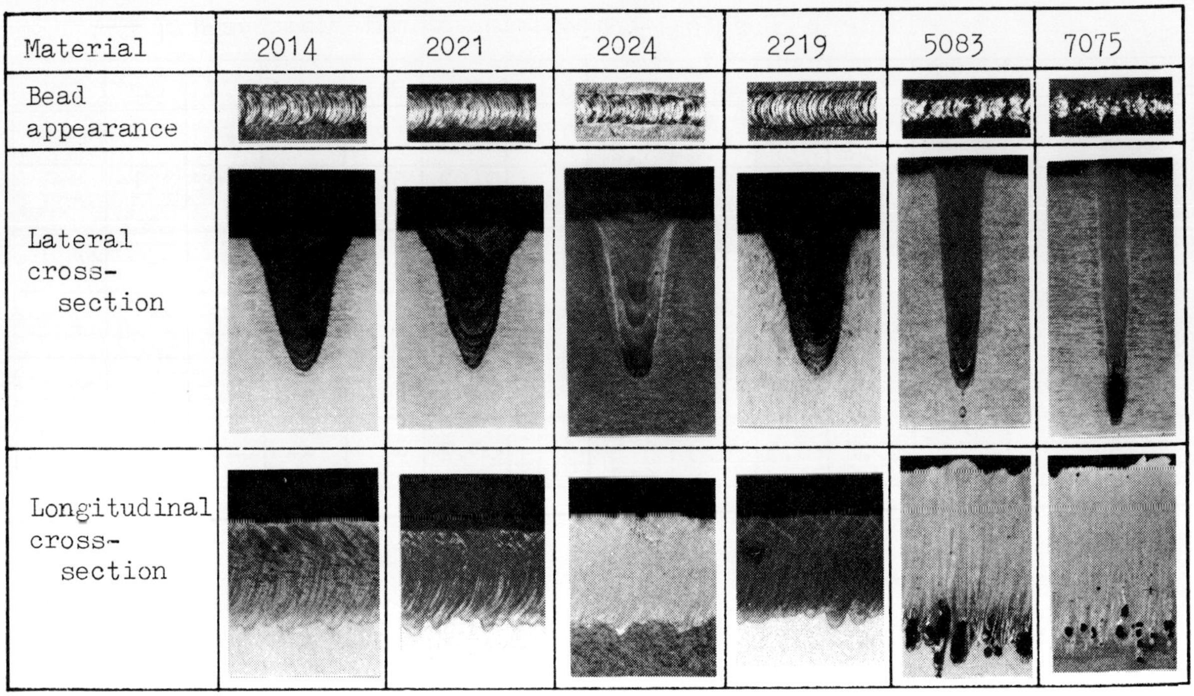

Photo. 1 Penetration geometries of aluminum alloys

B: base element
M_A:the atomic weight of alloying element
M_B:the atomic weight of base element

In Fig. 2, it is found that the vapor pressures of Zn and Mg are approximately 10^2 and 10 Torr respectively at 1000°C, which are extraordinalily higher than $4x10^{-5}$ Torr of Cu, the principal alloying element of 2xxx grade aluminum alloys.

This vaporization phenomenon is proved by spectrum analysis on the longitudinal cross-section of weld fusion zones of 7075. The results are summernized in Table 3 (1), in which vaporization rate is greater for Zn than for Mg, and increases as welding speed is reduced, Zinc content becoming below the specification requirement at a welding speed of 1250mm/min.

Observation of weld bead formation during welding and the resulted bead appearance also implies that alloying elements vaporize at considerable

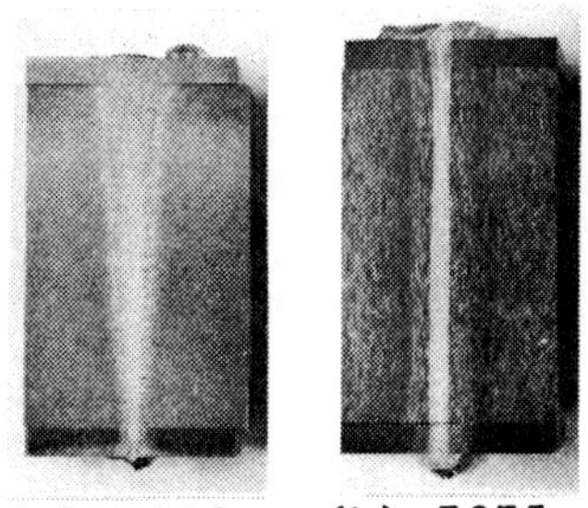

(a) 2219 (b) 7075

Photo.2 Lateral cross-section of welded area (full penetlated, $57mm^t$)

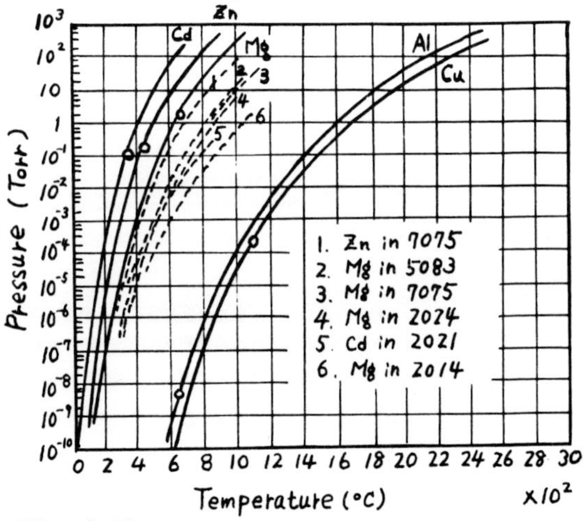

Fig.2 Vapor pressure vs. temperature diagram

REFERENCE

(1) M. Ohsumi, etal,: MITSUBISHI JUKO GIHO, (1970-5)

(2) A.H. Meleka: Electron Beam Welding: Principles and practice

Table 3 Chemical composition of weld fusion zone (%)

	Welding speed(mm/min)	Si	Fe	Cu	Mn	Mg	Cr	Zn	Ti	Al
Specification		<0.50	<0.70	1.2-2.0	<0.30	2.1-2.9	0.18-0.40	5.1-6.1	<0.20	Bal
Base metal		0.10	0.21	1.50	0.03	2.50	0.20	5.38	0.01	
Weld fusion zone	125	0.10	0.21	1.48	0.03	2.14	0.20	3.70	0.01	
	500	0.10	0.21	1.48	0.03	2.35	0.20	4.50	0.01	
	1250	0.10	0.21	1.48	0.03	2.45	0.20	4.28	0.01	
	1750	0.10	0.21	1.48	0.03	2.50	0.20	5.10	0.01	

rate in 7075. Distinct differences in the colors of electron beam(spectra) also exist.

According to the concept that inside pressure and the surface tension of molten metal is balanced in molten pool capillary (2), it seems that the pressure in molten pool capillary of 7075 reaches around the order of 100 Torr, too.

3. Mechanical properties of EB welded 7075

3.1 Static Strength

Plates of 7075-T6, 12.5mm thick, were EB welded at several welding speeds, from which 8mm thick JIS No.5 test specimens were machined and subjected to tensile test two months after welding. Tensile test result is shown in Fig.3. There exists the range of welding parameters within which a joint strength as high as 45kg/mm^2 is obtained in the as welded condition, but fracture mode appears not to be ductile enough with very small elongation.

Fracture frequently occurs at weld fusion zone slightly apart from weld bond.

Bending test result is shown in Fig. 4. Test specimens, 5mm thick 25mm wide and 180mm long, were machined from the previously welded plates of 12.5mm thickness, and were bend tested with a bending radius of 19mm and a bending span of 45mm. All of the as-welded specimens fractured at very small angles.

This brittleness can be remarkably improved if postweld solution treatment is applied.

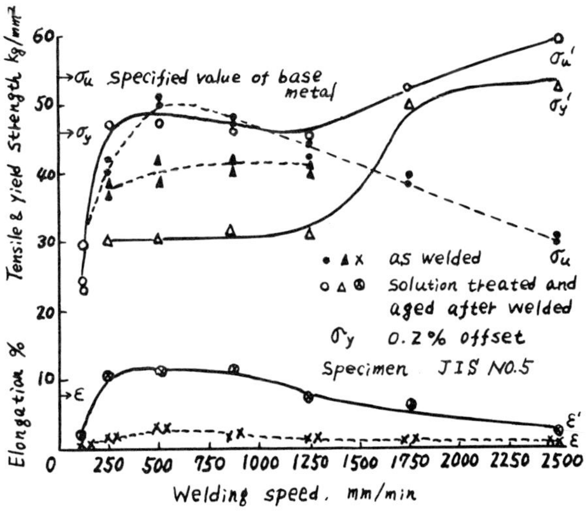

Fig. 3 Tensile test result (7075-T6)

3.2 Fracture Toughness Test

Plates of 7075-T6, 30mm thick, were butt welded according to the welding schedule shown in Table 4.

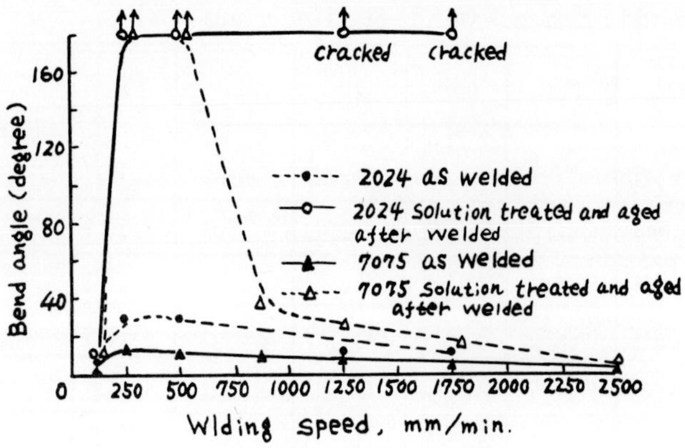

Fig. 4 Bending test result (face bend)

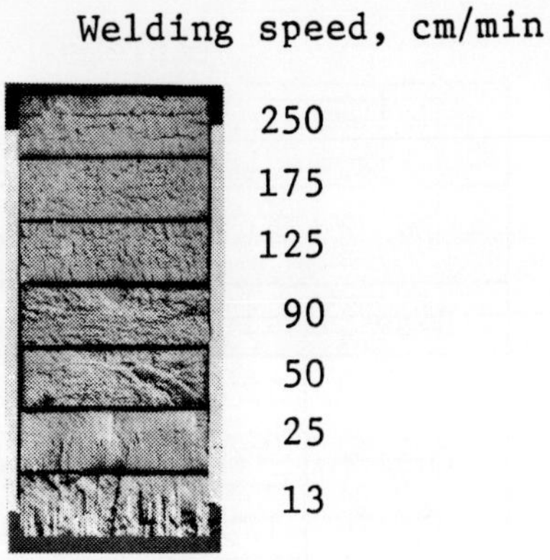

Photo. 3 Apperance of fracture (tensile test)

Non destructive and metallurgical inspections of welded area did not indicate weld defects at all. CKS type fracture toughness test specimens of 25mm thickness were machined from welded plates above mentioned. Tests were made both in the as-welded and postweld T6 treated conditions. Notch was located at the center of weld fusion zone or weld bond. Test pieces were precracked on a bend testing machine rated 1690 cps, and then tensile tested using a universal testing machine in LT direction.

Test results are listed in Table 5. Fracture mode is shown in Photo. 4.

Test results are as follows;

1) The fracture toughness of welded area exceeds that of base metal (84kg$\sqrt{mm}$/mm^2), independently of the conditions after welding and the location of notch. The steength in the as-welded condition, in particular, reaches 124kg$\sqrt{mm}$/mm^2, which is 1.5 times as high as that of base metal. Individual toughness values are considerably uniform.
2) The fracture toughness value of postweld heat treated specimen also exceed 100kg$\sqrt{mm}$/mm^2. In this case hardness distribution is uniform and there exist no soft zone of weld.
3) Some of the fracture surfaces shift from weld fusion zone to weld bond. Incomplete fusion is suspected at the shift point, but no decrease in fracture toughness is accompanied.
4) The fracture of test pieces notched at weld bond wholly propagates from heat affected zone to base metal.

Table 4 Welding schedule

Parameter	
Accelerating voltage	50 KV
Beam current	225 mA
Welding speed	625 mm/min
Work distance	125 mm
Focus	defocus(ab= 1.24 at small current)
Atmosphere	1x10^{-4}Torr

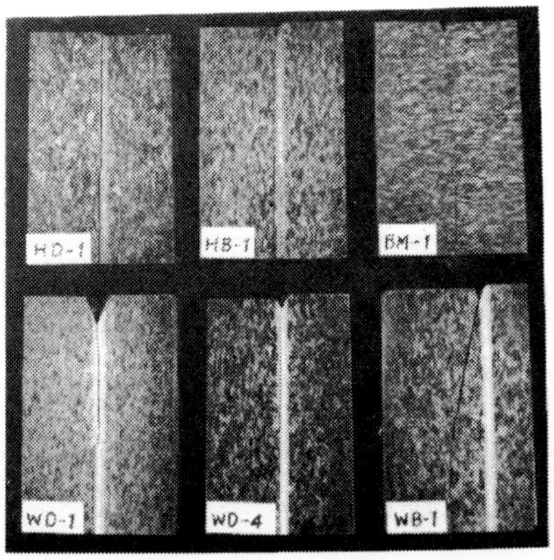

Photo. 4 Fracture mode (fracture toughness test)

Table 5 Fracture toughness test result

Specimen NO.	Condition	Location of notch	Location of crack point	Fractured at	Kc $Kg\sqrt{mm}/mm^2$
WD-1	as welded	bead center	→bead center	→center→center/bond	139
-2	"	"	→ "	→ " → "	127
-3	"	"	→ "	→ bead center	127
-4	"	"	→ bond	→ bond	110
-5	"	"	→bead center	→ bead center	135
WB-1	"	bond	→ HAZ	→ base metal	108
-2	"	"	→ "	→ "	122
-3	"	"	→ "	→ "	126
-4	"	"	→ "	→ "	123
HD-1	T6 after weld	bead center	→bead center	→ bond	130
-2	"	"	→ "	→ bead center	102
HB-1	"	bond	→ bond	→ HAZ	112
-2	"	"	→ "	→ "	103
-3	"	"	→ "	→ "	105
Base metal-1	T6	center			84
-2	"	"			84
-3	"	"			83

Test pieces recovered in hardness through postweld T6 treatment has shown the same inclination, proving the superior toughness of weld fusion zone.

5) All of the fractured surfaces exibit normal plain-strain condition.

4. Summery

1) Partial penetration phenomenon of EB welded aluminum alloys is considerably affected by vaporizing alloying elements. In the welding of 7075 and 5083 which contain high-vaporization-pressure elements of Zn and/or Mg, the bead surface is irregular and rugged, and narrow penetration about twice as deep as 2xxx aluminum alloys is obtained with high inclination of root defects such as cold shuts or spikes.

2) Tensile strength of EB welded 7075-T6 reaches as high as endurance limit of base metal if proper welding schedule is employed, but the fracture appears to be brittle with very small elongation and bend test develops ~~the same~~ result. Fracture toughness of welded area, however, exceeds that of base metal, proving superior weld toughness.

Source: *Advanced Welding Technology*, The Japan Welding Society, 1975

Electron beam welding of cast copper alloys*

by R. L. Jones**

Summary

In this short presentation the practical details and metallurgical aspects of electron beam welding two cast alloys have been summarized. Also the possible benefits to be derived in terms of welding speed and reproducibility have been suggested.

A further important aspect, however, which has not been discussed concerns the welding of these alloys for service in the sea water environment. It is known that for both these complex alloys, microstructural changes produced by welding can significantly affect corrosion performance, and thus the narrow zone welding approach with reduced thermal influence has this additional benefit of microstructural control.

Introduction

The problems of producing defect-free complex castings are accentuated by film forming alloys, and those that have a narrow freezing range. Although much has been done to improve foundry technology of copper alloys, in order to achieve a 100% production of sound castings, excavation of defects and subsequent weld repair has become a necessary feature of foundry practice.

However, the uncontrolled residual stress patterns, modification of microstructure, both of the casting and by the introduction of a filler material as produced by welding is a cause of concern, and has led to this reappraisal of casting production philosophy.

A solution to the casting defect problem is the technique used extensively in the steel industry of part cast/weld fabrication. This approach involves the production of complex castings, by casting a number of sub-components of relatively simple shape, which gives control over directional solidification, and subsequently joining them together by welding. The benefits of such a concept are twofold, first, large castings are no longer outside the scope of the smaller foundry providing they have a welding facility, and secondly, the stress fields caused by random weld repair are replaced by those from welded joints in predetermined locations, which may be readily assessed and thereby give greater confidence in service performance.

Inert gas welding techniques have been shown to be perfectly satisfactory in producing cast/weld fabricated components and for a number of component shapes will be the only contender technique. However, electron beam welding is an attractive alternative for some situations, because of its low heat input which minimizes associated factors such as microstructural modification, level of residual stress and distortion. The necessary jigging and joint preparation is of course critical for electron beam welding, but there is an appreciable reduction in both welding time and the likelihood of welding defects which should be considered especially when a large number of similar shapes have to be joined.

Description of alloys

At CDL we are assessing the feasibility of using both inert-gas and electron beam welding as applied to the concept of cast/weld fabrication, and have concentrated our efforts on two alloy systems, namely Ni-Al-Bronze and complex cupro nickel. This presentation is, however, concerned only with the response to electron beam welding of these alloys and the requirements for post weld thermal treatment to achieve the required microstructural and mechanical properties. Table 1 illustrates the typical mechanical properties and corrosion performance of the two alloys in their cast form.

Under equilibrium conditions the Ni-Al bronze alloy exhibits an $\alpha + k$ structure. However the microstructure is dependent on the cooling rate and for those pertaining to sand casting a proportion of the higher temperature beta phase may be retained. For larger sections giving a slower cooling rate the beta phase transforms to an alpha plus kappa eutectoid. Fig 1 illustrates a typical cast microstructure showing proeutectoid alpha contained within a matrix of lamellar alpha plus kappa. A small amount of retained beta is also present, shown as the dark etching constituent.

Fig. 1. Typical cast microstructure with proeutectoid alpha contained within a matrix of lamellar alpha + kappa.

**Lecture presented to the Belgian Welding Institute, Brussels, 1976, and reprinted with their permission.*

***Central Dockyard Laboratory, Portsmouth, U.K.*

Table 1

Description	Nominal Composition					Typical Mechanical Properties			
	Cu	Al	Fe	Ni	Mn	0·1 % Proof Stress	U.T.S.	Elongation	Impact
Ni Al Bronze to BS 1400 AB2	REM	9·5	5·0	5·0	1·5	250-300 N/mm^2	640-700 N/mm^2	20%	20 joules

Description	Property	Value
Marine Corrosion Data	Corrosion potential in sea water	190 MV (SCE)
	Self corrosion rate	63 μm/year
	Crevice corrosion	not more than 508 μm/year
	Galvanic corrosion	compatable with most copper base alloys
	Form of galvanic corrosion	pitting and selective phase attack
	Impingement resistance	4·3 m/sec

Description	Nominal Composition						Typical Mechanical Properties			
	Cu	Ni	Mn	Ti	Fe	Al	0·1% Proof Stress	U.T.S.	Elongation	Impact
90/10 cupro nickel	REM	12	10	0·4	5·0	2·5	250 N/mm^2	460 N/mm^2	25%	41 joules

Description	Property	Value
Marine Corrosion Data	Corrosion potential in sea water	200 MV (SCE)
	Self corrosion rate	25 μm/year
	Crevice corrosion	38 μm/year
	Galvanic corrosion	cathodic to most copper alloys
	Form of galvanic corrosion	pitting
	Impingement resistance	4·0 m/sec

The microstructure of the complex cupro nickel-alloy is not dependent on the cooling rate. A typical as-cast structure, shown in Fig 2, contains a cored solid solution with the presence of precipitates both within the nickel rich dendrites and along grain boundaries.

Welding data

All the welding trials were undertaken in the downhand position in a vacuum of 10^{-2} torr employing a 150 kV, 6 kW Hawker Siddeley electron beam welding unit. The parameters obtained for a simple butt weld joint geometry are given in Table 2. Comparable to most single pass techniques it can be seen that for increased section thickness it is necessary to increase the beam current and to decrease the welding speed.

It was found that Ni-Al bronze was prone to weld bead undercutting, possibly as a result of higher vapour pressure. Although beam oscillation was found to improve the situation, the technique was only applicable to small section thicknesses because of the power limitations of the welding unit. Conversely the cupro nickel alloy readily formed a more acceptable weld bead profile although on welding thicker sections of both alloys there is a tendency for weld metal "drop out", it is considered that this could be overcome by employing a welding unit with a horizontal gun.

Microstructural effects of welding

Microscopic examination of transverse sections taken through the weldments revealed that the Ni-Al bronze weld consisted of fine pro-eutectoid alpha within a martensitic type phase designated beta prime. On traversing across the weldment a narrow HAZ adjacent to the fusion line is observed and evidenced by a coarse beta prime plus alpha structure which gradually merges with the typical "as cast" structure (Fig 3).

Table 2

Material Thickness	Material	Accelerating Voltage (Kv)	Beam Current (mA)	Welding Speed (in/min)	Focus Position	Deflection	Working height (in)
¼″	Ni Al Bronze	140	35	100/120	Surface	—	6
½″	Ni Al Bronze	140	40	50	Below Surface	X5mA Y5mA 2k Hz	6
1″	Ni Al Bronze	140	40	20	Surface	—	6
¼″	Cupro Nickel	140	35	120	Surface	—	6
½″	Cupro Nickel	140	40	70/75	Below Surface	—	6
1″	Cupro Nickel	140	40	15	Surface	—	6

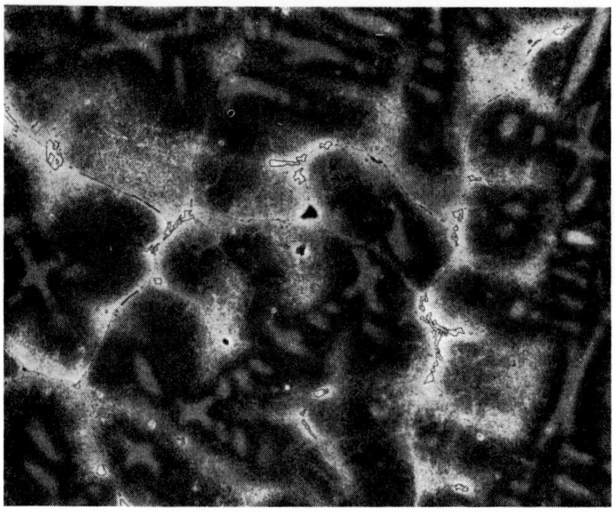

Fig. 2. A cast structure with precipitates within nickel rich dendrites and along grain boundaries.

The weld zone of the cupro-nickel sample is characterized by a fine cored dendritic structure. On traversing across the weldment there is no significant change in microstructure when compared with the parent material. However an indication of the presence of a HAZ is given by the difference in response to etching. The presence of the pale etching zone suggests that there has been some degree of solution treatment (Fig 4).

In view of the microstructural changes produced in both alloys on welding, hardness surveys were undertaken. These indicated that for Ni-Al bronze a significant increase in hardness results whilst for the cupro nickel softening occurs. Since these could have a deleterious effect on the mechanical properties, the response to post weld thermal treatment was also investigated. Microscopic examination of the Ni-Al bronze specimens tempered within the range 400—500°C show the structure of the weld to have modified due to transformation of the beta prime to beta plus Widmanstätten alpha. The coarse acicular beta contained within the HAZ is however retained (Fig 5).

The specimens tempered within the range 600—700°C show precipitation of a small directionally orientated kappa

Fig. 3. A narrow HAZ adjacent to the fusion line, evidenced by a coarse beta prime plus alpha structure.

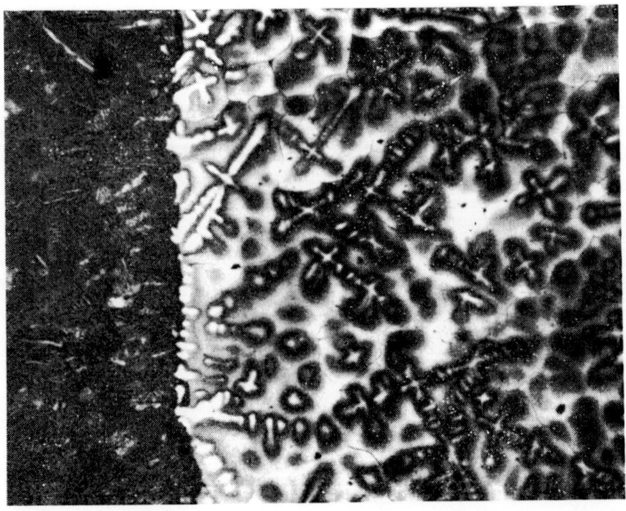

Fig. 4 A pale etching zone suggests some degree of solution treatment.

Table 3

Run No	Test Piece No	Accelerating Voltage	Beam Current	Welding Speed	Focus Position	Focus Setting	Working Height	Beam Deflection			Weld Time		
								Amplitude Setting		Frequency	Upslope	Weld	Downslope
		kV	mA	in/min			in	X	Y	Hz	Sec	Sec	Sec
1	A7/1	140	40	25	Below Surface	490	11¾	5	5	1000	3	42	5
2	A7/1	90	20	20	Above Surface	640	11¾	5	5	1000	3	42	5

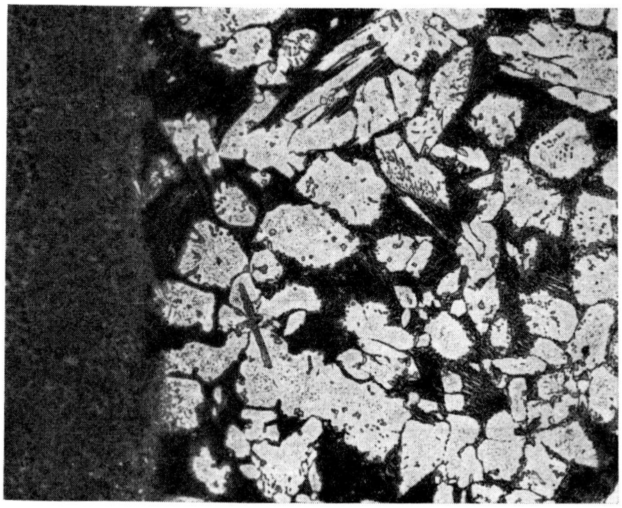

Fig. 5. Coarse acicular beta retained in the HAZ.

Fig. 7. A reversion of the lamellar alpha/kappa eutectoid to beta.

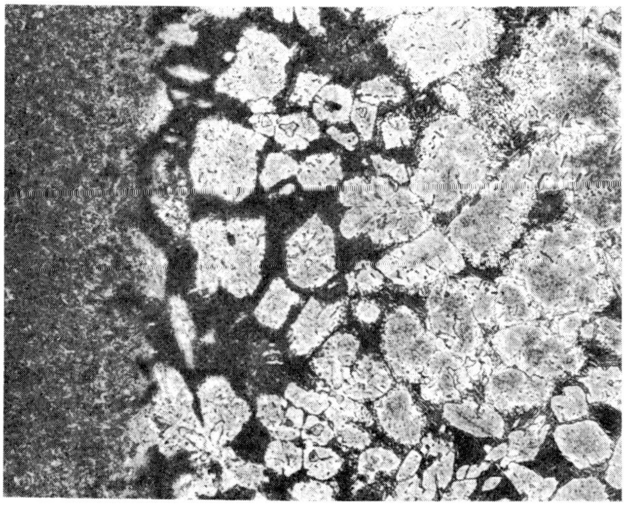

Fig. 6. Transformation of the retained beta to a fine alpha/kappa eutectoid.

Fig. 8. A broadening of the HAZ zone and an apparent thickening of the grain boundaries.

phase within the primary alpha, and transformation of the retained beta to a fine alpha/kappa eutectoid (Fig 6).

A coarsening of this structure occurs at higher tempering temperatures until approximately 850°C when there is reversion of the lamellar alpha/kappa eutectoid to beta (Fig 7).

In comparison very little microstructural change is observed following heat treatment of the cupro nickel weldments. Tempering within the range 400—500°C shows a broadening of the HAZ region and an apparent thickening of the grain boundaries (Fig 8). Tempering at higher temperatures shows little change until 800°C when recrystallization of the weld is observed together with removal of all indications of the HAZ (Fig 9).

Mechanical properties of weldments

Vickers hardness traverses taken across the "as welded" Ni-Al bronze weldments, as previously indicated show the weld metal to be harder than the parent material. Fig 10

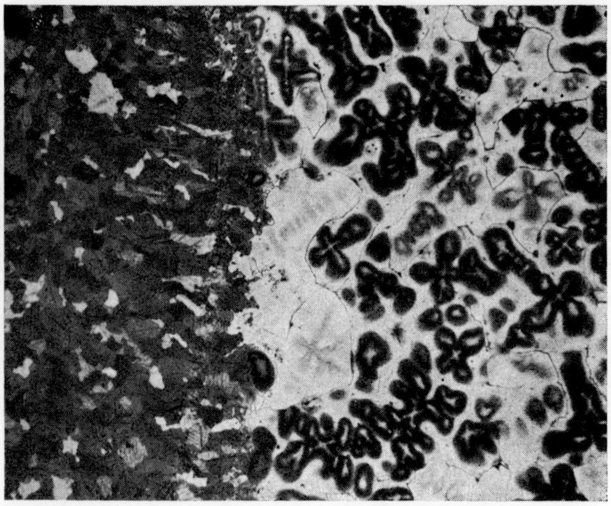

Fig. 9. Recrystallization of the weld with removal of all indications of HAZ.

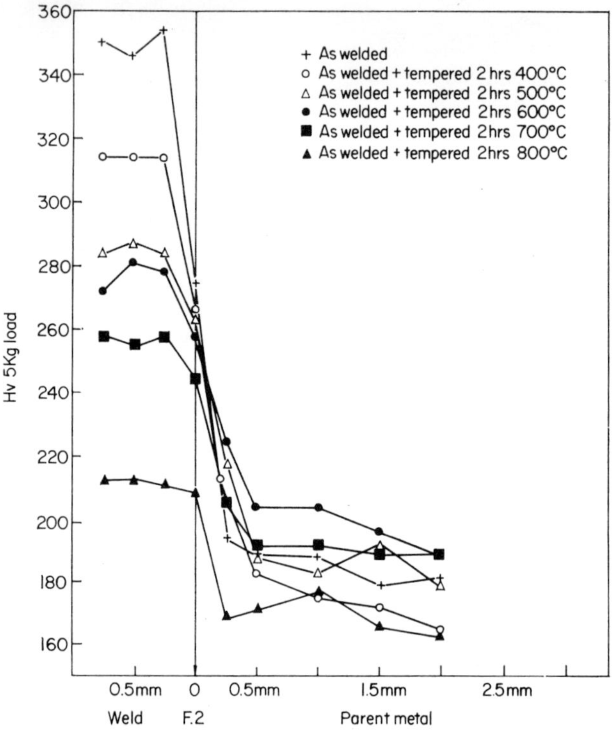

Fig. 10. Vickers hardness plotted against distance away from the fusion line for poor weld thermal treatments.

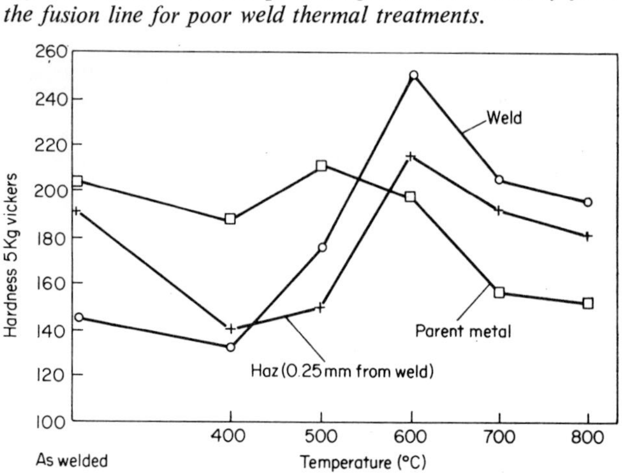

Fig. 11. Vickers hardness plotted against tempering temperature.

shows Vickers hardness (5 kg load) plotted against distance away from the fusion line for a series of post weld thermal treatments. As can be seen tempering gives a progressive softening of the weld metal. The parent material shows an initial softening on tempering at 400—500°C, followed by an increase in hardness to a maximum at 600°C, before finally softening to a minimum at 800°C. The hardening of the parent material at 600°C is consistent with precipitation of kappa within the primary alpha phase.

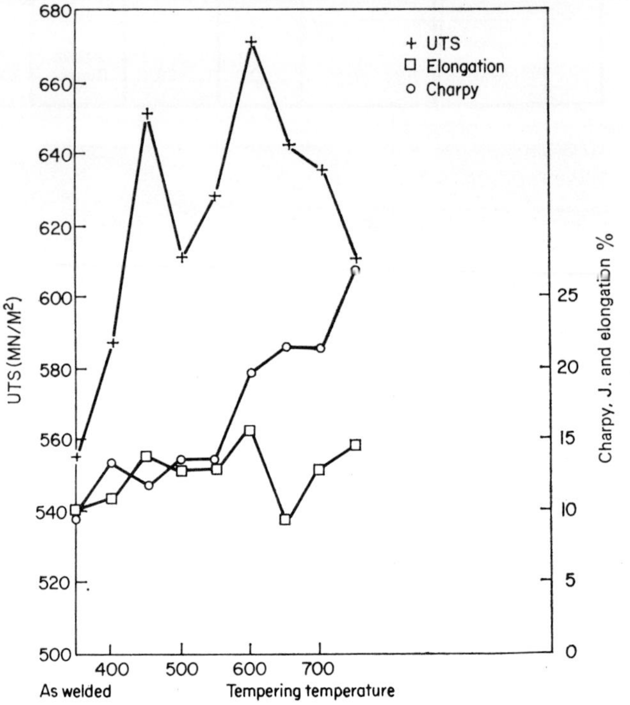

Fig. 12. Post weld thermal treatment can be beneficial in restoring ductility.

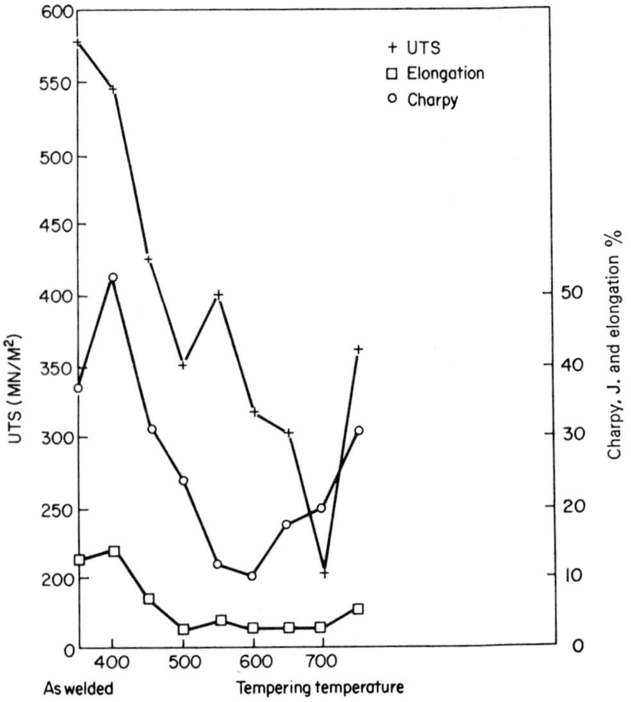

Fig. 13. Post weld thermal treatment has a deleterious effect on mechanical properties.

The response of hardness to heat treatment for the cupro nickel alloy is more complex and therefore to clarify the trends, Vickers hardness has been plotted against tempering temperature for three locations within the weldment (Fig 11). The results indicate that in the "as welded" condition both the weld metal and HAZ are softer than the parent material. Post weld thermal treatment as shown effects a hardening of the weld metal and HAZ, characteristic of a precipitation mechanism. The parent material also shows a further increase in hardness followed by overageing and softening. Micro-hardness testing revealed that the inter-dendritic regions are softer than the nickel rich dendrites, and that on ageing there is no significant increase in hardness of the white etching interdendritic regions. In fact it was observed that hardening takes place by a gradual enlargement of the brown staining area surrounding the dendrites.

The mechanical properties of the weldments were assessed from Hounsfield N° 12 tensile, and Charpy impact specimens. The weld was positioned in the specimens to coincide with the V notch of the Charpy and the centre of the gauge length of the Hounsfield.

Although the results of the tensile tests obtained from the Ni-Al bronze specimens are in doubt, because for all conditions except the "as welded" and 400°C heat treatment, failure occurred at the shoulder of the test piece, they do indicate that a post weld thermal treatment may be beneficial in restoring ductility (Fig 12). In fact the graph indicates that an improvement in both strength and ductility may be realized.

In comparison, the results obtained from the cupro nickel specimens indicate that apart from a low temperature anneal, post weld thermal treatment has a deleterious effect on the mechanical properties (Fig 13). In all instances the position of failure of the tensile specimens was situated within the weld. However, the Charpy impact specimens for the intermediate tempering temperatures failed outside the V notch, indicating that these heat treatments had resulted in embrittlement of the far HAZ.

Circumferential welding and fabrication exercise

Having established the welding parameters for a simple flat plate butt joint geometry, a machined Ni-Al bronze tube 152 mm diameter 19 mm wall thickness has been circumferentially butt welded to assess the feasibility of the approach to this more practical welding configuration. A support alignment jig was constructed. The butting surfaces of the tube were finely ground (8 μm CLA) and the welding undertaken in the downhand position. In order to achieve a continuous sound weld it is usual practice for the beam to be sloped out at the end of the weld run. It was unfortunate, however, that in this instance, interruption of the weld occurred before the welding sequence had been completed. This interruption was probably caused by metallic vapour condensing in the vicinity of the electron gun. In order to remedy this situation the defective region was rewelded at the same settings followed by a low power pass to diffuse the top bead. The top line of table 3 refers to the initial welding sequence and the second to the low power pass. It has been observed that such a low power pass gives rise to an improve-

Fig. 14. The welded tube.

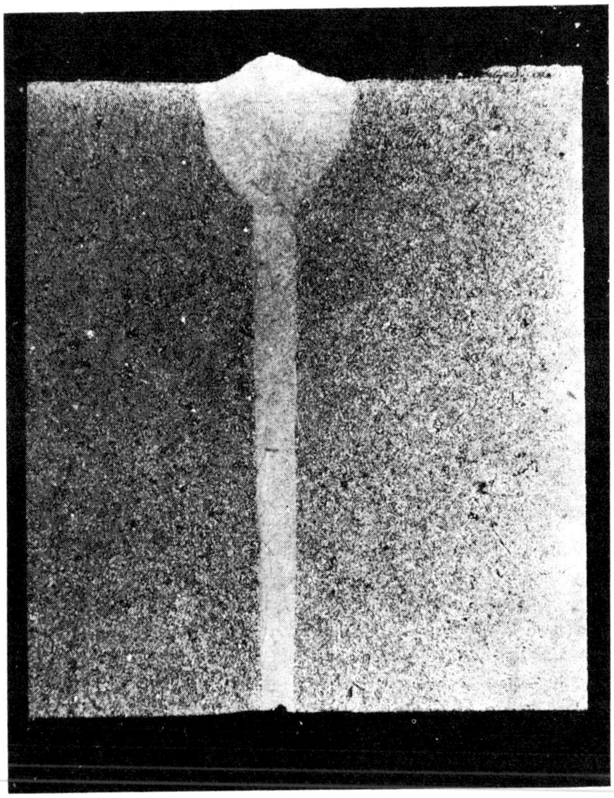

Fig. 15. Heat-treatment has transformed the HAZ.

Table 4

Condition	0·2% Proof Stress	UTS	Elongation	Charpy Impact
As welded	265·5 N/mm²	579·4 N/mm²	9%	2·7 Joules
Heat treated	288·8 N/mm²	566·5 N/mm²	5%	19·0 Joules
DG Ships 8520 F	250 N/mm²	620 N/mm²	13% min	

Fig. 16. The casting's support and alignment jig.

Fig. 17. The completed casting.

ment in the profile of the top weld bead and may alleviate the need for a post weld grinding operation. Fig 14 shows the tube after completion of the welding exercise.

Although an initial radiographic examination of the weld indicated the presence of a variety of pipes and pores together with large linear defects reminiscent of lack of fusion; by re-radiographing following a profile grinding operation it was ascertained that these defects could be attributed to undercutting of the underbead.

The tube was then longitudinally sectioned to undertake a comparison of the microstructure and mechanical properties of the as welded condition with a 2 hour 815°C/air cool heat treatment. The results obtained from 50 mm tensile test pieces (Table 4) indicated that the properties obtained for both conditions were below the specification requirements for cast Ni-Al bronze. It was observed, however, that failure occurred in the parent material away from the weld and that casting defects were present on the fracture surface. Necking, apparent either side of the weld zone in the as welded tensile specimen was not significant in the heat treated test piece indicating that the thermal treatment had homogenized the material. The macro section shown in Fig 15 indicates that the heat treatment has been effective in transforming the HAZ and also illustrates the profile of the underbead. It can also be seen that the low power pass has affected an approximate four fold enlargement of the weld zone.

Finally as a practical demonstration of cast/weld fabrication using electron beam welding an exercise has been undertaken to weld two flanges onto the body of a part casting. The support/alignment jig used is shown in Fig 16. In this particular instance because of the difficulties presented in aligning the side flange a step and recess (approximately 0·25 mm) was machined in the respective butt surfaces. The completed casting is shown in Fig 17.

SLIDING-SEAL ELECTRON-BEAM WELDING OF TITANIUM

R. H. Witt, J. G. Maciora and H. P. Ellison
Grumman Aerospace Corporation
Bethpage, N. Y.

Abstract

The sliding-seal electron beam (SSEB) welding process demonstrated capability for achieving aerospace quality weldments in titanium alloys equivalent to "hard" vacuum EB welding in chambers using production, nondestructive inspection techniques for inspection. For Ti-6Al-4V titanium alloy SSEB weldments, joint efficiencies of 100% of the base metal were obtained for both tensile and fatigue properties. Fracture toughness (K_{IC}) of one-inch-thick, SSEB welded Ti-6Al-4V titanium alloy was found to be 39.0 ksi$\sqrt{in}$. Values obtained were for weldments made in the flat and vertical up welding positions.

A selected simulated aerospace structure - an integrally stiffened wing skin - was successfully welded to demonstrate capability to fabricate production-type parts.

1. INTRODUCTION

The purpose of this investigation was to advance the electron beam welding process and establish the functional capability of a government-owned, sliding-seal electron beam (SSEB) welding system as a manufacturing method to fabricate aerospace structures. The information generated should benefit the defense and aerospace industries by providing an additional welding process that can be used to fabricate complex assemblies out-of-chamber in place of other more costly or inadequate processes. It has particular potential for welding structures too large to be contained in the high vacuum chambers presently available.

The objectives of this program were as follows:

- Produce suitable square butt weldments in various thicknesses of Ti-6Al-4V titanium alloy plate
- Determine quality and mechanical properties of titanium alloy SSEB weldments.
- Conduct weld repair studies on SSEB weldments
- Demonstrate the capability of the SSEB welding equipment to produce acceptable weldments on a simulated structure.

2. EXPERIMENTAL

2.1 EQUIPMENT

The sliding-seal electron-beam welding system used in these studies was designed and produced by Sciaky Bros., Inc., under Air Force Contract AF33(615)-5277. This unit, presently located in Grumman's Plant 2 Electron-Beam Welding Facility, Bethpage, N.Y., consists of a portable vacuum, moving electron-beam welding head that is mounted on a ram manipulator, back-up tooling and associated power supply and controls (Figure 1). The power capacity of this unit is 30kw(60kv, 500ma)

Source: *Materials & Processes for the 70's*, SAMPE, 1973

Figure 1. Sliding-Seal Electron Beam Welder Installed at Grumman

capable of welding at vacuum pressures ranging between 1 x 10^{-4} torr and 100 microns of Hg. The back-up table vacuum pressure is maintained between 10 and 50 microns of Hg. The vacuum in the moving welding head is obtained by pumping between two nonmetallic vacuum seals in the head that rides on the workpiece being welded.

A partial penetration seal weld (GTA) is used to join the components to be welded so that the vacuum level necessary for EB welding can be achieved. The ram manipulator, which supports the welding head and associated vacuum pumping equipment, is capable of traversing a distance of six feet in the flat position (X-axis) and five feet in the vertical position (Z-axis). The vacuum-sealed back-up tool is capable of being used in the flat position or remounted in the vertical position. The use of the SSEB welder permits welding large aerospace structures that cannot be welded in available chambers. The maximum butt weld lengths accomplished in these programs were two feet plus the length of end-tabs utilized.

2.2 MATERIALS

The Ti-6Al-4V titanium alloy plate was purchased per Grumman Material Specification GM 3103A which is more restrictive than Military Specification MIL-T-9046F in that it specifies a more uniform microstructure, ultrasonic testing and a 5,000-psi higher yield strength. The Ti-6Al-4V titanium alloy filler wire was purchased per Grumman Material Specification GM 3108. The chemical analyses for these as-received materials are shown in Table I. All base-metal material microstructural examinations were satisfactory.

2.3 PROCEDURES

2.3.1 Panel Preparation

Square-butt-joint weld panels (6 inches wide by 24 inches long) were machined from the as-received 48 x 48-inch plate materials. Starting and run-off tabs (9 inches long by 12 inches wide) were also machined. Titanium alloy plates had 0.030 inch per surface (0.060 total) machined off to insure removal of any mill alpha-case as well as any possible minor surface defects that could cause invalid mechanical properties test data. All the panels were machined on all four edges for square-butt and end-tab welding. It was necessary to weld nine-inch-long end-tabs to the test panel in order to over-lap the 37-inch outer seal. The 42-inch inch-long welded panel covered the weld groove and outer "0" ring in the back-up table and permitted clamping for vacuum sealing. The end-tabs also allowed an area for starting and stopping the SSEB welder so that a full 24-inch-long weld could be made on the panel material.

2.3.2 Cleaning Procedure

Cleaning procedures prior to SSEB welding were accomplished as follows for Ti-6Al-4V titanium alloy plate (machined surfaces):

- Immerse panels in non-etch alkaline cleaner Ridolene No. 53 for 5-10 minutes, hot water rinse, air dry.
- Immediately prior to seal-pass welding, wipe weld surfaces with clean acetone and air dry.

2.3.3 Specimen Configuration and Test Procedures

2.3.3.1 Tension. Both the base metal and welded titanium specimens were stress-relieved at 1200°F for four hours prior to testing.

The static tensile properties (yield and ultimate strengths and elongation) were determined on specimens of the configurations shown in Figures 2 and 3. The former was used for plate thicknesses of 0.500 inch or less, the latter for plate thicknesses in excess of 0.500 inch. The specimens were tested to failure using the procedures of ASTM E-8. Specimens were tested at a strain rate of 0.005 inch/inch-minute through yield.

2.3.3.2 Fatigue. Specimens of the configuration shown in Figure 4 were used to determine the fatigue charac-

Table I - Chemical Analysis Of Materials Evaluated

Material	Alloying Constituent, Percent							
	Ti	Al	V	Fe	C	N	O_2	H
Ti-6Al-4V titanium, 1/4 in. thick	Rem.	6.000	4.200	0.110	0.022	0.013	0.140	0.005
Ti-6Al-4V titanium, 1/2 in. thick	Rem.	6.000	4.200	0.130	0.022	0.008	0.120	0.006
Ti-6Al-4V titanium, 3/4 in. thick	Rem.	5.900	4.200	0.130	0.022	0.010	0.150	0.007
Ti-6Al-4V titanium, 1 in thick	Rem.	6.100	4.100	0.140	0.026	0.014	0.170	0.004
Ti-6Al-4V Specification (GM3103A)	Rem.	5.50-6.75	3.50-4.50	0.30 max	0.08 max	0.05 max	0.20 max	0.015 max
Ti-6Al-4V Welding Wire (1/16 in diam)	Rem.	6.17	4.20	0.13	0.014	0.009	0.057	0.0013
Ti-6Al-4V Wire Specification (GM3108)	Rem.	5.50-6.75	3.50-4.50	0.15 max	0.03 max	0.012 max	0.080 max	0.005 max

TABLE II GTA SEAL-WELDING PARAMETERS FOR Ti-6Al-4V TITANIUM ALLOY, SQUARE-BUTT SSEB WELDED PANELS

Material	Thickness, In.	Amperage, Amps	DC Volts	Carriage Travel Speed, In./Min.	Helium Shielding Gas Flow, Cu. Ft./Hr.	Argon Trailing Shield Gas Flow, Cu. Ft./Hr.	Helium Back-Up Gas Flow, Cu. Ft./Hr.
Ti-6Al-4V Titanium Alloy	0.190	70	15	30	40	180	5
	0.440	85	15	30	40	180	5
	0.690	85	15	25	60	180	5
	0.940	85	14	20	60	180	5

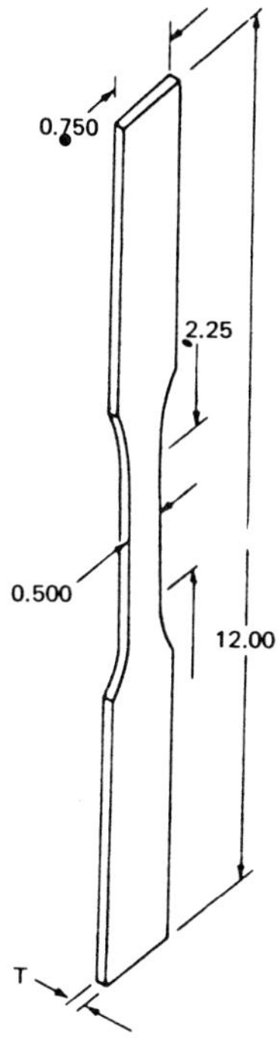

Figure 2 Flat Tensile Specimen

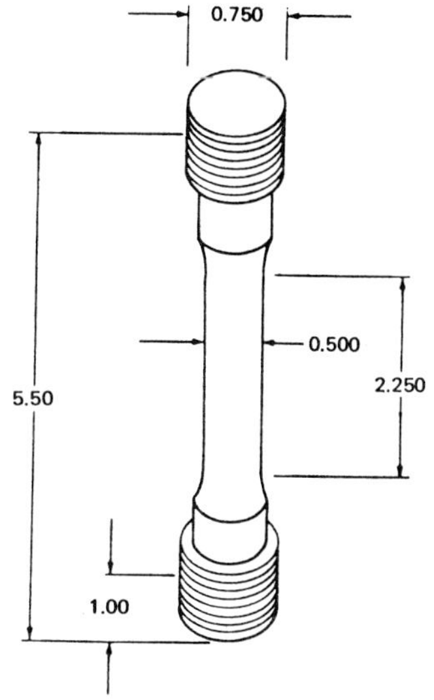

Figure 3 Round Tensile Specimen

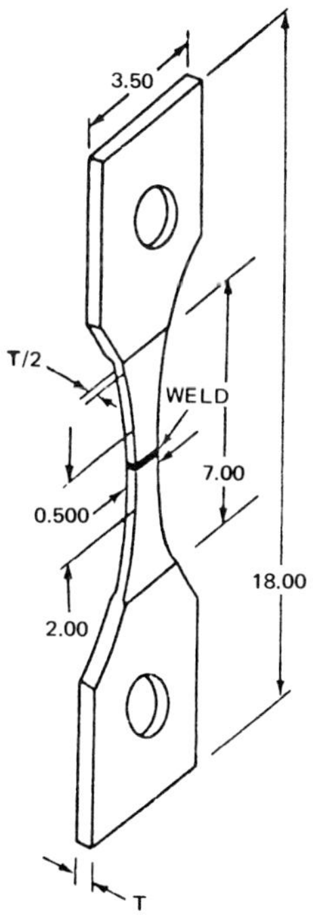

FATIGUE CONFIGURATION

Figure 4 Fatigue Specimen

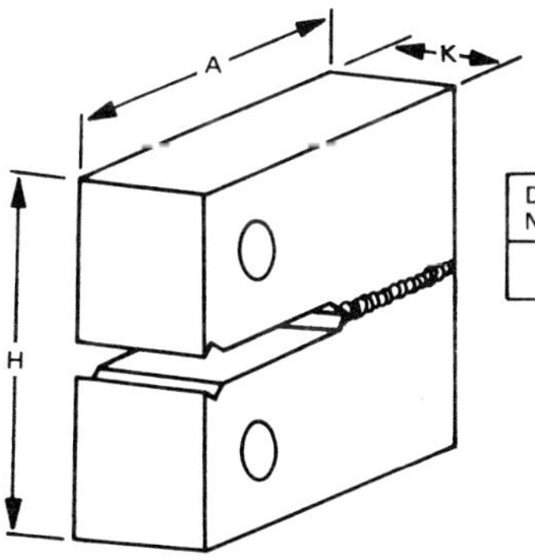

DASH NUMBER	NOMINAL DIMENSIONS A	H	K
3	3.125	3.000	1.000 *

*0.940 INCH FOR Ti-6Al-4V

Figure 5 Fracture Toughness Specimen Configuratior (Compact Tension) teristics. Each specimen was installed in a constant-force-type fatigue machine and subjected to constant-amplitude axial tension loading at a rate of 1800 cpm and a stress ratio (R) of 0.10. Stress levels were selected so as to define typical S/N curves with either catastrophic failure or an endurance limit of 10^7 cycles constituting test termination.

2.3.3.3 Fracture Toughness. Fracture toughness is the property which quantitatively determines a material's susceptibility to crack propagation.

Fracture toughness tests were performed on both base metal and welded Ti-6Al-4V titanium alloy straight-through compact tension specimens (see Figure 5) following the procedures of ASTM E-399. The specimens were initially installed in fatigue machines where cracks were initiated and grown to a crack-length-to specimen-depth (a/w) ratio of 0.6. Following cracking the specimens were removed from the fatigue machine and tested in static tension to failure. During static tests the crosshead speed of the testing machine was controlled so as to produce a stress intensity rate of increase of between 30,000 and 150,000 psi$\sqrt{\text{in.}}$ per minute. A compliance gage was used to measure the displacement across the crack. The output of the gage was autographically recorded as a function of applied load.

2.3.4 Investigation Sequence

The following investigations were conducted for titanium welding and are discussed under results:

- Optimization of welding parameters using the sliding-seal electron beam welding equipment for 1/4, 1/2, 3/4, and one-inch thick, Ti-6Al-4V titanium alloy plate, non-destructive inspection and test specimen fabrication.
- Determination of mechanical properties (i.e., tensile and fatigue strengths, and fracture toughness) for Ti-6Al-4V titanium alloy.
- Establishment of weld repair techniques and determination of ability to start and stop a weld on the assembly being joined.
- Evaluation of weldments by nondestructive techniques by means of dye penetrant, radiographic, and ultrasonic inspection.
- Fabrication and testing of component

3. RESULTS

Satisfactory GTA seal welding techniques were developed for 0.190,0.440, 0.690 and 0.940-inch-thick Ti-6Al-4V titanium alloy weld panels. Helium torch shielding gas gave better results than argon gas for GTA (direct current) welding (parameters are given in Table II). Radiography was used to examine several seal-pass weld panels for freedom of porosity/inclusions prior to SSEB welding and negative results were obtained. Helium torch shielding gas was used in all cases with satisfactory results (Table II). For all materials, current and voltage readings were found to vary by plus or minus five amperes and 1/2 volt, respectively, with satisfactory results. Weld penetration varied from 0.040 to 0.060-inch depending on thickness.

3.1 SSEB WELDING PARAMETER DEVELOPMENT

3.1.1 Flat Position Welding Parameters. The following summarizes the optimized SSEB welding parameters and compares them with the welding parameters determined using a Grumman "hard" vacuum EB welding chamber for the same gages.

3.1.1.1 0.190-Inch-Thick. Parameters were optimized using four bead-on-plate welds and four square-butt joint weld panels. The optimized parameters for square-butt joints were a beam voltage of 40 kilovolts, a beam current of 180 milliamperes, a gun travel speed of 60 inches per minute, a focus current of 6.40 amperes and a total heat input of 7.2 kilojoules per inch. For the development of these parameters, the 500-milliampere filament was used. The weld surface bead was 3/32-inch wide with a slight crown; the penetration was 1/16-

in wide and 0.020-inch deep. The underfill on the weld edge was about 0.005 inch deep. A photomacrograph of the weld cross-section is shown in Figure 6.

3.1.1.2 0.440-Inch-Thick. Parameters were optimized using four bead-on-plate welds and four square-butt joint weldments. Other square-butt welds were made subsequently with excellent results. The optimized parameters for square-butt joints were a beam voltage of 45 kilovolts, a beam current of 270 milliamperes, a gun travel speed of 50 inches per minute, a focus current of 6.40 amperes and a total heat input of 14.6 kilojoules per inch. The 500-milliampere EB gun filament was used for this investigation. The surface bead had a uniform crown about 1/8 inch wide with very little underfill (0.006 inch). Penetration was complete (about 1/16 inch wide and 0.020 to 0.030 inch deep). A photomacrograph of the weld cross-section is shown in Figure 7.

3.1.1.3 0.690-Inch-Thick. Bead-on plate welds were made prior to welding the square-butt panels. Square-butt weld panels were SSEB welded from 0.690-inch-thick material using the optimized parameters (50-kilovolt beam voltage, 300-milliampere beam current, 50 inches per minute gun travel speed, 6.40-ampere focus current, and 18.0-kilojoules per inch heat input). These parameters were obtained using the 500-milliampere filament. A photomacrograph of a typical weld cross-section is shown in Figure 8.

3.1.1.4 0.940-Inch-Thick. Starting parameters for flat-position SSEB welding were those used for bead-on-plate welds. These parameters were a beam voltage of 45 kilovolts, a beam current of 250 milliamperes, a gun travel speed of 40 inches per minute (which gave a total heat input of 17.0 kilojoules) and a beam focus current of 6.40 amperes. The 1/8-inch-wide surface bead was uniform with an underfill of only 0.010 inch. The penetration bead was about 0.060 inch wide and 0.030 inch deep with a few flush areas (P1). The welding parameters had to be changed slightly to give smooth, consistent penetration. The best top surface and penetration beads were obtained in Specimens P-4 and AFT-101 (square-butt joint) which were welded using a beam voltage of 50 kilovolts, a beam current of 330 milliamperes, and a gun travel speed of 45 inches per minute (which gave a total heat input of 22.0 kilojoules per inch). A typical cross-section of a 0.940-inch-thick quare joint is shown in Figure 9. Additional welds were made and were acceptable. Panels AFT-103 and AFT-105 were made while problems with the voltage control which regulates beam current were being experienced. Panel AFT-105 exhibited a varying weld beam shape but showed no porosity when radiographed.

3.1.2 Vertical Position Welding Parameters. Square butt joint panels were SSEB welded in the vertical-up position from 0.190-inch-thick material. Welding parameters used were a gun travel speed of 50 inches per minute, a beam voltage of 40 kilovolts, a beam current of 150 milliamperes, a focus current of 6.40 amperes and a heat input of 7.2 kilojoules per inch. These parameters gave a uniform surface bead 0.100 inch wide with a 0.010-inch-high crown with minimum underfill at the edges. The penetration bead was uniform, 1/16 inch wide and 0.020 inch deep. A photomacrograph of a typcial weldment is shown in Figure 10. The surface weld bead had a 0.020-inch-high crown and less edge underfill than the weld bead made in the flat position.

3.2 WELD QUALITY

The selection of titanium SSEB welding parameters was based on the following criteria:

- A properly shaped weld bead having a minimum amount of concavity or underfill.
- A consistent, uniform weld penetration bead.
- A weld deposit with a minimum of weld defects.

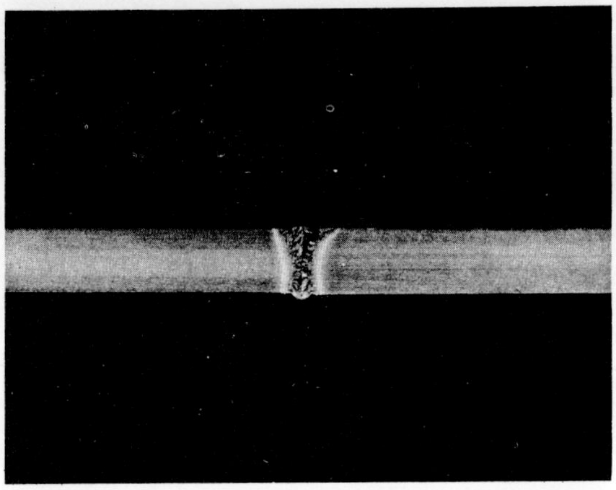

Figure 6 Photomacrograph of an SSEB Welded 0.190-Inch-Thick, Ti-6Al-4V Titanium Alloy Square-Butt Joint Weld Coupon Made with an EB Gun Travel Speed of 60 Inches per Minute in Flat Position (2X Magnification)

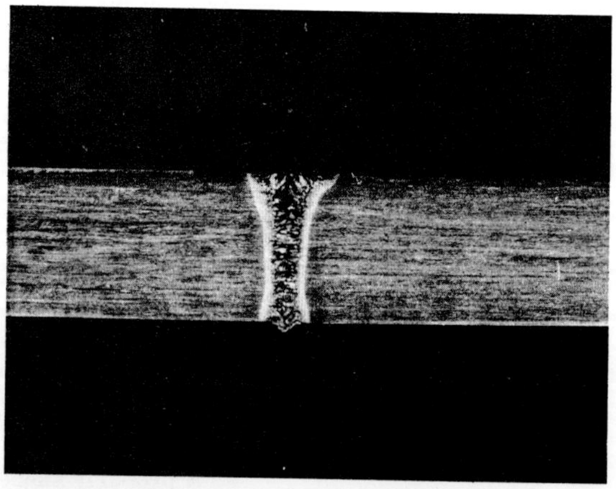

Figure 7 Photomacrograph of an SSEB Welded 0.440-Inch-Thick Ti-6Al-4V Titanium Alloy Square-Butt Joint Weld Coupon Made with an EB Gun Travel Speed of 50 Inches per Minute in Flat Position (2X Magnification)

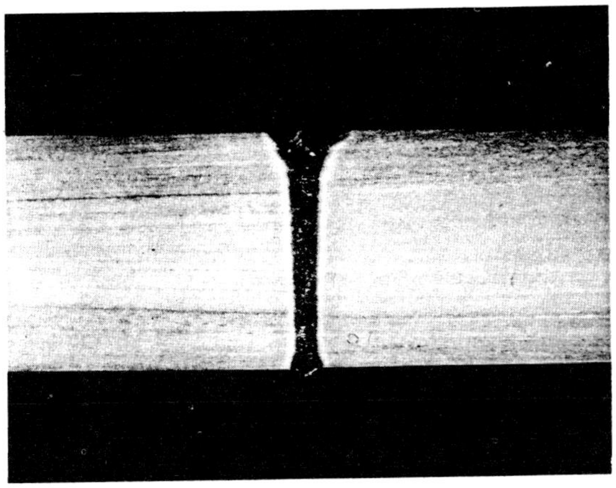

Figure 8 Photomacrograph of an SSEB Welded 0.690-Inch-Thick Ti-6Al-4V Titanium Alloy Square-Butt Joint Weld Coupon Welded at an EB Gun Travel Speed of 50 Inches per Minute in Flat Position (2X Magnification)

Figure 9 Photomacrograph of an SSEB Welded 0.940-Inch-Thick Ti-6Al-4V Titanium Alloy Square-Butt Joint Weld Coupon (EB Gun Travel Speed - 45 Inches per Minute - Flat Position - 2X Magnification)

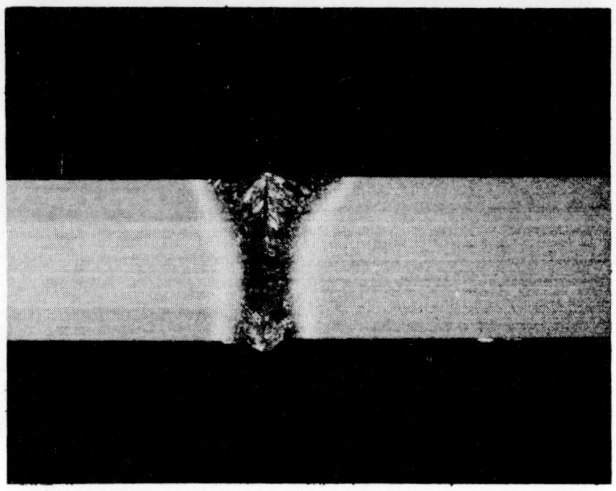

Figure 10 Photomacrograph of an SSEB Welded 0.190-Inch-Thick, Ti-6Al-4V Titanium Alloy Square Butt Joint Panel (AFT-125V) made with an EB Gun Travel Speed of 50 Inches per Minute in the Vertical-Up Postion (5X Magnification)

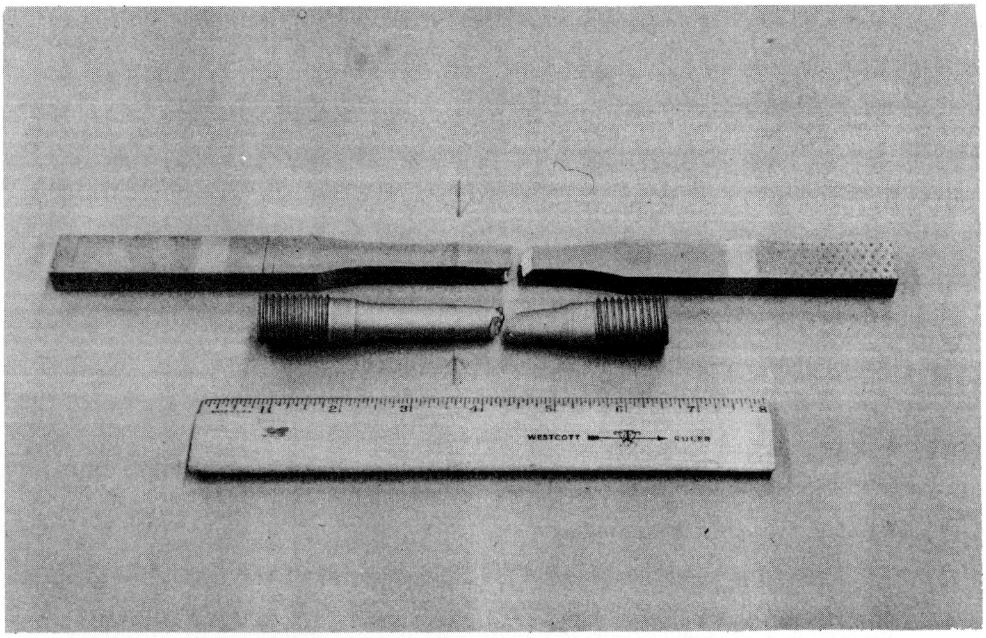

Figure 11 SSEB Welded Ti-6Al-4V Tensile Specimens After Testing at Room Temperature (Failures Occurred in Base Metal)

These criteria were selected to give a final weldment that would meet aerospace requirements for fabricated components or structures. Grumman Electron Beam Welding Specification GSS6205B was used to determine weld quality for Ti-6Al-4V titanium alloy SSEB weldments. This specification permits porosity based on 0.1T or 0.050 inch in diameter (whichever is less) with a spacing between defects based on three times the diameter of the smallest pore and a maximum of five pores per two inches of weld length. Underfill was not considered as a criterion for rejection of the weld, since it could be removed by machining or by using a "cosmetic" second pass. Radiography was the prime inspection method for the detection of internal weld defects. Ultrasonic inspection was used to check the results of radiographic inspection. Any surface weld defects were detected by visual and dye-penetrant inspection methods.

The SSEB welded titanium panels exhibited the typical, surface weld-bead center buildup and edge underfill characteristics of electron-beam, "hard" vacuum chamber and plasma-arc welded titanium panels. The titanium alloy SSEB weldments met the quality requirements of Grumman Specification GSS 6205 for electron beam weldments. Porosity was very limited and, if any existed, it was usually at the start of the weld. Lack of penetration was not a problem; this was confirmed by radiographic and ultrasonic inspection methods. Any lack of penetration detected by radiography, which was the cause of a missed seam, was repaired by subsequent rewelding. Underfill on the surface weld bead was about 0.005 to 0.007-inch deep with a maximum underfill depth of 0.010-inch. Subsequent cosmetic pass welding with the SSEB welder eliminated the underfill. Since Grumman Specification GSS 6205 requires that electron beam welds be machined to remove any underfill before final acceptance, underfill on SSEB weldments is not a criterion for acceptance of an SSEB weldment to the Grumman specification. The other requirements of the Grumman specification were followed for acceptance. The radiographic quality of Ti-6Al-4V titanium alloy weldments indicated general good quality with the weld defects confined to minor porosity or some lack of fusion due to weld seam joint misalignment with no evidence of cracking.

Ti-6Al-4V titanium alloy weldments (0.440, 0.690 and 0.940 inch thick) were examined ultrasonically. The 0.190-inch-thick SSEB weldments were not examined because the material was not thick enough to give the proper back reflection. For the three gages examined, the ultrasonic trace scans were negative.

3.3 COMPARISON OF "HARD" AND "SOFT" VACUUM WELDING PARAMETERS FOR Ti-6Al-4V TITANIUM ALLOY.

Several thicknesses of Ti-6Al-4V titanium alloy plate were electron beam welded in a prototype "hard" vacuum chamber for comparative purposes. Program material (0.190, 0.440, 0.690 and 0.940 inch thick) was used for this investigation.

Grumman EB "hard" vacuum large chamber welding parameters for 0.187, 0.660 and 1.030-inch-thick Ti-6Al-4V titanium alloy EB weldments are compared to SSEB parameters for the same relative thicknesses in Table III. It can be seen that the EB "hard" vacuum large chamber welding parameters are similar to the SSEB welding parameters. In some cases, the weld bead configurations may vary slightly due to the vacuum level used in the SSEB welder.

3.4 MECHANICAL PROPERTIES

3.4.1 Tensile

The longitudinal and transverse properties of the base metal exceeded the minimum procurement requirements and were used in establishing the joint efficiencies of the weldments investigated. The average tensile properties and joint efficiencies ob-

TABLE III COMPARISON OF EB "HARD" VACUUM CHAMBER AND SSEB WELDING PARAMETERS USED FOR WELDING Ti-6Al-4V TITANIUM ALLOY PLATE SQUARE-BUTT JOINTS

Actual Thickness, Inch	Beam Voltage, Kilovolts	Beam Current, Milliamperes	Gun Travel Speed, In/Min	Heat Input, Kilojoules/ Inch	Focus Current, Amps
LARGE EB "HARD" VACUUM CHAMBER					
0.187*	40	180	60	7.2	5.15
0.660*	45	275	45	16.5	5.25
1.030*	48	325	45	20.8	4.95
SSEB WELDING MACHINE					
0.190	40	180	60	7.2	6.40
0.440	45	270	50	14.6	6.40
0.690	50	300	50	18.0	6.40
0.940	50	330	45	22.0	6.40

*GRUMMAN ELECTRON BEAM CERTIFICATION (EBC) NUMBERS USED ARE:

0.187 inch thick EBC # 161

0.660 inch thick EBC # 14

1.030 inch thick EBC # 38

tained on the SSEB welded specimens are presented in Table IV. Visual examination of the failed specimen determined that those specimens containing transverse welds failed in the base metal (see Figure 11). Failure of those specimens having the weld in the direction of the major axis initiated in the weld.

The average tensile properties and joint efficiencies obtained on EB weldments performed in the prototype chamber using optimum SSEB welding parameters are compared to those of SSEB weldments in Tables IV and V.

With the exception of the 0.625-inch material welded in the production chamber using EB welding parameters the data reported was obtained on material purchased to Grumman Specification GM 3103. The tensile strength requirement of this specification is 5 ksi higher than that of MIL-T-9046 which was the procurement specification for the 0.625-inch-thick production welded material. In addition, this material was tested in the non-stress relieved condition. Therefore, the ultimate tensile and yield strengths are significantly lower than those of the stress relieved material.

These results show that SSEB welding is capable of producing joints in Ti-6Al-4V titanium alloy having the same efficiencies as EB welded joints provided the material is purchased to the same specification and stress relieved after welding.

The obtained average tensile properties and joint efficiencies for 0.190-inch-thick vertical-up position weldments are compared to those obtained on flat position welded specimens of the same thickness material in Table VI. These results show that the ultimate and yield strength joint efficiencies are about the same. However, the elongation efficiency of the vertical-up welded specimens is 4.5 percent higher than that for the flat positioned specimens. Both sets of specimens failed in the base metal.

3.4.2 Fatigue

The results of the fatigue tests on the base metal and SSEB welded plates are presented as S-N curves in Figures 12 and 13 for the 0.440 and 0.940-inch-thick plate, respectively. These results show that the constant amplitude fatigue life of SSEB weldments in 0.440 and 0.940-inch-thick Ti-6Al-4V titanium alloy are equivalent to that of base metal specimens.

3.4.3 Fracture Toughness

The determined fracture toughness values for base metal and SSEB welds are reported in Table VII. The average base-metal value of 55.4 ksi$\sqrt{\text{in.}}$ is well within the 40 to 80 ksi$\sqrt{\text{in.}}$ range reported in the literature and is, therefore, considered acceptable. In general, a minimum base metal K_{IC} value of 60 ksi$\sqrt{\text{in.}}$ is required for critical parts. Material having fracture toughness values below this level are used for less critical applications.

The average K_{IC} value of the SSEB welded (39.0 ksi$\sqrt{\text{in.}}$) specimens falls within the 30 to 45 ksi$\sqrt{\text{in.}}$ values obtained by Grumman on EB chamber weldments in annealed Ti 6Al 4V titanium alloy having a minimum base metal K_{IC} of 65 ksi$\sqrt{\text{in.}}$ As shown in Figure 14, both the base metal and SSEB welded specimens exhibited plane-strain fracture.

3.5 REPAIR WELDING

During SSEB welding of test panels, arc-outs made it necessary to stop and restart again on the weld panel. In order to again restart after a burn-through arc-out, a repair weld had to be made on the EB weld arc-out hole by gas-tungsten-arc (GTA) welding using filler metal before a suitable vacuum could be obtained in the

TABLE IV AVERAGE PROPERTIES AND JOINT EFFICIENCIES OF SSEB WELDED Ti-6Al-4V TITANIUM ALLOY

Specimen Thickness, inch	Average Tensile Properties			Average Joint Efficiencies		
	Ult Str, ksi	Yield Str, ksi	Elongation In 2 Inches, %	Ult Str, %	Yield Str, %	Elongation, %
0.190 Long.	143.2	138.3	9.7	99.2	98.0	70.7
0.440 Long.	138.2	134.1	11.8	100.0	99.7	87.5
0.440 Trans.	144.6	140.7	12.3	105.3	102.8	80.6
0.690 Long.	139.5	138.0	10.3	101.2	101.9	76.7
0.940 Tran.	159.1	146.4	9.3	113.5	107.9	66.4
0.940 Long.	144.1	137.7	11.0	99.0	97.2	82.7

TABLE V COMPARISON OF PROPERTIES AND JOINT EFFICIENCIES OF SSEB AND EB WELDMENTS IN Ti-6Al-4V TITANIUM ALLOY

Gage, in	Chamber	Avg. UTS, ksi	Avg. Y.S. ksi	Avg. Elong (2")%	Loc. of Failure
0.190	SSEB	143.2	138.3	9.7	B.M.
0.190	EB-Dev.-exp.	139.9	136.3	10.7	B.M.
0.187	EB-Tunnel	146.9	141.7	11.4	-
0.440	SSEB	138.2	134.1	11.8	B.M.
0.440	EB-Small	137.5	135.1	12.3	B.M.
0.690	SSEB	139.5	138.0	10.2	B.M.
0.690	EB-Small	141.1	139.4	10.2	B.M.
0.625*	EB-Tunnel	134.2	124.9	10.5	-
0.940	SSEB	144.1	137.7	10.0	B.M.
0.940	EB-Small	142.9	138.1	11.0	B.M.
1.00	EB-Tunnel	142.6	136.3	12.7	-

*Non-stressed relieved.

TABLE VI COMPARISON OF FLAT AND VERTICAL-UP POSITION WELDMENTS IN Ti-6Al-4V TITANIUM ALLOY (0.190 INCH THICK)

SSEB Welding Position	UTS/JE, ksi/%	YS/JE, ksi/%	EL/Eff, %/%
Flat	143.2/99.2	138.3/98.0	9.7/70.7
Vertical-Up	141.2/97.6	139.3/98.7	10.3/75.2

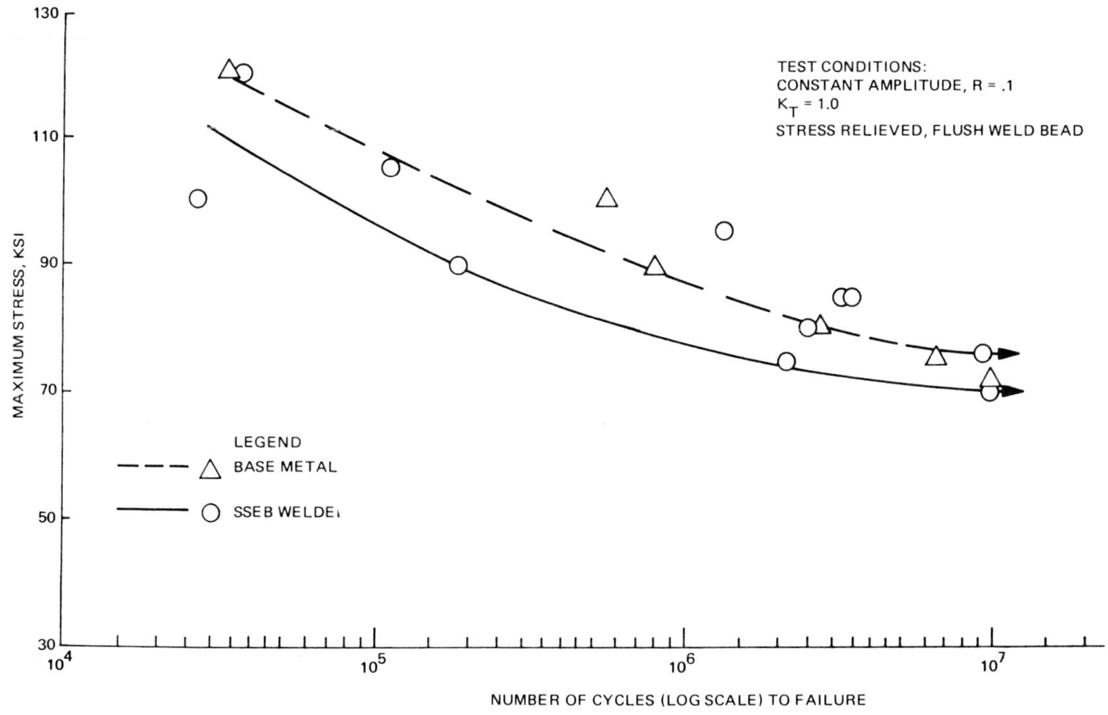

Figure 12 S-N Curves for Base Metal and SSEB Welded Ti-6Al-4V, 0.440-Inch-Plate

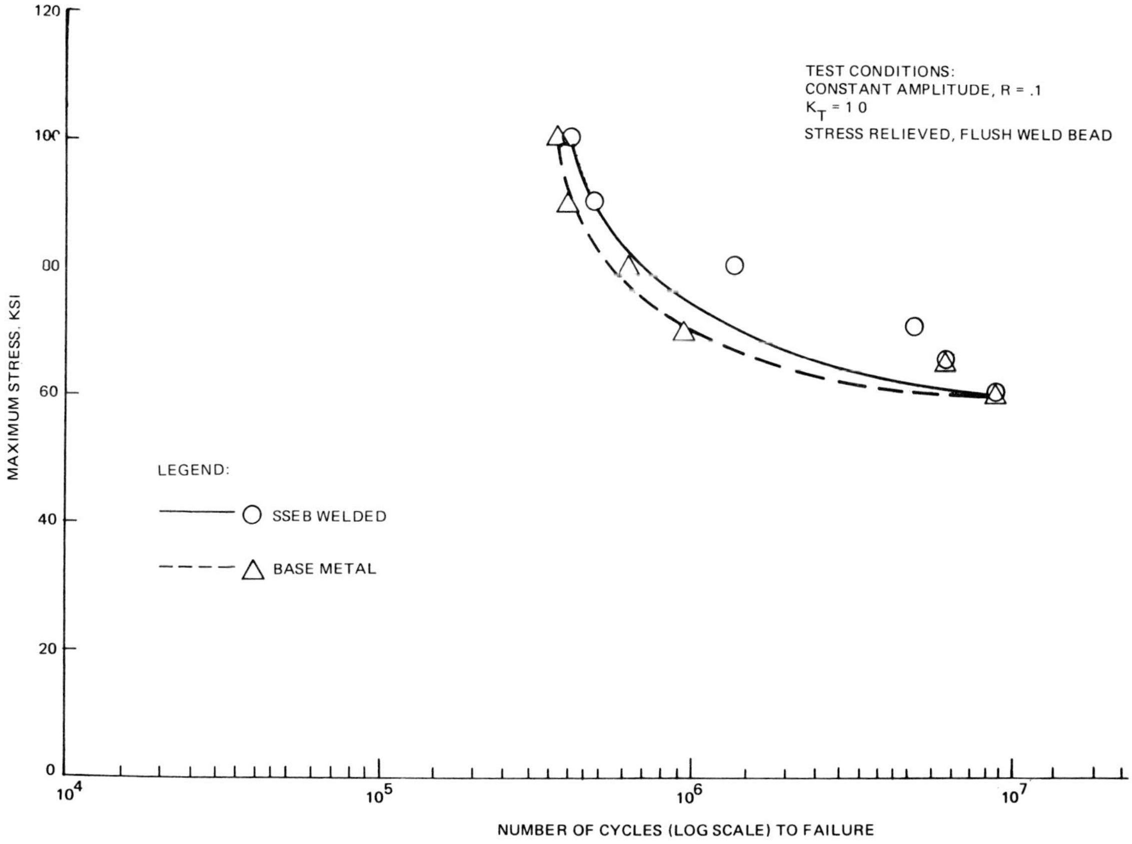

Figure 13 S-N Curves for Base Metal and SSEB Welded Ti-6Al-4V, 0.940-Inch-Plate

TABLE VII FRACTURE TOUGHNESS OF 0.940-INCH-THICK BASE METAL AND SSEB WELDED* Ti-6Al-4V TITANIUM ALLOY

Test Specimen Number	Fracture - Toughness Value (K_{IC}), Ksi $\sqrt{in}$
BASE METAL	
T100LC-1	56.4
-2	54.8
-3	47.0
-4	62.1
-5	65.5
-6	46.4
Average Value	55.4
SSEB WELDED	
T100LW-1	41.3
-2	35.5
-3	37.5
-4	45.8
-5	37.3
-6	38.9
-7	37.3
-8	37.5
-9	39.0
-10	39.0
-11	40.8
-12	37.5
Average Value	39.0

*Stress relieved, flush weld, weld and fracture transverse to rolling direction.

Figure 14 Base-Metal and SSEB Welded, 0.940-Inch-Thick, Ti-6Al-4V Titanium Alloy Fracture Toughness Specimens After Testing

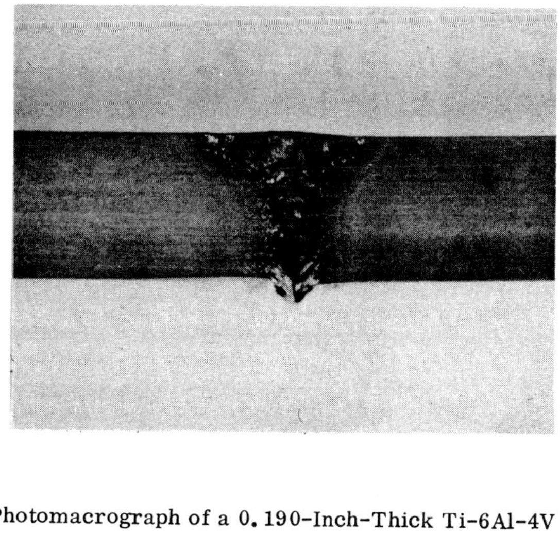

Figure 15 Photomacrograph of a 0.190-Inch-Thick Ti-6Al-4V Titanium Alloy SSEB Welded Panel Repaired by Mechanized GTA Welding with No Filler Metal (5X Magnification)

SSEB welding chamber. This procedure was necessary because of the differential in vacuum between the welding chamber and the back-up table. After the repair weld was completed, filed or machined smooth (to prevent seal damage) and cleaned with acetone, it was possible to restart again. If no burn-through arc-out hole occurred, the test plate was filed or machined smooth and cleaned with acetone, and the SSEB welding machine was restarted.

Repair of an SSEB welded, 0.190-inch-thick, Ti-6Al-4V titanium alloy panel for simulated lack of fusion, surface bead crowning and underfill was accomplished by rewelding with the SSEB welder without adding any filler metals. Radiographic examination of this repaired panel revealed it to be acceptable for both repairs. Tensile results show that rewelding again due to a lack of fusion or a missed seam has only a small effect on the ultimate and yield tensile strengths but considerably improves percent elongation or ductility. The effect of rewelding ("cosmetic pass") over a regular SSEB weld on ultimate and tensile yield strengths is slight with some improvement in percent elongation or ductility. About 40 percent of the uppermost portion of the repair was removed during machining. This could have resulted in lower joint efficiencies than those previously shown. Photomacrographs of SSEB rewelds to repair lack of fusion and underfill defects are shown in Figure 15.

Repair of an SSEB welded 1/4-inch-thick Ti-6Al-4V titanium alloy panel for simulated small surface porosity and underfill was accomplished by making a cosmetic pass using the mechanized GTA process (no filler metal additions) and the standard seal-welding parameters of 70 amperes, 15 volts and a carriage travel speed of 30 inches per minute (heat input of 2.1 kilojoules/inch). This resulted in a 3/16-inch wide surface bead and no underfill. The ultimate and yield tensile strengths are only slightly affected by the mechanized GTA "cosmetic pass". However, the percent elongation or ductility was greatly improved (by 37%). Visual examination of the test specimen revealed that about 40% of the uppermost portion of the GTA "cosmetic pass" had been removed during machining of the test specimen which may have resulted in lower joint efficiencies than those obtained. A photomacrograph of this GTA "cosmetic pass" repair weld is shown in Figure 16.

3.6 COMPONENT FABRICATION AND TESTING

3.6.1 Simulated Wing Panel Fabrication

An integrally stiffened, annealed Ti-6Al-4V titanium alloy wing-type skin was chosen as the component. A fitted rib block technique was developed for fixturing and vacuum sealing the component prior to conducting welding tests which were conducted to verify quality and strength. From this work, it is now possible to recommend approaches for vacuum sealing and SSEB butt-welding flat parts several feet long with complex cross sections. The SSEB welded panel was two feet wide. One-foot-wide fatigue specimens were machined from this panel. Figure 17 shows the wing skin panel before SSEB welding with fitted rib blocks in place and with starting/stopping tabs attached. The important dimensions of this longitudinally integrally stiffened test specimen were:

- Wing skin thickness - 0.130 inch
- Longitudinal rib thickness - 0.100 inch
- Longitudinal rib height - 0.735 inch
- Longitudinal rib length - 29.0 inches
- Overall length - 46.5 inches
- Overall width - 11.0 inches.

In the area where this panel was to be SSEB welded, the ribs were 0.030 inch heavier to allow finish machining after SSEB welding.

The sealing compound (RTV-90 silicone rubber) used for a practice wing panel was found to be very viscous

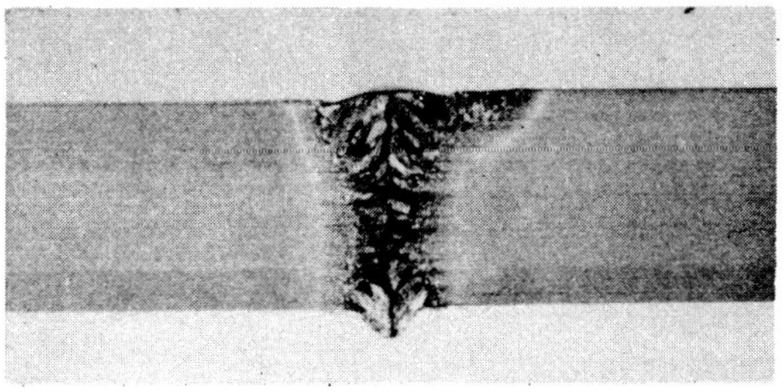

a. Lack of Fusion

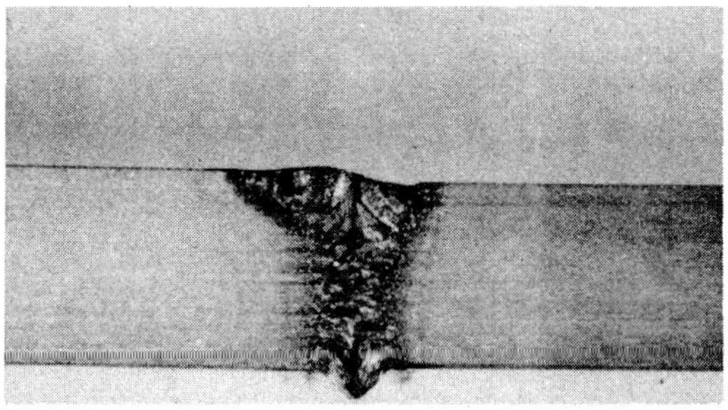

b. Underfill (slight misalignment resulted in some underfill remaining)

Figure 16 Photomacrographs of Cross-Sections of SSEB Repair Rewelds of 0.190-Inch-Thick Ti-6Al-4V Titanium Alloy Plate (5X Magnification)

Figure 17 Simulated Wing Skin Panel Showing Sealing Compound, Rib Blocks and Starting/Stopping Tabs Before SSEB Welding

and did not penetrate into tight areas with ease. For the larger skin panel, the sealing compound was changed to General Electric's RTV-60 silicone rubber with T-12 accelerator. This compound has a viscosity of about 500 poises, pours without difficulty and has good tensile strength, shear strength and bondability. It also has very low shrinkage, excellent firmness, a working life of about one hour and cures at room temperature in 12-48 hours without external heat (curing time was shortened by baking at 150°F. for 12-24 hours). It has Shore A Durometer hardness of 65. Figure 17 shows the wing skin panel with the solidified sealing compound in place. The panel was previously butted together and GTA seal pass welded. RTV-731 silicone rubber sealing compound was employed to hold the insert rib blocks to the panel and prevent the RTV-60 silicone rubber from leaking through and getting into the welding area.

The GTA seal pass welded skin panel with inserted rib blocks, attached starting/stopping tabs and vacuum sealant was then placed on the SSEB back-up table. The panel was held in place on the back-up with two prebent curved hold down bars placed far enough apart to permit the SSEB welding chamber to pass between them (Figure 18). "C" clamps were used on the starting tab and hold down clamps were used on the stopping tab as shown in Figure 18 to prevent them from lifting and shifting during the SSEB welding operation. Initially, the vacuum level attained in the back-up table was 150 microns which was not suitable for SSEB welding. Additional air leakage was stopped by applying glyptol around the edges of the panel and the back-up table. In order to insure that a suitable vacuum level (below 80 microns) was maintained during SSEB welding a 100-cubic-foot-per-hour mechanical pump (Figure 100) was installed in the back-up table vacuum line between the outer and inner rubber "0" rings. This mechanical pump also prevented air leakage from occurring as the SSEB welding chamber traversed the wing panel due to minute breaks in the sealing between the wing panel and the back-up table. With this arrangement the back-up table vacuum reading attained was 45 microns. After pumping over night, the back up table vacuum and welding chamber levels were 40 and 15 microns, respectively. The simulated wing panel was now ready for welding.

SSEB welding of the simulated wing panel was accomplished using the developed welding parameters for one-inch-thick Ti-6Al-4V titanium alloy in the flat position. These welding parameters were beam voltage - 50 kilovolts, beam current - 330 milliamperes and EB gun travel speed - 45 inches per minute (resulting in a heat input of 22 kilojoules per inch). The focus current setting was 6.40 amperes. While welding, vacuum levels in the back-up table and SSEB welding chamber rose to about 60 and 40 microns, respectively. No problems were encountered during SSEB welding; the weld looked very uniform. The top surface of the welded wing panel is shown in Figure 19. The bottom of the same wing panel after removal of the rubber sealant compound is shown in Figure 20. A close-up view of the penetration bead is shown in Figure 21.

The titanium alloy insert blocks were machined off and the penetration bead was machined flush with the inner wing skin or rib. The top surface bead was machined lightly to remove the crown and slight underfill. The panel was then cut lengthwise to produce two one-foot-wide fatigue test specimens. The two panels were radiographed per Grumman Specification GSS 6205 and found to be satisfactory. Ultrasonic inspection was not conducted because a good ultrasonic trace could not be obtained for the test specimen thickness of 0.0130 inch.

Following complete machining, the specimens were dye-penetrant inspected before further processing. After satisfactory dye penetrant examination, two one-half-inch-thick Ti-6Al-4V titanium alloy doubler plates were electron beam (hard vacuum) fillet welded with 1/8-inch-diameter filler wire (Ti-6Al-4V) to the grip ends of each

Figure 18 Simulated Wing Skin Panel On SSEB Back-Up Table Held in Place With Hold-Down Bars. Auxiliary 100 CFH Mechanical Vacuum Pump is Shown At Lower Right In Photo.

Figure 19 Top Surface Bead of SSEB Weld (Center) in Simulated Wing Panel After Welding

Figure 20 Bottom Penetration Bead of SSEB Welded Simulated Wing Panel

Figure 21 Close-Up View of Penetration Bead of SSEB Welded Simulated Wing Panel

specimen to prevent premature failure at these location in testing.

After the test specimen was acid-cleaned and Turco-Formkotc coated as described previously, it was placed into the stress-relief fixture; a hold-down plate was placed over the fatigue specimen. The plate had pressure applied to it by means of cross-bars to the stress-relief operation. The part and fixture were the stress-relief operation. The part and fixture were held at 1200^{0}F. for eight hours (time was increased from four to eight hours) for heating the fixture to assure uniform temperature in the part. After stress-relieving, the specimen was removed from the fixture and pickled to remove the protective Turco-Formkote coating. Six 3/4-inch-diameter holes were drilled in each grip end of the simulated wing panel fatigue specimen. The specimen then was submitted for fatigue evaluation.

3.6.2 SSEB Welded Fatigue Test Results

One SSEB welded wing panel fatigue test specimen was fatigue tested at the static limit stress level of 85 ksi. After 34,910 cycles, the specimen failed in the heat-affected, weld area.

Another SSEB welded specimen was tested at a design stress level (33 ksi) simulating stresses used for an EB weld in a wing-skin panel for a fighter aircraft. The operational stress level of 33 ksi for an EB square butt weld is based on the fact that the base-metal fracture toughness (K_{IC}) is not fully recovered after stress-relieving the weldment. The 33-ksi level assumes that if a given flaw size (~0.100) is present in the weld area, the flaw growth will not result in failure before twice the life of the aircraft is attained. This specimen sustained 1,102,820 cycles at the 33-ksi stress level. Further testing was done at an 85-ksi stress level and the specimen then attained 23,000 cycles additional and failed in the base metal inboard of the rib (Figure 22). Metallographic analysis of the first SSEB welded fatigue specimen showed that failure initiated in the heat-affected-zone/fusion zone boundary about 0.9-inch inboard of one of the stiffening ribs. The failure crack, initiated by fatigue at a cluster of pores (0.001 to 0.003-inch in diameter) that propagated from pore to pore, was approximately 0.050-inch long. A number of these micropores were open to the surface where the fatigue crack initiated. The source of the porosity in the heat affected-fusion zone was most probably hydrogen contamination. This was due to long processing required and not welding soon enough after cleaning. This can be circumvented in the future by using integral weld lands on the part instead of rib block inserts to minimize sealing problems.

Metallographic analysis of the second SSEB welded fatigue specimen showed that failure initiated in the base metal about 0.7-inch inboard of one of the stiffening ribs. The failure crack initiated by fatigue on the underside of the wing panel indicated that crack initiated from a machining defect that was improperly blended.

3.6.3 Analysis of Results

The results show that high-quality SSEB straight butt weldments in the flat and vertical-up positions are producible in Ti-6Al-4V titanium alloy gages ranging from 1/4-inch to one-inch thick with mechanical pro perties equivalent to those obtainable in high vacuum EB welding chambers. The repairability of the alloys employing GTA and SSEB rewelds is excellent. Additional work is required to develop SSEB welding techniques and tooling to fabricate long-length (> two feet) flat-position weldments in aerospace structures. Proximity control devices and techniques are also required for contoured/curved parts.

The results of the wing skin fabrication phase of the program showed that an aerospace structural component could be fabricated using the insert rib block method with a pourable sealing compound. However, it is felt that the processing time required limits the

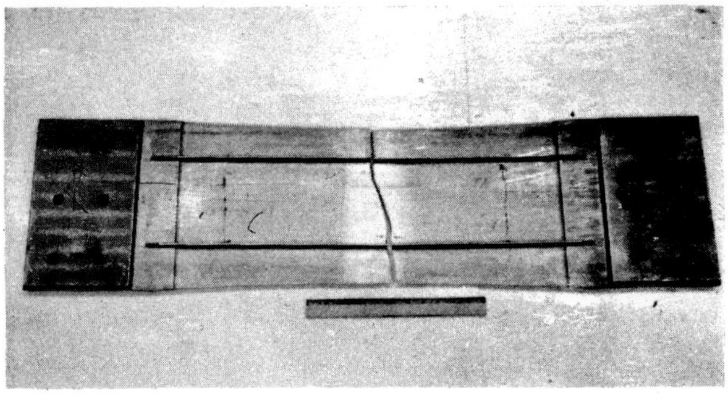

a. Overall View

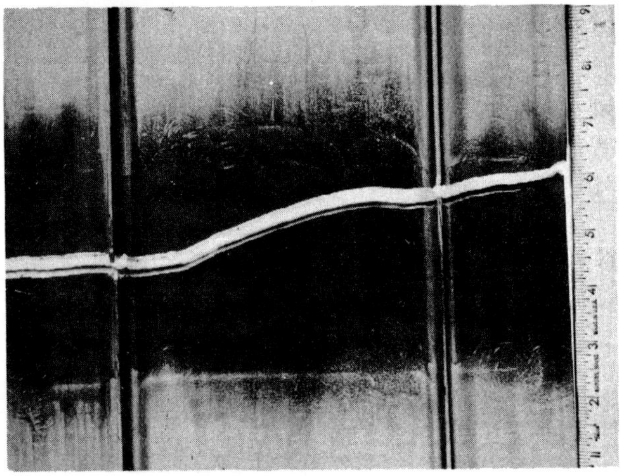

b. Close-Up View

Figure 22 Views of Failed SSEB Welded Fatigue Specimen (Note: Failure Occurred In The Base-Metal 0.7 Inch Inboard of Rib) use of this method to less critical parts, particularily where stringent high stress, low cycle fatigue strength is required.

Preparing the wing skin panel for SSEB welding was found to be difficult using the insert rib blocks and pourable sealing compound. Even though machining was carefully accomplished, some gap existed between the wing panel and the rib blocks. It was also difficult to keep the rib blocks in place while pouring the sealing compound. It was concluded that a better method would be to leave a weld land equal to the thickness of the rib block (in each wing skin segment being joined) to eliminate any gap between component and rib block. Sealing between the weld land and back-up fixture could then be accomplished adequately and quickly by employing "0" rings or a gasket that could be reused each time.

4. CONCLUSIONS

- Aerospace-quality, flat, square-butt panels can be SSEB-welded from 0.190-, 0.440-, 0.690- and 0.940-inch-thick annealed Ti-6Al-4V titanium alloy.
- Optimized SSEB welding parameters are similar to those for EB hard-vacuum chamber welding and can be used interchangeably with only minor adjustments
- Tensile strength, fatigue strength and fracture toughness of SSEB and EB high vacuum-chamber, annealed Ti-6Al-4V titanium alloy weldments are comparable. Tensile and fatigue joint efficiencies are 100 percent for both welding processes
- Fracture toughness of SSEB and EB large-chamber, annealed Ti-6Al-4V titanium alloy weldments are comparable. Fracture toughness (K_{IC}) is 73 percent of base metal for both processes
- The parameters developed for gas-tungsten-arc seal welding of straight butt joints prior to full-penetration SSEB welding produce porosity-free welds in thicknesses used in this program.
- Repair of lack of fusion and porosity defects in annealed Ti-6Al-4V titanium alloy can be readily accomplished by SSEB rewelding. Smoothing passes eliminated or minimized top surface underfill.
- Arc-outs can be repaired by gas-tungsten-arc welding
- The square-butt joint is the simplest weld joint for use with SSEB welding. Therefore, other joint types were beyond the scope of the program.
- Aerospace-quality SSEB welds can be accomplished on an aircraft structure such as a wing skin panel
- Radiographic, ultrasonic and dye-penetrant inspection of Ti-6Al-4V titanium alloy SSEB weldments were successfully employed to establish weld quality.

5. ACKNOWLEDGEMENT

The work reported herein was completed as part of Air Force Contract F33615-70-C-1806 initiated under Manufacturing Methods Project 828-9. The program was under the technical direction of Mr. Fred R. Miller of the Metals Branch (LTM), Manufacturing Technology Division, Air Force Materials Laboratory, Wright-Patterson Air Force Base, Ohio.

6. BIOGRAPHIES

Robert H. Witt, Advanced Welding and Metallurgy Development Group Head, Materials and Processes Department. Mr. Witt has 25 years of diversified experience in welding, metallurgy and fabrication of ferrous and non-ferrous alloys. He holds B. Met. E. and M. Met. E. degrees from the Polytechnic Institute of Brooklyn and did further graduate work at New York University. At Grumman he has worked on EB and SSEB welding of titanium and steel structures, manufacture of the LM descent-stage heat shield and coordination of materials studies for re-entry structures. Prior to this Mr. Witt was employed by Republic Aviation Corporation and Sylvania Electric Products Company where he investigated metallurgy, welding and brazing of high-strength, high-temperature materials for airframe and nuclear applications. Mr Witt has published over 30 technical articles on welding and metallurgical studies and is a member of SAMPE, AWS, ASM and RESA. In 1969-70 he was chariman of the Long Island chapter of ASM. He is presently a member of the AIA Plasma Welding Equipment Specification Committee and the Welding Research Council Reactive and Refractory Metals Committee.

Joseph G. Maciora, Advanced Welding and Metallurgy Development Welding Engineer, Materials and Processes Department. Mr. Maciora has 22 years of experience in welding processes, metallographic weld evaluation, metallurgy and materials applications. He holds a B. Met. E. from the Polytechnic Institute of Brooklyn and did graduate work at Rensselaer Polytechnic Institute. At Grumman he has been principal investigator on the sliding-seal electron beam welding Air Force Contract F33(615)-70-C-1806 which is partially reported in the present paper. He has also been associated with projects dealing with electron beam welding of titanium, aluminum and steel alloys, plasma arc and GTA melt through welding of titanium alloys and GTA, GMA, Flash and stud welding of aluminum alloys. Prior to this, Mr. Maciora worked for the Electric Boat Division, Groton, Conn.. and the Lycoming Division of the Avco Corporation. He is a member of the American Welding Society and is on the Aerospace Fabrication Committee. He has written numerous technical reports on welding processes and fabrication.

Harold P. Ellison, Advanced Welding and Metallurgy Development Welding Engineer, Materials and Processes Department. Mr. Ellison has 13 years of experience in welding and welding process development at Grumman and holds a BS in Mechanical Engineering from C.W. Post College. He is presently Project Engineer on a continuation of the SSEB welding development under Air Force Contract F33615-73-C-5030, Project No. 864-3. At Grumman he has worked on programs dealing with electron beam and sliding seal EB welding of titanium, aluminum and steel alloys, plasma-arc welding of titanium alloys and gas-tungsten-arc welding of aluminum alloys. Mr. Ellison was a certified production welder for five years and a welding engineering technician for several years prior to his present position. He has co-authored several technical papers.

ELECTRON BEAM SHAPE-WELDING TITANIUM STRUCTURAL SHAPES

By

Mario L. Ochieano
Senior Research Specialist
Lockheed-California Co.

ABSTRACT

A fabrication-feasibility assessement on EB welding titanium substantiated the ease of single-pass welding of square butt thick sections and the welding of other than constant thickness sections with fixed machine settings.

The first approach to EB square butt joining of shapes was accomplished by building-up of a constant through section by locally adding metal - called 'block-welding.' Improvements were made to effect near net 'shape-welding' by elimination of building blocks with resultant reduction in weld-through thickness. This methodology as applied to butt welding of engine pylon longerons on the Lockheed L-1011 commercial transport is related.

Over 280 engine pylons - more than 1680 weldments - have been made using the 'shape-welding' technique.

The application of the electron beam (EB) as a source for generating heat for welding has extended joint design thinking into areas heretofore not normally considered practical. That is:

- o The vacuum need for effective electron travel is an ideal environment for melting the air sensitive titanium, hence greatly increasing the confidence level for the fusion joining of the titanium alloys.
- o Highly localized beam concentration with power densities in the millions of watts/in.^{2} indicates the practicability of single pass square butt welding with minimal heat-affected zone and warpage.
- o Beam conformal and penetrating capability permits consideration of other than uniform constant-thickness-section type joints.

A fabrication-feasibility assessment substantiated the ease of single pass welding of thick sections at respectable speeds and the welding of other than constant thickness sections with fixed machine settings.

The coming-of-age of the electron beam technology coincided with our 2000 mph SST design activity where in-flight heating dictated a titanium structure. Of first concern in this endeavor was the strength properties of the EB welds in titanium. Generated data indicated:

- o Static tensile strength comparable to base metal.
- o Fracture toughness and ductility well within useability limits.
- o Fatigue strength better than base metal with a hole (K_t 3).

Source: Technical Paper AD76-150, SME, 1976

Figure 1 depicts the fatigue strength relationship of smooth base metal, smooth EB weld joint, and base metal with a hole. It should be noted that the weld joint appears to indicate a K_t of less than 1.5.

These findings were sufficiently encouraged as to support a design approach which would use the square butt weld construction rather than the lapped joints mechanical construction for permanent attachments.

Our first approach to the joint design for square butt welding of shaped structural sections was referred to as the 'block-welding' technique. Basically this approach involves the build-up of a constant cross-section thickness at the joint by locally adding metal. Figure 2 shows the block build-up method on "F", "H" and "T" sections. The photograph insert depicts an actual block-welded SST wing span "T" section. Note how the blocks were sculptured to fit the "T" geometry. The blocks were machined off after welding.

The first confidence shaker encountered was the missing of the joint on an "F" Section. X-ray and penetrant inspections accepted joints which were not fully welded. Confidence was restored by the adaptation of "witness lines" and ultrasonic inspection. The basics of the witness line technique is depicted in Figure 3. Here measurements are made after welding to ascertain that weld fusion is centered about the joint.

The nondestruction inspection (NDI) system as applied to present day production is shown by Figure 4. The NDI system includes indirect and direct controls. The indirect controls involve:

- o Machine qualification tests - welding through thick and thin square butt sections of aluminum and stainless steel alloys.
- o Operator qualification tests - academic training on machine operation and process specification requirements; practical training by demonstrating machine set-up capability and welding of joints.
- o Calibration checks - dial setting vs output.
- o Weld procedure - establishing of proven/tested weld parameters and documenting for production welding includes, joint geometry details, beam catch block positioning, weld sequence, destructive tests (tensile, bend and metallurgical), and weld size limitations.
- o In-process control checks - macro-sectioning and observing weld for quality and size requirements, and sign-off parameter/sequence check sheets.

The direct controls cover the actual inspection performed on the parts, namely:

- o Visual - external appearance, size consistency, undercut and through penetration.
- o Witness line measurements

- o Penetrant - fluorescent method
- o X-ray - multiple views
- o Ultrasonic - contact shear and longitudinal wave

The SST never came into being, but we did get the opportunity to apply our EB square butt joint weld smartness to a Ti 6Al-4V rotor hub on a helicopter program. Figure 5 depicts the EB welded rotor hub which is made up of four forgings having a thickness of 2-3/16 inches at the weld joint and a overall span of 10 feet 4 inches. Ten helicopters were made and flown, and about five additional rotors were destructively tested. This application, an industry-first, reinforced our confidence in the feasibility of square butt EB weld joining of critical structure.

The next utilization opportunity came on the engine pylons - predominately a titanium structure - for our L-1011 TriStar commercial transport. Figure 6 shows the complete engine package attached to the TriStar wing. The pylon is the connecting structure in between the engine and the wing.

During the pylon design stage a decision was made to apply EB square butt joint welding for the attachment of a forged Ti 6Al-4V fitting to a Ti 6Al-4V extrusion to fabricate the longerons. Thus providing a beneficial cost/weight savings. The geometry of the longeron sections were such that the block-welding technique approach would require welding through thicknesses greater than 5 inches, coupled with the need of machining numerous fill block inserts to accommodate the "H", "L" and "C" sections involved. Figure 7 depicts the framework of the engine pylon showing the EB weld joints and geometry, Note that the "H" sections would require welding through a thickness of about 5 1/4 inches (5" plus 1/8" beam catch plates top and bottom). There is a total of six joints.

So the necessity of technique simplification - the elimination of build-up blocks (a cost reduction), and the reduction in the weld-through thickness (a technical need) - was quite evident. These determinations lead to the development of the 'shape-welding' technique. Basically, this technique involved simple beam catch blocks, welding vertically up, angling gun and providing excess metal locally at joint. The longeron components are machined net prior to welding, except for the excess metal at joint and hole sizing. This application/weld method is an industry-first. This welding method is depicted by Figures 8 and 9. In reviewing the figures it should be noted that:

- o the electron beam is penetrating through a varying thickness and that the gun (EB source) to work distance also varies as the gun travels vertically up.
- o the maximum thickness penetrated is 3-1/2 inches and the minimum is 7/8 inches. These extremes are both on the same weld joint - WP #3.
- o the maximum beam travel (gun to work distance) is 9-1/2 inches -WP #3 and #4. On these procedures the minimum beam travel is 4 inches. Hence, welding is being accomplished between 4 and 9-1/2 inches with constant machine settings.

- the beam at some points along its travel path on WP #2, #3 and #4 penetrates and welds at two in-line places simultaneously, i.e., the beam enters near side of part-- exits-- travels through space -- reenters in-line portion on the far side of part-- welds-- exits.

- the longeron shape-welding is accomplished at speeds and power settings of 32 IPM and 39.2 KW for the "H" sections; 44 IPM and 30.2 KW for the "C" section; and 50 IPM and 29.9 KW for the "L" section.

- The shortest weld time is less than 10 seconds (WP #2) and the longest weld time is less than 20 seconds (WP #3 and #4).

Weld size range as determined from weld procedure test welds is documented for each weldment. The weld size is monitored in production through the in-process control macro-section. A macro-sectioned in-process control check weld for WP #1 is shown by Figure 10. Measurements for weld size are made at 1/8 inch below the as-welded surfaces and at a midpoint. The 1/8 inch below measurements are at the actual part surfaces after final machining.

Sequence-wise the components have the witness lines applied, are then aligned, fitted and beam catch blocks TIG welded in place in a preweld set-up fixture. The tacked assembly is then placed in EB chamber and clamped in a holding fixture and welded. After visual inspection and witness line measurements the weld joint area is machined flush. Then assembly is acid etched and penetrant, x-ray and ultrasonic inspected.

Figure 11 shows a close-up of WP #1 weld after final machining and acid-etch. It should be noted that the weld changes in size with the upper portion of weld being wider than the lower portion. This is due to the gun to work distance (beam travel) variation coupled with the focal point of the beam. On the lower part of the weld the beam travel is within 5 inches and on the upper 9.1 inches. See Figure 9, WP #1.

Weld Rework. Rewelding is permitted. Tests conducted on weldments over-simulating production repairs were evaluated for tensile strength and elongation, macro-structure, bend ductility and hardness properties. These data indicated:

<u>Tensile Strength and Elongation</u> (See Figure 12)

- the ultimate tensile strength (UTS) and yield strength (YS) and elongation (El) obtained on all specimens equal or exceed minimum values for the base metals, namely, 130,000 psi UTS, 120,000 psi YS and 10% El in 2.0 inches.

- the lowest values obtained on any of the weldments deviating from the single pass base line control weld was 136,120 psi UTS, 122,120 psi YS and 10% EI in 2.0 inches.

- Macro-structure - the weld structure and heat-affected zone were similar in appearance for all weldments. No notable metallurgical effects were apparent due to rewelding.

Bend Ductility (See Figure 13)

- o All weldments - base line, multiple reweld and wide weld - satisfactorily bent through an angle of 180 degrees over both a 10T and 8T radius without cracking.

Hardness Survey (See Figure 14)

- o the weld metal hardness was greater than that of the base metal for all welds.

- o the weld metal hardness on repeated rewelds was less than the hardness obtained on the single pass base line control weld.

In summary, all repair simulating weldments - multiple rewelds and wide welds - displayed strength, ductility and metallurgical structure comparable to the single pass base line control weld.

Production Welding. Over 280 engine pylon assemblies - more than 1680 EB welds have been fabricated to-date using the square butt 'shape-weld' technique.

LIST OF FIGURES

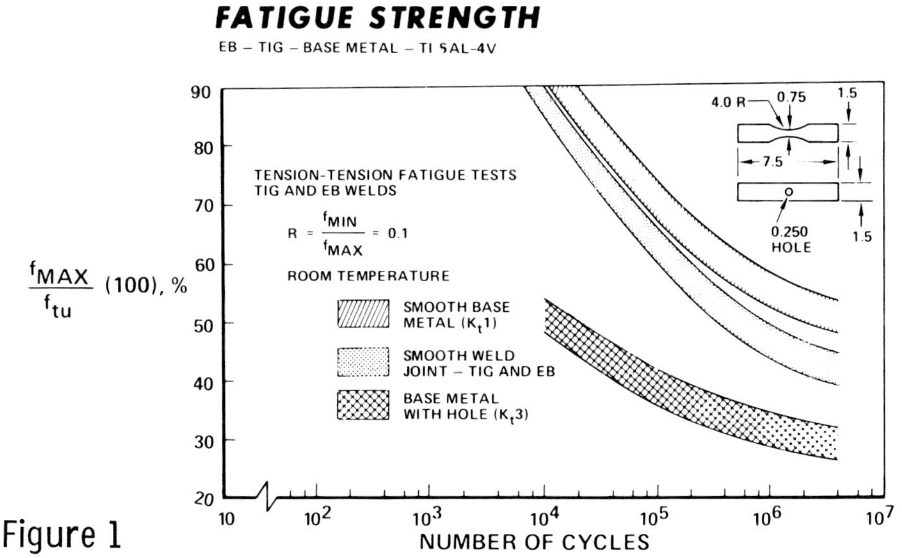

Figure 1

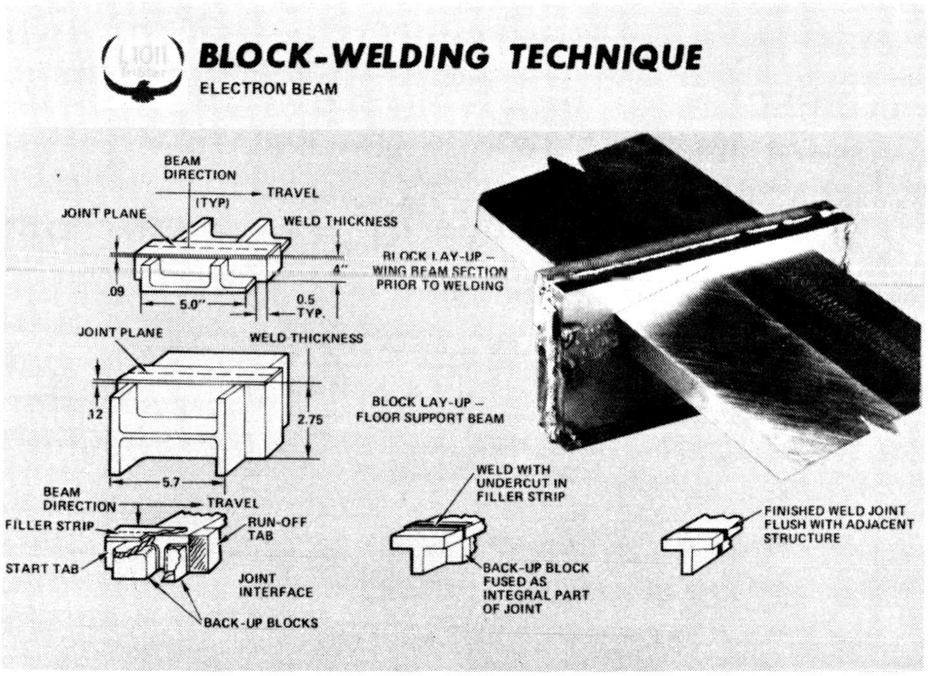

Figure 2

WITNESS LINE TECHNIQUE

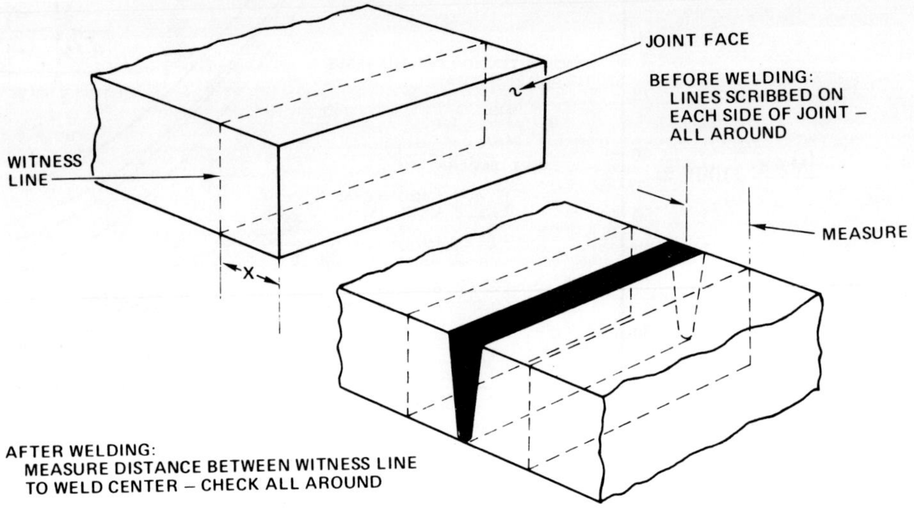

Figure 3

NDI SYSTEM - EB WELDS

INDIRECT:
- MACHINE AND OPERATOR QUALIFICATION
- CALIBRATION CHECKS
- WELD PROCEDURES (TENSILE, BEND AND MACRO'S)

IN-PROCESS CHECKS:
- MACRO-SECTION – QUALITY/WELD SIZE
- QUALITY CONTROL CHECK SHEET

DIRECT:
- VISUAL
- MEASUREMENTS – WITNESS LINES
- PENETRANT/MAGNETIC PARTICLE
- X-RAY – MULTIPLE VIEWS
- ULTRASONIC – SHEAR WAVE/LONGITUDINAL

Figure 4

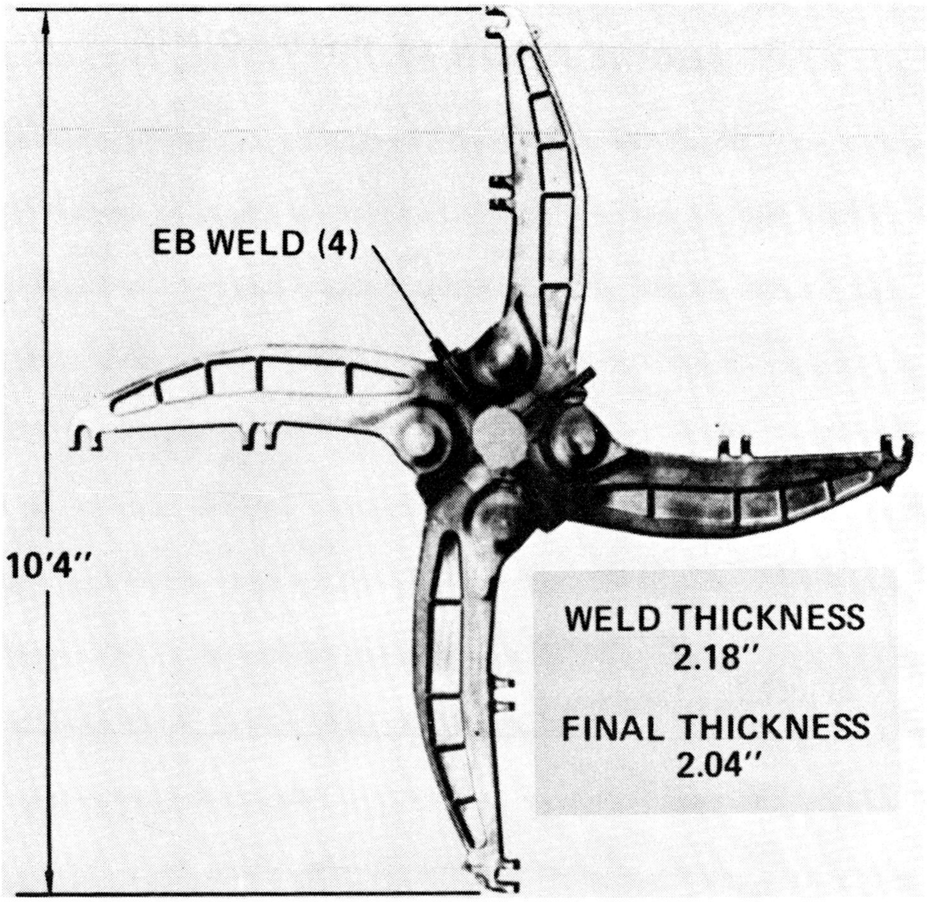

Figure 5

Figure 6

ENGINE PYLON EB WELD JOINTS

6 AL-4V TITANIUM

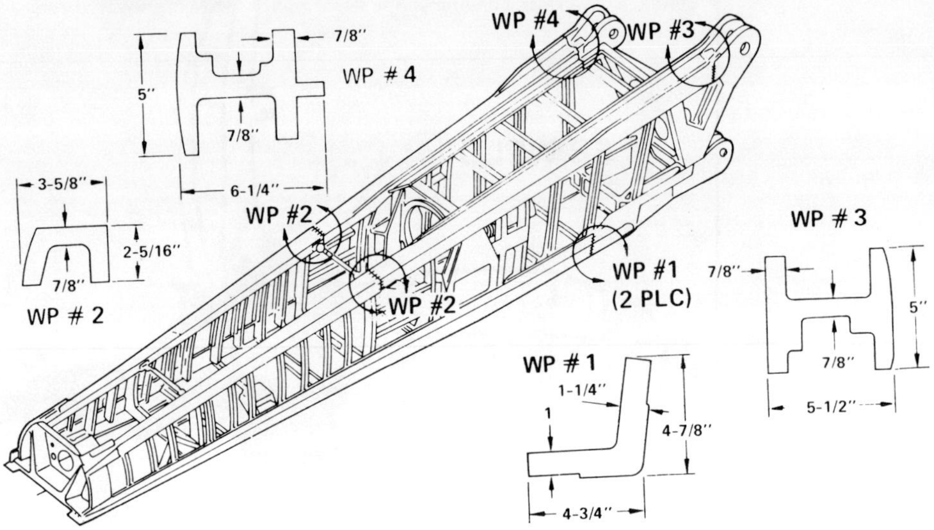

Figure 7

SHAPE TECHNIQUE - WELD METHOD-SECTION GEOMETRY

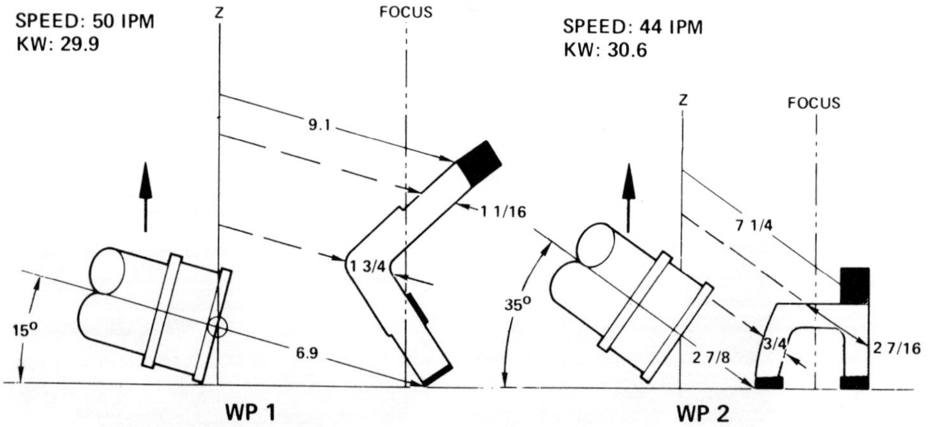

Figure 8

SHAPE TECHNIQUE
WELD METHOD/SECTION GEOMETRY

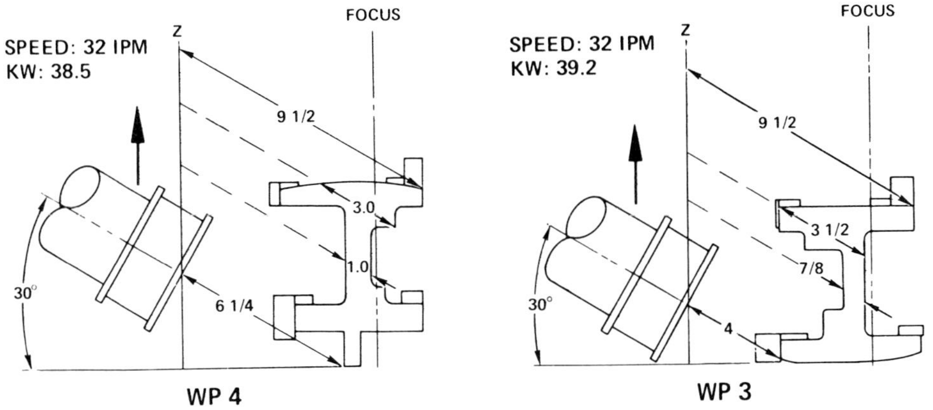

Figure 9

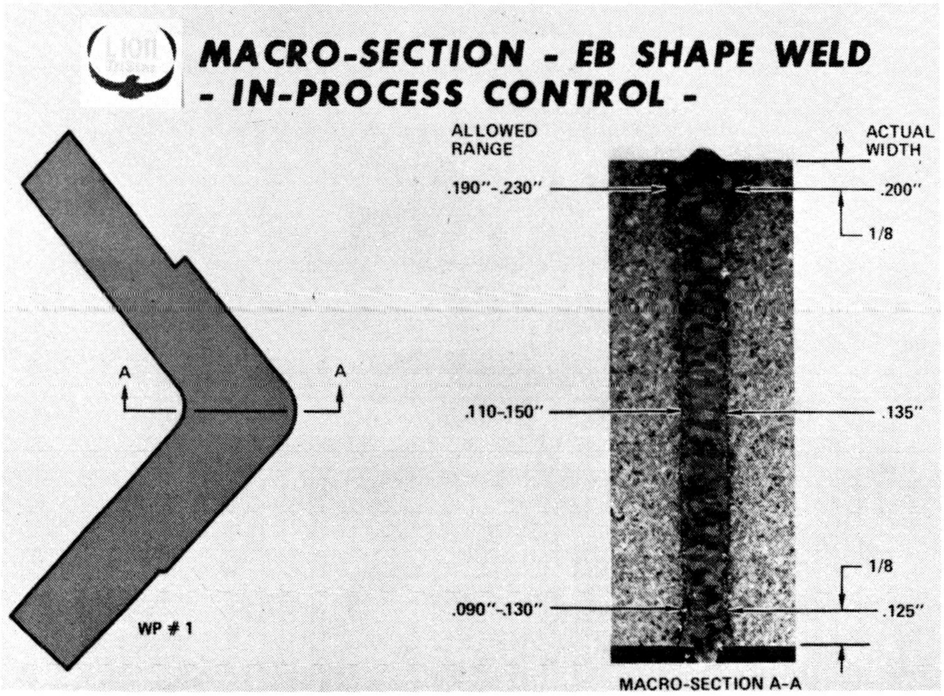

Figure 10

Figure 11

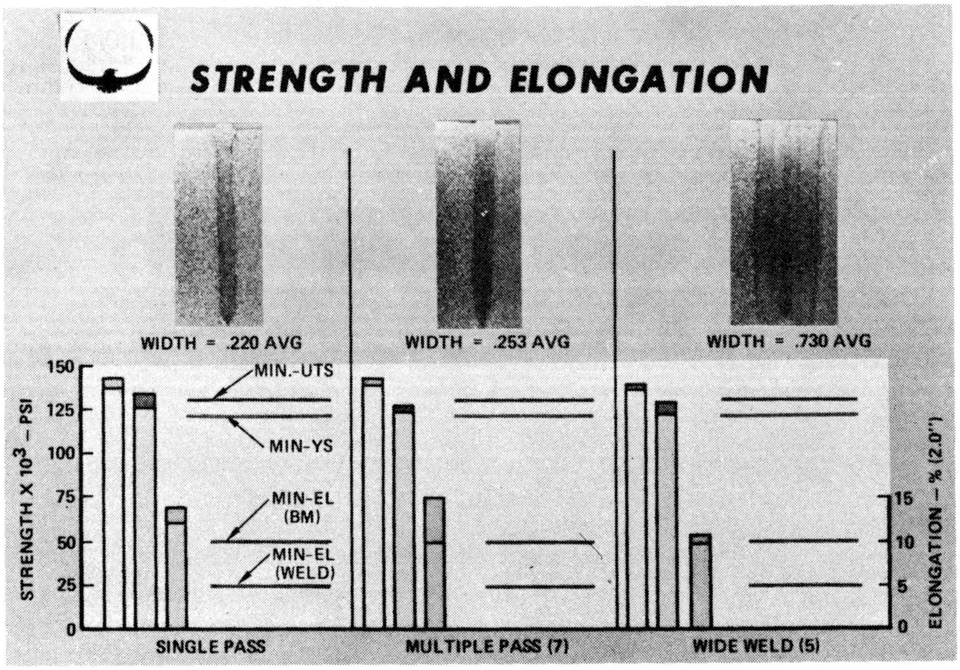

Figure 12

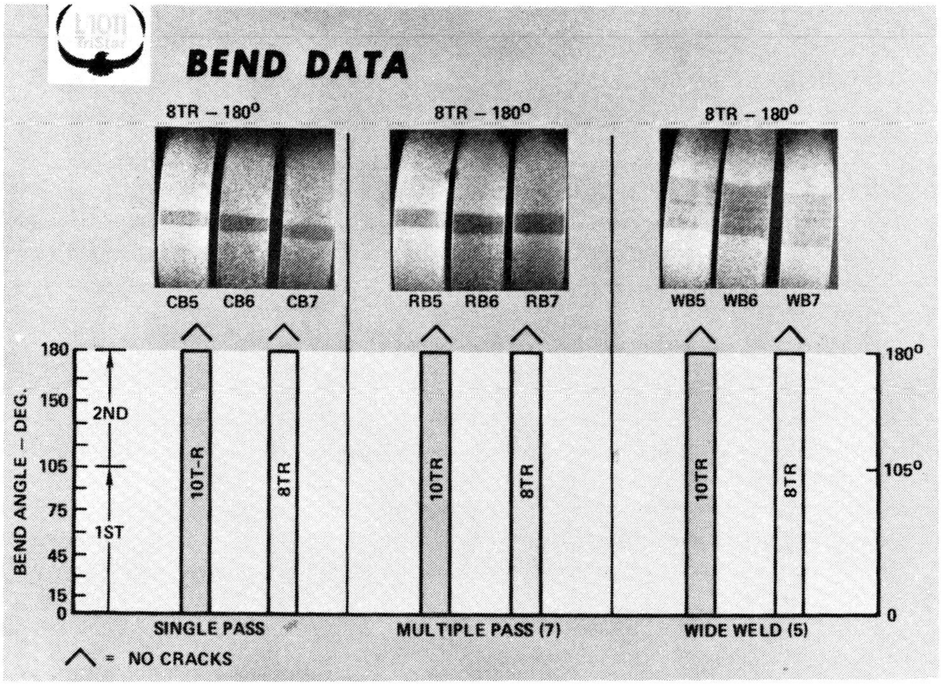

Figure 13

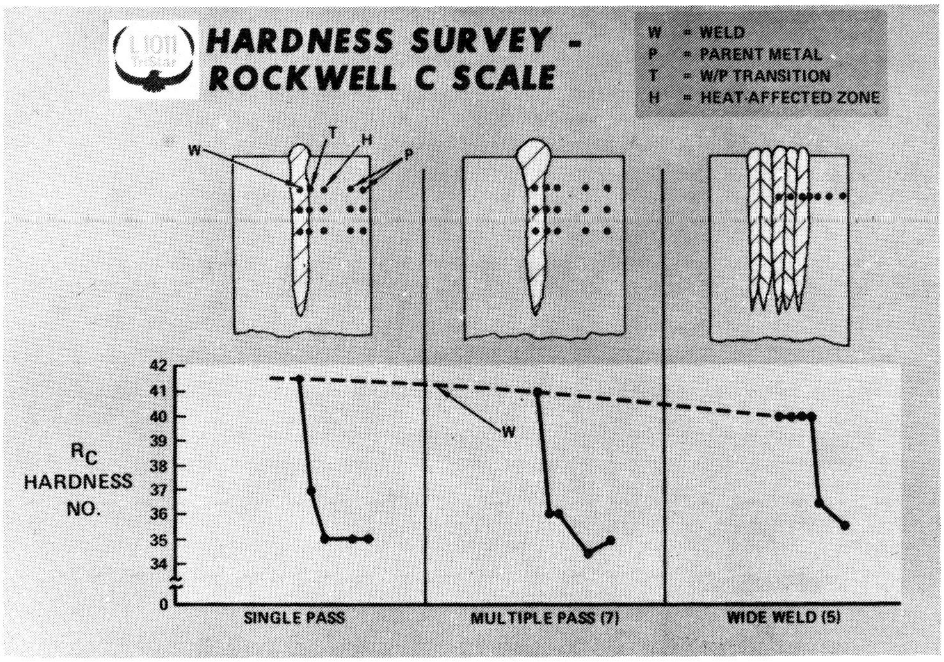

Figure 14

Electron Beam Welding of Titanium to Carbon Steel using Ag Insert Metal

Yoshiaki Arata, Fukuhisa Matsuda and Shōji Harada *

Synopsis

Electron-beam welding between low carbon steel and commercially pure titanium of 3 mm thick sheets has been investigated using Ag insert metal. The effects of the thickness of Ag insert metal and the location of electron beam displaced onto base metal on tensile strength of welded joint and on metallurgical behaviors were investigated. As a result, there were cleared, in order to make the sound joint of steel-titanium using Ag insert metal, that little melting of titanium base metal was one of the most important factor during welding and less than 15 wt% of Ti concentration should be kept in Fe-Ti rich layer of boundary between Ag- and Fe-rich zones in weld metal, and that for practical purpose the location of electron beam impinged was displaced by about the radius of electron beam onto the steel side from the edge of titanium base metal. The strength of steel-titanium joint using Ag insert metal showed about 30 kg/mm^2 in tensile test by an adoption of the optimum welding condition.

I. Introduction

Electron beam welding process has many obvious advantages as compared with conventional arc welding. Of these the heat source of high power density, fine focus and controllability in location of electron beam impinged are the most beneficial advantage for welding of dissimilar metals. That is, in spite of a great discrepancy in physical and thermal properties between them they are easily welded by means of electron beam welding having a symmetric fusion zone, and they are also easily welded with an odd fusion zone when the location of electron beam impinged is consciously displaced from the seam line onto either base metal.

In the meantime, fusion welding by conventional arc methods between steel and titanium has been tried already and most of direct welding without any filler metal have been not succeeded for the sake of occurrence of many cracks in weld metal during cooling after welding. The main reason for the cracking is considered that some brittle Fe-Ti intermetallic compounds which are formed in the weld metal due to mixing of molten iron and titanium in the puddle can not endure the shrinkage succeeding to welding.

Therefore, as regards conventional fusion welding between them, application of various insert metals has been examined, so far.[1-4] In this investigation electron beam welding is applied to make butt welded joints of low carbon steel and commercially pure titanium 3 mm thick sheets using pure silver insert metal. By changing the thickness of silver insert metal and displacing the location of electron beam impinged the weldability and the strength for the joints of steel-titanium were investigated.

II. Experimental Procedure

2.1 Materials used

The materials used in this investigation are as follows: (a) 3 mm thick plain low carbon steel sheet (C:0.03, Mn:0.19, Si:trace, P:0.008 and S:0.016%), (b) 3 mm thick commercial pure titanium sheet KS 50 and (c) commercial pure silver foils of 0.3, 0.5, 0.8 and 1.0 mm thickness.

---

* Professor, Associate Professor and Graduate Student,
Welding Research Institute of Osaka Univ., Suita, Osaka, 565, Japan

2.2 Welding procedure

A high voltage type electron beam welding machine (maximum power 6 KW, 150 KV-40 mA) was used throughout the investigation. The joint profile for butt welding between steel and titanium sheets using Ag foil is shown in Fig. 1. Ag foil, the dimension of which was 8 mm in width and W mm in thickness, was inserted between them. In order to make a welded joint without any defect in appearance there are protuberances of Ag foil on both top and back surfaces, the dimensions of which were decided from experimental results.

The locations of electron beam impinged which were represented in the arrows were in the center of Ag foil and successively varied to both base metals. The dimensions of the locations displaced were 0.3 + W/2 and W/2 onto the Ti side, the center of Ag foil and W/2, W/2 + 0.3 and W/2 + 0.6 mm onto the steel side which were illustrated as A, B, C, D, E and F in Fig. 1. The welding condition of electron beam is 150 KV in accelerating voltage, 9 to 13 mA in beam current, 100 mm/min in welding speed and 0.96 mm in beam diameter which was measured in slope-type welding test piece.[5]. All the welding were done with a straight single run under a high vacuum of 10^{-4} Torr or less.

2.3 Testing methods

The welded joints were cut and machined all sides of welds in order to perform metallurgical and mechanical tests. Optical microscopic examination, hardness test and analyses of elements in the weld metal with EPMA were done as the metallurgical test and tensile test as the mechanical test.

III. Results and Discussions

3.1 Metallurgical test

All the welded joints were fractured within weld metal. General schematic illustration of the weld metal in cross-section is shown in Fig. 2. The weld metal is generally composed of three regions, that is Ⓐ zone in Fig. 2 is mostly from Ag containing 2 to 15 wt%Ti and 1 to 2 %Fe, Ⓑ zone is mostly from Fe containing 1 to 2 %Ti and Ag and Ⓒ layer which is located between Ⓐ and Ⓑ zones is several to ten-odd μm in thickness, and the composition of which is composed of Ti and Fe containing a small amounts of Ag, but the ratio of Ti and Fe is widely varied for each case.

Moreover Ⓒ layer is generally higher in hardness than Ⓐ and Ⓑ zones, and Ⓐ is the most soft because of Ag-rich zone. A typical example of metallurgical structure and hardness in knoop test is shown in Fig. 3. The metallurgical characteristics of Ⓒ layer is the most important factor in order to make a sound steel-titanium joint using Ag insert metal. Ti element is the most concentrated in Ⓒ layer of the weld metal though the layer is apart from titanium base metal by Ag insert metal. An example of distribution of Ti and and Fe elements near and in Ⓒ layer is shown in Fig. 4. The peak concentration of Ti element is shown in Ⓒ layer. This peak concentration is increased with an increase of the concentration of Ti element in Ag-rich Ⓐ zone. That is, the more the melt of Ti base metal by electron beam, the higher concentration of Ti element in Ⓐ zone and moreover Ⓒ layer in general.

If the peak concentration of Ti element is high enough and some brittle intermetallic phase occur in Ⓒ layer, the welded joint may be brittle. There are two intermetallic phases $TiFe_2$ and TiFe in Fe-Ti binary alloy system.[6]

3.2 Mechanical test

Effects of the location of electron beam impinged on tensile strength of welded joints are collectively shown in Fig. 5 (a),(b),(c) and (d) for various thicknesses of Ag insert metal. The arrow marks on the axis of abscissa in Fig. 5 represent the location of electron beam impinged, respectively. Mark ● in the figures represents the tensile strength of the welded joints which were fractured in Ⓒ layer. Mark X represents that of the welded joints which were fractured in Ⓒ layer where penetration was poor because the location of electron beam impinged was far from Ⓒ layer. Mark ■ represents that of the welded joints which were fractured within Ag insert metal, that is, Ⓐ zone. Moreover mark ⊗ in (d) represents the welded joints which were fractured in the poor penetrated Ag-Ti interface.

In Fig. 5 (a) using 0.3 mm thick Ag insert metal the joints were weakened at 0.45 mm on Ti side due to poor penetration in Ⓒ layer, and were widely scattered in strength at 0 and 0.15 mm on steel side, but became stable in strength to about 25 to 35 kg/mm^2 at 0.45 mm on steel side. In Fig. 5(b) a stable strength for the welded joints is seen at 0.25 mm on steel side in case of 0.5 mm thick Ag insert metal. Moreover a stable strength is also predicted at between 0 and 0.4 mm on steel side for Fig. 5(c) of 0.8 mm in thickness and near 0 mm for Fig. 5(d) of 1.0 mm in thickness.

Considering from these results and the beam diameter of 0.96 mm, it can be concluded that, in order to have a stable joint strength, the location of electron beam impinged should be apart from the Ag-Ti interface by a distance of the beam radius or a little more. This means that the melting of titanium base metal should be as little as possible because the peak concentration of Ti element in Ⓒ layer is strongly affected. The effect of the peak concentration of Ti element in Ⓒ layer, which was measured with EPMA having about 2μm diameter of electron beam, on tensile strength of the welded joints is shown in Fig. 6. It is concluded from Fig. 6 that, in order to make the welded joint having more than about 25 kg/mm^2 in strength, the peak concentration of Ti element in Ⓒ layer should be kept in the value less than 10 to 15 wt% by taking care of excess melting of titanium base metal, and that when the value reach to 20 % the strength of the joint would be reduced and become brittle.

IV. Conclusions

Electron beam welding between steel and titanium 3 mm thick sheets is investigated using Ag insert metal. The main conclusions obtained are as follows:

(1) Weld metal with electron beam is mainly composed of three regions, that is, Ag-rich zone, Fe-Ti-rich layer and Fe-rich zone as shown in Fig. 2 schematically.

(2) There is the most metallurgical feature in the Fe-Ti-rich layer between Ag-rich and Fe-rich zones. The layer shows the hardest in hardness in the weld metal, often appears as intermetallic compound from its concentrations of Ti and Fe and shows brittle in mechanical property. When the peak concentration of Ti in this layer exceeds about 15 wt%, the welded joint shows brittle and low strength in general. In order to make a sound joint, therefore, the peak concentration of Ti in this layer should be kept less than 15 wt%.

(3) So as not to increase Ti concentration in the Fe-Ti-rich layer, titanium base metal should be melted as little as posibble. For the purpose of this it is recommended that a proper metal foil such as Ag is inserted between steel and titanium sheets and the location of electron beam impinged is displaced onto steel base metal.

(4) In this investigation using Ag insert metal it was the optimum, in order to make a stable joint in strength, that the distance which was displaced from the edge of titanium base metal was about the radius of the beam, about 0.5 mm or a little more. However the excess distance in displacing caused the lack of fusion in the boundary between titanium base and Ag insert metals. The

thickness of Ag insert metal is recommended about 0.5 mm in case of 3 mm thick base metals in this investigation from economical and reliable standpoints. In the above case, therefore, the location of electron beam impinged is at the edge of steel base metal.

(5) The tensile strength of 3 mm thick steel-titanium welded joint using Ag insert metal is considered about 25 to 35 kg/mm^2 when it is adopted the optimum electron beam welding condition.

References

1) F. W. McBee, Jr., et al. : W.J., 35 (1956) 10, 481-S
2) D. R. Mitchell, et al. : W.J., 40 (1961) 12, 546-S
3) V. S. Golovchenko, et al. : Svar. Proiz., 1971, 12, 36
4) D. Canonico, et al. : W.J., 38 (1959) 2, 71-S
5) Y. Arata, et al. : Trans. JWRI, 2 (1973) 2, 1
6) M. Hansen: "Constitution of Binary Alloys" NcGraw-Hill, 1958

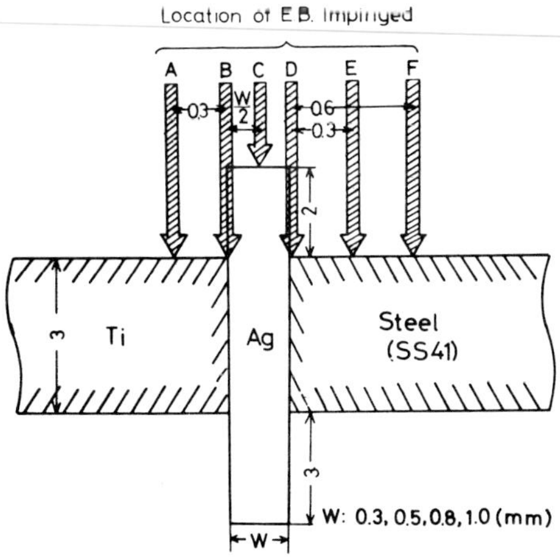

Fig. 1 Schematic illustration of joint profile.

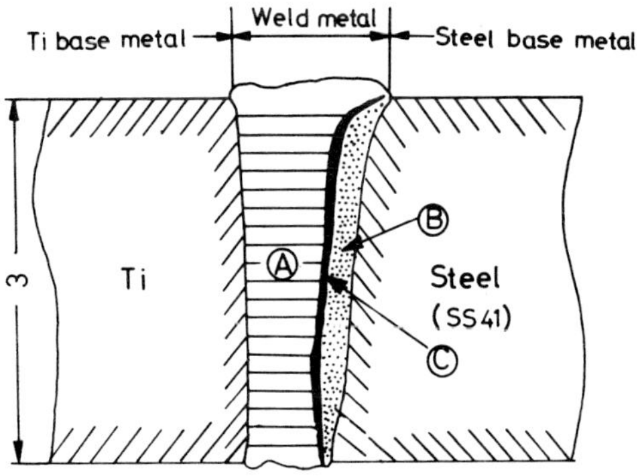

Fig. 2 Schematic illustration of welded joint in cross-section.

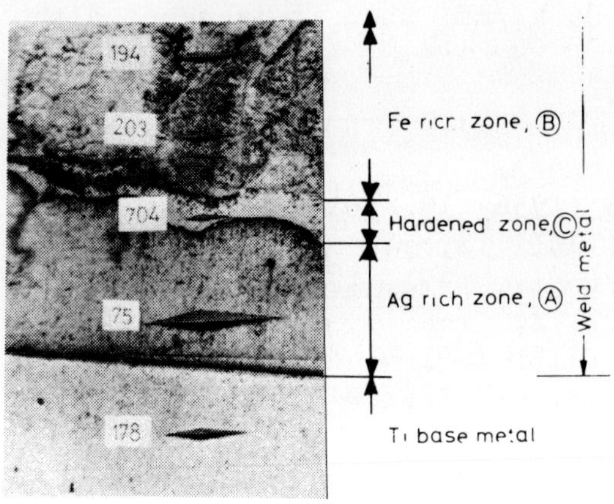

Fig. 3 Knoop hardness in welded joint.

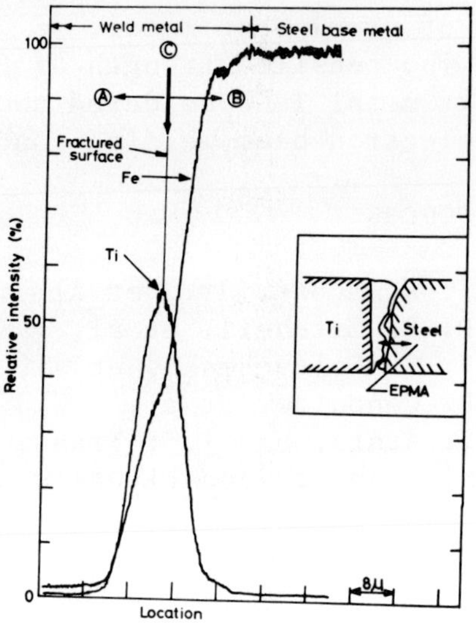

Fig. 4 Distribution of Ti and Fe near Ag-Steel interface.

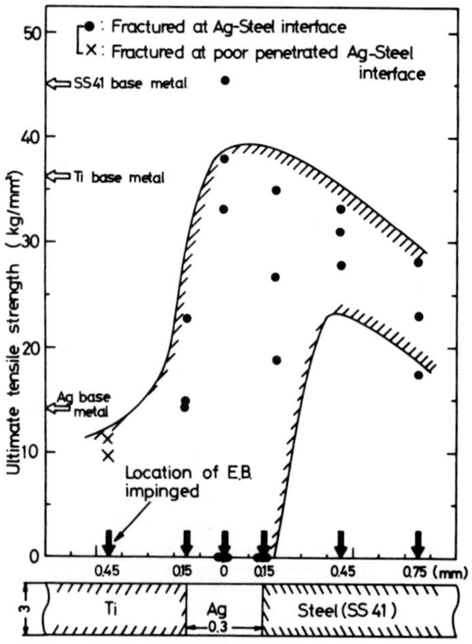

(a) Ag : 0.3 mm

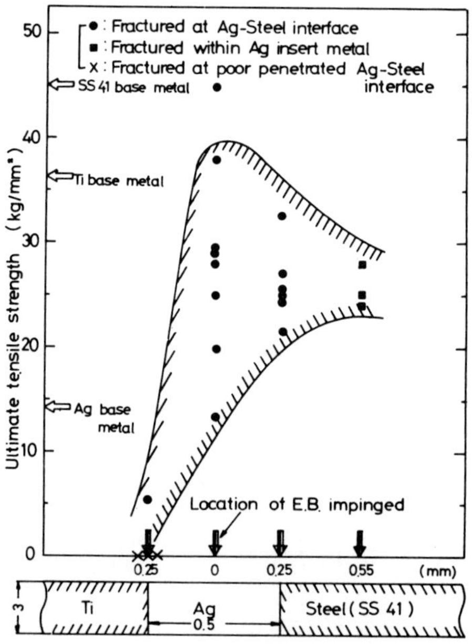

(b) Ag : 0.5 mm

Fig. 5 Tensile strength vs. location of electron beam impinged.

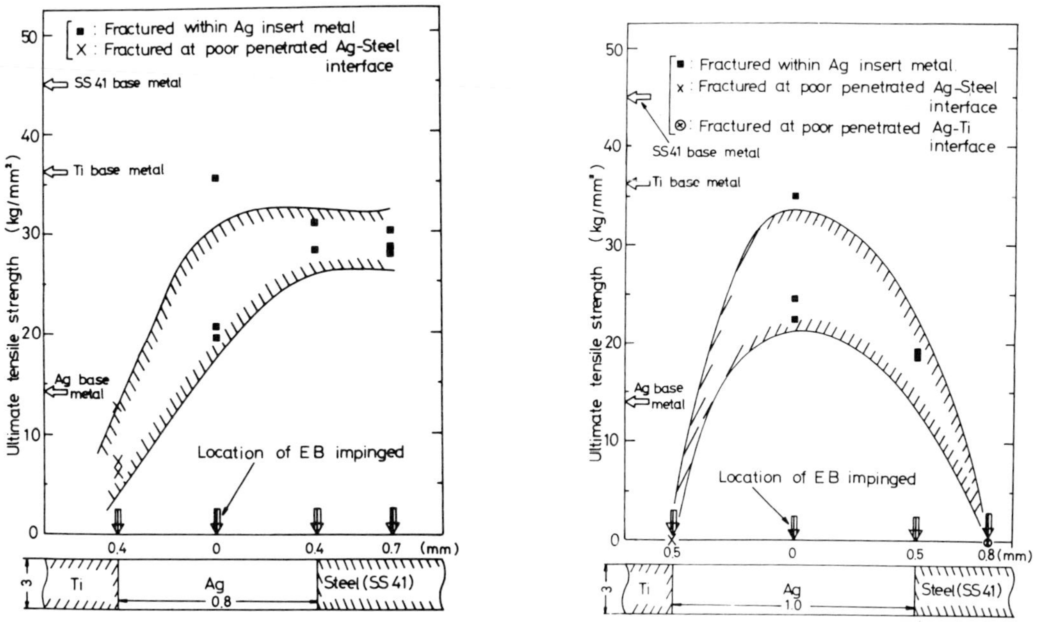

(c) Ag : 0.8 mm

(d) Ag : 1.0 mm

Fig. 5 Tensile strength vs. location of electron beam impinged.

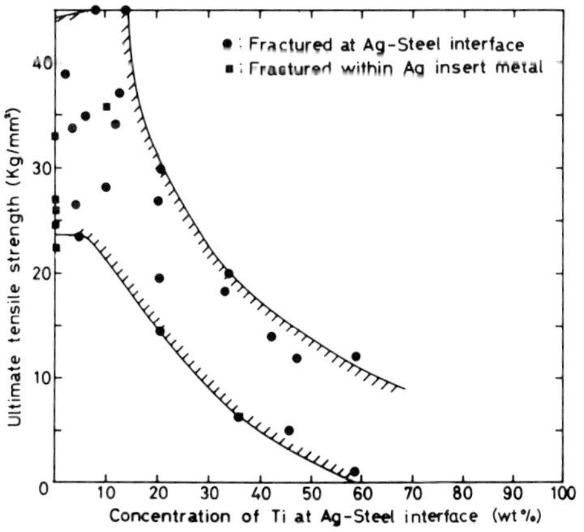

Fig. 6 Tensile strength vs. concentration of Ti at Ag-Steel interface

EB welded dual metal

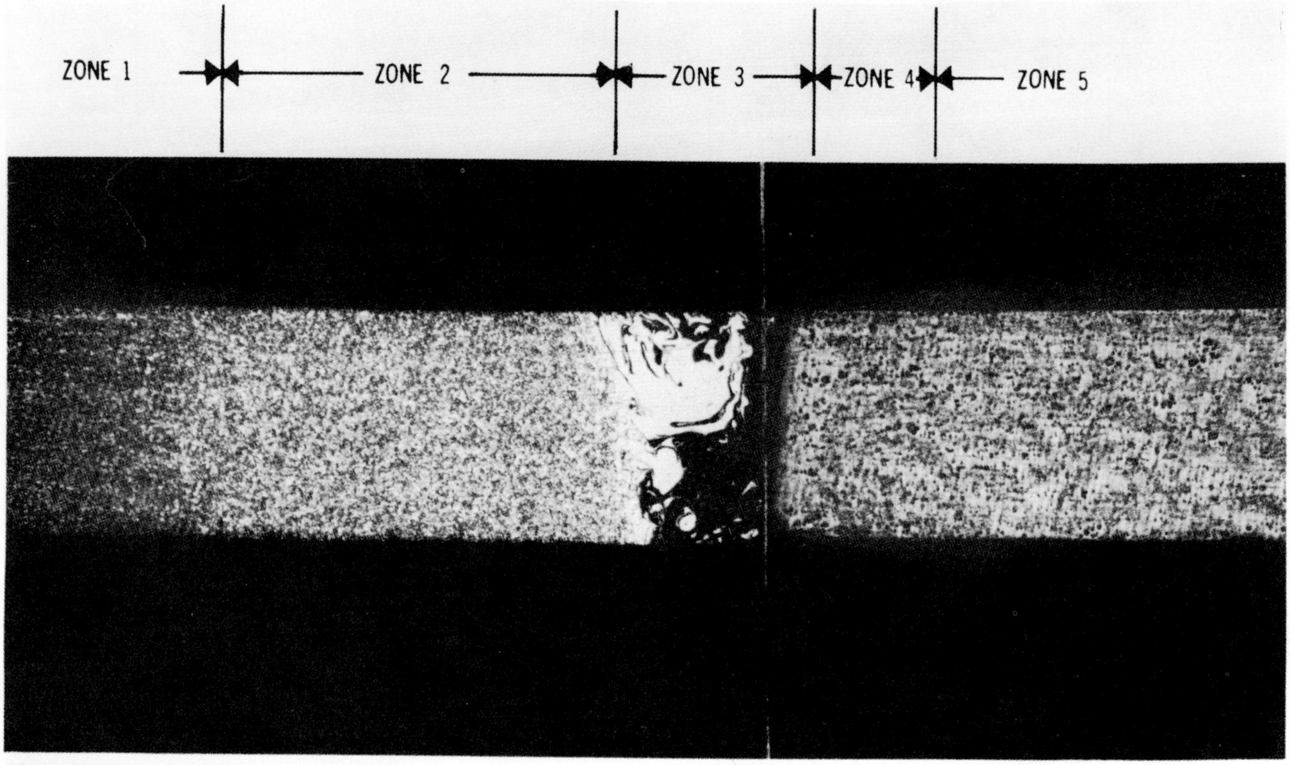

This photomicrograph at 50X magnification shows an electron beam weld of silver and Alcoloy. Zone 3 is the fusion zone and has a thickness of 0.010" to 0.015". Zones 2 and 4 are the heat-affected areas where some annealing has taken place. Zones 1 and 5 are unaffected metal.

he Micro Switch Div of Honeywell, Freeport, IL, was about to introduce a new series of pushbuttons, but conventional production methods and material constraints threatened cost over-runs and impaired product integrity.

The solution employed was to change from hand-welded contacts to electron beam (EB) welded strip. For Micro Switch, the benefits of changing to the dual metal strip are predicted annual production/material savings of $164,000; total cost reduction amounting to 30 percent; elimination of three manufacturing steps; simplified materials handling, inventory and documentation; improved product integrity; improved production capability; and faster delivery.

It is said that conventional welding and joining techniques cannot be controlled as closely as electron beam welding. There is a danger that the two metals—base and contact—will be detrimentally affected. The resistance type weld joint area is less consistent, often more porous, and less dense, resulting in functional characteristics which are too variable for many applications.

Electron beam welded dual metal strip, jointly developed by Micro Switch

strip saves money

and Technical Materials, Inc (TMI), Lincoln, RI, combines dissimilar metals and/or tempers in one long, continuous coil. The possible combinations of metals are many and the weld "joint" has little effect on both metals. As an added bonus for switch applications, EB-welded strip conserves precious metals, in this case silver.

Actually, the switch material is an electron beam welded combination of Olin Alcoloy 688 and fine silver. The Olin alloy contains 73.5 percent copper, 22.7 percent zinc, 3.4 percent aluminum and 0.4 percent cobalt. TMI supplies the 0.025″-thick by 1.680″-wide material—width of fine silver is 0.250″—in coils to several thousand feet in length.

Value improvement

At the outset, the company was looking for ways to reduce manufacturing time and costs for contact lead frames without sacrificing quality. According to Curtis Stoops, supervisor of engineering, "The most important change to meet our need would have to occur in our welding processes. This would require more expensive lead frame material—silver-brass."

Analysis indicated that although the new strip would cost more than the combined metal costs used in conventional joining techniques, recovery of those costs, plus additional savings, would come through reductions in production steps.

"This," continues Mr Stoops, "is what prompted us to make the change. Actually, when you compare the two methods, value improvement becomes readily apparent."

Conventional methods awkward, costly

Conventional methods for making contact lead frames for precision snap action switches were awkward, and costly in terms of both time and labor. One conventional method, small welded contacts on a base metal strip, requires three time-consuming steps: 1) individually welding silver slugs onto blanked lead

MATERIAL COMPARISON FOR EB WELDING

Metals	Best suited application	Manufacturing benefits	Cautions	Price/cost benefits
Beryllium copper/ copper	Crimp-type connector contacts	Combines good spring properties (be cu) with good formability (copper)	Avoid forming within 0.030″ of weld zone	Eliminates costly secondary operations such as zone annealing
Silver/ copper	Switch contacts	Eliminates need to apply silver contact through welding or staking	Design tooling to minimize and collect precious metal scrap	Extremely cost-effective with silver at $4.50/oz vs copper at $1.00/lb
Invar/Stainless steel	Thermostatic bimetal components	Finely defined weld zone with minimal heat-affected zone	-----	Availability in coil stock eliminates need to weld individual pieces and costly secondary handling operations
Copper/ silver/ copper	Switch contacts	By welding silver down center of copper, additional economies are available to conserve precious metal and increase tooling productivity	Design tooling to interdigitate parts wherever possible and provide scrap separation stations in die	Allows designing of a minimum silver width of 0.025″
Steel/ copper	Lead frames	Allows combination of different tempers using steel in area of part not requiring high conductivity and providing good lead strength of steel vs copper	Design tooling to facilitate cutting through differing tempers	Consumes higher priced copper while providing economical solution to design problem

Source: *Tooling & Production*, Oct 1978

frames, 2) trimming excess silver to form the contacts and 3) inspecting to insure excess silver was removed.

Another alternative, manual welding of base and contact strips, was discarded as being totally unusable. Says Mr Stoops, "Hand welding or joining two ultralong strips is really a little overtaxing, especially in what is supposed to be a mass production situation. Manual or semiautomatic welding increases the margin for error. Welds cannot possibly be consistent enough and, obviously, just take too much time."

Although Mr Stoops admits that either of the conventional methods are adequate for single or larger contact operations, they simply were not practical for small, multicontact operations.

The Micro Switch component is a 4-unit lead frame, having two sets of contacts on each unit. Four joined units comprise a workpiece. Hand welding small individual contacts on the workpiece confines operators to too small an area with too many small welds.

Formerly, the operator welded eight slugs on one four-unit-long lead frame, trimming in between each to make a total of 16 separate contacts per workpiece. So the conventional operation usually took between 1½ to 2 hr per thousand contacts. When the time and labor required to trim excess silver was considered, costs were excessive.

One step replaces three

Taking advantage of TMI's EB welded dual metal strip, Micro Switch eliminates the three separate operations. The entire lead frame is blanked out of the strip in one operation. The required thickness and width of silver for contacts is thus precisely controlled. And because the contacts are now "continuously" formed in a punch press, they're more dimensionally consistent. Precision blanking and trimming are done simultaneously. As a result, pushbutton switches are now more reliable as well as less costly to make.

According to sources at TMI, electron beam produces a strip that is easier to form (bend, stamp etc) through the weld area. Any loss in temper of the metals is restricted to a very narrow band and does not migrate into the two metals. This is important where the final component requires bending through the welded zone, or adds support when encapsulated in the body of a switch.

TMI's EB welding process has minimum impact on the material hardness. Usually, the metals are processed individually to provide the exact characteristics required, then joined. This is especially important when strips are narrow. During conventional welding, narrow strips ½"-wide, for example, would be completely annealed, losing more than half of their hardness. Recovery of temper after welding would not be feasible.

EB welding allows a multitude of metal combinations, including copper base, nickel alloys, steels, and all precious metals and alloys. In addition to being able to join dissimilar metals, the process permits the joining of different tempers. For example, mill-hardened beryllium copper can be joined to ¼-hard phosphor bronze, which provides an excellent combination of spring characteristics and formability for connector parts. Thus, the designer has a wider range of material configurations from which to choose.

In the past, lead frames (left) were made by blanking the body, individually welding silver slugs to it and trimming off the excess silver. The lead frame on the right was blanked out of the dual metal strip in one operation.

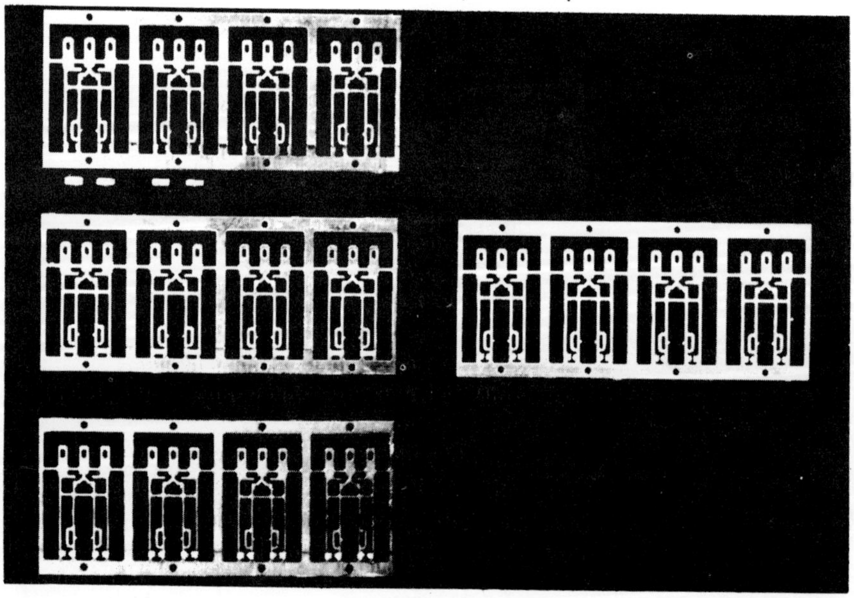

This 1.5-million watt resistance heater will boost helium temperature from 1,400° to 2,250° F. for a helium gas loop in the Advanced Test Reactor at the National Reactor Testing Station, Idaho Falls. Arrow points to cold-wall insulation.

Electron beam welding of molybdenum

by **W. P. McGREGOR** *and* **H. MARQUEZ**, *Solar Division of International Harvester Co., San Diego, Calif.*

If you've welded molybdenum, or tried to weld it, you know the hang-ups: grain growth, brittleness. In spite of this, there are applications where a brittle weld is OK.

Like a molybdenum electrical heating element for a closed cycle gas loop of an advanced test reactor (ATR) for Idaho Nuclear, Arco, Idaho. The heating element raises the temperature of helium from 1,400° to 2,250° under steady state conditions while maintaining a purity of less than a part per million of contaminants. To do this, 1.5 million watts power the heater from 2,800° to 3,000°F. Battelle Memorial Institute, Pacific Northwest, Richland, Wash., made the ATR with Solar Division of International Harvester Co., San Diego, electron beam welding.

The loop tests metals and other materials that may be used in reactors that are expected to operate at 2,000° F.

One good reason why we selected electron beam welding over all others: high operating temperatures of ATR. Another good reason: EB welding was the only way. Without it, we couldn't have done the job. We considered Mig and Tig but neither has a great enough depth-to-width ratio for the weld bead. So, we designed for EB welding. Our results prove the soundness of our choice.

As insurance against failure, we made two elements, one for a backup. Each called for 75 EB welds, 72 of them on joints between element tube and connectors, 3 between connectors and electrodes. Tube diameter: 0.670 inch, 36 of them for each assembly. We welded each tube on each end of the assembly to a fitting. To weld each joint without rotating the assembly, an eccentric positioner moved the part so the joint was always tracking concentrically. We designed and built the positioner to rotate the part so the electron beam could track each joint. We moved the gun remotely from one joint to another.

Whenever we'd move the EB gun

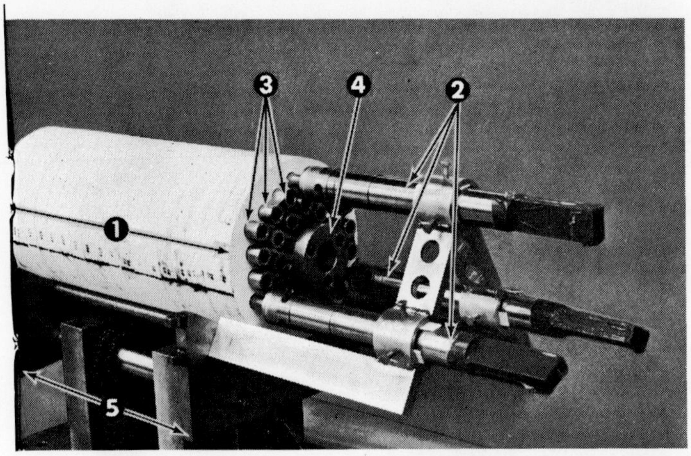

Right: eccentric positioner rotates heating element so electron beam can track each joint. Left: 1. Nineteen ceramic insulators. 2. Molybdenum electrodes. 3. Molybdenum tubes and connectors. 4. Connector ring. 5. Weld fixture.

for welding on either end of the assembly, we couldn't see the weld joint, so we had to set up mirrors. We used regular glass first, but the mirror coating caused the electron beam to arc along the edge of the glass. Polished stainless steel prevented the arcing, but vaporized metal coated the mirror forcing us to replace it every 6 or 7 welds.

Before welding, we cleaned, then assembled the parts.

Cleaning:

- Wear white cotton gloves.
- Immerse weld area in 1 part HF, 1 part HNO_3 1 part water for 5 seconds at room temperature.
- Desmut—probably molybdenum oxide—in 95 percent H_2SO_4, 4.5 percent HNO_3, 0.5 percent HF, and 18.8 gm/l CrO_3 at room temperature.
- Rinse in water, follow with an acetone rinse.

After cleaning, we assembled the tubes, fittings, and 19 ceramic insulators. We had already welded tubes to inserts, inserts to the three electrodes—three large pieces on the right end of the assembly.

We tested weld samples—which we later submitted to Idaho Nuclear—visually, ultrasonically, and

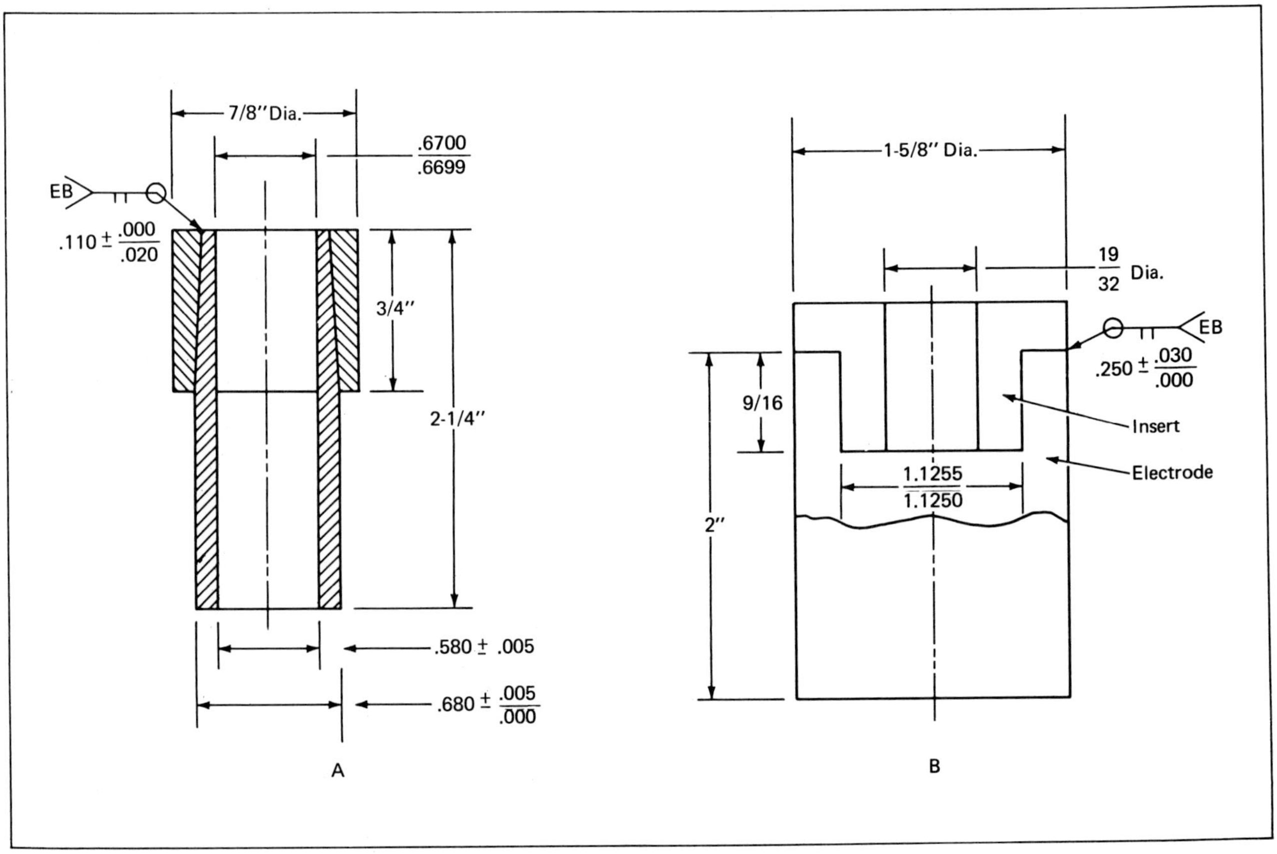

Joint configurations for joining tubes and connectors (A) and insert to electrode (B).

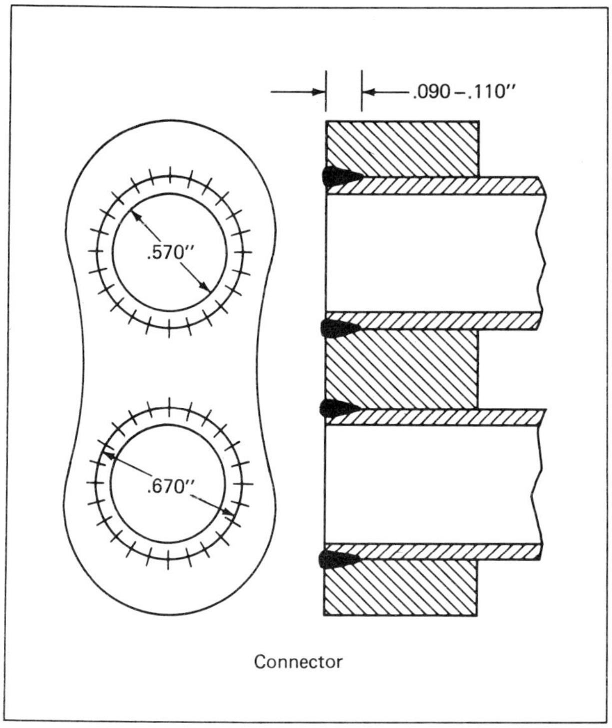

Subassembly: electrode, insert, and tube.

metallographically. Test results helped us plan acceptable welding methods.

WELDING

Before welding any of the joints, we welded a simulated part and inspected it visually, metallographically, and ultrasonically. Idaho Nuclear's customer rep approved these welds. Customer approval was SOP whenever 8 hours or more lapsed between welds, whenever equipment broke down—anything from a blown filament to a burned-out transformer, or whenever we changed welding operators.

SUBASSEMBLY WELDING

Welding three tubes to inserts.

Welding the insert and tube assembly to the electrode. We welded horizontally, penetrated 0.250 inch minimum, 0.280 inch maximum, then inspected.

Welding tubes to connector rings, one ring per assembly. Three welds join three tubes to the ring.

FINAL ASSEMBLY

Each assembly has 19 ceramic insulators with holes to take the 36 tubes. We pushed the three tubes welded to the connector ring and the three tubes welded to the electrodes and 30 unwelded tubes through the holes in the 19 insulators. We assembled tube end connectors, one for each two tubes, on the end of the tubes.

We checked each of the three circuits for continuity (there are three electrical circuits in ATR, one for each of the large electrodes). A stainless steel shield protected the ceramic insulators from damage which might be caused by the electron beam.

The EB gun was then positioned remotely in the vacuum and each joint—tubes to connectors—spotted and welded. Minimum penetration: 0.085 inch. Better: 0.090-to 0.100 inch. Actually, penetration varied from 0.090- to 0.110 inch because of joint tracking and heat difference: sometimes the electron beam was a few thousandths of an inch off the joint caused either by machining or joint misalignment. Also, while welding, the part rotated several times to preheat, weld, and taper off: some fluctuation in power or speed would cause a difference in weld penetration. Any of this, we picked up ultrasonically. We had to reweld some joints to correct lack of fusion from the tracking situation.

After we welded on one end and had met our inspection requirements, we reversed the part, made adjustments, welded and inspected as we did before.

Idaho Nuclear inspected all welds ultrasonically using a pulse-echo unit designed and built especially for this program by Battelle. Battelle too inspected and accepted all welds before the unit left us.

SECTION A-IV:
Applications

The application of electron beam welding – Part 2

A general review based on an IIW paper, by **Kiyohide Terai*** and **Hiroyoshi Nagai***, from a Japanese company for which The Welding Institute built an EB welder

Nuclear structures

Electron beam welding has a number of possible applications in the nuclear industry including reactor pressure vessels, piping, heat exchangers etc. The growth in nuclear power installations is shown in Fig.17 and the increasing thickness of the constructional materials in Fig. 18. This growth has brought with it an increase in the use of high tensile steels which, when welded by conventional multipass arc welding processes, are liable to weld defects, cracks, etc, as well as a reduction in the toughness of the HAZ. Using the 180kV, 108kW equipment shown in Fig.19, steel up to 320mm thick can be welded in one pass, thereby reducing the metallurgical effects (Fig.20) even when welding at 100 mm/min.

Turning to pipework for nuclear reactors, weld joints showing little distortion and with good quality, can be readily produced (Fig. 21). Single side welding was used to produce these joints in a large vacuum chamber and the resulting welds had a good appearance with zero defects when examined by NDT. The deep sharp penetration of the electron beam, with its low heat-input characteristics, reduces the risk of weld decay and stress corrosion cracking experienced with austenitic stainless steel tubing fabricated by MIG or TIG welding for use in light water reactors. A comparison of EB, MIG, TIG and SAW (submerged arc) welds on SUS 304 stainless steel is shown in Fig. 22. There was no trace of $Cr_{23}C_6$ precipitation in the EB weld and hence no risk of weld decay.

EB welding has been applied also to the manufacture of the in-

*Dr Eng Terai is Manager and Dr Eng Nagai is Research Engineer, Welding Research Department, Kawasaki Heavy Industries Ltd, Japan

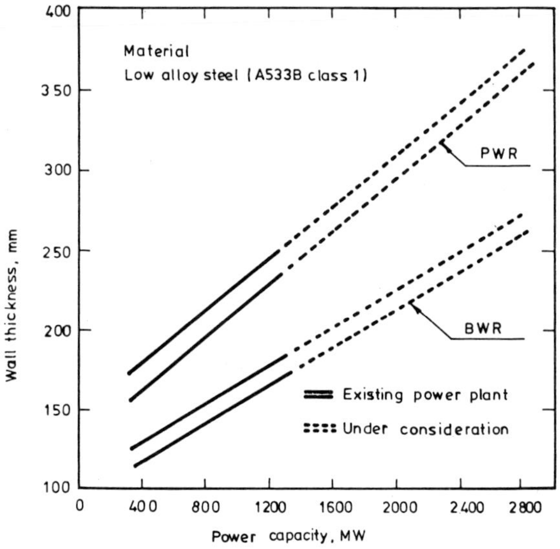

18 Relation between wall thickness of reactor and power in nuclear power plant.

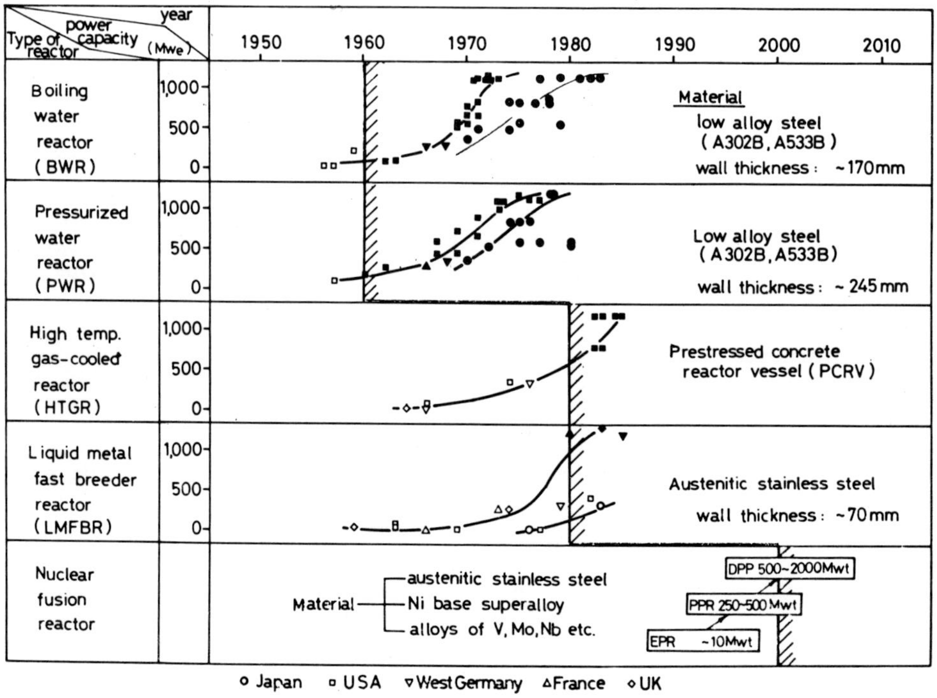

17 The international situation in the development of nuclear power plant.

Above **19** *High power EB welding equipment (180kV-108kW type).*

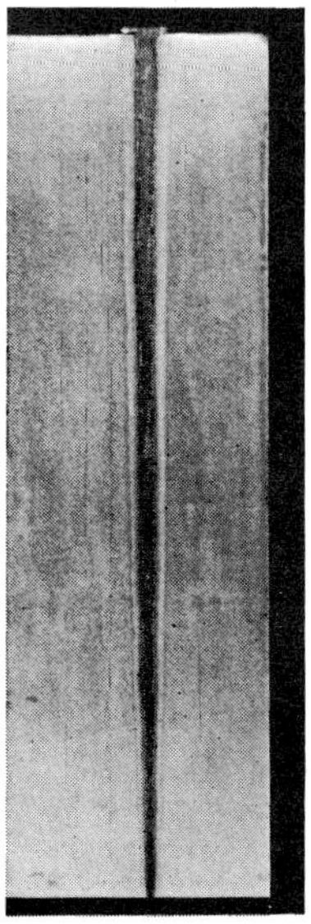

Right **20** *Single pass weld in mild steel plate 320mm thick made on the EB equipment shown in* **19** *at 100 mm/min using 108kW at 2 × 10^{-4} torr.*

Below **21** *EB welding of pipe for nuclear power plant (power: 12kW; speed: 1 m/min; vacuum: 6 × 10^{-4} torr. Material: stainless steel SUS 304).*

Below **22** *Comparison of austenitic stainless steel welding by various welding processes.*

Item \ Process	EBW	TIG and SMAW	SAW
Base plate	Stainless steel (sus 304)		
Welding sequence	10	10	25
Welding condition	Vb(kV) 150 Ib(mA) 40 υb(mm/min) 800 D_F (mm) 170 D_O (mm) 205	pass No. TIG 1, SMAW 2~4 I (A) TIG 75, SMAW 90~150 V (V) TIG 10, SMAW 22~24 υ(mm/min) TIG 55, SMAW 180	I (A) 450~500 V (V) 30~32 υ(mm/min) 320~350

23 *EB welding of an in-pile tube for radiation test (material: Hastelloy X+ stainless steel SUS 316).*

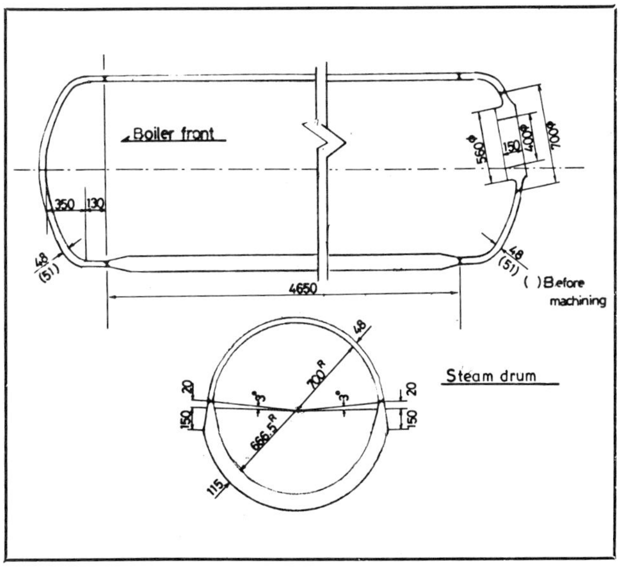

24 *Shape of trial boiler drum for local chamber EB welding (dimensions in mm).*

pile tubes for radiation tests for the high temperature, high pressure helium gas loop for a high temperature gas cooled nuclear reactor. Manufacture was carried out using a conventional vacuum chamber, but for welding the in-pile tube a special sealing jig was required outside the main vacuum chamber (Fig. 23).

Each 8m long in-pile tube consists of four large diameter double tubes made from thin plate and is required to have an accuracy of straightness of better than 1 in 2000. By using conventional arc welding it proved almost impossible to achieve the required degree of straightness. However, EB welding produced excellent results on these complex structures made from Hastelloy X and stainless steel SUS 316.

Boiler drums

Electron beam welding has been applied to boiler drums and the successful results suggest that nuclear power plant heat exchangers could be fabricated by the same techniques. Figure 5 shows welding using an EB machine with a local chamber, while Fig. 24 shows the shape and dimensions of the trial boiler drum.

In this application, it is necessary to use a backing shield assembly and three techniques are possible (Fig. 25), although use of a backing frame is the most common. Figure 26 compares a

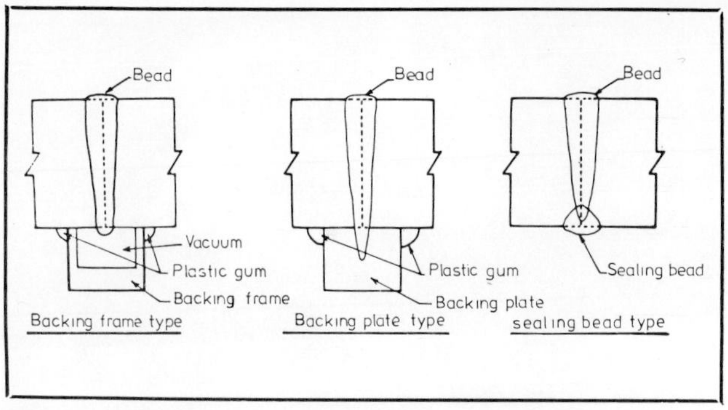

25 *Backing methods in the local chamber EB welding process.*

Item	Submerged arc	Electron beam
Welding time	Reverse side 8 Hr, Surface side 18 Hr) 26Hr	0.5 Hr
Number of pass	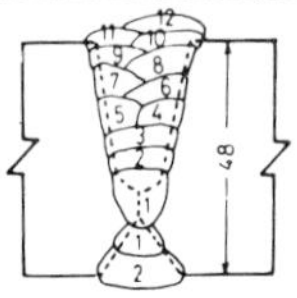	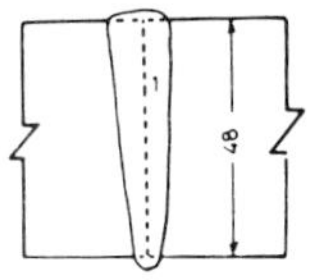

26 *Comparison of submerged arc welding and EB welding of the boiler drum shown in* **24.**

27 *Local chamber EB welding of short tube.*

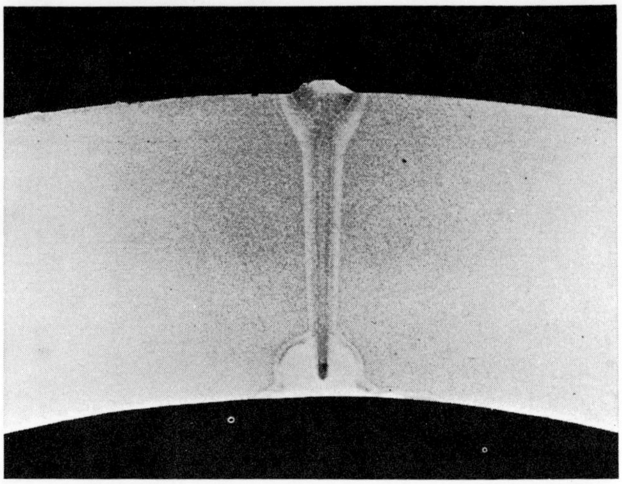

28 *Macrosection of joint obtained by local chamber EB welding of short tube, 60mm thickness mild steel.*

29 *Local chamber EBW equipment for long welds.*

30 *Hydraulic jack to prevent deformation of the long frame shown in* **29.**

submerged arc weld made using 14 passes in 26hr with an EB weld made using a backing plate (Fig. 25b) in only 0.5 hr. Figure 27 shows a set up for the local chamber EB welding of 60mm wall thickness short mild steel (SS41) tubes using a backing shield (Fig 25c) and Fig. 28 is a section through the weld, indicating that the weld made from the external surface adequately penetrates the sealing bead.

For the welding of long boiler drums, a local chamber EB machine with a long welding frame is under development (Fig. 29). A problem with this approach is that the pressure difference between the atmosphere and the vacuum can lead either to misalignment of the EB gun or to seizure of the moving carriage thereby stopping welding.

To overcome this, the authors' group developed equipment where the welding frame supporting bar is operated in sections by hydraulic jacks (Fig. 30); thus only the supporting bar under the EB gun is under pressure at any one time.

Submarines and deep submersible rescue vessels

High strength steels are being specified for these applications to enable greater depths to be reached but these steels can cause problems when welded with high heat input processes as a result of degradation of properties and cracking in both weld metal and HAZ. Tensile properties and Charpy test results for welded joints in HY90 and HY130 steels showed that optimum results were

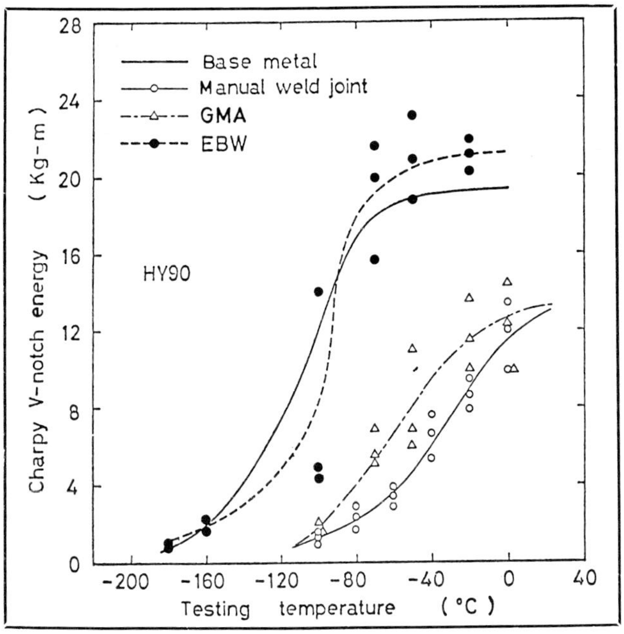

Above **31** *Charpy results for welded joints in HY90 steel made by various processes.*

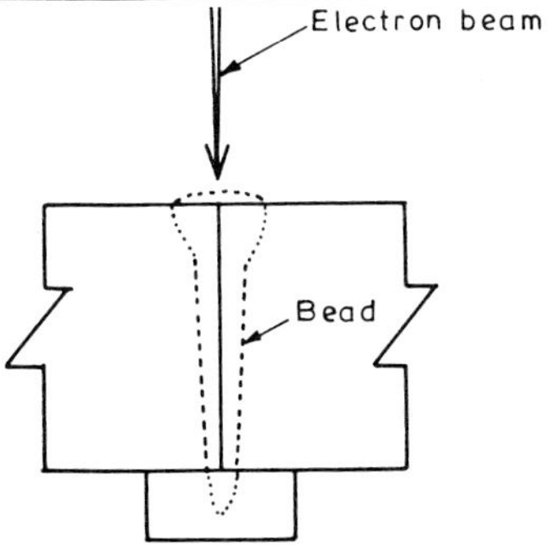

Left **32** *EB welding with metal insert (schematic).*

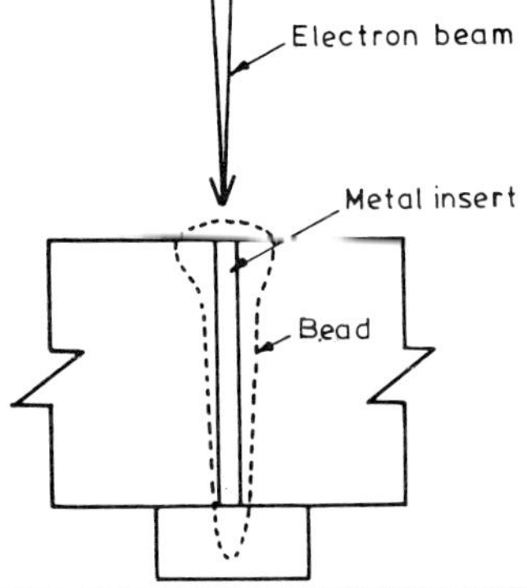

Below **33** *Schematic drawing of the application of local chamber EB welding to the inner hull structure of a submarine butt joint.*

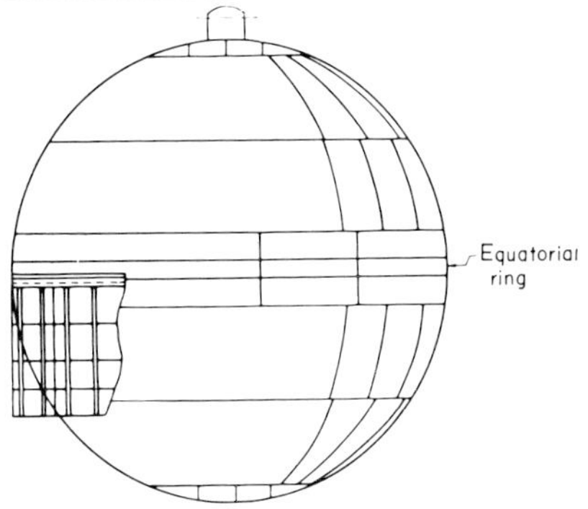

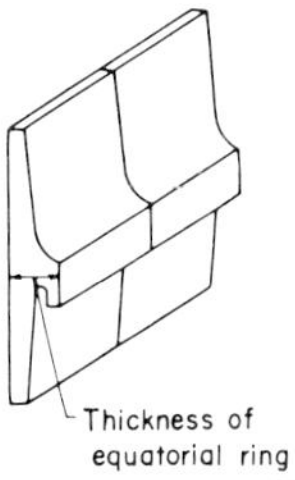

Capacity (m³)	Diameter of tank (m)	Number of tanks	Thickness of equatorial ring (mm)	
			5083Al alloy	9%Ni steel
128,000	37	5	200	80
160,000	40	5	230	90
200,000	43	5	260	105
250,000	46.3	5	300	120

34 *Relationship between thickness of equatorial ring and capacity of spherical tank for LNG marine carrier.*

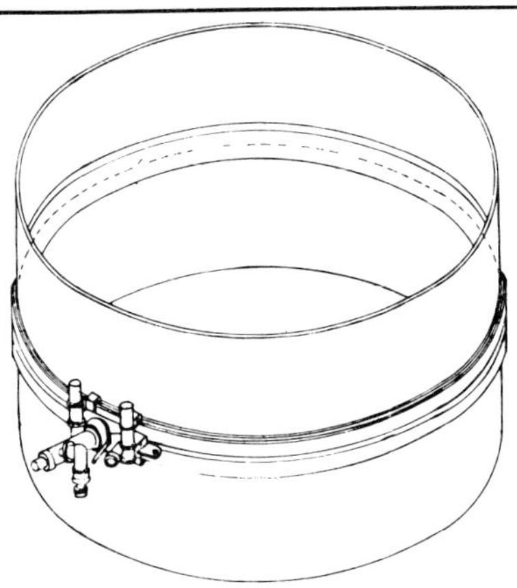

obtained with the EB process compared (for example, Fig. 31) with other processes. With some of the HY130 samples, EB welds were made with Inconel 600 inserts, as indicated in Fig. 32 and these gave excellent results.

The application of local chamber EB welding to the manufacture of the inner hull structure of a submarine is shown schematically in Fig. 33; macrosections of the joints produced in high tensile strength steels were satisfactory and had excellent mechanical properties.

Spherical tanks for LNG carriers

The increasing size of LNG marine carriers, necessitated by the growing worldwide market for gas products, is likely to cause problems in the not too distant future. In the case of a Moss-type spherical tank made of aluminium alloy, the diameter and equatorial ring thickness must be increased to allow for an increase in volume of the sphere (Fig. 34). However, at present the greatest thickness of rolled aluminium alloy plate which can be supplied is 200mm, whereas for future requirements, plates up to 250 or even 300mm will be required with good mechanical properties.

With this in mind, the trial manufacture of thick plates was attempted using EB welding. Figure 35 shows the method of fabrication while Fig. 36 shows two completed ring segments with apparent plate thicknesses of 275mm fabricated from 80mm and 100mm thick plates. After welding, tensile, Charpy, COD and fatigue crack propagation tests all gave satisfactory results. From these results, and using a system similar to that shown in Fig. 37, single sided manufacture of 10m long thick composite plates using a high power EB gun can be considered a practical proposition. A single pass weld carried out by the authors' group on 5083 aluminium alloy in 590mm thickness material was found to be satisfactory. Butt welding the equatorial ring will pose the next problem since it would require about 100 passes to complete the 200mm thickness ring using the most advanced form of conventional MIG welding. If the plate thickness increases above 200mm, the problems of producing MIG welds free from blow holes and lack of fusion, let alone the enormous number of weld passes required, will increase disproportionately. The authors' group has devised a local chamber EB system (Fig. 38), for welding these joints in only two passes.

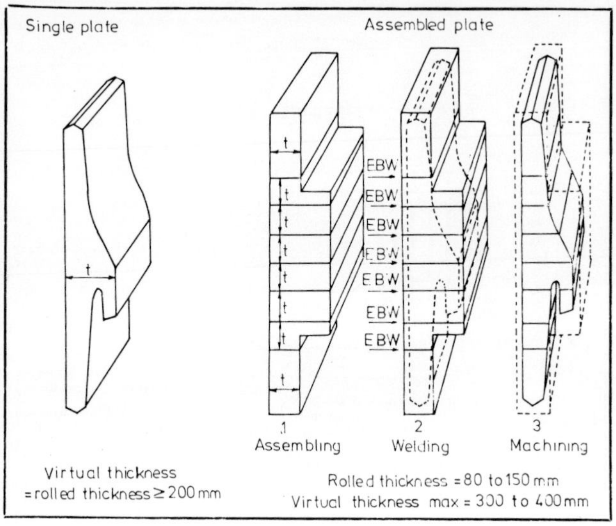

35 *Fabrication of integrated plate for ultra-thick equatorial ring by EB welding.*

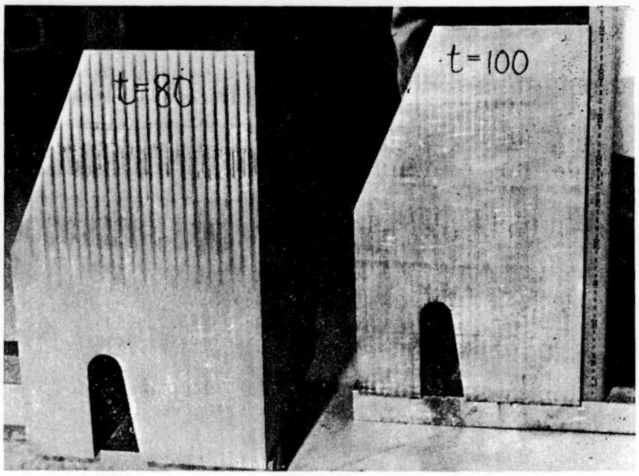

36 *Integrated plates made by EB welding (virtual thickness: 275mm) (thickness of each rolled plate, t).*

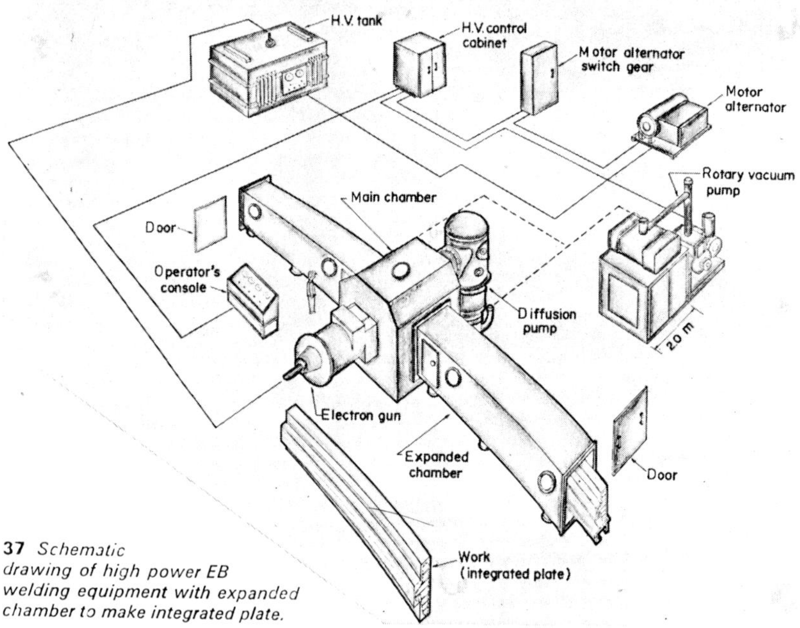

37 *Schematic drawing of high power EB welding equipment with expanded chamber to make integrated plate.*

38 *Vertical type local chamber EB welding equipment.*

As a result of the trifurcated shape of the ring, the joint is difficult to weld even using EB welding and so a special flange was manufactured to fit tightly over the ring to provide a weld of equal section thickness; the flange was then machined off after welding (Fig. 39).

Conclusions

The examples outlined in this contribution show clearly the work done, particularly by Kawasaki Heavy Industries, in the industrial application of electron beam welding to the fabrication of structures such as pressure vessels. Further efforts are necessary in the development of restraint jigs, positioning equipment, etc, as well as in the sliding and shielding mechanisms of the electron gun.

One of the big problems still to be overcome is the acceptance and approval of the technique for the fabrication of pressure vessels, a topic which is mentioned only briefly in cases 1461 and 1471 of ASME Boiler and Pressure Vessel Code. Even these cases imply that EB welding is intended for small parts rather than structures such as boiler drums and pressure vessels. In the light of these possible applications, the revision of fabrication and acceptance standards will become necessary and to this end, sufficient test data should be collected.

39 *Equatorial ring welded by vertical type local chamber EBW.*

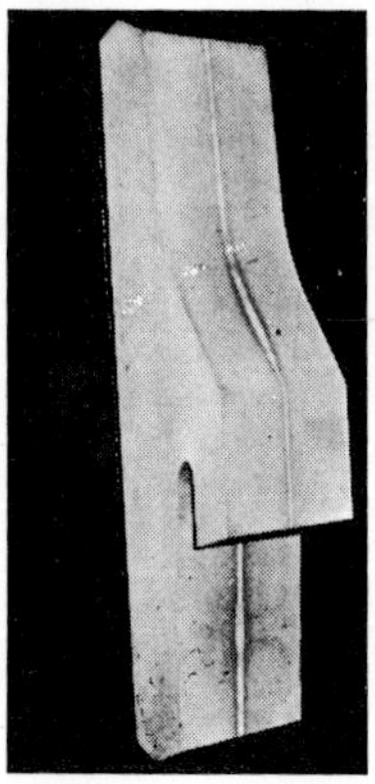

Electron-beam welding thick-walled pressure vessels*

A study of the practical engineering problems involved in applying EBW to a production requirement for uniform-diameter constant wall-thickness vessels

by **J. D. Russell,** B.Sc., A.I.M., M. Weld.I., **K. W. Brown,** T.D., C.Eng., F.Weld.I and **M. J. Adams,** B.Sc. (Eng.), A.R.S.M., The Welding Institute.

Problems will arise from the employment of electron-beam welding in the fabrication of pressure vessels considerably larger than constructions currently using this welding process. The engineering problems concerned with the provision of good joint fit-up, mechanical handling and a vacuum environment have been considered from a design feasibility point of view. Practical work has been carried out where necessary to define unknown limits such as weldability of materials and process tolerance for the inevitable joint gaps.

An examination of the feasibility of fabricating a particular type of large cylindrical pressure vessel with a wall thickness range of 25—75 mm (1—3 in) by electron-beam welding, has been carried out by The Welding Institute. The pressure vessel, of the order of 6 m (20 ft) diameter, is constructed of a high-quality low-alloy steel.

The EB process makes very deep, narrow welds for which the ratio of depth to width is typically 10 : 1. It is usually a single pass process and the highest power machines can penetrate up to 75—125 mm (3—5 in) in steel at travel speeds of 130—500 mm/min (5—20 in/min). EB welding is also a mechanised process, lending itself to automated situations. The low heat input and resulting low distortion, combined with the high rate of joint production, make EB welding very attractive for joining thick sections. On the other hand, the complexity and high capital cost of the equipment, and the need (at present) to weld in a vacuum, are obstacles to its use for welding large structures. In addition, as EB welding is a mechanised process, and because the focused electron beam is extremely small [0.20—1 mm (0.010—0.040 in) dia.], the quality of preweld joint preparation must be very high.

Although very large structures have been welded by the EB process, in general, these have been much lighter than the present proposed application. Mainly aircraft structures have been welded; these are extremely costly components and it is usually possible to justify, either on economic or technical grounds, the use of the highest standards of machining to achieve the fit-up necessary for successful EB welding. The situation is very different in general engineering. The problems of applying electron-beam welding to large thick-walled vessels are mainly concerned with joint fit-up, and the need to weld in a vacuum. These vessels are much heavier, although not necessarily bigger, than anything previously EB welded. It will thus be necessary to manipulate and machine to close tolerances, very heavy pieces of metal, which may distort under their own weight. It is necessary, therefore, to devise means of increasing the tolerance to fit-up of the EB process while still making reproducible welds.

Problems associated with materials

Weldability. Before more costly and lengthy process technology work is carried out, the electron-beam weldability of the materials of interest must be determined. However, in practice EB welds are subject to most of the defects found in conventional fusion welds, including weld-metal cracking, porosity and lack of fusion. Here we are concerned with the weldability of one type of material only, a high-quality low-alloy steel (see Table 1).

Service performance. Welding procedures which produce sound welds with good external profiles may be established with the aid of conventional proving techniques, metallography, radiography, ultrasonics, tensile, impact and bend tests. Electron-beam welding is a low heat-input process, causing minimum metallurgical effects in the parent materials. However, the cooling rate after welding is high and martensitic fusion and heat-affected zones (HAZs) are produced in low-alloy steels. Since EB welding is usually a single-pass process there will be a fully hard structure through the full material thickness, unlike the conventional multipass welds in which each run is tempered by a subsequent pass. The hardness of the fusion and HAZs may, in itself, be harmful as regards crack propagation. However, the narrowness of the hard region, which is surrounded by relatively soft material, may result in satisfactory joint properties.

Engineering and welding technology problems

Joint fit-up. This is a problem in applying any mechanised welding process, since such a process normally depends on the maintenance of an equilibrium in the weld zone. Anything which tends to upset this equilibrium, including variation in the joint gap or the degree of mismatch of the joint, cannot be accommodated as easily in a mechanised as in a manual process. Such fit-up defects can be more easily tolerated in manual welding because:

a) the weld beads are much wider;
b) the operator can change his welding technique or power parameters to accommodate the fit-up

* *A summarised version of Research Report P/47/70. The Welding Institute.*

TABLE 1 Materials

Material	C%	Si%	Mn%	S%	P%	Ni%	Cr%	Fe%	Cu%	Nb%	V%	Mo%	Ti%	Al%
25 mm (1 in) thick air-melted degassed steel	0.11	0.29	0.75	0.003	0.007	4.96	0.54	Rem.	0.05	—	0.065	0.57	0.035	0.016
25 mm (1 in) thick vacuum arc remelted steel	0.10	0.20	0.75	0.003	0.003	4.80	0.54	Rem.	0.09	—	0.05	0.59	0.04	0.008

tolerance.

In addition, it should be noted that because the focused electron beam is generally only 0.25—1 mm (0.010—0.040 in) dia. it can pass through relatively small joint gaps without heating the metal on either side. Since it is difficult to produce very good close butt joints between two large structures, new techniques must be developed to permit welding with the largest possible gaps between faces. These techniques all tend to produce a rather wider fusion zone. Filler metal may then be added to compensate for the gap. Alternatively, new machining and jigging techniques must be developed which will improve considerably on present practice in general engineering. The types of misfit likely to be found in making butt joints between large and heavy cylindrical components are shown in figure 1.

Vacuum requirements. Deep welds may be made with a range of pressures in the working area, depending on the accelerating voltage used. With voltages in the range 30—40 kV the pressure in the workchamber must be around 10^{-3}—10^{-4} mm Hg. At 60—150 kV a pressure of 10^{-1} mm Hg can be used without serious loss of welding performance.

Systems operating with the workpiece at atmospheric pressure have been developed in the USA and USSR with accelerating voltages of 175 and 1000 kV respectively. However, the reported applications of non-vacuum electron-beam welding systems are for material not more than 19 mm (¾ in) thick.

Therefore, the welding of thick-walled vessels must be carried out *in vacuo* at present.

Alignment of beam and joint. The narrowness of the electron beam, and the high depth-to-width ratio of the fusion zone produced, make it imperative that joint faces are perpendicular to the work surface, or at a constant known angle to it. If this is not the case, a beam aimed accurately at the joint on the top surface of the material will miss the joint face at some depth below the surface (see figure 2).

Joint tracking. Once the beam is aligned with the joint, alignment must be maintained throughout the welding operation. In a large component, it is unlikely that the jointline will be straight and the electron beam must be made to follow any deviations of the joint from a straight line. This can be done by moving the electron gun sideways when the angle of the beam with regard to the jointface is kept constant. Alternatively, the beam may be deflected to follow the jointline while the gun is stationary. A low-voltage electron gun, which can be mounted inside the vacuum chamber, can be moved more easily than a high-voltage gun. On the other hand, if the beam is deflected to follow the jointline at the upper surface of the material, there is a danger that the joint will be missed near the lower surface. This danger is lessened by the use of a larger gun-to-work distance. Thus seam-following by beam deflection is best carried out using a high-voltage electron gun, which has a long working distance. The position of the jointline with respect to the electron gun must be sensed in order to correct for jointline deviation. Mechanical tracking systems, using a stylus riding in a grooved jointline, have been used for this purpose and techniques using a scanning electron beam have been developed recently.

Fig. 1. *Fit-up problems.*

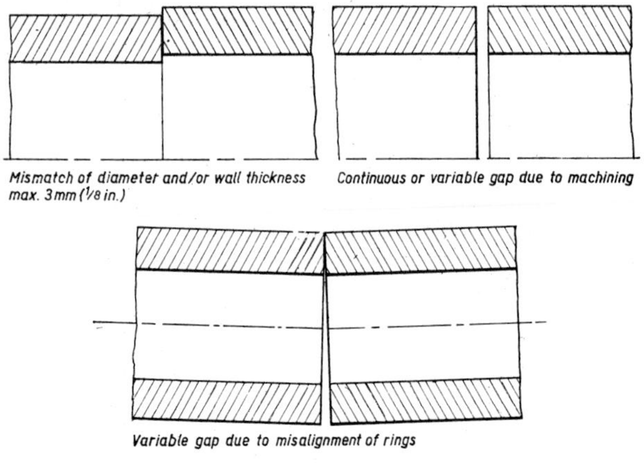

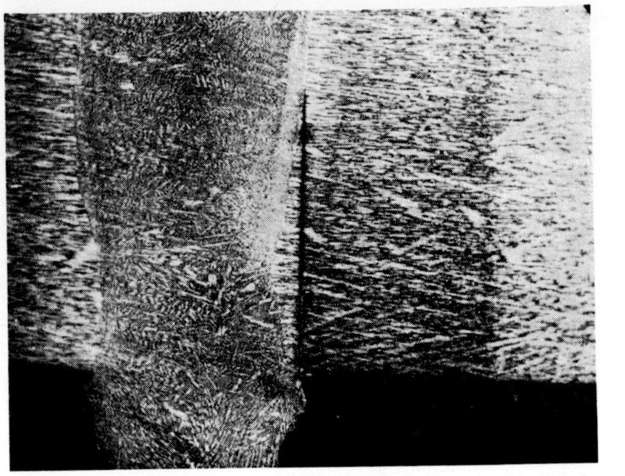

Fig. 2. *Transverse section of electron-beam weld showing lack of fusion over lower 10% of joint face because of misalignment of electron-beam axis with respect to joint face.*

Weld profile control. The extremely high power density of the focused electron beam makes it impossible to use copper or ceramic backing bars to control underbead shape and to support the weld pool. Any normal backing bar is penetrated by the beam and fused to the workpiece. For this reason backing bars used in EB welding must be of similar composition to the workpiece and would normally be removed after the welding operation. The backing materials may be attached by welding, adhesion or by conventional vacuum sealing material.

Experimental work

Welding machines. The electron-beam machine used for most of the experimental work was a Hamilton Standard

Fig. 3. *Wire-feed pass in progress.*

W2, 150 kV, 6 kW machine with a fixed gun for welding in the downhand position.

The workchamber, approximately 900 x 750 x 600 mm (36 x 30 x 24 in) is pumped to full vacuum of 5 x 10^{-4} mm Hg in about 4 minutes. At this power level it is possible to produce welds up to 50 mm (2 in) deep in steel at the low travel speed of 100 mm/in (4 in/min). In order to have some flexibility in choice of welding parameters, our work was limited to 25 mm (1 in) thick material.

During the later stages of the investigation a low-voltage gun and chamber (30 kV, 9 kW) were modified to provide positional welding facilities, and a limited amount of downhand welding was also carried out with this equipment.

Wire feed unit. When necessary, filler wire was added in the conventional way using a small wire-feed unit built at The Welding Institute. The unit incorporates a swivelling device to allow the wire to be fed into the path of the beam accurately. The wire feed nozzle is positioned in front of the beam with the wire emerging at about 5° to the horizontal (see figure 3).

Electron beam sampling probe. This produces a display on an oscilloscope screen, the shape of which is a function of the distribution of electrons, and, hence, energy, within the beam. The probe was used to define the position of the beam focus in relation to the workpiece surface accurately, and also to delineate the beam energy di3tribution whcn thi3 wa3 modificd, in an attempt to reduce the incidence of defects in the root of non-penetrating welds.

Beam spinner. The Welding Institute beam spinner was used extensively in this work. The device is used to move the beam in a small circle [0.25—1.2 mm (0.010—0.060 in) dia.] or to produce similar longitudinal or transverse deflections at up to 100 000 Hz during welding, so modifying the distribution of energy in the welding zone. The spinner was used in this investigation to improve the weld bead shape, to widen the fusion zone, and in an attempt to reduce the incidence of root defects in non-penetrating welds.

Material weldability

The material was a high-quality low-alloy steel, produced by air melting followed by either vacuum degassing or vacuum-arc remelting.

A preliminary assessment of the weldability of these materials was made using melt runs in preference to butt joints in order to avoid inconsistencies due to fit-up and surface preparation. All the experiments were made on 25 mm (1 in) thick material.

The preliminary melt runs and close butt welds which were made following the melt run experiments showed that, as expected from a vacuum-deoxidised steel, porosity was not a problem. Full penetration welds were free from cracking but wide, blind welds showed a tendency to produce solidification cracking. The presence of solidification cracks in material of this quality was not expected and may be due to adverse segregation effects peculiar to this steel under the welding conditions used for producing wide blind welds. Cracking was only found in welds which had been made deliberately wider than normal by beam oscillation or defocusing. The majority of the work was carried out on the vacuum degassed material and only a limited number of experiments was performed at a later stage on the vacuum-arc remelted material. It was noted that cracking only occurred in the vacuum-degassed material although the two materials were apparently of almost identical composition, the only obvious difference being in the number of inclusions present. Both materials contained manganese sulphide, alumina and unidentified duplex inclusions, the vacuum-arc remelted material containing much lower quantities of each type.

Another defect associated with blind welds was found to be a series of cold shuts in the weld root — ' necklace defect ' (see figure 4). This has been found in the majority of blind welds and, therefore, must be regarded as a likely defect. It is a serious difficulty since one of the methods of overcoming the vacuum problem involves the use of a blind weld (see figure 5). It was noted that necklace defects have a greater tendency to form in the root of sharply tapering welds and were sometimes completely absent in welds with rounded roots. Unfortunately, such welds are normally wide and, as mentioned above, prone to solidification cracking. Various techniques, including defocusing and beam oscillation or spinning over a wide range of frequencies and amplitudes, were investigated. On the high-voltage equipment, these techniques failed to produce welds consistently free from necklace defects. The 30 kV welding machine has an electron beam with hollow, as opposed to the normal gaussian, energy distribution and it was found that this type of beam is capable of producing

Fig. 4. *Nooklaoo dofoot (x23).*

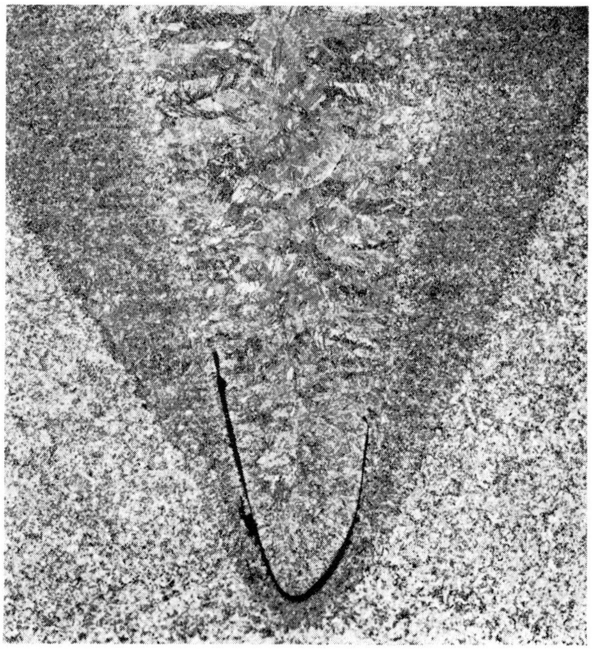

defect-free welds with rounded roots.

A further difficulty encountered in welding this steel was in producing good top and bottom bead shapes at the same time. It was shown that good top beads could be produced by sacrificing the quality of the underbead and *vice versa.* Therefore, the use of a blind weld into a backing material is a valuable method of simplifying the welding procedure to prevent weld drop-through. If backing materials have to be used then it is necessary to ensure that all the root defects are in the backing member which is subsequently removed from the fabrication. The problem of weld drop-through, which occurs when good underbeads are produced, applies to melt runs and close butting joints and is aggravated by the presence of any joint gap. Therefore, if substantial joint gaps are to be tolerated, it would probably be necessary to have an underbead support which could take the form of an arc weld deposit or pre-placed backing bar.

Effect of joint fit-up

Even using the best possible joint preparation methods it must be accepted that some degree of joint gap is to be expected with the large structures under consideration. We aimed, therefore, to predict the maximum possible joint gap which could be tolerated, consistent with acceptable weld quality.

As mentioned above, weld top-surface depression is produced when welding conditions are adjusted to produce a good underbead profile. This necessitates the addition of filler wire using a two pass procedure. The second pass is a non-'keyhole' type in order to avoid the formation of 'necklace' defects. An alternative procedure which uses a conventional single pass weld into either a backing bar or arc weld deposit was investigated. This produced satisfactory welds without top bead sink in close butt joints, provided the 'necklace' defect was arranged to be in the backing material which would subsequently be removed. When these experiments with both unsupported and backed welds were extended to investigate joint gaps up to 1 mm (0.040 in), it was found that top bead depression occurred with both techniques. Thus, if joint gaps are encountered in practice, a double-pass procedure will probably be necessary. It would be possible on backed welds with small gaps to eliminate the weld sink by filler-wire addition during the first pass. However, where substantial gaps are involved, 0.5 mm (0.020 in) upwards, inconsistent results were obtained. In most cases filler wire was fed so as to be coincident with the electron beam at the plate surface, whereas it would probably be preferable to feed wire into the gap. This can only be achieved if a consistent gap without periodic bridging exists. This can be produced if the first pass is made without filler wire, producing a trough of consistent width into which filler wire can be fed during a second filling pass (see figure 6). A 'hot wire' procedure can be used to pre-cast material into the joint gap or can be fed into the pool during welding to improve deposition rate and fusion.

An alternative procedure for gap filling has been developed using a metal powder mixture of parent-metal composition. This procedure is of greatest value for laps large enough to allow dense compacting to be carried out on the pre-placed mixture before welding. The investigation covered gaps of 0.4—0.9 mm (0.015—0.035 in) with a

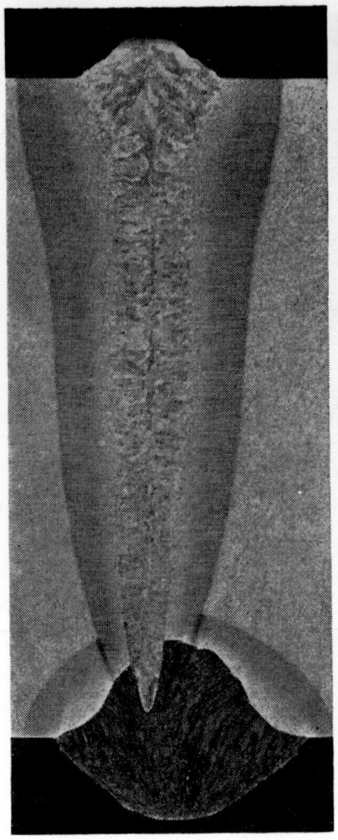

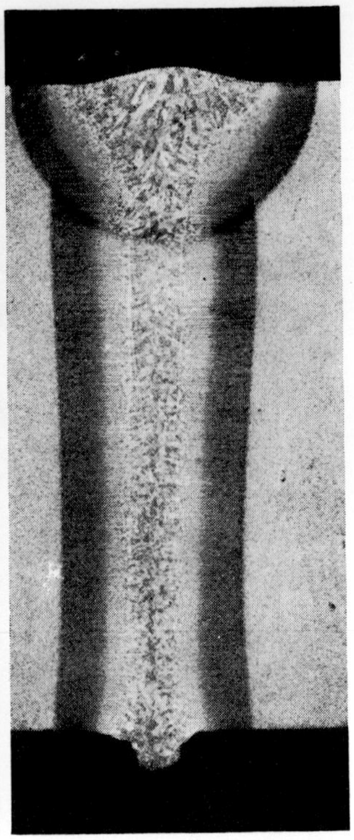

Fig. 5. *(left) single-pass electron-beam weld into MIG backing weld (original, x4).* Fig. 6. *(right) two-pass electron-beam weld with filler-wire addition, 0.38 mm (0.015 in) gap (original, x4).*

backing bar and it was found that the powder procedure still produces weld sink (owing to the low density of compacted powder), requiring a second shallow filler pass using filler wire or, possibly, powder fed from a hopper (figure 7). Pre-placing and compacting the powder are possible sources of inconsistency and it would probably be worthwhile investigating the use of a hopper-fed powder system. Although the use of metal powder mixture permits the filler-metal composition to be adjusted easily, it also introduces a possible source of trouble because of inhomogeneous and incorrect mixtures.

The following ranges of gap tolerance have been drawn up together with suggested methods of welding.

Range A: 0—0.25 *mm* (0.010 *in*) *maximum clearance*

This fit-up situation can be welded satisfactorily without backing material using a two pass procedure, the second pass being made with filler wire addition. Widening of the fusion zone by beam deflection would be an advantage. An alternative procedure would be to add filler during the first pass using a backing bar, in which case the weld could be completed in one pass. Orbital welding could probably be carried out satisfactorily.

Range B: 0—0.5 *mm* (0.020 *in*) *maximum clearance*

These require transverse beam deflection or beam spinning in order to widen the fusion zone. Two passes would be required, the first being made without filler and the second with filler wire or metal powder. Metal powder could also be pre-placed in the joint before welding in order to reduce the amount of filler required subsequently.

Range C: 0—0.9 *mm* (0.035 *in*) *maximum clearance*

This range requires similar techniques to B. Pre-placing of metal powder is necessary to reduce the amount of weld sink. A second pass is applied using filler wire or metal powder. With larger gaps it is possible to provide more effective compacting of the metal powder. It is

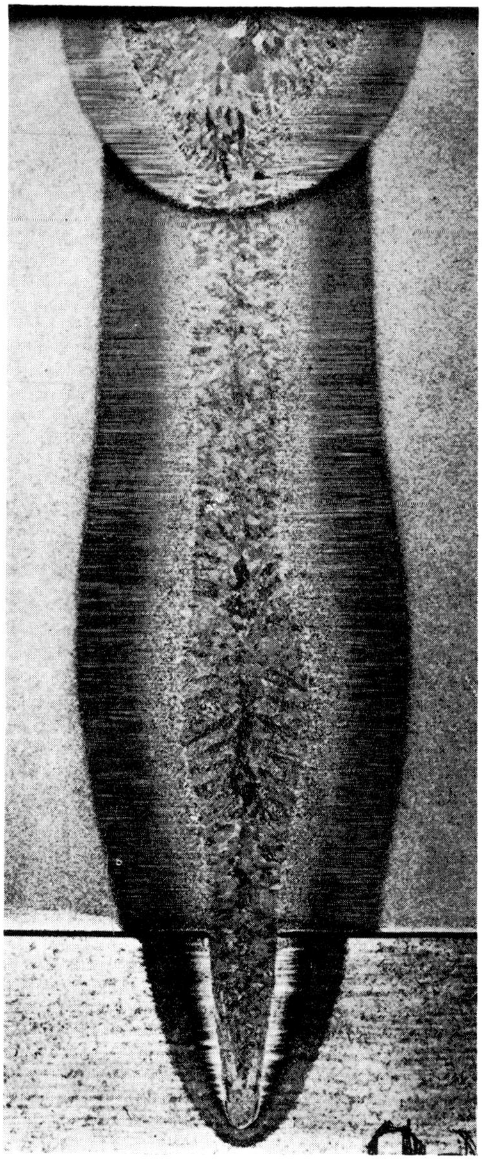

Fig. 7. Two-pass electron-beam weld with metal-powder addition, 0.79 mm (1/32 in) gap (x5).

doubtful if welding in the overhead position could be carried out with gaps of this width.

The general conclusions on welding with joint gaps were that beam deflections large enough to provide a wide weld pool to allow adequate filling of the joint should be used; this is also aided by low travel speeds of 100—250 mm/min (4—10 in/min).

Although the use of metal powders has been found to be very effective, it is probable that filler-wire procedures could be developed to accommodate the large gap ranges. The success of welding in the horizontal/vertical position with close butting joints (reported below) suggested that vertical upwards welding could also be carried out and might be more tolerant of gaps.

Although pre-placed powders have been used so far, it is probable that powder-hopper systems could be developed which would feed sufficient filler into the weld pool to allow complete filling in one pass; this would be confined to the downhand position only. The powder-paste system would be used for positional welding.

It was noticed during the gap experiments that the gap closed ahead of the beam, although shims and tack welds were placed at approximately 100 mm (4 in) apart along the testpiece. For instance, an original cold gap of 0.5 mm (0.020 in) had closed to 0.25 mm (0.010 in) just ahead of the beam. after switching off the beam and cooling the sample. It therefore appears possible that a thermal technique could be developed for gap closing ahead of the weld.

From a consideration of the metal flow during welding giving rise to sunken top beads, it is apparent that downhand welding might not be the best welding position from the weld profile point of view. Accordingly, a low-voltage gun was mounted on the side of its chamber in order to facilitate welding in the horizontal/vertical position. Few welds were made, but those produced had superior top and bottom beads to those produced in the downhand position. Radiographic and metallographic examination showed that the welds were free from defects. From these preliminary experiments it appears likely that welding in the horizontal/vertical position, and probably also vertically upwards in the 3 o'clock position, would be more attractive from the weld shape point of view than conventional downhand procedures.

Service performance

Within the scope of the present study it has only been possible to carry out simple tests for the determination of weld properties, including tensile, bend, hardness and V-notch Charpy tests. The test specimens in most cases were taken from welds produced with close butt joints but bend tests were used extensively to assess joint quality using various filler wire and powder filling techniques.

The test results are recorded in Table 2.

The V-notch Charpy tests carried out at —18°C and 196°C on transverse samples with the notch in the weld metal show promising results, comparable with parent material (Table 2 and figure 8). At —18°C, fracture occurred in a ductile manner through the plate material after passing at right angles to the notch through the weld metal *via* the shortest path. Thus little information on weld-metal properties was obtained at this temperature and, therefore, repeat tests were carried out at —196°C. At this temperature fracture occurred in a brittle manner through the weld metal at 13.6—16.4 J (10—12 ft lbf) which is again comparable with parent-plate performance. The nature of the failure at —18°C is encouraging from the point of view of performance in service and in large-scale fracture tests. 25 mm (1 in) wide plate tensile specimens produced failure in the parent plate as would be expected from the hardness values of weld metal and parent material (see figure 8 and Table 2). Conversion of hardness values into tensile strength indicates 1050 N/mm^2 (155 000 lbf/in^2) for the parent material, which is in good agreement with the tensile test results: 1190 N/mm^2 (176 000 lbf/in^2) for weld metal and 1310 N/mm^2 (190 000 lbf/in^2) for heat-affected zones. The weld area, therefore, probably overmatches the parent-material strength by approximately 30%. The bend test results from several joints, including autogeneously welded single-pass close butt joints and wide two-pass welds with filler addition, were satisfactory. However, on wide welds produced with 0.9 mm (0.035 in) gap joints and pre-placed metal powder, the bend test samples failed in the weld metal after approximately 15°. This is probably a result of incorrect chemistry of the weld deposit resulting in

TABLE 2 Mechanical test results

	Tensile test at room temperature				Hardness V.P.N. 20 kg load			V-notch Charpy J (ft lbf)			
Material	0.1% proof N/mm² (lbf/in²)	Tensile strength (lbf/in²)	Location %	Elongation %	Weld metal	HAZ	Average parent plate	−18°C	−196°C	−180°C	RT
25 mm (1 in) vacuum degassed steel transverse weld test	924 (134 000)	1054 (152 800)	parent plate	21% on 100 mm (4 in) gauge length	376-383	401-415	327-329	170, 172 (125, 127)	14, 17 (10, 12.5)	—	—
Vacuum degassed parent plate	—	—	—	—	—	—		152-247 (112-82)	—	14 (10)	—

low weld-metal ductility, since metallographic examination revealed no weld defects. A further factor was the lower hardness of the weld deposit than either HAZ or parent material, owing to the fact that carbon was not added as a constituent of the filler powder.

NDT techniques

Although gross defects are unlikely to occur in the materials of interest when welded with the correct procedures under careful control, it would still be necessary to carry out radiographic examination, at least in the early stages of production.

Deliberately defective welds containing porosity and missed joints have been produced in order to assess the relative merits of radiography and ultrasonics. The joints were missed partially at the root by positioning the plate at angles up to 4° with respect to the beam, at a working distance of 150 mm (6 in).

Examination of these welds by conventional radiography revealed areas of porosity and lack of fusion lines.

Ultrasonic examination using a 5 MHz shear-wave probe at 45° was found to be a more effective method of finding lack of fusion defects and it was also possible to make an estimate of their size with this technique. The ultrasonic technique would need to be developed to define the smallest defect which could be found with confidence; this work should be backed up with appropriate tests to determine the critical defect size for the material and application. Visual observation of the under surface, together with dye penetrant, will detect either lack of fusion or lack of penetration defects, provided careful weld-bead dressing is carried out.

Engineering and production considerations

A general appreciation of EB welding requirements will be necessary to obtain a balanced comparison of the EB welding process with, say, multi-pass submerged-arc or manual metal arc welding procedures.

Equipment

Vacuum pumping machinery, EB welding equipment and ancillary devices (such as for gun manipulation, traversing, seam tracking, etc.) are commercially available to meet the technical specification of the present proposed application. However, the user with engineering facilities available may wish to undertake the construction of the vacuum chamber himself to restrict capital outlay, and possibly save time as well as money.

Design

Heavy engineering firms not using EB welding, are unlikely to have design staff or welding engineers/metallurgists fully conversant with the process, but specialist information can readily be obtained from equipment manufacturers and The Welding Institute, and experimental facilities are available to establish the feasibility of any new project. In a short time, a firm's own engineering staff would be able to handle general EB welding design problems.

Vacuum environment

The interelectrode region of the EB gun must always be maintained at a hard vacuum of 10^{-4} mm Hg to protect the gun filament, to provide insulation and to increase the electron mean free path. In some instances the workpiece may also be required to be contained at this pressure. This is not essential and for many of the large fabrication applications, a soft vacuum in the range 10^{-1}—10^{-2} mm

Fig. 8a. *V-notch Charpy testpiece, tested at —18°C notched in weld-metal.*

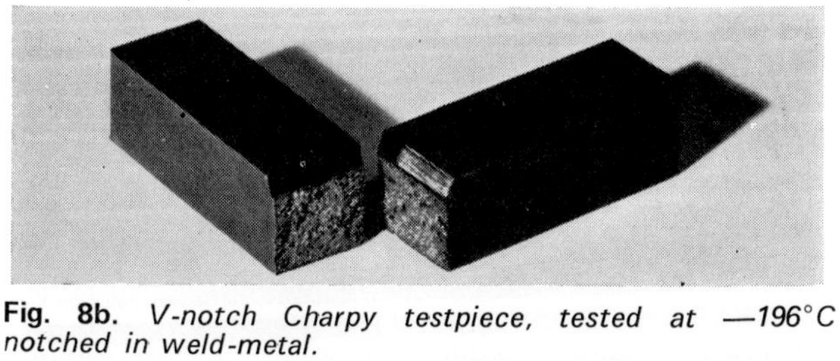

Fig. 8b. *V-notch Charpy testpiece, tested at —196°C notched in weld-metal.*

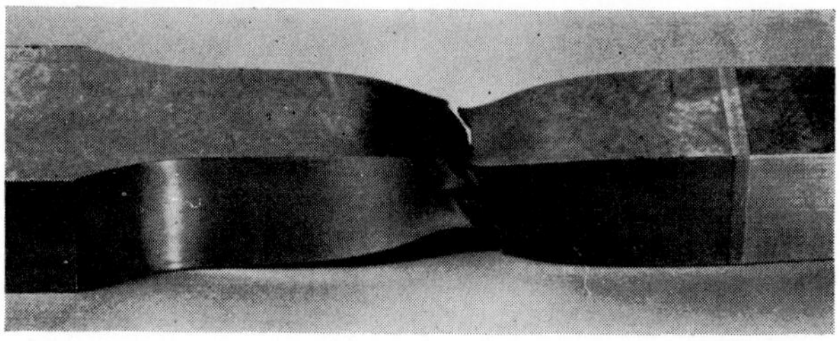

Fig. 8c. *Tensile test specimen in 25 mm (1 in) thick steel.*

Hg will be adequate and is much easier to pump and maintain. The workpiece may be either fully enclosed in a hard or soft vacuum chamber or protected locally by a hard or soft vacuum contained by a static or sliding seal system.

Distinct types of vacuum pumps exist and are characterised by high pumping speeds in particular ranges:

Pump type	Range for high pumping speed (mm Hg)	Maximum capacities commercially available (m^3/h)
Mechanical rotary	760—1	850
Mechanical booster	1—5 x 10^{-2}	50 000
Vapour booster	10^{-1}—below 10^{-4}	70 000
Oil vapour diffusion	5 x 10^{-2}—below 10^{-4}	150 000

Vacuum sealing

The capacity of the vacuum pumping system must be sufficient to evacuate the welding chamber to the required pressure within the allowable process time, and to maintain that pressure against the rate of outgassing (release of absorbed or occluded gas) of the chamber/workpiece surfaces and leakage from outside. Sealants are available which may be painted on the surface of the chamber to seal pinholes or porous walls and reduce outgassing. The sealing against leakage from outside will present the major problem and considerable work has been carried out in developing suitable sealing materials. The efficient functioning of seals must be unquestionable over a reasonable working period without the need for frequent maintenance or replacement. The properties of seals must include resistance to heat, radiation and chemical corrosion. Demountable, re-usable seals (chamber or chamber-to-workpiece closure) will be particularly vulnerable to damage. Seals used for chamber lead-ins (providing low voltage electric or hydraulic power for actuators, etc., or angular/linear shaft movement) have all been in use for many years without giving trouble. High-voltage kinetic cable lead-ins for high-power guns still need development and tend to impose limitations on some moving gun concepts.

For many large, totally enclosing chambers it is considered worthwhile to incorporate a guard vacuum system into the seal configuration for the chamber joint faces necessarily separated for workpiece removal. This entails providing an evacuated volume between an inner and outer seal, thereby considerably reducing the leak rate, and offering means of testing the flange sealing by quickly evacuating this enclosed pocket. It will be difficult to guarantee that all demountable mating flange faces will retain their original close limits in heavy duty use, so the use of solid gasket material is not the first choice as it is only capable of accepting about 25% portion compression. Inflatable seals are therefore attractive for the sealing of gaps with a wide range of tolerances. Sealing techniques for localised sliding seals present a greater design problem.

Joint fit-up

Although it is little more than a decade since the electron beam was applied to a commercial welding application, the initial problems of clinical cleanliness, precision joint fit-up, alignment and need for overall hard vacuum have been overcome to an extent that make its proposed use in heavy engineering realistic. However, meticulous joint-gap cleaning is still a vital requirement, and visible dirt and metal turnings, if left on the faces to be butted, will negate efforts made to ensure a close fit-up. All contaminants physically adhering to the surfaces must be removed. With newly machined surfaces, solvent cleaning will probably suffice if carried out before surface deterioration becomes apparent, with temporary protection for the cleaned surface until welding. The use of hot or cold filler-wire or metal-powder techniques permit gaps of up to 0.9 mm (0.035 in) to be tolerated.

It is proposed that the components be first rough-machined to overall dimensions sufficiently oversize to permit parallel-gap re-machining *in situ* after the mating components have been assembled and faired. After parallel-gap machining, the sized components would be cleaned and re-positioned to butt joint faces to within Class A fit-up, if such is required. Machining may be carried out using single-point cutting tools, with the workpiece or cutting tool traversing. Alternatively, rotating milling cutters, slitting saw/grinding wheels or abrasive disks could be used, depending upon the application and amount of material to be removed.

Magnetic effects

The electron beam can be deflected by magnetic fields. Therefore components to be welded must be demagnetised to an acceptably low level before welding. This level depends on the accelerating voltage in use. Alternatively it is possible to 'stiffen' the beam against magnetic fields by using the deflection coils to maintain the beam aligned with the electron-gun axis. This system, however, is able to cope with external fields only and cannot correct for beam deflections as the beam passes through the workpiece.

Discussion

The potential advantages of electron-beam welding in this particular application are:

1. to reduce welding time in comparison with conventional multi-pass arc welding procedures;
2. to reduce incidence of significant defects and, therefore, repair work;
3. to produce weld properties as good as, or better than, those associated with arc welding without the use of preheat and/or post-weld heat treatment.

The desirable characteristic features of electron-beam welding—deep narrow welds, low distortion and high welding speeds—are offset in the context of large structures by the high capital cost involved and the many heavy engineering problems posed. Our programme considered these problems from a practical fabrication viewpoint and used laboratory investigations to define such unknowns as weldability of material and joint fit-up requirements as a preliminary study, in order to justify the more costly development programme leading to a production facility.

It has been established that sound welds can be produced in the particular steel under investigation. This does not mean that defects would be produced if the specified procedures were not adhered to. It is essential to develop the welding procedure to suit the application and situation, e.g. the extent to which gaps and misfits have to be accommodated.

It is inevitable because of handling and machining difficulties, that some degree of joint gap and mismatch should occur. The welding procedures provisionally developed are able to cope with gaps up to 0.9 mm (0.035 in) in 25 mm (1 in) thick welds. However, these procedures become more complicated as the gap increases and in order to simplify the welding process every effort should be made to produce the best possible fit-up, i.e. of the order of 0.25 mm (0.010 in) maximum gap.

It is considered that the proposed parallel gap machining methods would be capable of achieving this figure for the majority of fabrication procedures developed. Where larger gaps are inevitable it would be necessary to add substantial amounts of filler material either in the form of metal powder or filler wire. The ability to produce good joint fit-up would, in addition to simplifying the welding procedure, give more freedom of choice of welding position, i.e. downhand roll or vertically upwards welding, with either workpiece rotation or orbiting gun.

The amount of joint gap which has to be accommodated by the welding process affects the occurrence of defects owing to either inadequate filling of the joint or weld cracking. Where the use of a backing bar is required, the danger of 'necklace' defects occurs. The welding procedure must guarantee that such defects will occur in the backing which, subsequently, will be removed for inspection of the weld root.

The work reported here has been carried out on 25 mm (1 in) thick material, although the principles established are applicable to greater thicknesses. The engineering design study has shown that a number of different fabrication procedures could be devised based on electron-beam welding. Two procedures have been selected to cover both the welding of sub-assemblies or relatively short vessels in a totally enclosing chamber, and an adding-on procedure for a vessel of any length using small cylindrical sub-assemblies. One probably relies least on the development of new electron-beam and vacuum techniques but would involve a fairly high capital expenditure on the large vacuum chamber, whereas the other requires extensive development of local and sliding seals with the sliding part confined to a relatively small diameter near the axis. Various alternative schemes are possible using large chambers with horizontal assembly and local and sliding seals clamped externally. The choice of procedure is thus very wide and is affected by several factors, for example: size of vessel—length and diameter; standard size or range of sizes to be considered; number of vessels required per year; quality requirements; type of weld—longitudinal or circumferential; and the need for other welds such as nozzle attachment.

In all cases where orbital guns are employed it would probably be necessary to use intermediate voltage guns in the 60 kV range to minimise the electrical problems.

The basic choice of welding gun lies between 150 kV and 60 kV systems with powers in the range 25—30 kW, in order to provide some latitude on choice of parameters for welding 25—75 mm (1-3 in) thick material. In all applications involving large structures, it would be necessary to provide a joint-following device which could take the form of a mobile gun following a stylus or a beam-deflection system with fixed or mobile gun, provided a long working distance was used and large deflections were not required.

Although it is possible to carry out electron-beam welding with the weld area at atmospheric pressure, there is a serious loss of performance. Interaction between gas molecules and electrons causes degradation of the beam, with the result that 3—4 times as much power is required for a given penetration depth compared with in-vacuum welding. Also, even at very short working distances of the order of 6 mm ($\frac{1}{4}$ in) the effective beam diameter is degraded and considerably larger weld-widths result. Therefore, for the materials and thickness involved in this investigation, the present generation of out-of-vacuum electron-beam welding machines cannot be considered. In the future, as more power becomes available and weld depths of several inches are possible , there will still be doubts as to the effect of the increased weld width on cracking tendency and the ability to control the weldbead shape.

Vacuum pumping facilities would be designed to suit the chamber size and anticipated leak and outgassing rates, using rotary and mechanical booster pumps for the partial vacuum areas and oil diffusion pumps for the electron gun area. It is usual to provide full 10^{-4} mm Hg vacuum in the weld and gun areas where gun movement is involved, but with the large structures involved in this case it is considered that a mobile gun housing pumped by a small diffusion pump suspended from gimbals would be preferable. From a consideration of the weldability of materials, machining and handling methods, and vacuum techniques it may be concluded that electron-beam welding could be adapted for use in heavy-vessel construction, provided sufficient effort is put into developing the doubtful areas after decisions have been made on the preferred fabrication procedure, e.g. whether orbital welding is required.

Within the limits of this investigation it has been shown that tensile and bend test results are satisfactory on 25 mm (1 in) thick material of the composition examined. The Charpy values at —18°C and —196°C are comparable with those of the parent plate, which is encouraging. However, no assessment of service performance can be made without first carrying out more realistic tests, such as COD and K_{1c} tests in order to gain a better understanding of the performance of the narrow basically single-pass welds under the expected service loading conditions. Correlation of the minimum defect size detectable by radiography and ultrasonics with these test results, would be a vital factor in the acceptance of electron-beam welds in any situation where the possibility of brittle failure exists.

The solution of the technical problems does not, however, necessarily mean that electron-beam welding could be successfully introduced into a heavy engineering situation because of the many environmental problems involved, for example:

1. recruitment of new labour or re-training of existing personnel;
2. disruption of present shop functional layout;
3. need for more accurate workmanship and equipment and, possibly, heavier lifting gear;
4. new forms of material to be handled;

5. unusual health hazards;
6. more exacting standards of cleanliness and protection of machined surfaces required.

If these and other local problems can be resolved and sufficient incentive exists to develop a large scale electron-beam welding facility, then it is suggested that a development period of five years would be required from a given welding requirement to production welding of large cylindrical structures, at a cost of between £200 000 and £500 000 depending on the size of the vessel involved and the procedure selected.

Conclusions

This feasibility study has defined the problem areas in applying electron-beam welding to large cylindrical structures. Before electron-beam welding can be applied to pressure vessels in general, it is necessary to carry out further development and evaluation work covering the following aspects.

1. Techniques for welding thicker sections and development of higher power gun systems.
2. Tolerance for joint gaps, especially for sections greater than 25 mm (1 in) thick.
3. Assessment of the weldability of a wider range of materials.
4. Evaluation of joint properties for relevant materials and service performance.
5. Machining methods for good joint fit-up on large thick-section components.
6. Positional welding procedures.
7. Seam tracking systems.
8. Detailed consideration of vacuum requirements and sealing techniques.
9. Assessment of health hazards.
10. Economic comparison of electron-beam welding procedures with alternative fabrication methods.

Acknowledgement

We gratefully acknowledge the valuable assistance of Mr. A. N. Taylor and Mr. R. Stearn.

Electron beam welding — now doing the unbelievable

With the help of computerized control and material handling systems, EB welding makes a place for itself in automotive manufacturing

by TED B. JEFFERSON, *P.E., consulting editor*

Probably the most astonishing examples of electron beam welding are those done on the front and rear gear carriers of the Chrysler A904 transmission at the Chrysler Corporation Transmission Division plant, Kokomo, Ind. This welding is done on a Sciaky CV welder equipped with a sliding seal. So productive is this machine that the engineers equipped it with a six-station feed, a 12-station load table, and an automatic unloader. A computer controls loading, orients the work, aims the beam, controls the welding, and unloads the part after welding.

The first weld on the front carrier is an annular butt weld of a stamping in aluminum-killed 1008 steel plate to a 1½-inch- (38 mm) diameter cylindrical hub of 1030 steel. Production: 800/h. The weld penetrates 100 percent and does not project above the plate. To complete the frame the machine makes four ¾-inch-long (19 mm) welds that fasten a fine-blanked bent plate, also of aluminum-killed 1008 plate, to the plate carrying the cylinder. Again the penetration is 100 percent and the welds do not project above the plate surface.

The dial feed table and the computerized welder together make a proved and versatile combination that saves manufacturing cost and yet provides parts of high and assured quality.

The Sciaky computer control system with this E-B CV equipment includes a Data General Nova unit. Input is by punched tape or through on-line teletypewriter. The welder can accept other types of controllers or NC control. The system controls all parts positioning and translations, scanning, monitoring, and weld setting functions. The sequence of operations for the gear carrier welds be-

EB WELDS GEAR CARRIER

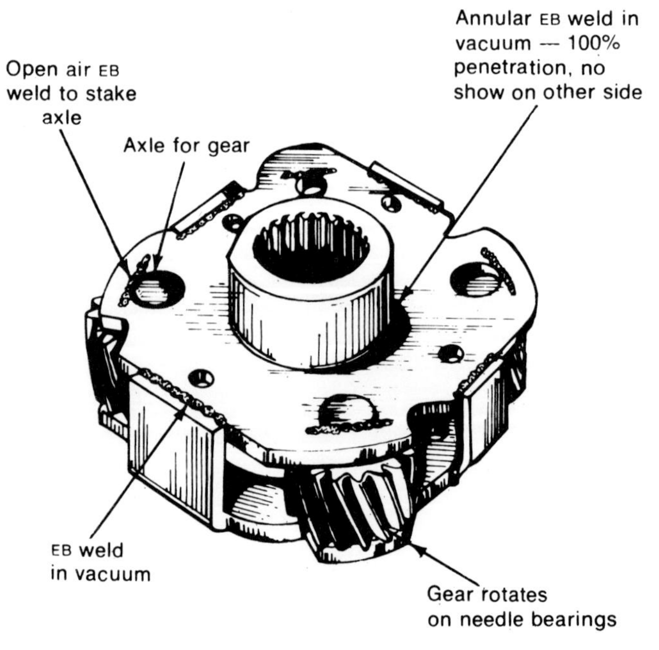

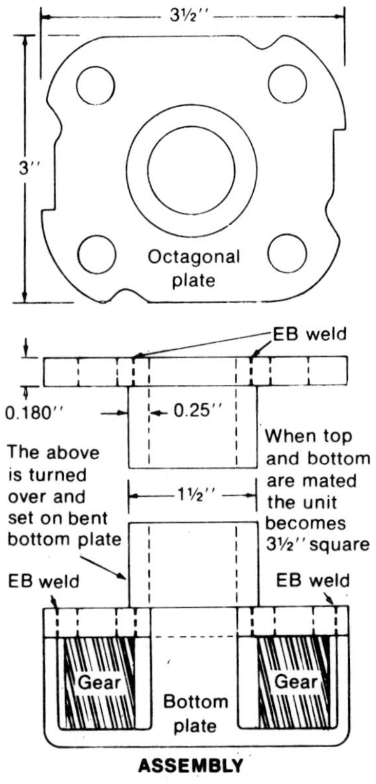

Production line for electron beam welding of automobile transmission parts. Welders are automatic Sciaky CV type.

Twelve-station load table with welder controls.

gins at the load station. The load table has 12 stations. It is synchronized to feed parts to the main dial-feed table, which has a sequence of six dwell and six index periods. An index period occurs between each dwell period, and it is initiated by the operator as he pushes the button to start welding. Parts are manually loaded at the load station. All other operations are automatic.

In one example — the front gear carrier — a 3-inch (76.6 mm) octagonal flat stamping 0.180 inch thick (4.6 mm) with a 1½-inch-diameter (38 mm) hole in the center is to be welded to a 1½-inch-diameter (38 mm) hub 1 inch (25 mm) long, 0.25-inch (6.5 mm) wall with a machined shoulder at one end which mates with the plate when fitted in the hole. The flat plate mates with the cylindrical hub, making a step-butt joint, and the electron beam joins the two parts in an annular weld. The advantage of the two-piece construction: material savings and cost savings over machining from one solid piece, and cost savings over a forging or casting.

Here's the sequence

During the welding operation this sequence takes place on the main feed table:

Station 1, load. The operator loads the machined hub into the tooling nest. A locating pin fits into the hole of the hub. Then the operator loads the flat stamping into the tooling nest to fit onto the hub. The inside and edge surfaces of the stamping sit on

PLAN VIEW OF ELECTRON BEAM WELDER AND LOADING SYSTEM

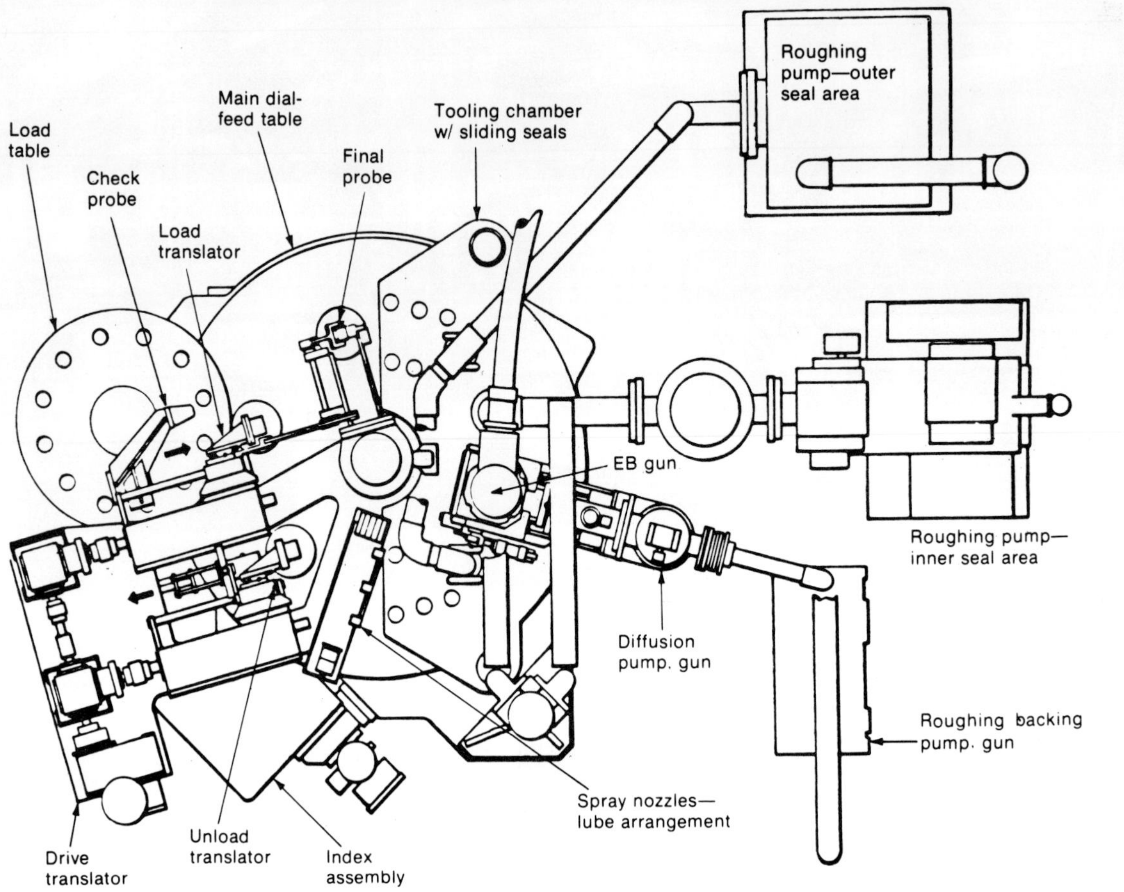

locating blocks. The operator pushes a button, and the machine runs automatically by using its built-in grasping, shifting, and placing devices. For this assembly, the pieces are positioned on a rotary loading table that feeds them to a loading translator equipped with a mechanical holding device that lowers 1¼ inches (32 mm) to grasp the part. The loader lifts the part up 1¼ inches (32 mm), carries it 8 inches (203 mm) transversely, then lowers the part 5 inches (127 mm) into the tooling cavity. After release, the loader lifts 5 inches (127 mm) and returns over an 8-inch (203 mm) path to a stop position to accept more parts. At the same time, an unloading translator at station 6 performs the same operation to remove the welded part.

Station 2, position. The forward motion of the loading translator mechanically actuates a probe that pushes the work into proper position for welding. A limit switch, mounted on the probe, does not make contact if the station is empty and stops the indexing sequence.

Station 3, prepump. The part in the tooling nest cavity, now under the sliding seal plate, is evacuated to a rough vacuum of 500 microns (66.7 Pa) or less.

Station 4, weld. The part moves under the electron beam gun. At this point the tooling nest cavity evacuates to 50 to 100 microns (6.7 to 13.3 Pa). The tooling nest rises, and spring-loaded clampdown fingers in the upper tooling housing press the stamping against the hub to hold the parts in exact alignment. The EB gun scans the joint by oscillating across it at 60 Hz/sec with a low energy beam, locating the joint electronically to confirm its location. Scanning power is 10 percent of welding power. The weld joint line reflects differently than do other areas, a difference discerned by a circuit in the controller. The controller remembers where the weld joint is, and with the scanning done, the controller directs the electron beam along the path. Should the scanner find the joint to exceed a set amount to either side of the nominal position, the machine would shut down. A 60,000-volt, 125 mA elec-

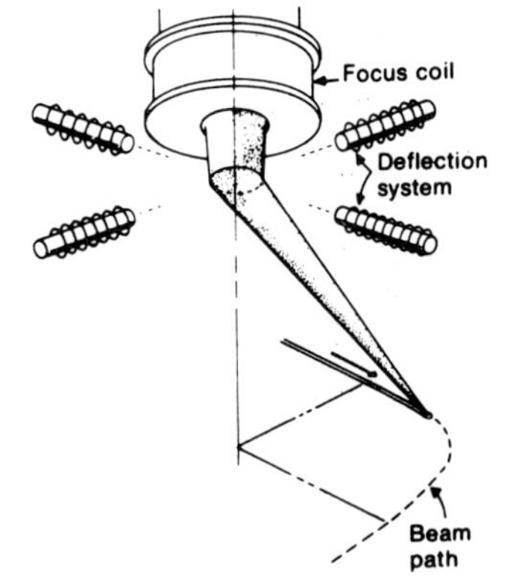

Deflection coils move electron beam along complex weld path.

New bottom delivery EB welder (left), fitted with Mark VII controller, will heat-treat as well as weld.

Unloading setup (below) is automatic.

tron beam moves along the joint to make the weld. With the weld completed, the lift cylinder retracts, withdraws the tooling nest from the tooling housing, and moves it into the tooling nest cavity.

Station 5, idle.

Station 6, unload. Each welded unit lifts out of the tooling nest and, released from the translator fingers, is placed onto unloading guide rails. The next cycle of the unloader causes mechanical ejecting fingers to push the piece onto an unloading chute. As the piece moves, a wand-actuated switch indicates all clear, and the cycle continues. If the wand switch does not signal, the operator knows that the work has not been properly removed. The machine will stop.

The gear cage is completed by welding the assembly described above to a bent plate stamping in a smaller welding sequence. This operation makes four ¾-inch (19 mm) straight-line welds. The spacing tolerance between the top and bottom stampings is held to 0.005 inch (0.13 mm). Assembly of four planetary gears, each with 18 needle bearings, on pins within the cage, completes the carrier. Open-air EB welds hold the pins in place. The welds made in vacuum are 0.015 inch (0.38 mm) wide; the open atmosphere weld is 0.020 inch (5.1 mm) wide, even with the EB gun within ⅝ inch (16 mm) of the target.

Similar procedures weld the rear gear carrier. Here again, the success of the EB operation is due to the built-in actions of the total welding machine. The go-no-go monitoring system, for example, automatically and continuously monitors the critical functions of the welder to check machine operation. This system measures the welding variables, namely, the beam power and current, the vacuum level of the EB gun chamber, and the focussing current. The engineers may select and set a high or low tolerance band for the variables. Should the variables fall outside the limits, a warning light will show which variable is out of tolerance and whether it is high or low. At the end of the weld sequence, the welder will lock out. It will not start up again until the out-of-tolerance variable is corrected.

A movable beam

Much of the versatility of the EB welder comes from the beam deflector. By following a program, the deflector makes the beam follow a prescribed welding path without outside help. The beam axis normally passes through the center of the electron gun and the magnetic focus coil. An array of two or four magnetic coils deflects the beam. The deflecting coils are placed below the focus coil around the beam path. Applying a selected amount of current to each coil deflects the beam as desired.

The machine also offers time-variable patterns. Signal frequency sets the speed of the beam along its path; voltage sets the welding pattern.

More to come

In addition to the five slide-seal machines that have been running for 5 years, Sciaky recently installed Chrysler's new dial-fed machine, one fitted with a bottom delivery. This machine satisfies Chrysler production requirements for welding components for a new transmission, even though a bottom delivery machine has a slower cycle. Its pump-down time is longer but the seal is easier to maintain.

This new machine, unique in many ways, offers capabilities mostly the result of the Mark VII welder control. It brings a new generation of controls for soft vacuum EB welding that offers the productivity of an advanced minicomputer-based control system that gets away from a lot of hardware. The reliability of the computer has been proven out in telephone switchgear. The minicomputer displays all machine settings on a CRT, so the operator — really more of a monitor in this case — can check machine settings and prevent blunders before they happen. This equipment does double duty at the transmission plant: it is part of the first commercial EB heat-treating installation. ■

EB welding takes on the catalytic converter

Faced with the prospect of making a new product out of a new alloy, AC Spark Plug is putting its faith in nonvacuum EB welding.

General Motors has a name for it. They call it the turtle. And, GM is mass producing these turtles like there was no tomorrow at the Milwaukee Operations of its AC Spark Plug Division in Oak Creek, Wis.

The turtle, of course, is the all-stainless-steel catalytic converter that is now standard equipment on 1975 automobiles from General Motors and American Motors.

The most crucial station on the production line is the welding station where leakproof, integral seams are applied to the exteriors of the four main components of the converter. In a bold decision, AC management has chosen the nonvacuum electron beam process to make these welds.

"The converter was designed," points out Francis J. Buckley, Jr., plant manager—AC Milwaukee Operations, "with this joining method in mind. It's a life-of-the-car component," he adds, "designed for 50,000 miles."

Industry is perhaps more familiar with the hard vacuum version of electron beam welding, the process that was used extensively to provide the leak-tight structural seams for the Saturn booster rocket.

Why EB welding? True, it's a high precision process, but isn't it very expensive and slow?

Counters Mr. Buckley: "The speed of electron beam welding is certainly fast enough to warrant serious consideration for this kind of application. It's also more economical in automotive volume than other methods requiring filler metal. A leakproof weld is mandatory to insure the integrity of the converter for 50,000 miles."

The converters themselves are produced from a titanium-stabilized grade of Type 409 stainless. The addition of titanium is included to improve the weldability as well as the corrosion resistance of the weld. This basic alloy was developed a number of years ago as a material for hot water tubing, but it did not become a big tonnage alloy until the auto industry decided to use it for exhaust emission control devices.

Furnishing this alloy to AC Spark Plug are Allegheny Ludlum, Armco, McLouth, United States and Republic. For the 1975 car models alone, GM will need well over 75,000 tons of Type 409 stainless steel. The same alloy is also being used for emission control devices by Ford and Chrysler.

At AC Spark Plug the turtle-shaped converter is produced in two basic sizes. Measured in terms of volume, the sizes are 260 and 160 cu in., respectively. All converters have round openings at both ends to accept short lengths of exhaust tubes which are Mig welded into place at a later stage in manufacture.

With tubes inserted, the larger converter weighs about 27 lb, while the smaller one weighs 18 lb.

Francis J. Buckley, Jr.

"The converter was designed with EB welding in mind."

GM's catalytic converter is made out of Type 409 ferritic stainless steel.

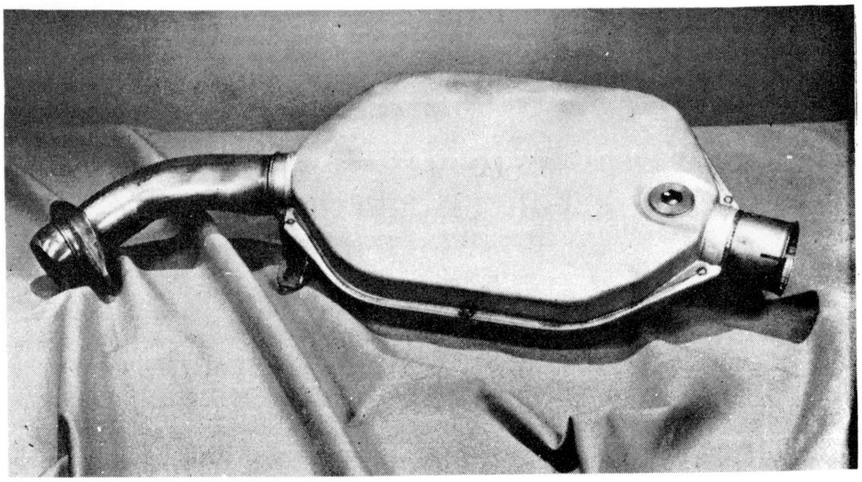

AC Spark Plug is also a beehive of Mig welding activity.

The electron beam process is used to make continuous seam welds along the circumference of the converter from tube opening to tube opening. Welds along both sides are made concurrently. Length of the weld will be either 42 or 46 in., depending on the model of converter.

The workpiece moves; the EB gun remains stationary. Actual welding is performed in the horizontal position and the beam is trained onto a joint configuration that represents a sandwich construction. The sandwich consists of four layers of stainless sheet, each one 0.050 in. thick. Thus, the total thickness to be welded is 0.200 in.

As the electron beam escapes from the orifice of the gun and enters the atmosphere, its diameter is about 1/16 in. By the time the stream of electrons reaches the workpiece, the beam diffuses to a diameter of about 3/16 to 1/4 in. This distance, known as standoff distance, is a constant 1⅛ in.

So obviously, when the beam strikes the workpiece it has more than enough magnitude to weld the four layers of sheet together. The beam also penetrates to a depth of 0.050 in. into the metal.

In the present production setup at AC Spark Plug, a converter is welded in 13.4 seconds. Despite the fact that these welds are made at better than 200 ipm, electron beam welding is still "the slow poke" of the manufacturing line. However, efforts are under way to increase the speed with an ultimate goal of 300 ipm.

Six separate production lines are in operation at the Milwaukee Operations to weld the flow of converters using the electron beam process. Each line is equipped with two 36 kw machines.

Eight of these units were built by the Hamilton Standard Division of United Aircraft Corp., Windsor Locks, Conn. The other four were constructed by Westinghouse Electric Corp., Sykesville, Md. There is an additional machine, a Hamilton Standard unit, on the plant's prototype line, which is used for repair work and R&D evaluations.

"Lots of people are interested in what we have here," remarks A. Frank Banyas, staff production engineer at the Milwaukee Operations. "Properly engineered, nonvacuum electron beam welding has great potential as a means of conserving starting stock. Relatively speaking, it also requires a small amount of floor space."

For safety reasons, all of the electron beam welding operations are carried out inside a lead-lined chamber. This enclosure, however, does not interfere with production speed or flow.

Mr. Banyas cites the integrity of the welds as the greatest selling point of the process. "If a part looks welded, it is welded."

That does not mean that visual inspection is the sole means of quality control. Not by a long shot. During Congressional hearings in the spring of 1973, the EPA saw AC Spark Plug's new manufacturing plans. In addition, GM was just as interested in seeing to it that there would be enough quality built into the system so the product would hold up for more than 50,000 miles on the road.

A plant tour of the facility might remind the visitor of an aircraft plant. A lot busier, but just as clean.

Back in the tryout months, major qualification tests had to be performed on all equipment prior to use. On the beam welders, for example, one parameter had to be held constant while the other two were varied. The parameters were velocity, standoff distance and power.

The weld settings are continually monitored by quality control to make certain that the equipment is operating within the pre-qualified range. Whenever a parameter is changed, the process has to be re-qualified. And, records are kept on all of these readings.

AC Spark Plug engineers see several other advantages in using the EB process. The energy input of any welding process, they reasoned, is proportional to the cross-sectional area of its fusion zone and the width of an EB weld is only about 1/25th that of a conventional weld.

Thus, the low energy input required in electron beam welding minimizes distortion in the workpiece. And, the low input will also have minimal effect on the properties of the material being welded.

In the final analysis, electron beam welding was selected over Mig welding as the main production method. To Mig weld the converters, however, AC Spark Plug would have been forced to consume annually more than 500,000 lb of stainless wire and more than a million cubic feet of shielding gas.

Despite the setback on the main application, the Milwaukee Operations is still a beehive of Mig welding activity. Two exhaust tubes are Mig welded to every converter. Also, four stainless studs are circumferentially welded to the base of each turtle.

AC Spark Plug operates 48 Mig welding stations to weld 24 million studs per year. These operations draw current from pulsed-arc power sources. The stainless filler metal is 0.045 and 0.060 in. diam, while the shielding gas is a 2 pct oxygen-argon mixture.

What about laser welding? Was that process considered?

"You must remember," says Mr. Banyas, "that we began to look at nonvacuum EB welding in early

Typical chemical composition of Armco 409 stainless

Carbon	0.05
Manganese	0.30
Phosphorus	0.02
Sulfur	0.010
Silicon	0.50
Chromium	11.00
Titanium	0.50

Typical mechanical properties in the annealed condition

Ultimate tensile strength, psi	70,000
0.2 pct yield strength, psi	41,000
Elongation, pct in 2 in.	30
Hardness, Rockwell B	75

970. At that time the state-of-the-art if laser welding was not as well advanced as it is today. We had to decide on the process several years ahead of time."

The partial or soft vacuum version of EB welding was also considered. Although test speeds of 160 ipm were recorded, there were problems with pumpdown which would have interfered with production flow.

From the standpoint of quality, resistance seam welding looked acceptable, but the test speeds never exceeded 20 ipm.

All of the electron beam welding machines are under the guidance of computer numerical control. The CNC system is the Vega III, furnished by Vega Servo Control, Inc., Troy, Mich. In this operation it's essential that three parameters—standoff distance, part velocity and angle of the beam to the workpiece—remain constant at all times.

According to Lee R. Herndon, Jr., president of Vega Servo Control, "The most demanding specification, and something new to us at the time, was the requirement for constant velocity. The speed had to be maintained within 5 pct.

"As it's operating now, our equipment controls the three axes. It also maintains a constant surface speed, including peripheral speed, and provides for proper workpiece orientation."

The Vega system has digitizing and editing capabilities. Thus, AC Spark Plug can develop a new program whenever needed or they can alter a present one, if needed. At present there are three programs available for loading.

Formed stock comes to the welding stations after having passed through a battery of 24 transfer presses, which range in capacity from 250 to 1800 tons.

The major stampings for the converter are produced on the new 1800 and 1200 ton transmat presses. Included on these presses is a quick die change system which incorporates hydraulic die clamping in the slide and T-rails for die roll-out.

There is also a constant energy drive system with programmed speeds to provide maximum press cycle rates as well as a Tri-Axial handling system with finger clamping to permit precise positioning of parts in the die nests in addition to maximum transfer speeds.

Press overloads are detected by an electronic load monitor system. If preset tonnage is exceeded, the press is shut down automatically.

Trouble-shooting instrumentation is carried a step further with a fault identification graphic display panel. This panel lights up the reason for press stoppage, pinpointing problems and part transfer faults for fast correction response.

Stainless stock is fed into the press forming operations from 12,000 lb coils.

After the press operations, the components are shuttled through a washing cycle. Here the water-soluble lubricants applied during press forming are removed. At the same time the surfaces of the parts are cleaned in preparation for electron beam welding.

Long live the stainless turtle! May it have a long, useful life.

Robert R. Irving

Stainless steel studs are attached to the converter housing by circumferential Mig welding. AC Spark Plug has 48 units in operation for this application.

Coiled stainless stock is stamped and formed into components for the catalytic converter by a battery of transfer presses. Later, after a wash cycle, the parts are ready for welding.

The Chrysler Die Cast and Welded Aluminum 6-Cylinder Intake Manifold*

JOHN LEE
and LEO GOLIN
Chrysler Corporation
Detroit, Michigan

Abstract

Chrysler's Kokomo Casting Plant is welding die cast 380 aluminum intake manifolds on a high volume production basis. The 75 inch (191 cm) peripheral weld is made by a computer controlled non-vacuum electron beam welder at 450 ipm (19 cm/sec.). The development of the die casting design is discussed. Various fastening and welding processes are evaluated and compared.

Introduction

When Chrysler first designed the 225 CID (3.7L) 6-cylinder intake manifold in 1960, sand cast gray iron was the optimum material for high volume production. Recently that approach was re-examined in the interest of increasing vehicle fuel economy. One method to improve fuel economy is to reduce the weight of the automobile. Significant weight reductions can be achieved by the substitution of sand cast aluminum for cast iron components, but at substantial cost penalties. However, the die cast aluminum intake manifold resulted in greater weight reduction over sand cast aluminum and a cost savings over the cast iron component. Hence the reason for developing the die cast and welded aluminum intake manifolds as described in this paper.

Mechanical Fasteners and Adhesives

The Chrysler 225 CID (3.7L) 6-cylinder intake manifold has curved passageways and therefore cannot be die cast in one piece. (See Fig. 1.) Die casting the curved passageways with expendable cores was considered and rejected because of costs and the corrosive nature of the expendable core material. In order to take advantage of the high productivity of the conventional die casting process, it was decided to split the manifold in the horizontal plane into two die castable pieces. The first prototypes were cast from plaster moulds in August 1972 using a production cast iron intake manifold sawn in half for patterns. The resultant two piece 319 aluminum alloy intake manifolds had the same wall thickness and surface finish as the cast iron counterpart. (See Fig. 2.) Metal screws and RTV sealant were used to join the first prototypes which were subjected to a 500-hour engine dynamometer test in March, 1973 without failure of any kind.

As a result, Chrysler started die fabrication for the

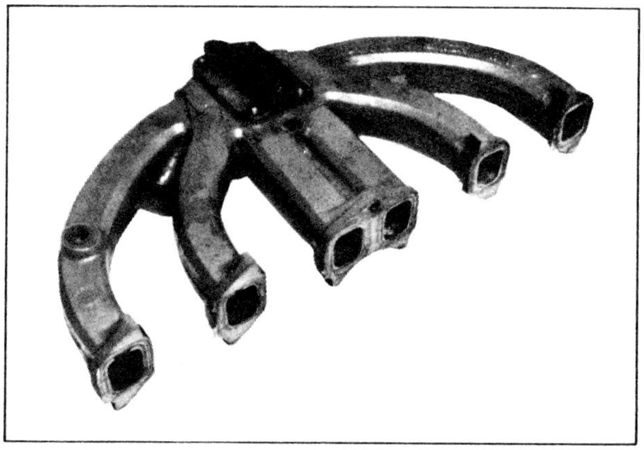

Fig. 1. Conventional one piece cast iron 6-cylinder intake manifold.

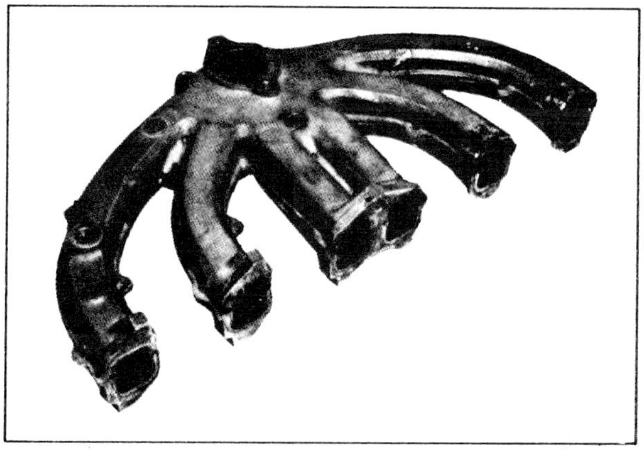

Fig. 2. Prototype two piece 319 aluminum intake manifold, plaster mould castings with metal screws and RTV sealant.

manifold halves to take advantage of the thinner wall thicknesses (less weight) and high productivity (less cost) that are possible with the die casting process. The method of joining the die cast halves was yet to be determined. Initially, epoxy adhesive and screws were rejected because of high temperatures (350°F or 177°C) present around the exhaust gas heated plenum area. Steel wire staples and RTV sealant were tried but the staples could not consistently penetrate the double aluminum layers at the joint line. It was then decided to use self-tapping screws with RTV sealant in a grooved track to fasten the assembly together. (See Fig. 3.) This construction tested satisfactorily, but it was unable to withstand the machining stresses when processed over existing tooling used for cast iron manifolds. The next design modification was to raise the split line as far as possible above the exhaust gas heated plenum area and to substitute epoxy adhesive for the RTV sealant which together with the self-tapping screws made a much stronger assembly. Nonetheless, extensive testing proved unsuccessful and the mechanical fasteners plus adhesives approach was abandoned in favor of welding.

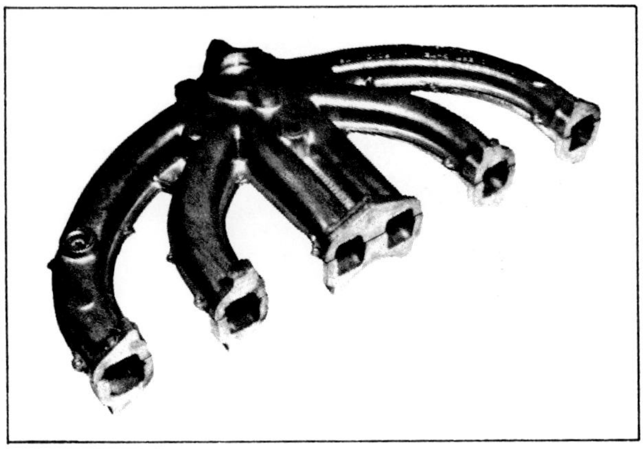

Fig. 3. Two piece die cast 380 aluminum intake manifold fastened by self-tapping screws and RTV sealant.

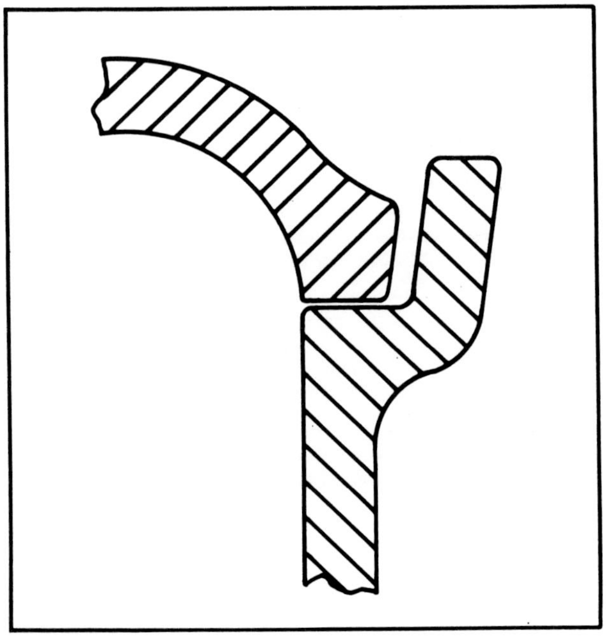

Fig. 4. Cross section of the weld flange showing the nested joint configuration (4x).

Welding Development

Preliminary electron beam weldability tests were made as early as 1973 on both die cast 380 and sand cast 319 aluminum alloy samples. However, electron beam welding of manifolds was not intensively evaluated until it became evident that adhesives and mechanical fasteners were not satisfactory for our 225 CID (3.7L) intake manifold. At a later date, a number of die cast aluminum manifolds were electron beam welded on our laboratory high vacuum welder. The experience gained by welding those manifolds enabled us to rule out the burn-through and lap joint configurations for production use. A modified butt joint was then designed and the split line on the assembly moved to the uppermost limit to provide maximum access for welding. (See Figs. 4 and 5.)

In addition to high vacuum electron beam welding various processes were evaluated including CO_2 cw laser, non-vacuum and partial vacuum electron beam, plasma arc, gas tungsten arc, and gas metal arc welding. Round test samples were specially fabricated for welding after which they were pressure tested and eventually sectioned and metallographically examined.

Conventional Fusion Welding

Test specimens were gas tungsten arc welded with adequate weld quality, but the welding speeds were very slow (15 to 20 ipm, 6.4 to 8.5 mm/sec). In addition, that process requires precise electrode to work positioning and the tungsten electrode is easily contaminated. All those factors combined to eliminate gas tungsten arc welding as a high volume production process. Plasma arc welding was also used successfully on die cast 380 aluminum samples at welding speeds up to 60 ipm (2.5 cm/sec). However our assessment is that plasma arc welding equipment must be further developed before it can be selected for high volume production. Gas metal arc welding of the round test samples was successful at speeds greater than 100 ipm (4.2 cm/sec). But high operating costs involving the welding wire and gas consumables were the reasons for rejecting that process for the 225 CID (3.7L) intake manifold.

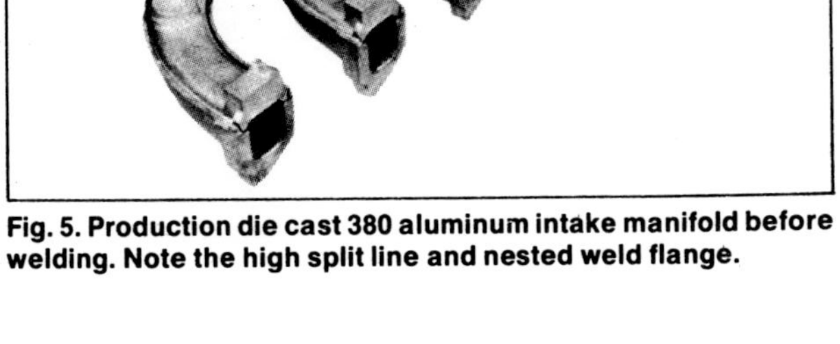

Fig. 5. Production die cast 380 aluminum intake manifold before welding. Note the high split line and nested weld flange.

Laser Welding[1]

High power (over 5 kw) as well as low power CO_2 lasers were used to weld test specimens and prototype manifolds at speeds in excess of 200 ipm (8.4 cm/sec). The resultant weld quality was adequate as determined by leakage tests. However the durability and the manufacturing feasibility of laser welders were not demonstrated to our satisfaction in time to purchase that equipment for production of the die cast aluminum 6-cylinder intake manifold. Nonetheless, laser welding will be considered for future die cast and welded aluminum assemblies.

Medium Vacuum EB Welding[2]

A number of electron beam equipment manufacturers welded test specimens and manifolds on their medium vacuum electron beam welders (25 to 100 Torr). Weld quality of those samples, determined by leakage rates, was adequate. However the 100 ipm (4.2 cm/sec) welding speeds were relatively low and aluminum vapors were generated during welding which caused arcing in the electron gun as well as work chamber contamination. Those conditions and the difficulty in pumping down the large work chamber were the reasons for rejecting vacuum electron beam welding as the production process.

Non-Vacuum EB Welding[2]

After considerable development work by Chrysler Engineering and Leybold-Heraeus Vacuum System Inc., test specimens were welded at speeds greater than 400 ipm (17 cm/sec) that were virtually leakproof. The key factor to our success appears to be the use of helium effluent instead of the usual compressed air. However, it was not possible to weld manifold assemblies with the process because a computer controlled non-vacuum electron beam welder and X-Y table combination was not available. Nonetheless, after reviewing the above information on alternate processes, Chrysler Engineering specified non-vacuum electron beam welding as the production process for the 6-cylinder intake manifold and its design was further developed with that process in mind.

Fig. 6. Inside the non-vacuum EB production welder showing the inclined welding head, conveyor and X-Y table. (Photograph courtesy of Leybold-Heraeus Vacuum Systems Inc.)

Die Casting Design

The key features of the 225 CID (3.7L) die cast manifold are the high single plane split line and the nested weld flange. (See Figs. 4 and 5.) The top casting has a flat surface which seats on a horizontal shoulder in the bottom casting and both pieces have matching 10 degree slanted walls which are self-piloting when pressed together. An off-line hot stake machine is used to press and stake the castings together before the assembly is presented to the non-vacuum electron beam welder. There is a designed 0.03 inch (0.7 mm) clearance between the castings to allow for manufacturing tolerances. The bottom casting is identical for both single and double barrel intake manifolds whereas the top castings have either a single or double carburetor port opening. There is no access problem in welding the single barrel manifold, unlike the two barrel configuration where the rear of the carburetor mounting pad interferes with the line-of-sight from the welding head to the weld seam. As a result the non-vacuum electron beam welding head had to be fixed 10 degrees backwards to ensure that line-of-sight access. (See Fig. 6.) During the welding cycle, this inclination causes a leading or lagging angle of beam impingement. It was found that the leading angle condition was detrimental to weld quality causing weld porosity or skips which were corrected by slowing the weld speed down.

Production Welding

The production welder made by Leybold-Heraeus moves the pre-staked and palletized manifold through a complex contour in the X-Y plane under the stationary welding head. The exit nozzle on the welding head is approximately 1.2 inches (3 cm) from the weld seam. The X-Y table motions are driven by hydraulic motors and ball screws in a closed loop system under the control of a dedicated minicomputer. The computer also modulates the electron beam power by varying the beam current and turns it on and off 7 times during the welding cycle. A total of 75 inches (191 cm) of weld is made on each manifold which is done at varying speeds between 300 and 450 ipm (12.7 and 19 cm/sec). No filler wire is utilized which avoids extra expense and process complication, however helium effluent gas is used. The entire welding cycle is over in 22 seconds and an additional 4 seconds is utilized for handling and transfer which makes for a 26 seconds floor-to-floor cycle time. At 100% utilization, production output is rated at 135 manifolds per hour.

Helium Effluent Gas

As mentioned earlier, helium effluent gas is used on the non-vacuum electron beam welder instead of the usual compressed air. The helium appears to perform two functions. It definitely reduces dispersion of the electron beam as it exits and thereby permits a more concentrated power distribution at greater work to nozzle distances. The resultant weld on 380 aluminum alloy has a narrower fusion zone than that produced under air effluent and the same

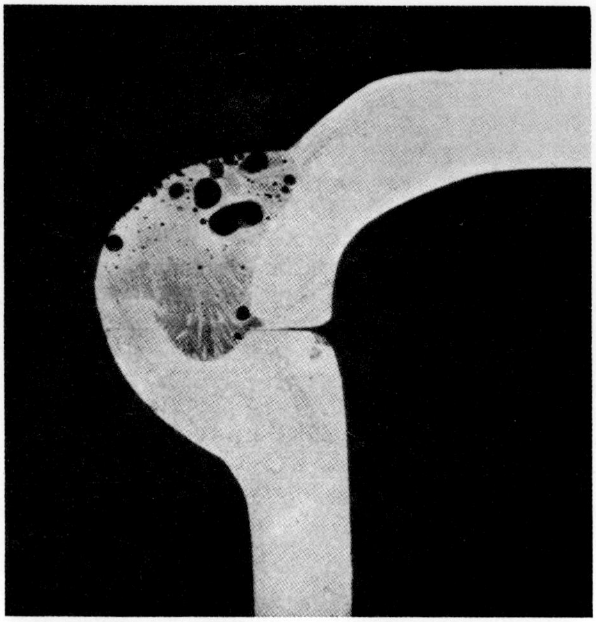

Fig. 7. Cross section of non-vacuum EB weld made by production welder on die cast 380 aluminum manifold showing relatively dense root and porous top (5x).

stand-off distance. Although not firmly established, the helium also appears to protect the molten weld metal from reaction with atmospheric gases. A typical weld, as shown in Fig. 7, has a relatively dense root structure with a porous top. The manifold assembly made by joining the halves by such a weld generally satisfies the leakage rate of 200 atm - cc/minute at 5 psig (35 kPa). However the assembly may require additional sealing which is easily accomplished. (See Fig. 8.)

Summary

Chrysler's Kokomo Casting Plant has started production of the above manifold as a die cast and welded aluminum assembly as a running change for mid 1978 model cars. Each manifold produced in that manner represents a 14 to 15 lbs. (6.4 to 6.8 kg), reduction over its cast iron counter part. Equally important is an estimated 30% cost reduction compared to the conventional cast iron manifold. In addition, preliminary evidence indicates that gas mileage, engine emissions, and performance are all improved by the die cast and welded manifold.

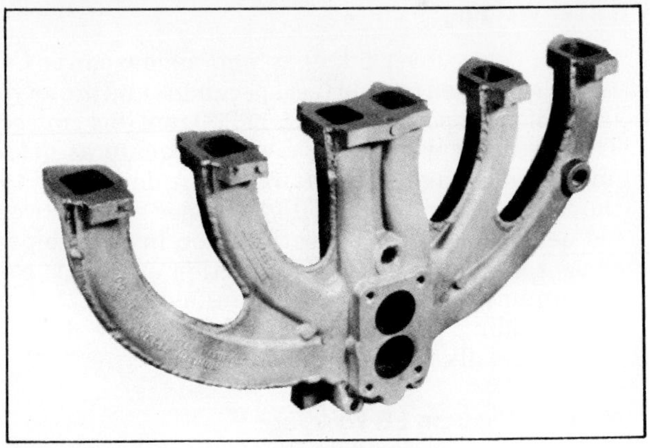

Fig. 8. Production die cast and welded 380 aluminum 6-cylinder intake manifold.

Conclusion

The Chrysler intake manifold is the first high volume welding application for die cast aluminum and is a highly significant advance in manufacturing technology. The manufacturing process combines the advantages of high productivity die casting and non-vacuum electron beam welding to fabricate assemblies at cost savings over permanent mould and sand castings. Chrysler is developing additional applications that could lead to more weight reductions that would improve fuel economy in future models.

References

1. W. C. Ball and C. M. Banas, "Welding with a High Power CO_2 Laser." SME paper 740863 presented at the National Aerospace Engineering and Manufacturing Meeting, San Diego, California, October 1-3, 1974.
2. D. H. Sundermeyer, "An Overview of Electron Beam Welding." SME paper AD77-701.

APPLICATION OF ELECTRON BEAM WELDING TO ROTATING GAS TURBINE COMPONENTS

S. M. Silverstein, V. Strautman and W. R. Freeman
Lycoming Division, AVCO Corporation

Introduction

Continuous advances in the aircraft and aerospace technology impose ever increasing demands on the materials used in the components and structures and also impose ever increasing demands upon the engineers and metallurgists engaged in the work to develop new joining techniques. The higher stress, higher temperature, as well as environmental requirements have led to the development of a score of new alloys which cannot be satisfactorily joined using the conventional joining processes. A demand was created to develop new joining techniques to satisfy the aforementioned requirements and, as a result of this, in the last ten to twelve years great advancements have taken place in the joining technology. One of the new processes developed for practical applications was electron beam welding.

Electron beam welding is being employed successfully at Lycoming in the fabrication of several rotating gas turbine components such as compressor rotors and power shafts. The application of electron beam welding to rotating gas turbine components was initiated for the following reasons: the uniqueness of this process permits the joining of fully heat treated components attaining a very high level of joint efficiency, a very low distortion in electron beam welding due to the very narrow weld and heat affected zones permits the joining of almost fully machined components, and finally the very low heat input in the joint permits the joining of some of the alloys which are very difficult to weld using the conventional joining processes. In this paper the new approaches used in the fabrication of these components will be described, the methods of evaluating the joints and the results of tests of the actual components presented. All welding was done in a Hamilton Zeiss high voltage electron beam welder having a maximum accelerating voltage of 150 KV and either 3 or 6 KW capability. An electromagnetic deflection system can oscillate the beam using alternating current in two axes or the beam can be swept in a circular, elliptical or sine wave path as desired. The 6 KW facility is also equipped with high precision drives and complete weld cycle controls, including pre- and postweld condition and slope control.

Compressor Rotor

One of the interesting applications of electron beam welding was the development of a welded axial compressor rotor for Lycoming's T53 gas turbine engine. The compressor rotates at 25,600 rpm. The current production rotor consists of a mechanical assembly of five discs, five spacers and a sleeve, or a total of eleven details bolted together (see Figure 1). AMS 6382 and 2014 aluminum alloy is used for disc material, and AMS 6832 and A286 alloy is used for spacer material. The sleeve material is AMS 6360. A new gas producer rotor system was designed to improve the integrity, the life and the corrosion resistance of the component. The new design called for an integral axial compressor spool made of heat treated Ti-6Al-4V alloy joined by electron beam welding at five circumferential joints in three thicknesses of 0.175 in., 0.220 in., and 0.375 in. (see Figure 2).

In the first phase of the development program, laboratory tests were performed in order to investigate the mechanical properties of electron beam welded joints in Ti-6Al-4V alloy (AMS 4928). The material was welded in the solution treated (1750° F/2 hr.) and aged (1000° F/8 hr.)condition and subsequently stress relieved at 900° F/4 hr. after welding. Flat plate Ti-6Al-4V weldments were prepared following this procedure and machined into tensile specimens. The results of tensile tests are shown in Table I. Eight out of ten specimens failed in the parent material and one specimen failed partially through the weld. One specimen with a known 0.060 inch diameter void in the weld was tested in order to evaluate the effect of such internal defects on uniaxial tensile strength. The specimen failed in the weld at the defect location with lower than the average elongation; however, ultimate tensile strength and 0.2% yield strength were of the same average values as the other specimens. This was also observed for the partial weld failure. The results indicated a 100% joint efficiency in short time tensile properties.

For the investigation of the fatigue strength of electron beam welded joints rotating beam test specimens of parent material and weldments were prepared and their fatigue behavior determined. The results were plotted, and, as shown in Figure 3, the decrease in fatigue strength of the weldment relative to parent material did not exceed 12%.

Of interest for this, as well as another development program, was the behavior of Ti-6Al-4V weldments in 2:1 biaxial stress field. For this purpose pressure vessels were designed, fabricated, and hydrotested to failure, the ultimate hoop stress for each vessel was determined. The pressure vessels were designed to be 4 1/2 in. I. D. by 7 in. long and 0. 050 in. thick. The size was selected for the hydraulic pressures required to burst the vessels in the laboratory rig to be within reasonable levels.

To establish the base line data, pressure vessels were designed for fabrication without welding. The welded vessels were designed using three different joint designs as shown in Figure 4. Joint design No. 1 was selected to evaluate electron beam welded joints which are finish machined on both inside and outside surfaces. Joint designs No. 2 and No. 3, were prepared to evaluate electron beam weldments with the underside (inside of the cylinder) left in the as-welded condition. The second design incorporated a premachined pilot in which the electron beam would terminate. The third design had a full penetration weld permitting evaluation of an unmachined underbead as produced by an electron beam. Six 5 1/2 in. O. D. by 4 in. I. D. by 16 in. long forgings of Ti-6Al-4V alloy from a single heat of material and complying to AMS 4928 were obtained in the annealed condition for the fabrication of twelve pressure vessels. Subsequent to rough machining and heat treating the twelve blanks were then machined into hydrostatic pressure test vessels: three unwelded parent material vessels for baseline properties and nine for evaluating the properties obtained from electron beam welds. Three vessels were machined for each of the three different joint design variations.

To optimize the weld settings for the desired joints in pressure vessels, rings of Ti-6Al-4V alloy were machined to the same thickness and joint configuration as those which were used in the actual pressure vessels. It was found by metallographic sectioning that weld settings using high travel speeds and thus relatively low heat inputs could be employed for joint designs No. 1 and No. 2 since this would produce narrow and desirable weld and heat affected zones. A relatively slow, higher heat input setting was developed for the joint design No. 3 vessels because this type of setting produced the smoothest underbead configuration possible. The detailed weld settings developed and finally used on the actual vessels are given in Table II. Typical cross sections through the welds produced by these welding parameters are shown in Figure 5. All nine vessels were assembled and welded using the optimized parameters, and subsequent to welding, all vessels were stress relieved at 1000° F for 4 hours. This stress relief was performed on the three unwelded vessels also in order to subject them to an identical thermal treatment. The vessels weldments were fluorescent penetrant inspected, radiographed and finish machined, as required, to meet the final joint design configurations and pressure vessel geometry required for hydrostatic testing.

Typically a pressure vessel with a No. 2 joint design subsequent to welding is shown in Figure 6. The finish machined test vessels were given a full dimensional inspection to permit calculation of ultimate burst strength subsequent to testing. Each pressure vessel was assembled in the test rig as shown in Figure 7 and hydrostatically tested to failure. Internal pressure was monitored during test by a Teledyne Model 176 pressure transducer which was calibrated prior to bursting each pressure vessel. Calibration was made against a 10,000 pound range Helicoil test gage. From the burst data the average burst strength or ultimate hoop stress (S_h) for each pressure vessel was calculated from the formula:

$$S_h = \frac{Pr}{t}$$ where:

P = Ultimate pressure at failure
r = Mean radius of the pressure vessel
t = Wall thickness of the pressure vessel

A tabulation of burst strengths and calculated joint efficiencies in a biaxial stress field obtained from the test program is shown in Figure 4. Typical failure for welded and unwelded pressure vessels are shown in Figures 8, 9 and 10. As may be seen, the failure propagated in a longitudinal manner as expected. However, one of the three welded test vessels with an unmachined pilot (joint design No. 2) failed in a circumferential manner as seen in Figure 11 with the failure initiating at the root of the electron beam weld and the unmachined pilot. The failure propagated along a 45° shear plane completely around the pressure vessel in the heat affected zone and unaffected base metal. This failure, however, occurred at an internal pressure in excess of that experienced by the unwelded vessels.

The fracture surfaces of all of the test vessels were examined. Although some of the failure of the welded pressure vessels appeared to initiate in the area of the electron beam weld, this is to be expected since the specimen was designed so as to place the weld in the area of greatest deflection and strain. Results of hydrostatic testing showed that thetest vessels with the joint designs No. 1 and No. 3, i.e. those with an electron beam weld machined on both root and crown and those with the weld machined on the crown only, had essentially the same biaxial burst strengths as unwelded test vessels. Thus the joint efficiency of the electron beam welds in biaxial stress field in heat treated Ti-6Al-4V alloy was essentially 100%. Additionally, the scatter in test results within triplicate tests was greater than the difference in averages between the unwelded specimens and the electron beam welded specimens of joint design No. 1 and No. 3. The welded test vessels with the unmachined internal pilots (joint design No. 2) showed a mean ultimate hoop stress of 217.6 KSI without correcting for the geometry of this type of test vessel. This was 8% higher than the strength exhibited by the unwelded vessels and was attributed to the reinforcing effect of the pilot.

Based on the results of the aforementioned test programs it was decided to utilize a joint machined on both root and crown after welding (joint design No. 1) for the compressor rotor weldment. Subsequent to the establishment of weld settings by welding a number of test rings simulating the rotor joints, a tacking and welding sequence was developed using test rings to: (1) minimize the transverse shrinkage, and (2) balance the welding distortion and effect a uniform transverse shrinkage around the circumference.

Rough machined forgings for the rotor details were solution treated and aged, then the details machined to the finished dimensions as shown in Figure 12. A welding fixture (Figure 13) was designed and made for welding the rotor assemblies. It includes an internal spring to provide a 50 lbs. axial end load on the rotor assembly which allows for expansion during welding and shrinkage after welding. Figure 14 presents a transverse microsection of a typical electron beam welded rotor joint and a close-up of a welded rotor in the as-welded condition prior to removal of the internal pilot shoulder and the machining allowance on the O. D. Subsequent to welding the assemblies were stress relieved in vacuum then finish machined on the I. D. and the O. D. in the joint areas and inspected by radiographic and fluorescent penetrant methods. The rotors were then finish machined on the end flanges and bladed. A typical bladed rotor is shown in Figure 15. The experimental rotor units were then programmed for a series of basic rig test evaluations for determinations of the rotor dynamic characteristics as follows:

1. Determination of radial static flexibility.
2. Determination of rotor natural frequency.
3. Determination of mechanical integrity.
4. Engine evaluation.

The complete rotor assembly was mounted on rigid supports at the positions of engine main rotor bearings. Varying loads were applied to the rotor in a radial direction at the fourth stage of the compressor and the deflection was measured by displacement indicators radially opposite the loading point. The results of these tests were compared with the previously acquired data on a standard T53 rotor and are as follows:

Standard T53 compressor:	1. 16 to 1. 56 x 10^6 pounds/inch
Welded Titanium T53 compressor:	1. 05 to 1. 25 x 10^6 pounds/inch

The full compressor rotor with its turbine wheel in place was mounted on a variable frequency shaker table and excited by a sinusoidal force input through the support cradle. The natural frequency of the rotor system was determined and compared with the frequency of the standard rotor:

Standard T53 compressor:	805 cps
Welded Titanium T53 compressor:	465 cps

In preparation for actual engine operation and testing a welded rotor was tested in the high speed balancing and spin-test facility. The full compressor rotor was initially balanced in a low speed Gisholt balancing machine at approximately 1700 rpm and the imbalance condition corrected in two planes to levels below 0. 2 in. -grams. Subsequent to installation in the high speed balancing facility, the rotor was checked out at 12, 000 rpm, and a minor correction made at the aft position to lower the residual imbalance from approximately 2 in. -grams to less than 1 in. -grams. After balancing, the rotor was checked out at 26, 000 rpm for mechanical integrity and found to be satisfactory. The rotor was then installed in a test engine and subjected to various test cycles including 2000 start cycles. The performance of the rotor was very satisfactory.

Power Shaft

The power shaft of Lycoming's T53 gas turbine engine shown in Figure 1, was another component which was redesigned for electron beam welding. The standard shaft is made by induction brazing a carburized AMS 6260 (SAE 9310) alloy steel forward end to a AMS 6382 (SAE 4140) alloy steel aft end using a silver base brazing alloy. The power shaft is driven by the power turbine wheel and transmits the power to the reduction gears in the helicopter or the prop versions of the T53 engine. It rotates at 21, 000 rpm. A program was initiated to develop a procedure for electron beam welding this power shaft to replace silver brazing, evaluate the properties of the resulting electron beam welded joint, and prepare power shafts by this method for engine test.

For the development of the electron beam welding technique AMS 6382 forgings were normalized at 1625° F for one hour, solutioned at 1550° F for 30 minutes followed by oil quenching, then tempered at 1025° F for two hours to a hardness of Rc 34-37. AMS 6260 forgings were normalized at 1750° F for one hour and air cooled, hardened at 1485° F for one hour followed by oil quenching, then tempered at 250° to 275° F for two hours. This treatment gave hardness in the specified range of Rc 32 to 40 for the uncarburized areas on the forward end of the power shaft. The development of weld settings was accomplished by first utilizing flat plates for establishing preliminary settings, and then welding full size test rings machined to the same joint configuration as the actual shaft details as shown in Figure 16. A series of electron beam welds were made to find the lowest energy input (joules per inch) which would give full penetration. This condition was desired since it would produce the least distortion of the component and minimize the requirement for balancing. A machine setting of 149 KV, 18.5 MA and 34 ipm with sharply focused beam and no oscillation was chosen as the optimum setting. Sample welds were inspected by radiographic, magnetic particle and fluorescent penetrant methods and no defects found. Metallographic examination of sections made through the welds showed no detrimental defects. Figure 17 is a typical photomacrograph of a weld made with the selected parameters in a test ring. Additional welds made with these settings were stress relieved at 900° F for four hours. This temperature was selected to negate the possibility of overtempering AMS 6382 material and was 125° F below the tempering temperature. Transverse microhardness surveys were performed on both as-welded and stress relieved weldments. Figure 18 shows the average hardness curves of four surveys in each condition. The average hardness of the as-welded fusion zone, which was 0.045 in. wide, was Rc 45. Adjacent to the fusion line in the AMS 6382 alloy there was a rehardened zone of untempered martensite approximately 0.010 in. wide where the critical temperature was exceeded and the hardness was Rc 55. Immediately adjacent to this zone was a very narrow overtempered area, i.e. high ferrite region, approximately 0.005 in. wide with a minimum hardness of Rc 27. No evidence of a rehardened zone existed in the AMS 6260 material, however there was an overtempered zone with a minimum hardness of Rc 33.

After the 900° F stress relief the hardness of the weld nugget was reduced to Rc 35 and the rehardened zone in the AMS 6382 to Rc 45. A general reduction in hardness was affected in the SAE 9310 material because the stress relief temperature of 900° F was considerably higher than the tempering temperature required in the carburizing cycle. The hardness of the parent material dropped 5 points to Rc 35 which, however, was still within the desired hardness range of Rc 32 to 40. Since the use of a 900° F stress relief was beneficial in tempering the fusion zone and the rehardened zone of AMS 6382 without detrimentally affecting the AMS 6260 material, this stress relief cycle was adopted.

In order to substantiate the fatigue reliability of an electron beam welded power shaft, a special test shaft simulating the power shaft was designed as shown in Figure 19 for torsional fatigue tests. It is identical in dimensions to the power shaft except that two joints are made for attaching special ends to fit the fatigue test rig. The materials and heat treatments used were the same as for the power shafts and described above. The test shafts were welded using the same procedure and weld settings optimized on the test rings. The completed fatigue test shafts were then thermocoupled and prepared for testing at elevated temperatures in the fatigue test rig shown in Figure 20. Fatigue tests were run at 500° F to 600° F in order to simulate the operating temperature of the component and the results compared with brazed test hardware failed under the same conditions. In cases where there was a run out, i. e. the test shaft did not fail after 2 x 10^6 cycles, the load was increased and cycling was continued to failure. Figure 21 is a plot of vibratory torque vs. cycles to run out or failure for the electron beam welded shafts. Also included in the plot are data obtained previously on brazed shafts. All failures in the electron beam welded test shafts occurred in the tube section of the specimen in an area unaffected by the weld, while all failures of the brazed shafts occurred in the brazed joints. Figure 22 shows a typical failure of an electron beam welded shaft and the location of the welds.

Because of the significant improvement in the component strength of electron beam welded fatigue specimens over brazed specimens, it was decided to proceed with the fabrication of actual electron beam welded power shafts. A typical set of power shaft details prior to welding are shown in Figure 23. In order to minimize the equalize the distortion due to welding a controlled tacking and welding sequence was utilized when welding the shafts. This was necessary for maintaining the run out of the shaft after welding to less than 0. 05 in. because (1) the front end of the shaft was carburized and semi-finish ground prior to joining, and (2) in dynamic balancing the maximum allowable unbalance is 0. 5 in. -gram. After welding the power shaft joints were to be subjected to a stress relief cycle of 900° F. However, since the front end spline was carburized, hardened and tempered prior to welding, it could not be heated to over 265° F. Therefore, a stress relieving rig was designed and built for locally stress relieving the power shaft joint after welding. The facility is shown in Figure 24 with a shaft used for calibration purpose in the holding fixture. The sleeve type furnace is six inches long and is made to fit the power shaft contour in the joint area. A controlling thermocouple is located in the center of the furnace. During the stress relief cycle the shaft is rotated and water cooling is provided for the carburized spline. Subsequent to the stress relief the shafts were ground in the joint area and subjected to radiographic and magnetic particle inspection. No defects were detected. The completed shafts were submitted for engine verification and all successfully passed extensive evaluation.

Turbine Rotor:

Electron beam welding is also being applied in salvaging Lycoming's T53 engine's second stage power turbine rotors shown in Figure 1. The rotors were being removed from service after extended periods of running if they had excessive wear on the front internal spline. A procedure was developed to replace the front end by electron beam welding and to restore the rotors to an essentially new condition. Timken 17-22AS forging stock, which is the material of the power turbine rotors, was heat treated to the same condition as the rotor material, then machined into front end replacement details. The front end of the rotors was removed and the rotors machined in preparation for electron beam welding. The front end replacement, assembled and electron beam welded to a rotor, is shown in Figure 25. Subsequent to welding, the rotors were stress relieved in vacuum at 900° F for four hours. In order to evaluate the repaired joints prior to engine evaluation, it was decided to fatigue test a number of the assemblies in a test rig in both torsion and bending. Unwelded rotors were also tested to provide baseline information. From these tests, it was concluded that the fatigue strength of the electron beam weld repaired power turbine rotor was of the same order of magnitude as that of the original new rotor in both torsion and bending. Subsequent engine testing of the repaired components also substantiated this finding; consequently, such repair procedure is now being used in salvaging power turbine rotors with worn front end splines.

Conclusion

The experience at Lycoming has shown that electron beam welding is an excellent process for the fabrication of high performance components. Significant also is the fact that in the alloys investigated, the mechanical properties of the joints produced efficiencies of essentially 100% without a complete reheat treatment of the component. The advantages offered by this process will result in a more extensive application of electron beam welding in the future.

TABLE I

Mechanical Properties of Ti-6Al-4V Material Electron Beam Welded in Solution-Treated and Aged Condition and Stress-Relieved After Welding

Specimen No.	0.2% YS psi	UTS psi	Elong. % 4D	Location of Fracture
1	159,500	178,600	8.0	PM*
2	162,400	179,400	7.5	PM
3	161,900	178,300	10.0	PM
4	164,500	186,400	5.5	PM & Weld
5	157,400	179,700	10.0	PM
6	162,000	180,500	7.0	PM
7	167,100	181,700	9.5	PM
8	151,200	175,700	10.0	PM
9	159,100	174,900	9.5	PM
10	160,400	174,300	2.5	Weld (0.060" void)
Average	160,550	178,950	8.0	

*Parent metal

Heat Treatment: 1750° F for 2 hours, water quench
1000° F for 8 hours, air cool

TABLE II

Electron Beam Welding Parameters
Employed on Pressure Vessels

	Joint Design #1	Joint Design #2	Joint Design #3
Accelerating Voltage (KV)	150	150	150
Tracking Bead Current(1) (ma)	7.5	7.5	No tacks used
Welding Bead Current(2) (ma)	13.0	13.0	3.75
Travel Speed (ipm)	100	100	15.0
Location of Beam Focus	Part Surface	Part Surface	Part Surface
Deflection Used	None	None	None
Distance Heat Shield to Part (in.)	12	12	12

(1) Eight tack approximately 1/4" long equispaced in staggered sequence about joint circumference with beam current given. All other parameters identical with those used for welding joint.

(2) All welds ended with beam current decay.

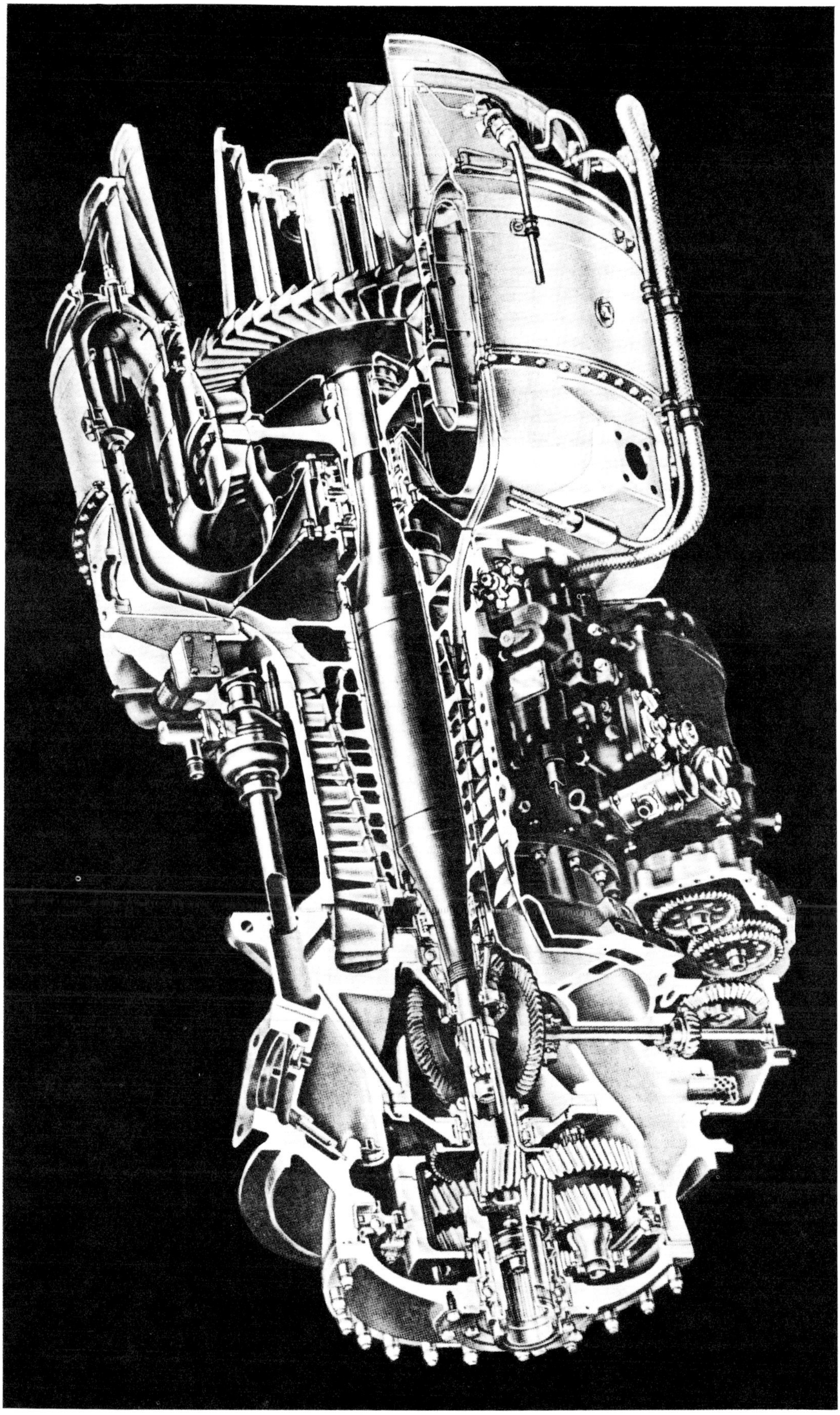

FIGURE I - LYCOMING T53 GAS TURBINE ENGINE

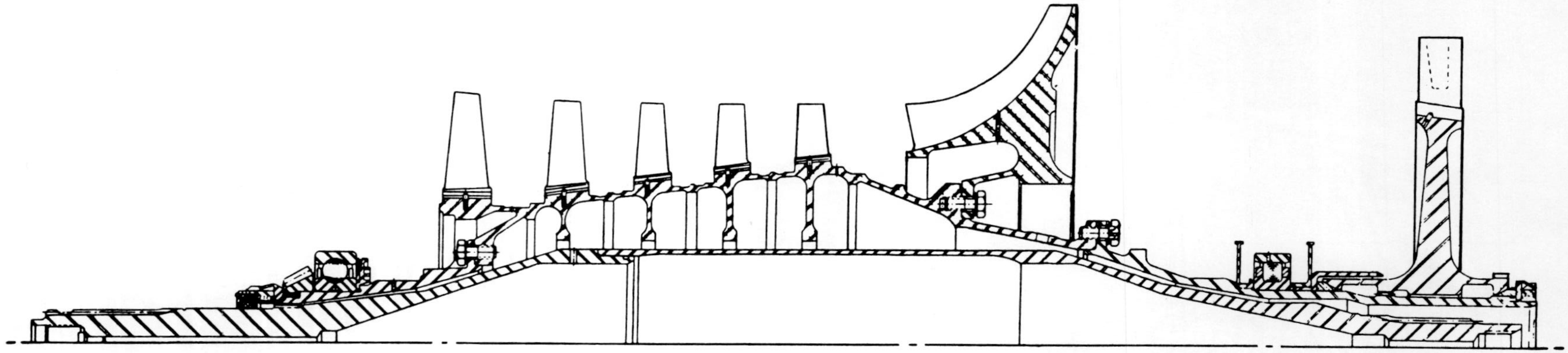

FIGURE 2 - T53 COMPRESSOR ROTOR AND POWER SHAFT ASSEMBLY

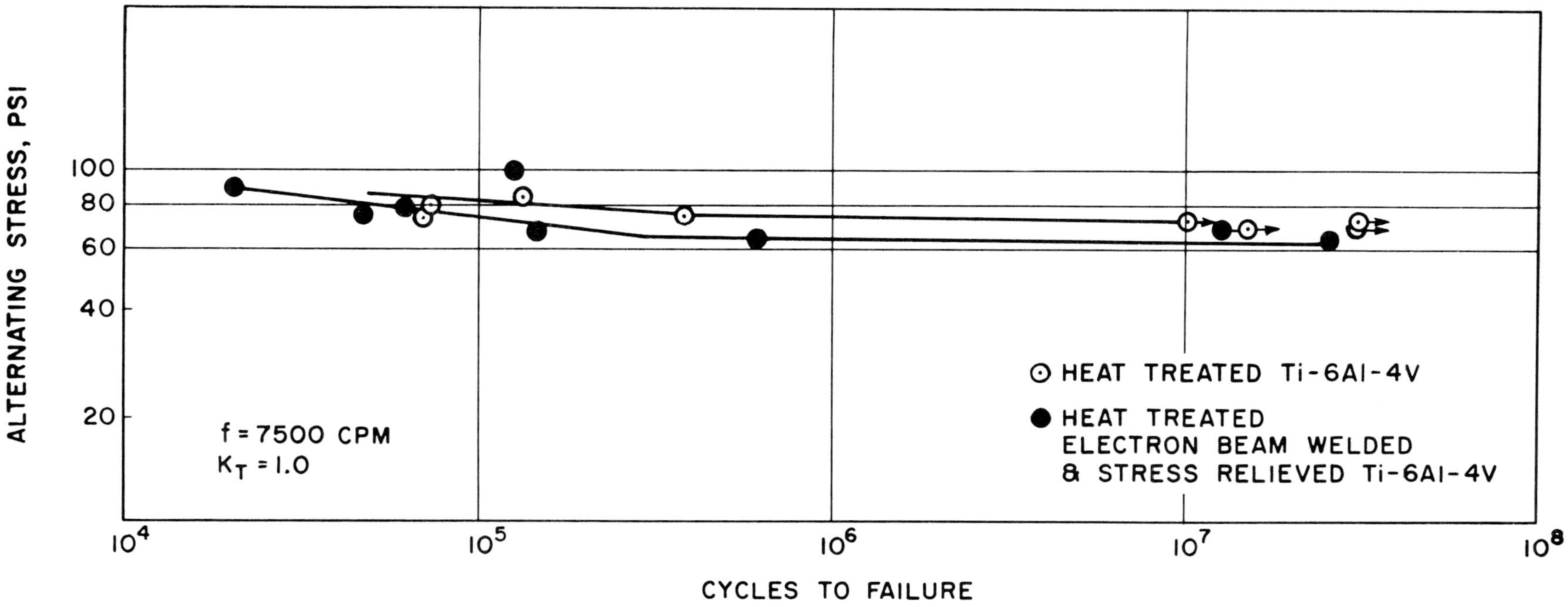

FIGURE 3 - ROOM TEMPERATURE ROTATING BEAM FATIGUE LIFE OF UNWELDED AND ELECTRON BEAM WELDED Ti-6Al-4V TITANIUM ALLOY IN FULLY HEAT-TREATED CONDITION

JOINT DESIGN		ULTIMATE HOOP STRESS, KSI	JOINT EFFICIENCY, %
BEFORE WELDING	AFTER FINAL MACHINING		
UNWELDED PARENT METAL	0.055 0.045 	208.3 198.9 197.0 MEAN 201.4	---
JOINT DESIGN NO. 1 0.190 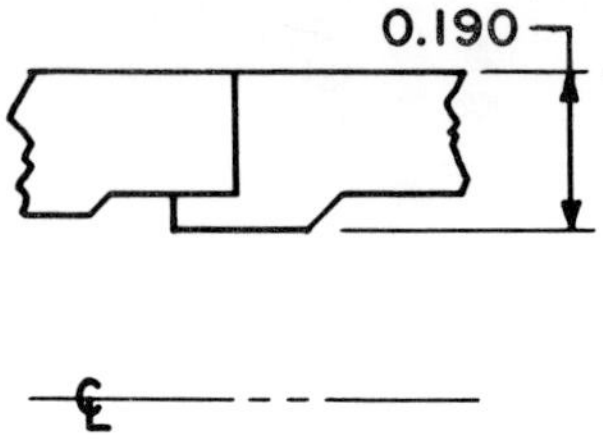	E.B. WELD 0.055 0.045 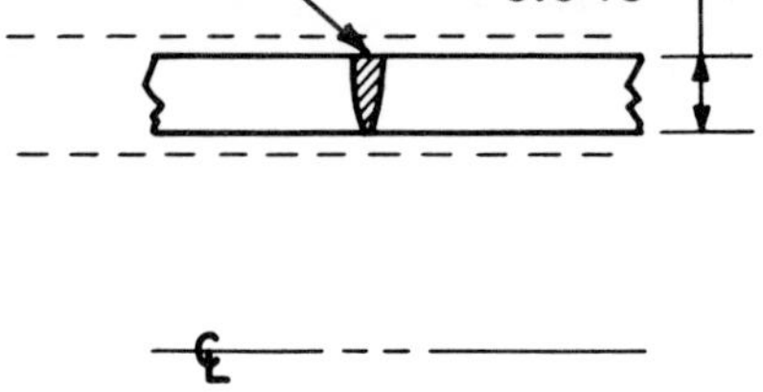	195.2 203.9 201.3 MEAN 200.1	99.4
JOINT DESIGN NO. 2 0.205 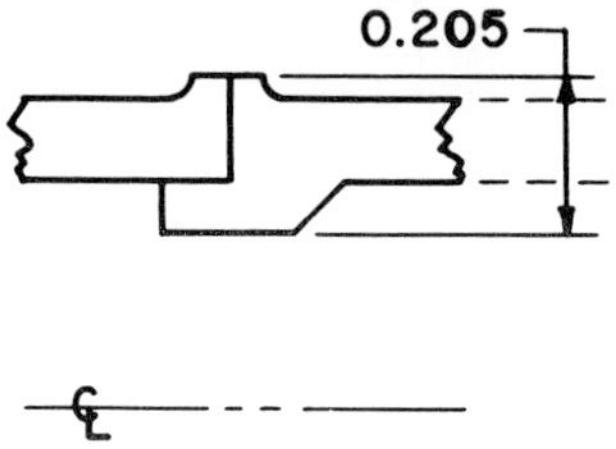	E.B. WELD 0.0505 0.0502 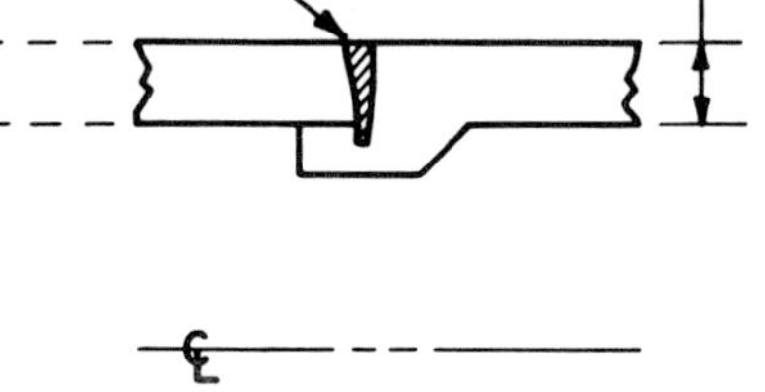	222.8 214.1 216.0 MEAN 217.6	100(1)
JOINT DESIGN NO. 3 0.130 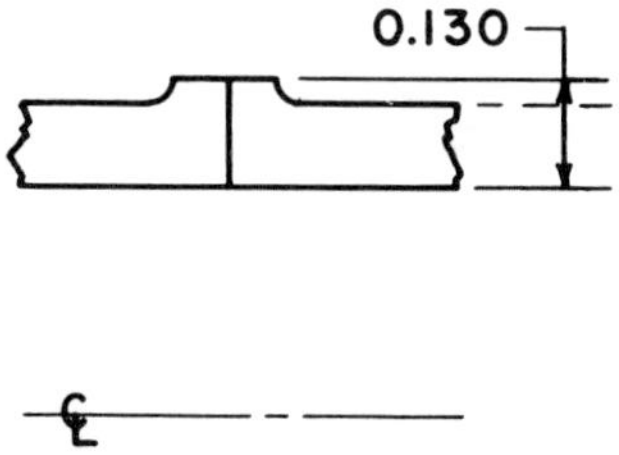	E.B. WELD 0.045 0.055 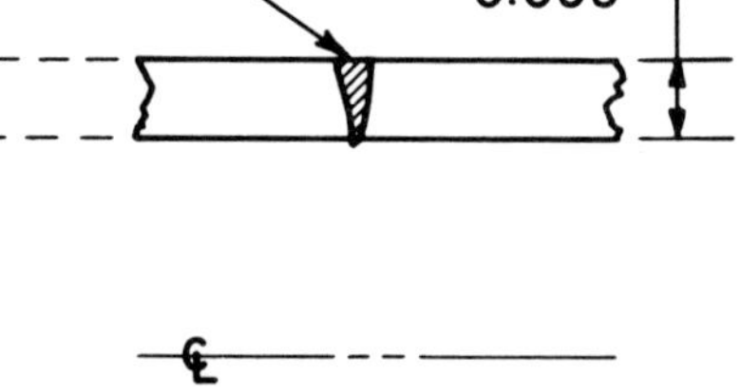	200.0 195.5 201.1 MEAN 198.9	98.8

(1) NO CORRECTIONS HAVE BEEN MADE TO INCLUDE EFFECT OF INCREASED AREA (UNMACHINED PILOT).

FIGURE 4 - SUMMARY OF BIAXIAL PRESSURE TESTS OF ELECTRON BEAM WELDED TITANIUM 6Al-4V ALLOY

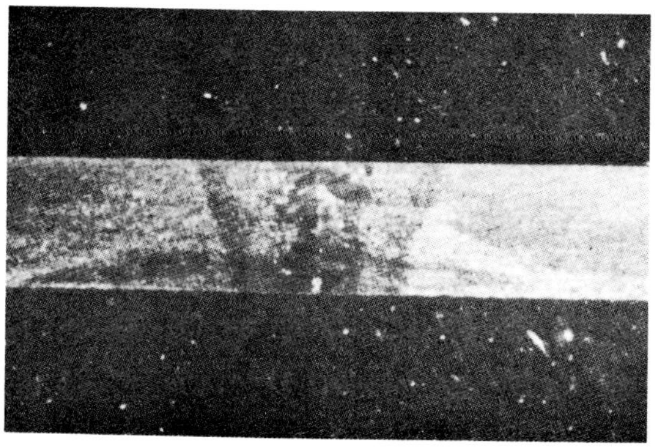

Neg. No. D115-7 Mag: 10X
Joint Design No. 1

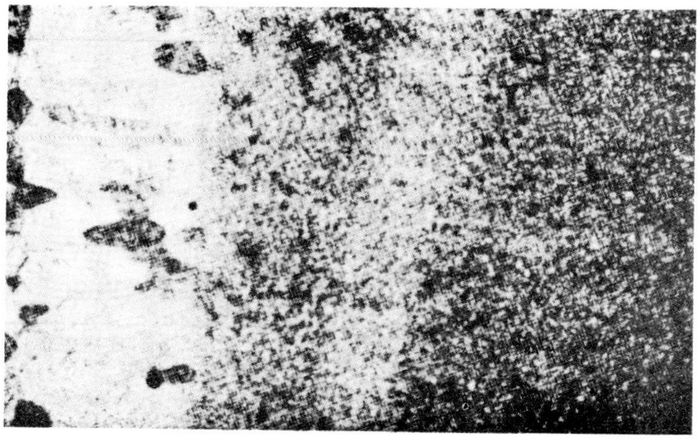

Neg. No. D115-2 Mag: 50X

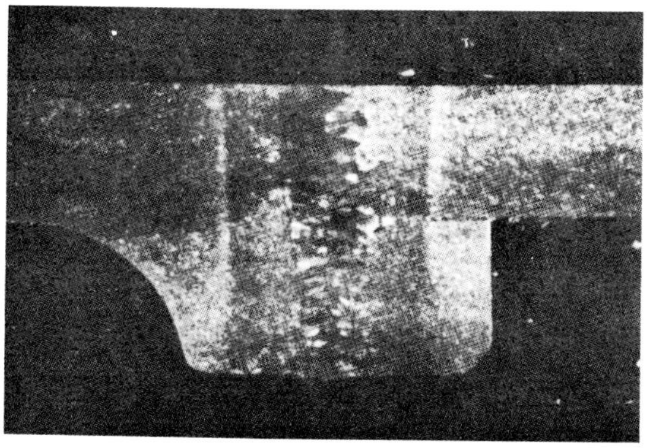

Neg. No. D115-8 Mag: 10X
Joint Design No. 2

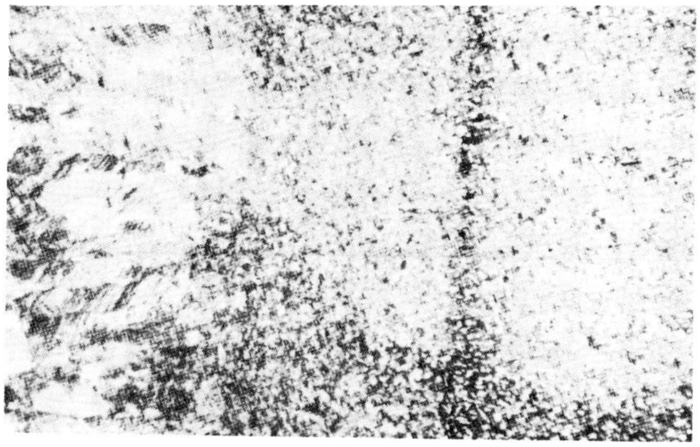

Neg. No. D115-5 Mag: 50X

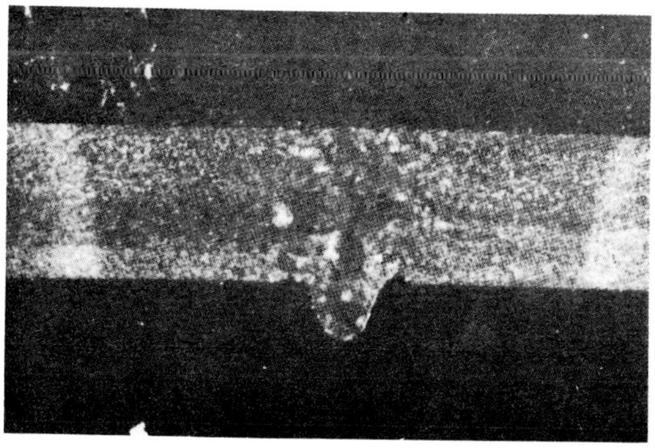

Neg. No. D115-6 Mag: 10X
Joint Design No. 3

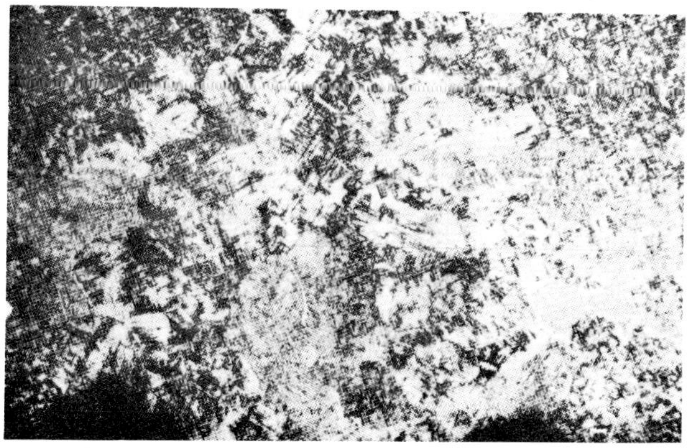

Neg. No. D115-4 Mag: 50X

Krolls etchant used for all photomacro and micrographs

Figure 5. Cross sections through electron beam welds used in the fabrication of pressure vessels subsequent to finish machining.

Neg. No. C8164-3 Mag. IX

Figure 6. Joint Design No. 2 Pressure Vessel After Electron Beam Welding.

Neg. No. 9220

Figure 7. Assembled Pressure Vessel Installed in Hydrostatic Test Stand.

Neg. No. C-8178-8 Mag. 1X

Figure 8. Typical unwelded parent metal pressure vessel failure.

Neg. No. C-8178-7 Mag. 1X

Figure 9. Typical electron beam welded joint design No. 1 pressure vessel failure.

Neg. No. C-8178-3 Mag. 1X

Figure 10. Joint design No. 2. pressure vessel after failure. Failure mode is typical of two of the three bottles fabricated to this joint configuration.

Neg. No. C-8188-2 Mag. 1X

Figure 11. Electron beam welded joint design No. 2 Pressure vessel after failure. Vessel shown was one which failed circumferencially.

Neg. No. B2021-1

Figure 12. Unassembled components of electron beam welded titanium compressor rotor.

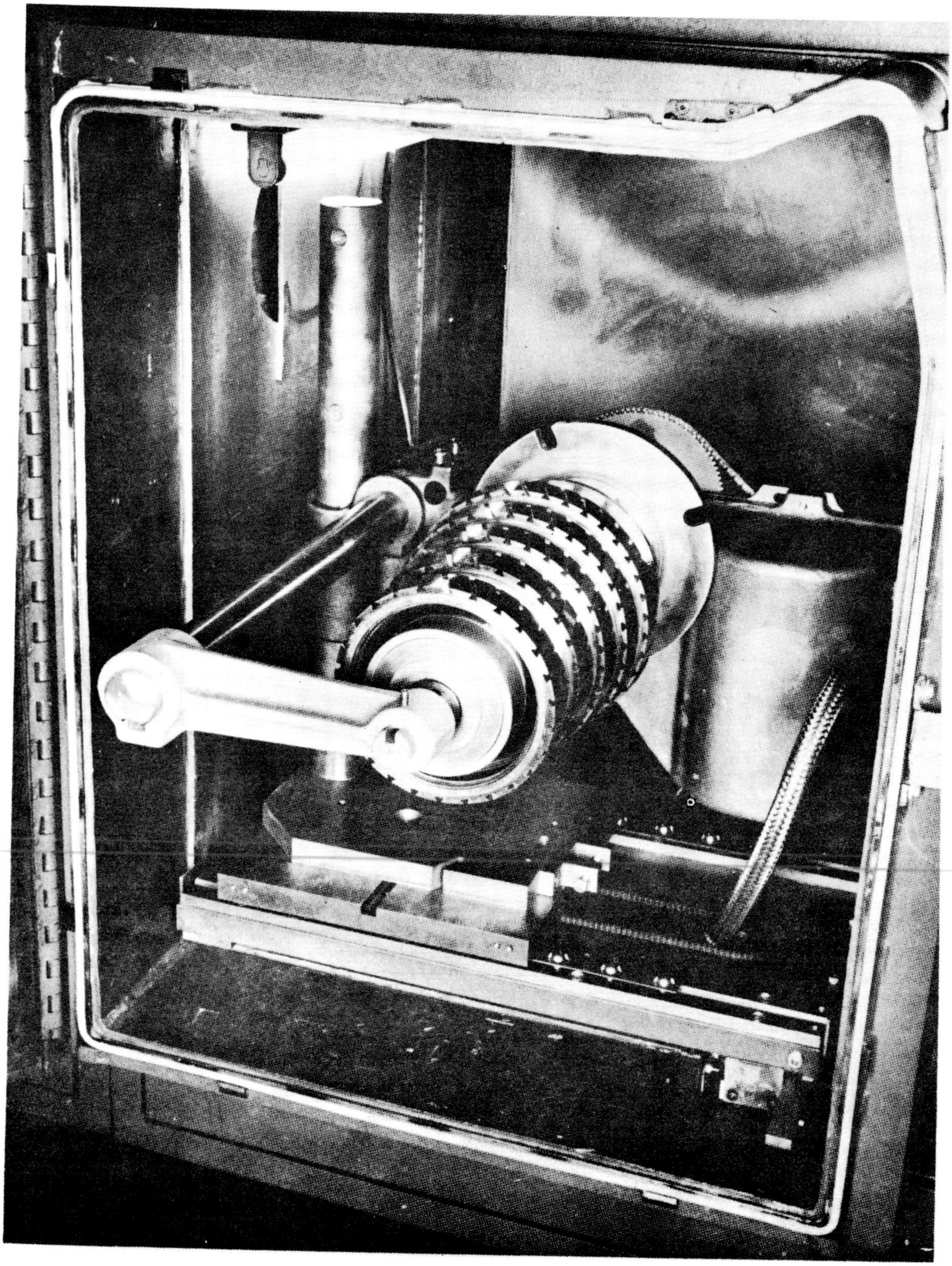

Neg. No. B9042

Figure 13. Fixture for welding compressor rotor assembly in the chamber of the electron beam welder.

Neg. No. B2030-1 Neg. No. B8053-A Mag. 20X

Figure 14. Portion of electron welded titanium rotor and cross section through weld joint.

Neg. No. D8115-1

Figure 15. Bladed Electron Beam Welded Titanium Rotor.

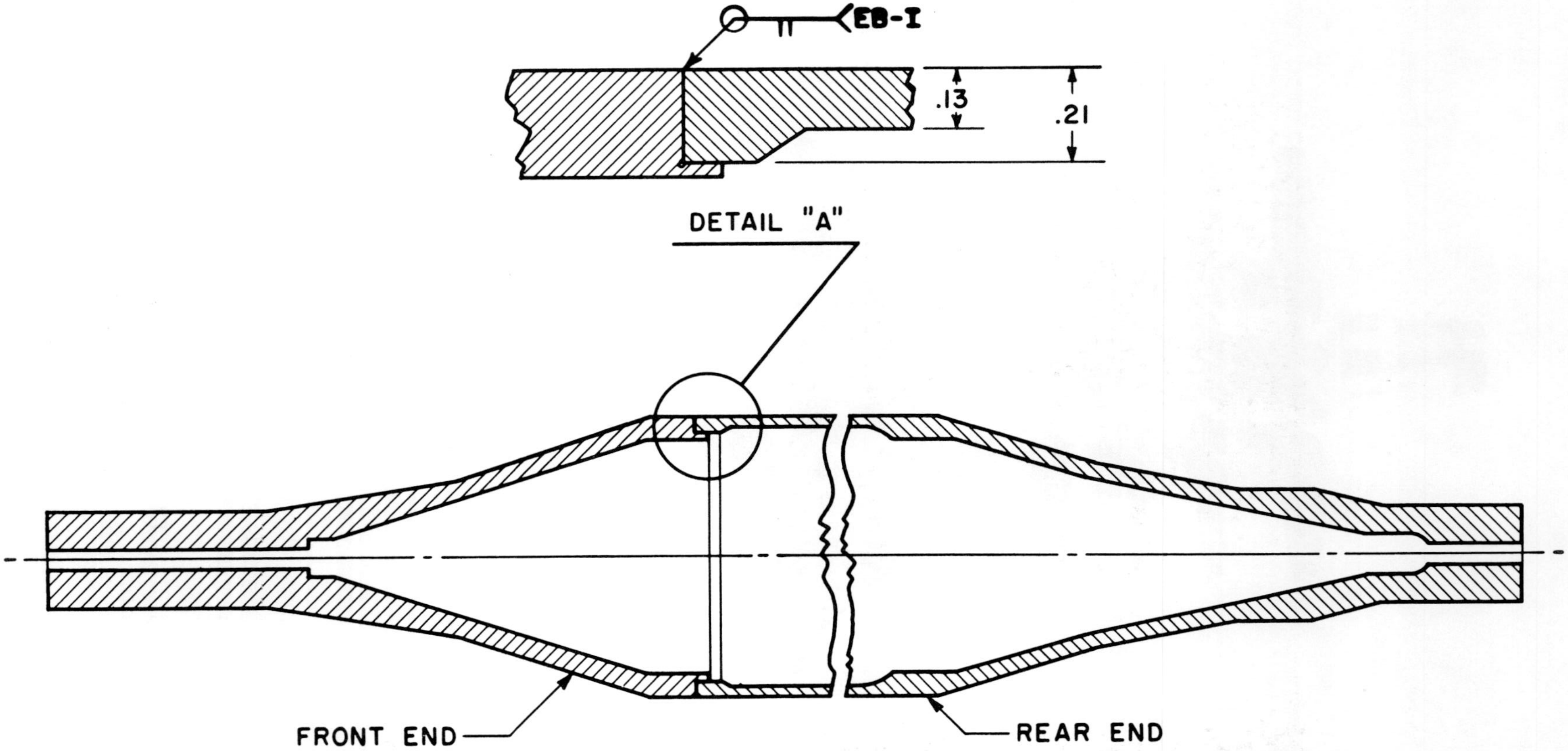

FIGURE 16 - ELETRON BEAM WELDED POWER SHAFT

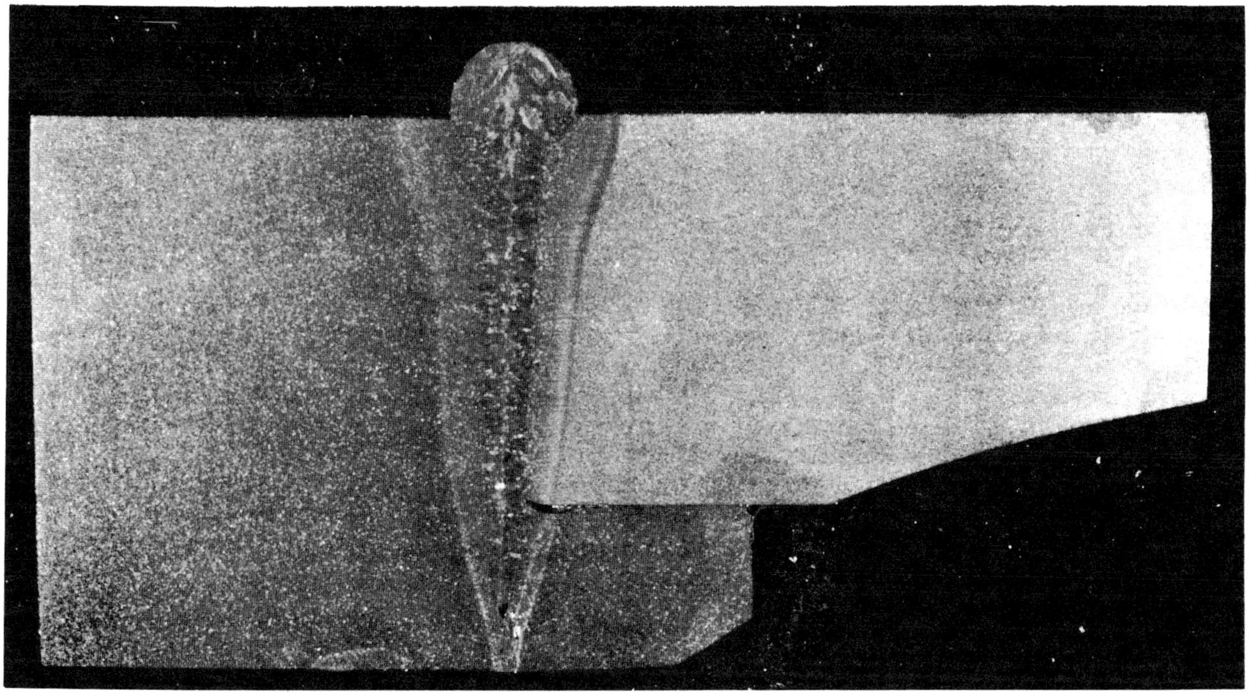

Neg. No. B8053C   Mag. 10X

Figure 17. Photomacrograph of electron beam weld in simulated power shaft joint.

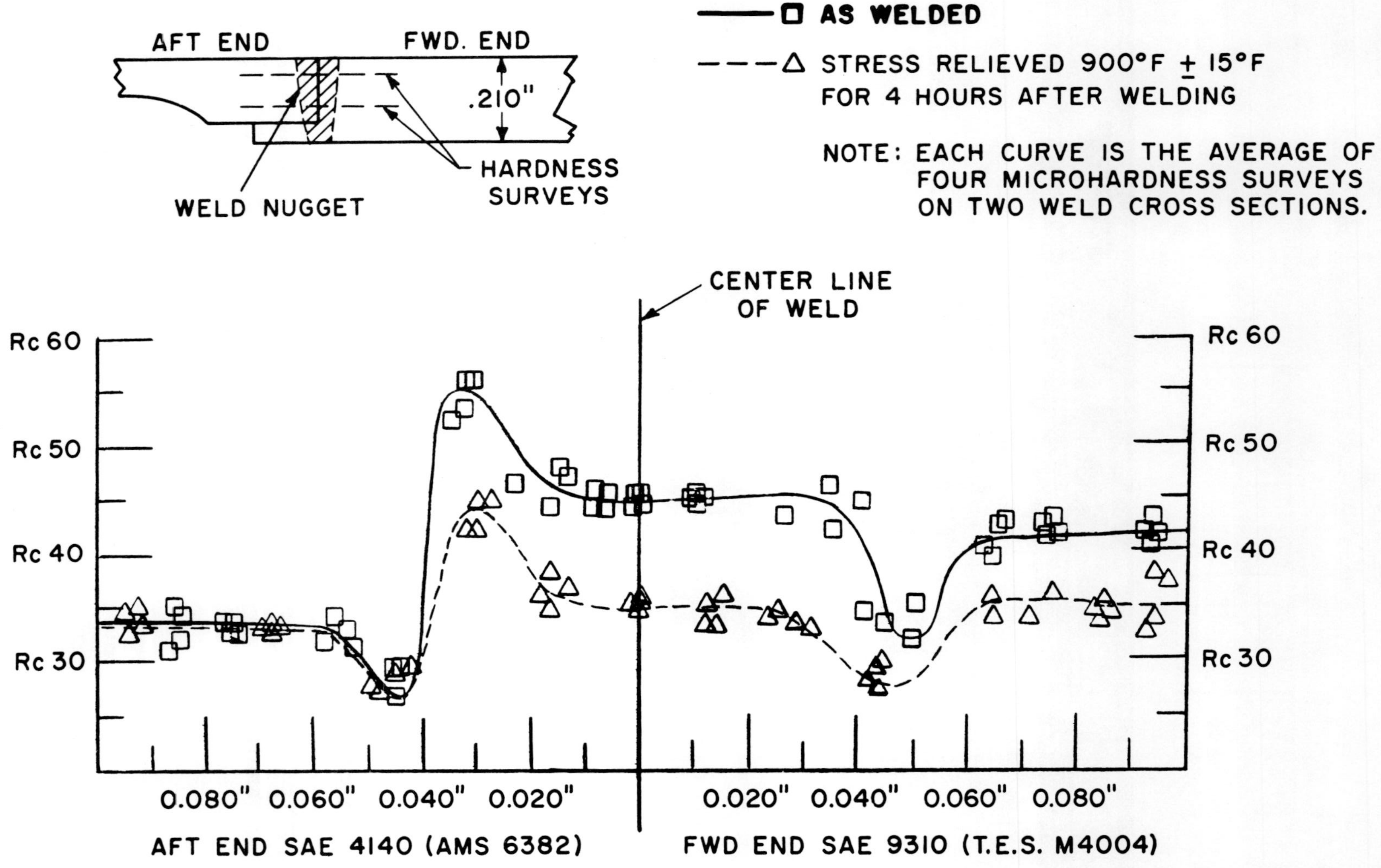

FIGURE 18-MICROHARDNESS SURVEY OF ELECTRON BEAM WELDED POWER SHAFT JOINT

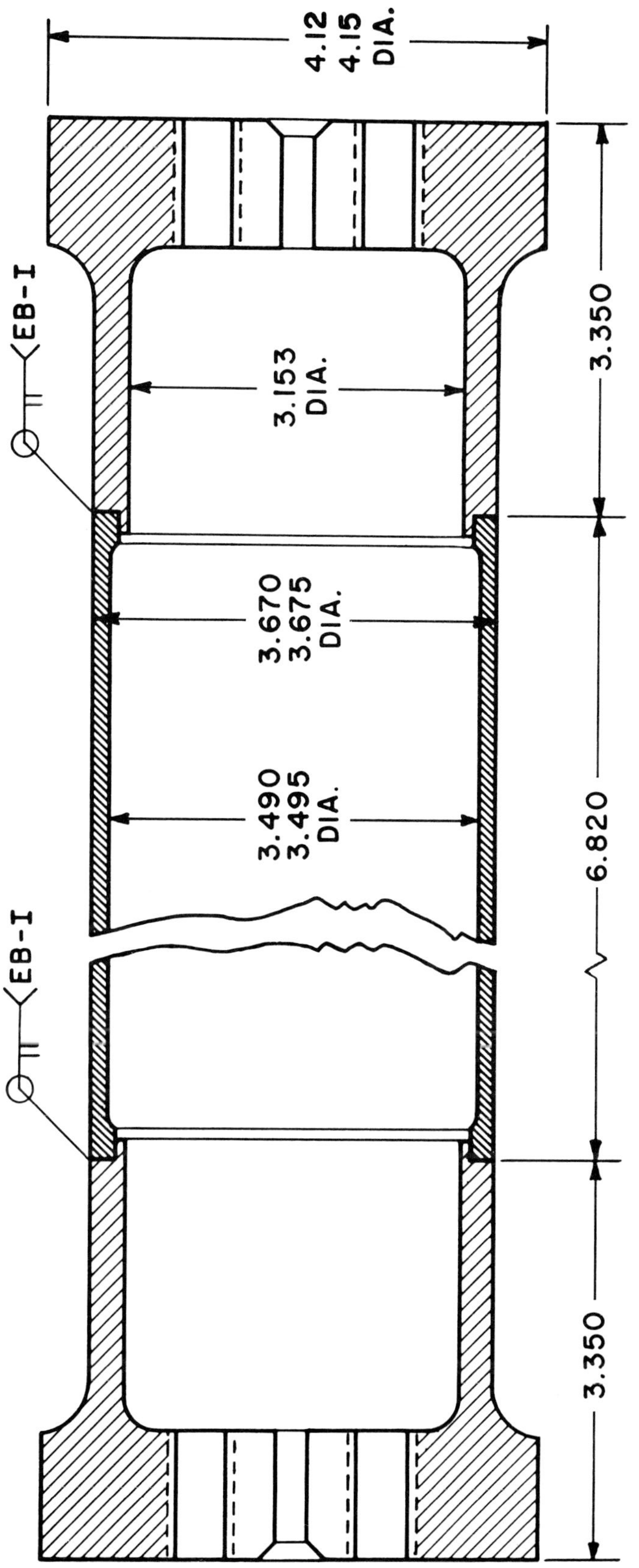

FIGURE 19 - FATIGUE TEST SPECIMEN SIMULATING AN ELECTRON BEAM WELDED POWER SHAFT FOR ENDURANCE TESTS AT ELEVATED TEMPERATURES

Neg. No. C9023-2

Figure 20. Instrumented fatigue specimen in testing rig.

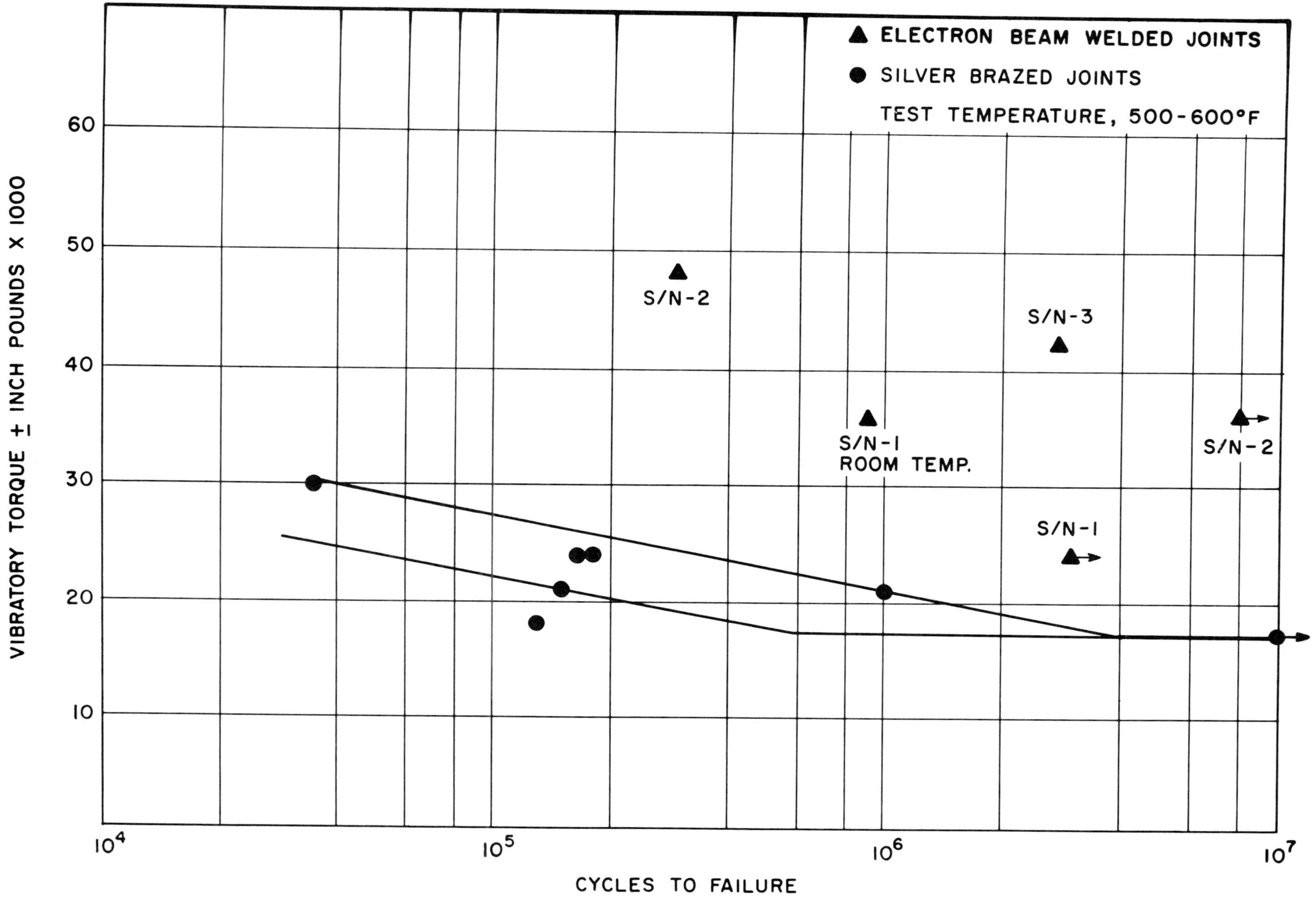

FIGURE 21-T53 POWER SHAFT FATIGUE RESULTS ELECTRON BEAM WELDED VS BRAZED SPECIMENS

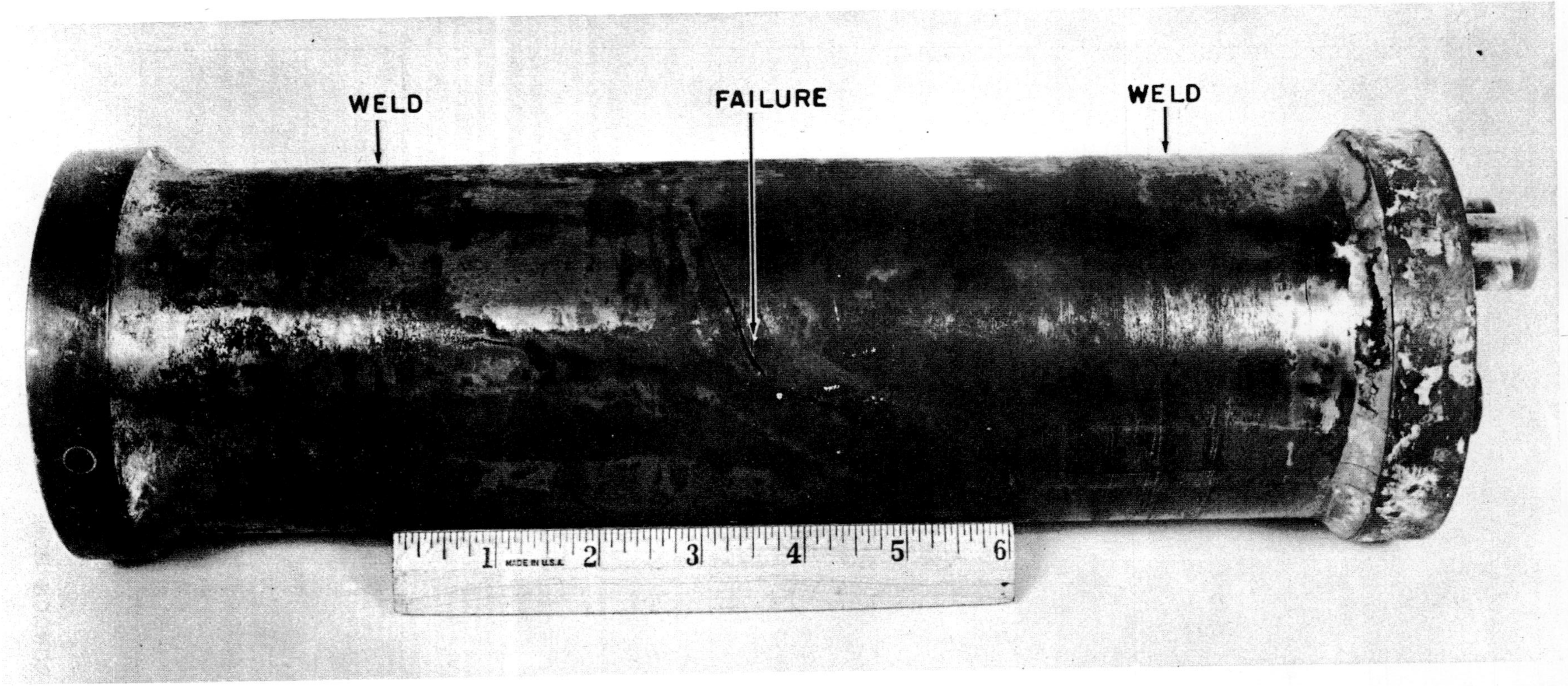

Neg. No. C-1069

Mag. 5/8X

Figure 22. Electron beam welded fatigue test specimen S/N 3 showing location of failure.

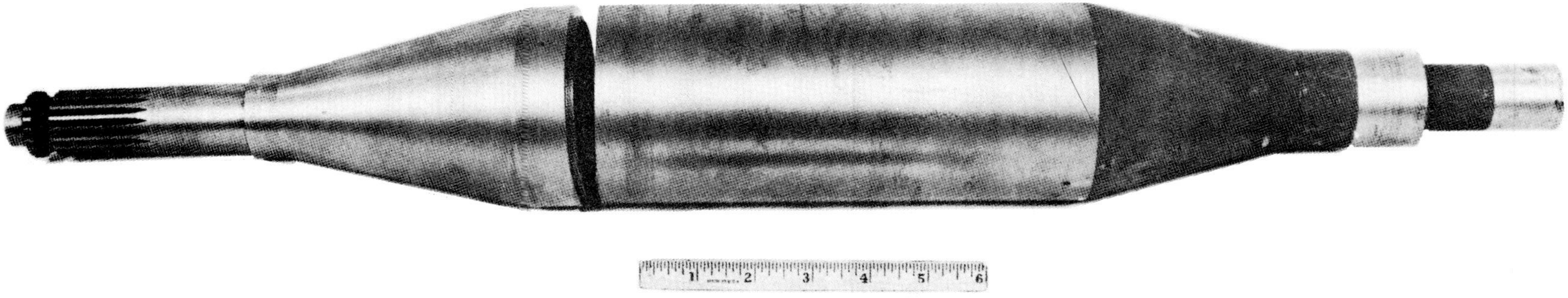

Neg. No. B2021-3

Figure 23. Power shaft details prior to electron beam welding.

Neg. No. B9043

Figure 24. Local Stress Relieving Fixture for Electron Beam Welded Power Shafts.

Neg. No. C2049 Mag. 1.2X

Figure 25. Power turbine disc with new spline blank after electron beam welding.

Electron Beam Welding— Applications and Design Considerations for Aircraft Turbine Engine Gears

By
N. F. BRATKOVICH
Section Chief
Joining Processes Laboratory
W. L. McINTIRE
Senior Design Engineer
Reduction Gears
ROBERT E. PURDY
Senior Project Engineer

Introduction • *Allison Division, G. M. C. is taking full advantage of a new welding process for the fabrication of precision mechanical components—Electron Beam welding. EB welding provides the design engineer with a joining process that results in a new degree of freedom from many of the limitations of present day machine tools. Too frequently design criteria are established on the basis of what can be produced readily with existing equipment rather than the best use of the functional material for maximum reliability, minimum size and weight, and minimum cost. EB welding allows the engineer to design precision parts that can be economically joined together without the distortion that is inherent with more conventional welding processes, and without resorting to cumbersome splines, locknuts, locktabs, bolts and flanges. On aircraft gear components, for example, Allison experience has shown that finished gears and shafts can be successfully assembled by EB welding without subsequent finishing operations. In other cases, EB welding can be used for joining semi-finished components with subsequent finishing by grinding, honing, etc. Since the degree of distortion caused by EB welding is related directly to the design of the weld fixtures and holding tools, each application may require a brief development program to check out the adequacy of the welding fixtures and the welding machine parameters.*

Design improvements by the use of the EB welding process are threefold: (1) significant design limitations are eliminated (2) cost is reduced and (3) reliability is increased. This report discusses some typical applications and design considerations for EB welding and the processing techniques involved and shows how the advantages of EB welding can be realized.

Applications

The variety of applications of EB welding to the fabrication of precision mechanical components is endless. Some typical aircraft gear design improvements made by Allison utilizing EB welding are discussed herein, but the engineer is limited only by his imagination in applications of EB welding to gain the advantages of this low distortion joining process.

DOUBLE HELICAL GEARS

Current design stress levels for double helical aircraft gears require that the gear teeth be hardened after cutting. The desirability of maintaining the smallest possible gap between helices for minimum size and weight precludes finish grinding of the gear teeth after hardening due to the large gap required for grinding wheel clearance. The designer is limited in his choice of the gear material to a nitriding steel to control hardening distortion. In cases where the

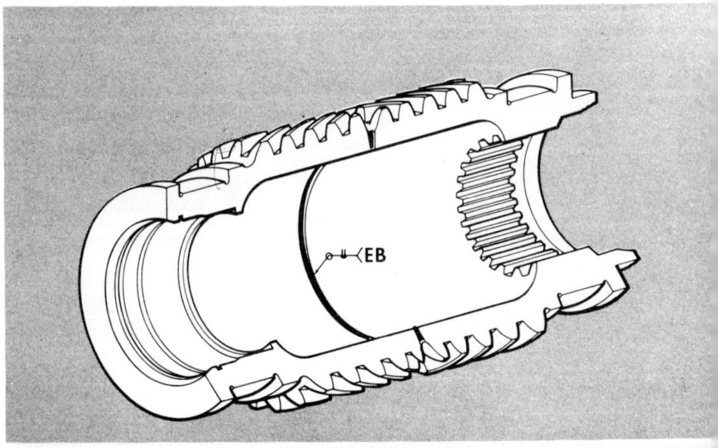

Figure 1—Double Helical Gear—Electron Beam Weld at Center

gear must have integral bearing raceways, experience with nitrided steel as a bearing raceway material is limited as compared to the more conventional thru-hardened or carburized bearing steels.

Using the EB welding process, the design of the double helical gear may be improved in two ways: (a) the double helical gear teeth can be manufactured in two pieces from carburizing steel, cut, case-carburized, hardened, and ground separately, and EB welded at the center as shown in Figure 1; or (b) the gear teeth can be made from nitriding steel, cut, shaved, nitrided and the bearing raceways made from carburizing steel, and EB welded to the gear tooth section as shown in Figure 2. In the case of alternative (a) the necessity of a large gap for grinding wheel clearance is eliminated and in alternative (b) the desirable properties of two different steels are utilized on an integrated piece.

On large double helical gears, with large diameter webs, the gear teeth and web section can be EB welded together as shown in Figure 3. This results in great weight savings and increased reliability over the bolted assembly shown in Figure 4.

CLUSTER GEARS

In many cases, accessory cluster gears requiring high levels of accuracy are joined together with splines, locknuts, or bolt flanges in order to avoid the necessity of providing space for grinding wheel clearance. Allison has eliminated the splines and bolts in engine accessory gears by finishing the gears separately and then EB welding the finished gears to their shafts. The advantages are obvious: (a) the inherently low reliability bolts, locknuts, and lockwashers are eliminated, (b) the design is simplified, and weight reduced, (c) the individual gears can be manufactured in "stacks" which results in greatly reduced cost in production. The cluster gear for a modern turbine engine accessory gearbox shown in Figure 5 is an example of how Allison uses EB welding to manufacture a gear that would be extremely difficult to make in production by more conventional means. A splined version of the same assembly is shown in Figure 6.

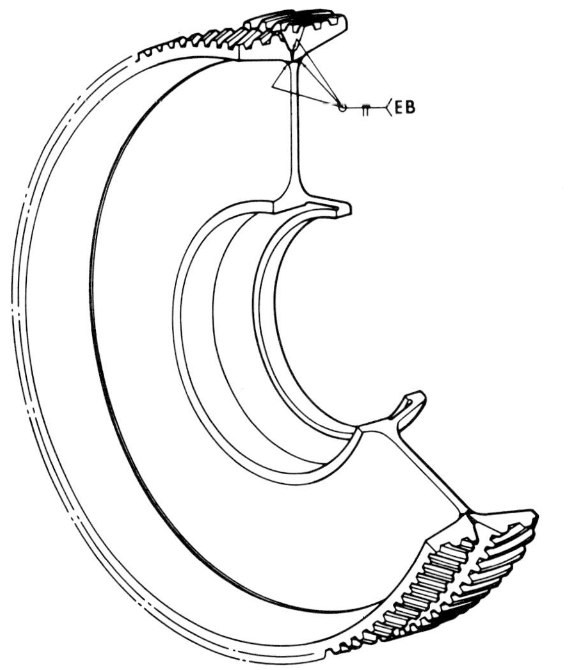

Figure 3—Double Helical Gear—Electron Beam Welded at Web

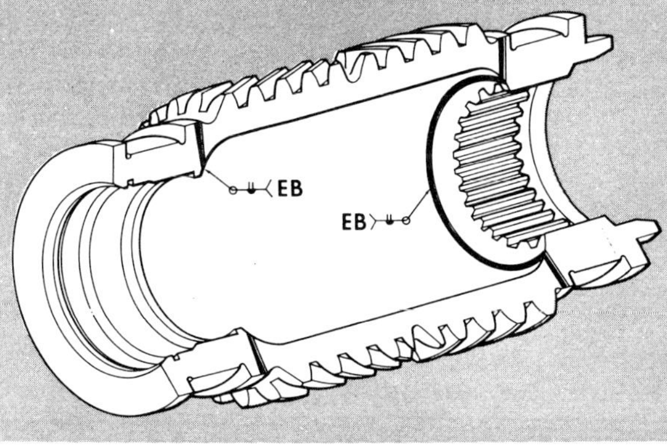

Figure 2—Double Helical Gear—Electron Beam Welded Hubs

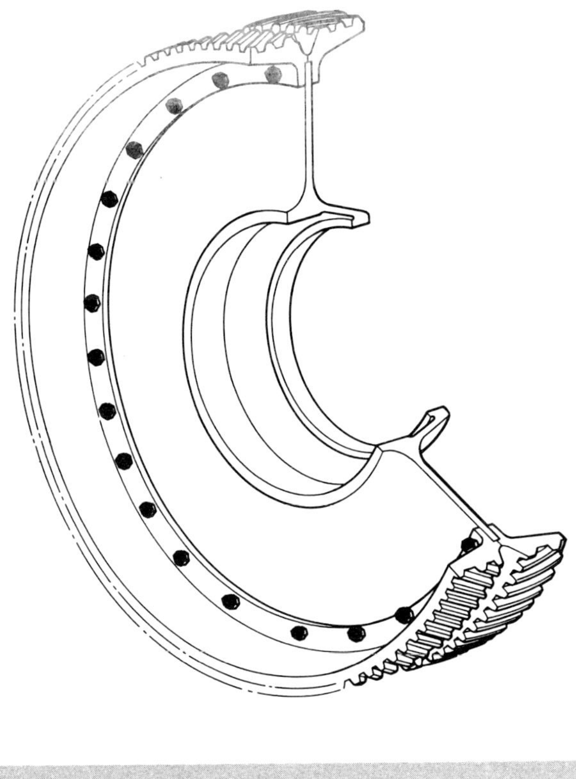

Figure 4—Double Helical Gear—Bolted Assembly

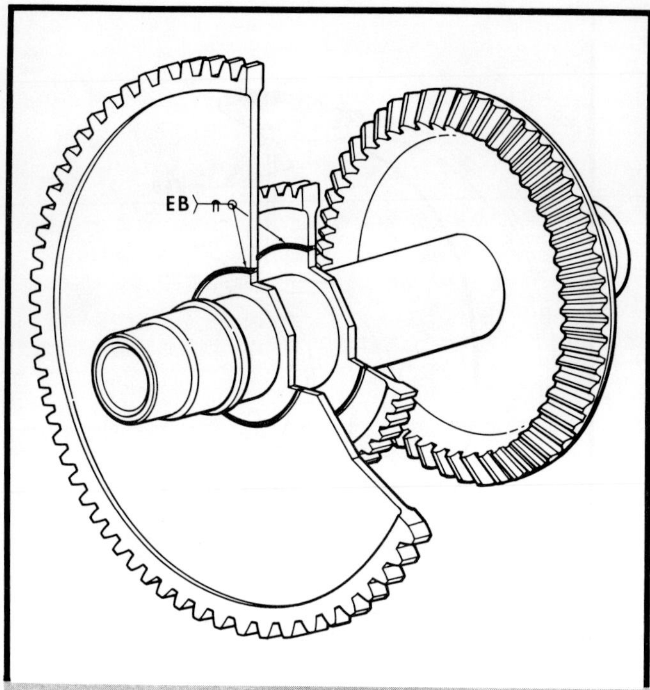

Figure 5—Accessory Cluster Gear—Electron Beam Welded at Webs

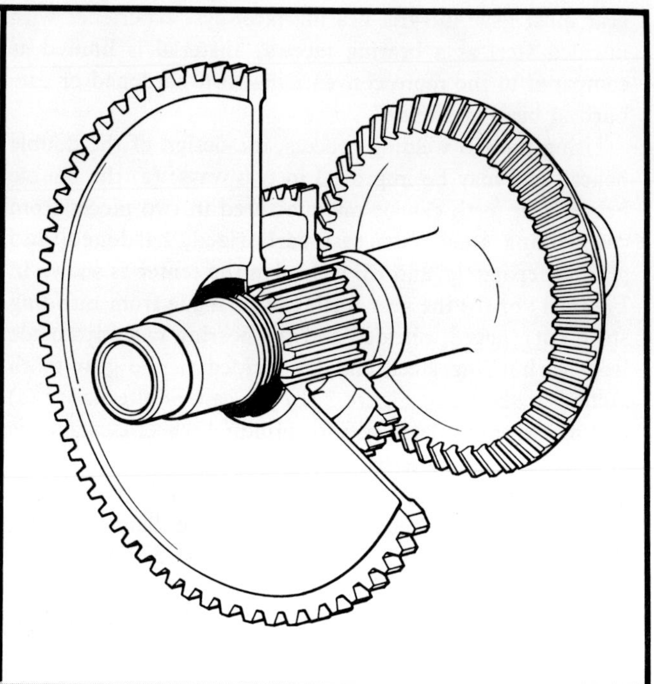
Figure 6—Accessory Cluster Gear—Splined Assembly

SHAFT GEARS

The EB welding process is ideally suited to manufacturing simple shaftgears similar to the one shown in Figure 7. Electron beam welding the gear blank to the shaft results in great savings of material. A costly one-piece die forging has been replaced by a two-piece construction consisting of a simple "pancake" and a piece of bar stock. Further cost reduction can be attained by making the shaft from a less expensive material than the more costly gear material.

DESIGN CONSIDERATIONS

The successful application of EB welding to aircraft components requiring close control of welding distortion is, to a great extent, dependent upon the configuration of the weld joint. It is up to the design engineer to choose a weld joint configuration and location consistent with the welding machine requirements in order to obtain satisfactory results. Weld joint location is usually dependent on electron beam accessibility since weld joint efficiency or strength can be considered in stress evaluation.

In general terms, the basic requirements of an electron beam welded joint to obtain minimum distortion and good weld quality are described below:

Weld Geometry

Basically, the thickness of the weld joints should be as small as possible. The greater the thickness of the weld, the greater the width due to the converging shape of the electron beam. Since welding distortion is related to the volume of metal heated to fusion temperature, it is obvious that thick welds inherently result in greater distortion than thin welds. The deep penetrating characteristics of EB welding allow straight abutting and tightly fitted surfaces to be successfully joined even in thicknesses exceeding 1 inch.

Tooling

Many weld joints can be made "self-tooling," i.e., the pieces to be joined are designed to fit together with interfaces that assure squareness and concentricity before welding without the necessity for elaborate holding fixtures.

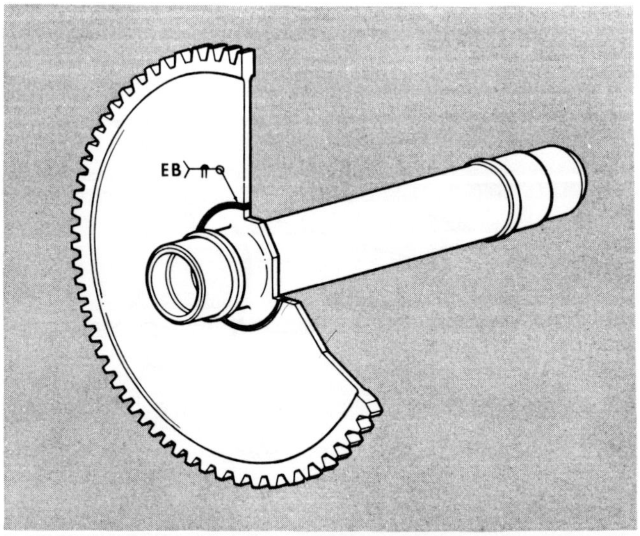

Figure 7—Accessory Shaftgear—Electron Beam Welded at Web

Anti-spatter Backup Lips

In cases where welding removes access for finishing the welded surface and to protect adjacent surfaces (see Figure 8), an anti-spatter lip should be provided to trap the weld spatter in a cavity as shown. This eliminates loose weld particles causing oil contamination and an unbalanced condition of a high speed rotating part. Wherever possible, however, the weld joint should be finished to permit accurate evaluation of the weld by nondestructive testing.

Processing Electron Beam Welded Gears

(1) The EB welding process is one of the more recent additions to the fusion welding field. A simple and basic explanation of the electron beam welding process is that the operation requires a vaccuum chamber, diffusion, roughing, and hold pumps, high voltage power source, electron gun, and control console. The diagram shown in Figure 9 depicts a hot filament generating electrons that are directed through an anode. A focusing coil focuses the fast moving stream of electrons that impinge the workpiece. The kinetic energy converts to thermal energy as the electrons strike the workpiece melting and fusion welding the metal joint. The most significant features of the process are given:

- The welding operation is accomplished in a vacuum of 1 x 10^{-4} mm of Hg or greater which eliminates the problem of interstitials or absorbed gases, and produces extremely clean welds. This means that the molten weld site is exposed to one (1) to two (2) ppm of contaminant as compared to 25 to 100 ppm of contaminants when welded with the inert gas tungsten arc (TIG) welding process.

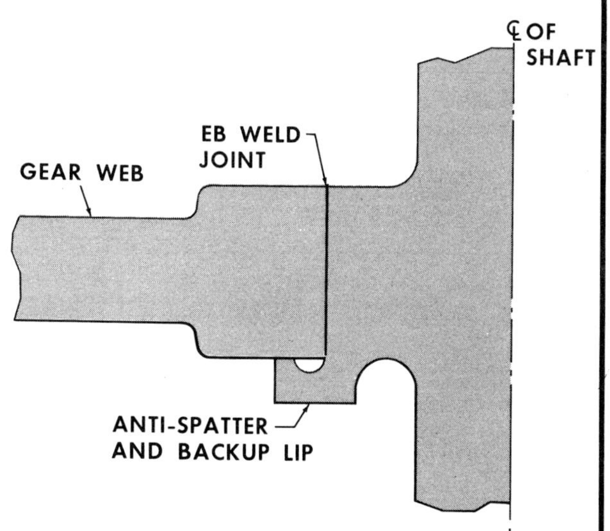

Figure 8—Anti-Spatter Lip Joint Design

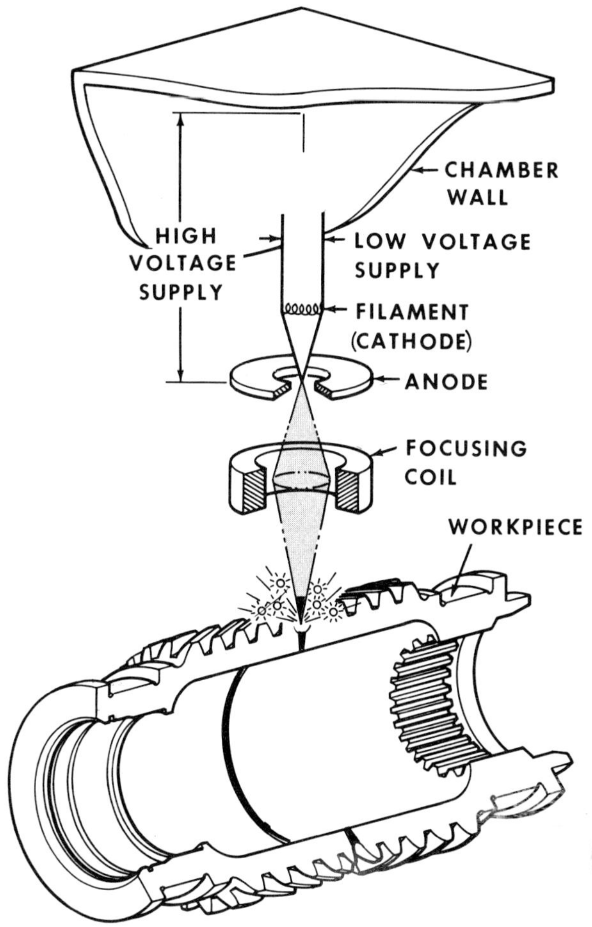

Figure 9—Electron Beam Welding of Double Helical Gear

- Inside a vacuum chamber, which is necessary for EB welding, the electron gun generates the electron beam for the welding and its power is increased or decreased in intensity by adjusting the beam current and accelerating voltage. The ideal weld parameter will result in a deep narrow weld nugget section up to a depth to width ratio of 20 to 1 as compared to a depth to width ratio of 1 to 1 or less for TIG welding. Since welding distortion is related to the volume of metal heated to fusion temperature, it is obvious that the greater depth to width ratio weld will result in less distortion.

- In reviewing all of the fusion welding processes EB welding is one of the highest in welding efficiency. Efficiencies of over 90% are realized when compared with approximately 50% efficiency for TIG and MIG welding. Experience has shown that, once a welding schedule is established, results are highly reproducible within close tolerances.

Source: Allison Research and Engineering, Vol 4, 1964

PROCESSING SEQUENCE

Processing for a typical EB welded shaftgear is outlined in the sequence shown in Figure 10. The gear cluster weldment is illustrated in Figures 11 and 12. The gear web thickness for the illustrated gear is 0.250 inch at the weld joint allowing stock to be removed from the face and root of the weld. This finishing operation is desirable in that weld bead irregularities are removed to facilitate nondestructive testing including radiographic and magnetic particle inspection.

An important processing feature is the grinding of the O.D. of the large gear to within close tolerances to position it accurately within the welding fixture shown in Figure 13. The fixture, as designed, accurately maintains the gear face step dimensions.

Upon completion of the EB welding process the gear eccentricity was 0.0015 inches and the gear face runout was 0.0006 inches, well within the 0.002 inches tolerance required.

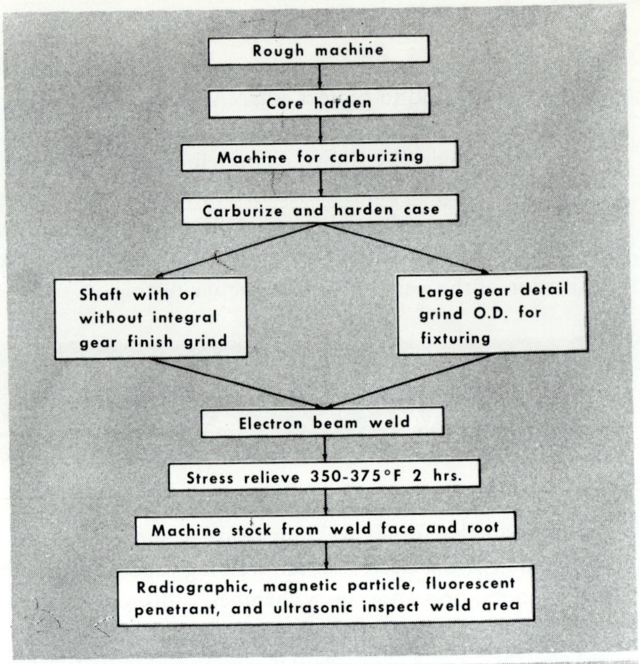

Figure 10—Typical Processing Sequence for Electron Beam Welded Gears

MATERIALS AND HEAT TREATMENT

Turbine engine gears have been successfully fabricated from the carburizing grade SAE 9310 material procured to AMS 6260 and AMS 6265 specifications. Both specifications have the nominal composition ranges shown below:

C	Mn	Si	P	S	Cr	Ni	[illegible]	Fe
0.07- 0.13	[illegible] 0.70	0.20- 0.35	0.025 max.	0.025 max.	1.00- 1.40	3.00- 3.50	0.08- 0.15	Bal.

AMS 6260 is air melted material while AMS 6265 is premium quality consumable electrode vacuum melted material subjected to more rigid magnetic particle inspection standards.

Forged details are conditioned for rough machining by normalizing at 1750°F and tempering to produce a hardness range of BHN 229-269. Intermediate heat treatment to produce a good machined surface finish is accomplished by heating to 1500-1550°F in a neutral atmosphere, oil quenching and tempering to produce the required core hardness of Rockwell C30 minimum. After a normal carburizing treatment of the required areas including gear teeth, bearing journals, or splines, the parts are heat treated to produce a case hardness of Rockwell C58 minimum.

WELDING PARAMETER DEVELOPMENT

In developing welding procedures, problems other than the establishment of welding parameters can affect the weldment results. Some of the major items are discussed:

- Weld undercut, regardless of varying weld parameters, is inherent in the process. Therefore, finishing stock (.030 inch) should be allowed at the face and root of the weld. Usually .005 to .025 stock removal is adequate to remove weld undercut.

- In achieving complete weld penetration, metal spatter and stray overshooting electrons can pit and damage adjacent finished surfaces. A variety of maskants such as copper and tungsten were tested and the best maskant was found to be a back-up ring of the same material as the base material. An integral back-up lip has been found very satisfactory in assuring weld penetration and preventing root spatter.

- Irregular weld bead formation and actually missing the intended weld joint can be caused by residual magnetism in parts and fixtures. To avoid erratic beam deflection, assured demagnetization of fixture and parts is essential.

Figure 11—Gear and Shaft Finished Details Prior to Electron Beam Welding

Figure 12—Electron Beam Welded Cluster Gear Assembly

Initial electron beam welding parameters are established by welding straight-line specimens of the chemical composition and joint thickness of the intended gear weldment design. Optimized weld parameter specimens are tested for metallurgical and mechanical properties.

The final phase of development is achieved by EB welding simulated gear detail blanks that are accommodated by the actual gear welding fixture. These tests ascertain the tooling approach as regards concept, back-up condition, concentricity, step-dimensions and ring retainment. Further tests entail nondestructive and destructive micro and macro-examinations for weld soundness.

The following is a typical welding schedule developed for the gear assembly shown in Figure 12.

Accelerating Voltage	29.5 KV
Beam Current	110 mil amps
Focal Distance	1.5 inches
Welding Speed	23.8 IPM

Specifications are established as are normally required for fusion and resistance welding.

Evaluation of Electron Beam Welds

NONDESTRUCTIVE TESTING

Evaluation of electron beam welds comprises nondestructive as well as destructive tests. It was found that the weld face and root finishing (besides removing weld undercut and stress raisers) is necessary for proper interpretation during radiographic, magnetic, fluorescent penetrant, and ultrasonic inspection. The very narrow fused zone may contain

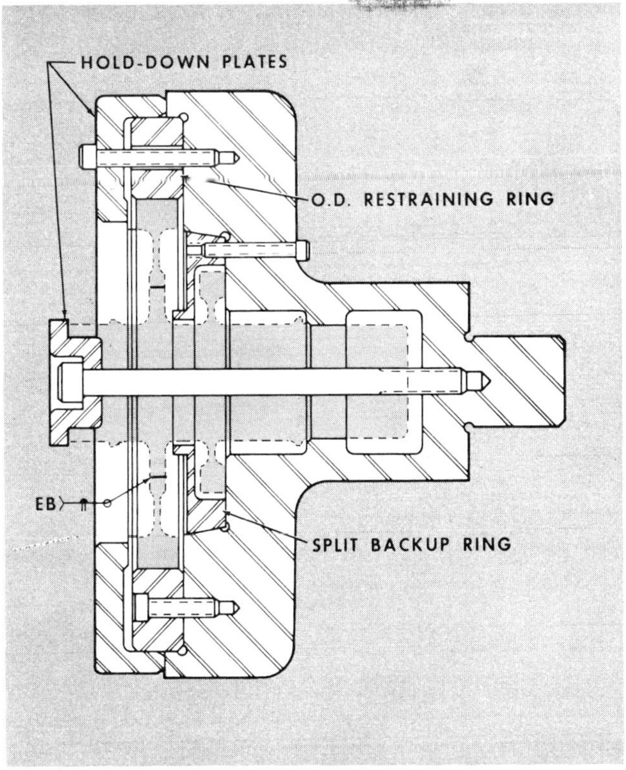

Figure 13—Welding Fixture for Cluster Gear Assembly

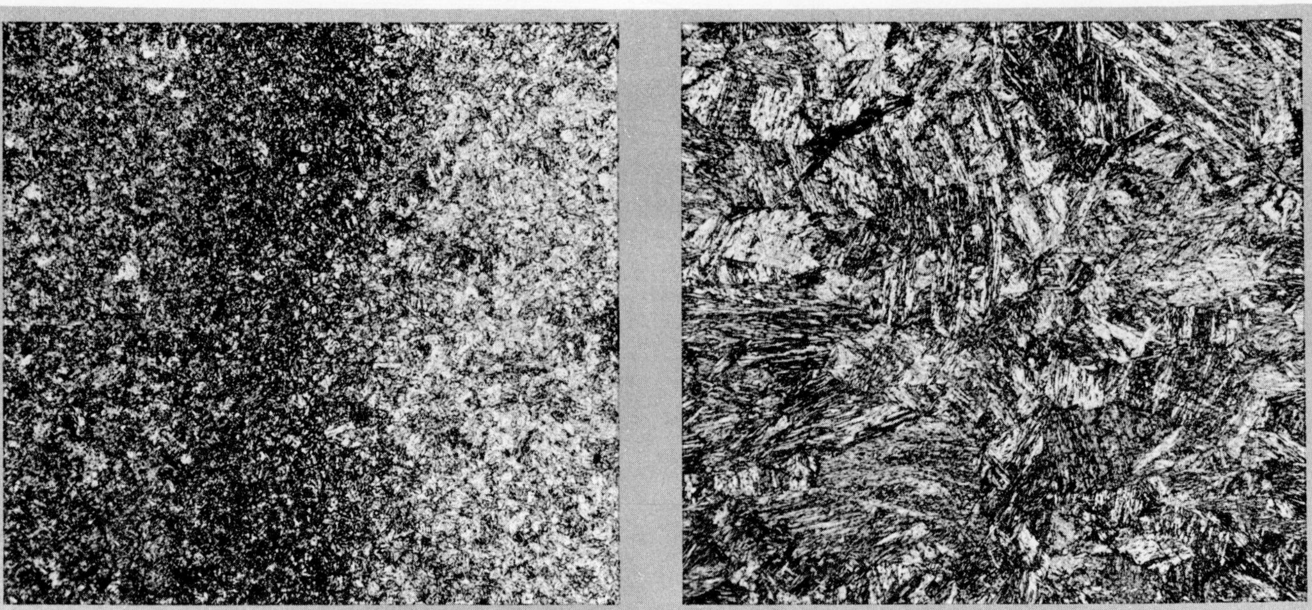

Figure 14—Typical Weld Microstructure of Electron Beam Welded AMS 6265 Gearshaft

microscopic fissures and/or very tight insufficiently fused areas along the abutted machined joint. Ultrasonic inspection methods have been successful in their detection.

METALLURGICAL EXAMINATION

Metallurgical tests have included micro and macroexamination, hardness evaluations, tensile and fatigue tests.

Typical weld and parent metal microstructures are compared in Figure 14. The cast weld structure is normal and coarser than the core treated parent material; however, it is not exceptionally columnar in grain orientation.

The macrostructure and hardness survey prepared from actual gearshaft specimens is shown in Figure 15. Note that hardness in the weld zone is comparable to the parent metal

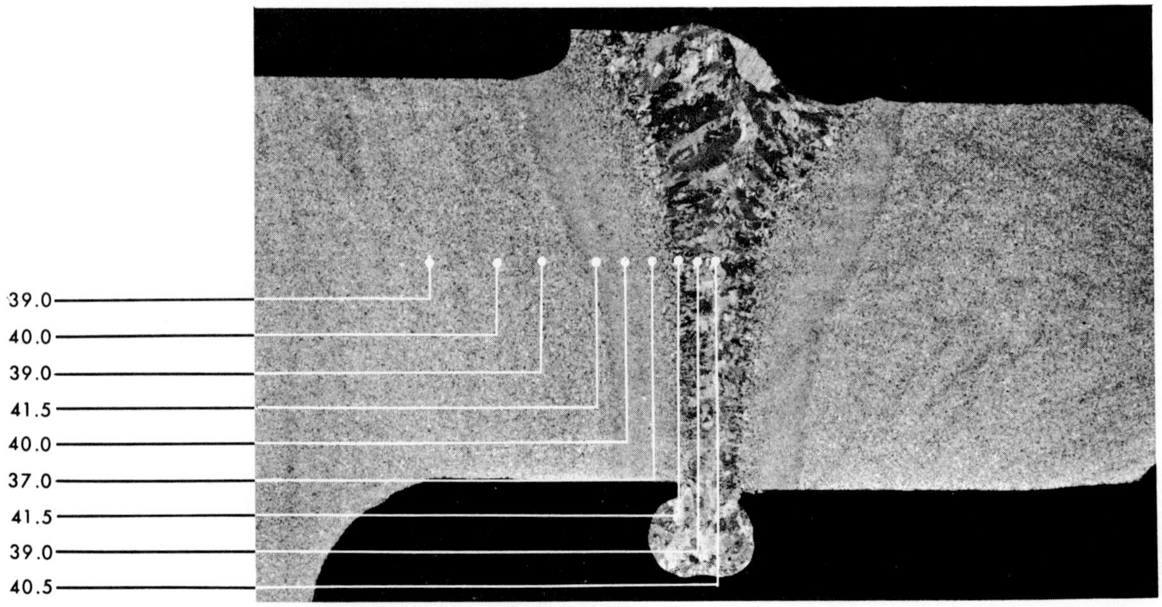

Figure 15—Typical Weld Macrostructure and Rockwell C Hardness Survey of Electron Beam Welded AMS 6265 Gearshaft

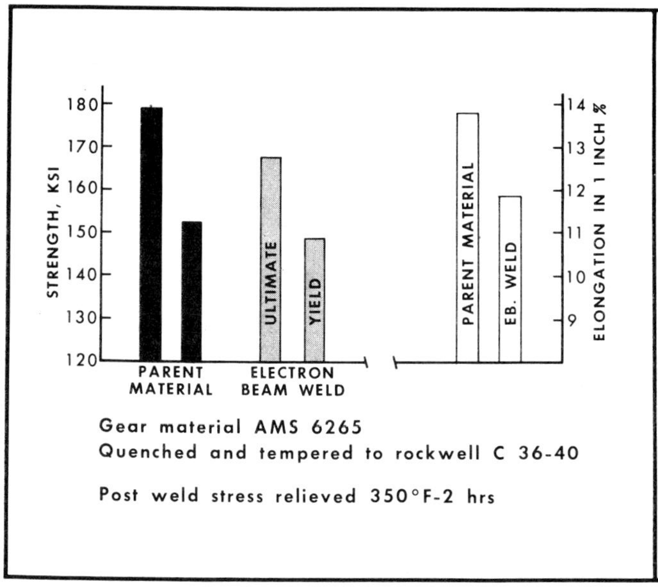

Figure 16—Tensile Test Properties of Electron Beam Welds

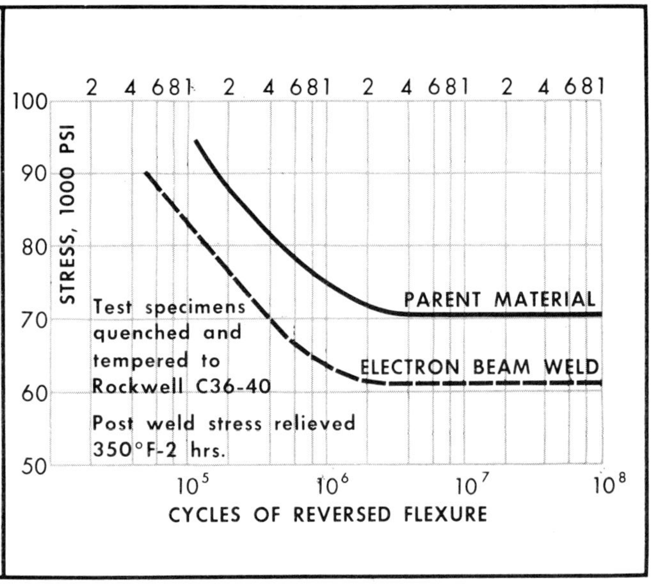

Figure 17—S-N Diagram for AMS 6265 Parent Material and Electron Beam Welded Specimens

hardness. Welding of the AMS 6265 material does not induce any localized zones of high hardness.

MECHANICAL PROPERTIES OF EB WELDS

Uniaxial tensile test properties are within normal expectancy with joint efficiencies of 92% ultimate strength and 97% yield strength. All of the tensile failures occurred in the weld zone (Figure 16).

Comparison of fatigue strength for electron beam welded AMS 6265 to that of the parent material is shown in Figure 17. The welded specimens failed in the weld zone at 86% joint efficiency.

In the weld evaluations made, excellent mechanical properties were found. Other gear materials tested resulted in comparable weld joint efficiencies. In general, it was demonstrated that electron beam mechanical properties were comparable or better than welds made with other fusion welding processes such as gas tungsten arc and metallic arc welding.

A variety of gearshaft electron beam welded designs were evaluated in extensive engine and gearbox tests and found to be successful. Nondestructive tests subsequent to these runs found no peculiarities in the gearshaft assemblies.

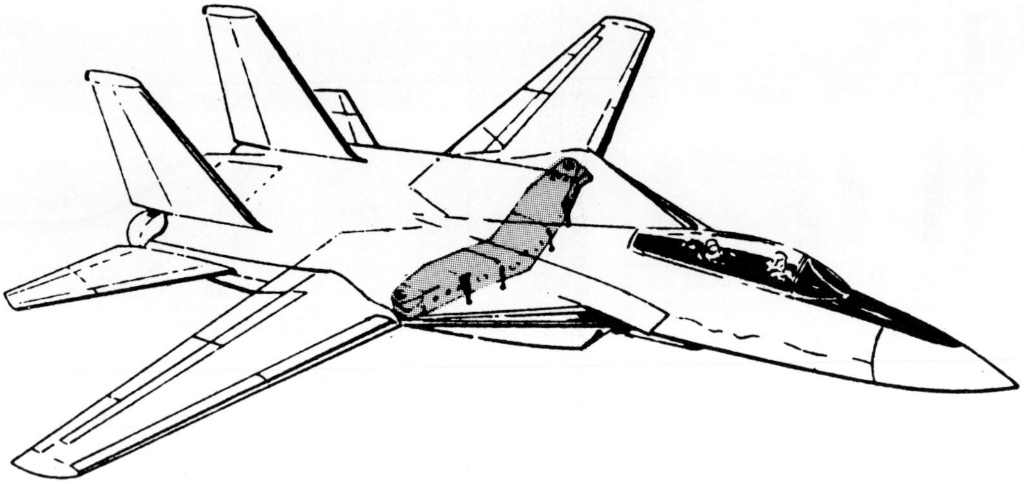

The superstrong wing box of Tomcat

Grumman Aerospace Corp. use advanced design and EB welding in building the world's hottest Navy fighter plane.

Drawing shows where the box goes in the Tomcat. Upper photo shows wing-box construction (open for inspection), and stiffener ribs on box cover standing behind the box. Lower photo show cover welded on.

The supersonic, variable-sweep-wing F-14 Tomcat can fly blurringly fast or as slow as a World War II transport. The F-14 can climb like a sky-rocket and turn on the aeronautical equivalent of a dime at g-loads high enough to pull your cheeks back to your ears. If you force it into a stall, it recovers almost instantly. It's probably the most maneuverable dogfighter flying today. Flying high or low, fast or slow, the F-14 performs at or close to maximum efficiency, always under full control. The variable-sweep wing gives it this versatility.

You're forced to design a fixed-wing fighter for optimum performance at a specific point in its speed/altitude envelope. Not because you want to, but because aerodynamic requirements are different at, say, Mach 2 than at subsonic cruise or during landing approach. Outside the design point, performance falls off. You settle for the best compromise.

But suppose the plane can change its wing configuration during flight.

Then you can have optimum performance at all points. The Tomcat does this trick. And it goes a giant step further. The pilot no longer takes the split-second actions necessary to vary the wing geometry. A computer-controlled hydraulic system does it automatically and continuously. The system can sweep the wings fore or aft 7° a second while the F-14 is pulling maximum g's.

It takes backbone

This kind of performance tremendously stresses the structure. The swinging back and forth of the moveable wing panels creates large, variable bending and torsional moments. To support these loads the plane requires a massive, superstrong wing center section or backbone or wing box.

In the F-14, aerodynamic considerations required that the wing box follow aerodynamic air passage contours. In practice, this box became a 20-foot-long, gull-wing structure. Translating this long, complex aerodynamicist's dream from paper to production was no easy matter.

This wing box demanded to be exceedingly stable because it was to be the structural heart of the aircraft—and to carry the fuselage, nacelles and outer wings. Bolting added weight and took space. Welding usually meant D6AC steel, with difficulties in process control and manufacturing, plus cost and weight penalties.

If the F-14 was to be the Navy's air-superiority fighter for the 1970's and 1980's the job would need advanced material and construction technology.

Enter EB

As far back as the early 1960's Grumman exprimented with electron beam welding of titanium alloys in a small vacuum chamber. They liked the unusually good welding characteristics of 6Al-4V titanium, and they like the ability of EB to join thick sections with very little distortion. One-pass EB also pleased because it could weld thick sections in a single pass. Eventually Grumman research proved that titanium could save substantial structural weight and that high-efficiency welds on the wing center section could save still more.

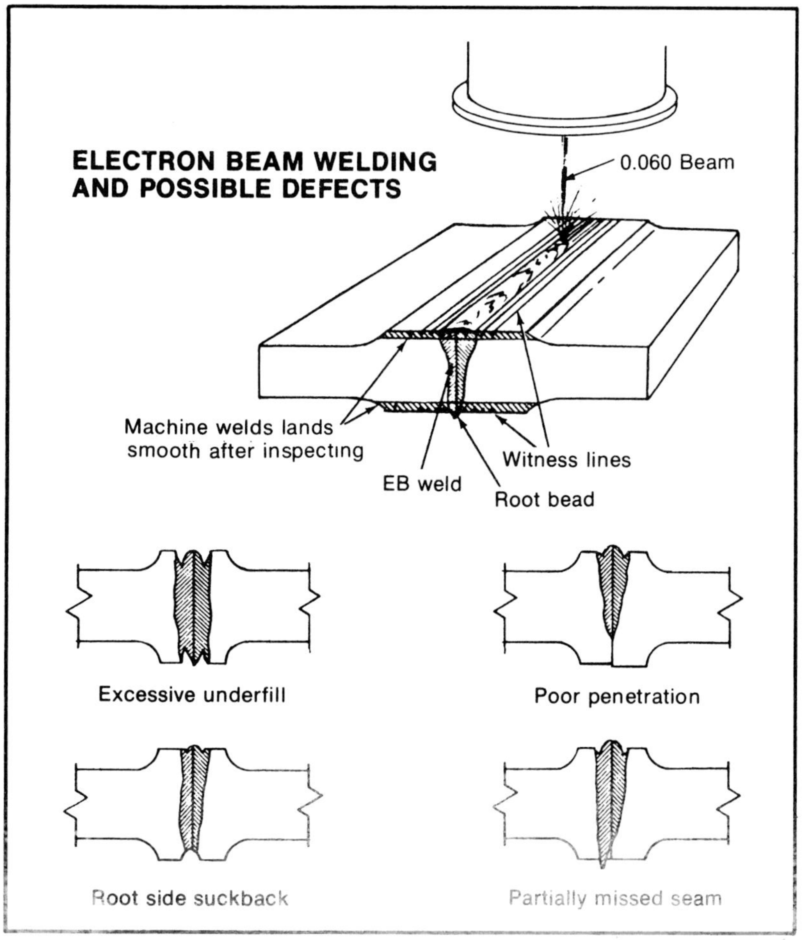

By 1968 Grumman had designed, fabricated, and tested a thick-section EB-welded 6Al-4V titanium box beam, representative of the wing-box beam of the fighter. This beam was the first aircraft structure to include welds in tension. The experimental beam successfully withstood severe fatigue testing. It finally failed when the test loads were increased 12 percent to reduce the test time—but the failure came at a tool mark on the lower cover; no welds failed.

Detail design, accurate knowledge of internal stresses, good material properties, tight process control, and high quality manufacturing control the fatigue life of structures.

The EB equipment for welding wing boxes cost about $3 million and turns out five F-14 wing boxes a month with ease. It consists of three large welding chambers—two end-loaded tunnels, 25 by 9 by 11 feet, and one clamshell, 32 by 10½ by 8 feet. The electron guns in the chambers are designed to operate at voltages up to 60, current up to 500 mA and speeds up to 100 i.p.m. Pumps evacuate the chambers to a pressure of 10^{-4} mm. of Hg in 20 minutes.

Wing box design

The wing center section, a single-cell box beam, consists of four basic weld assemblies fabricated from 33 machined parts. It transmits wing outer panel loads (applied to wing pivots) into fuselage nacelles and centerbody.

Each wing box needs 70 welds in 25 setups. Fifty-seven welds are square-butts and 13 are scarfed joints. Welds join metal thicknesses from ½ to 2 inches. The longest weld is 65 inches. As the photo shows, the box configuration is open for inspection during manufacture.

Structural and manufacturing considerations led to an instructive point at the intersection of the full depth

WING BOX WELDING JOINTS AND PROCEDURES

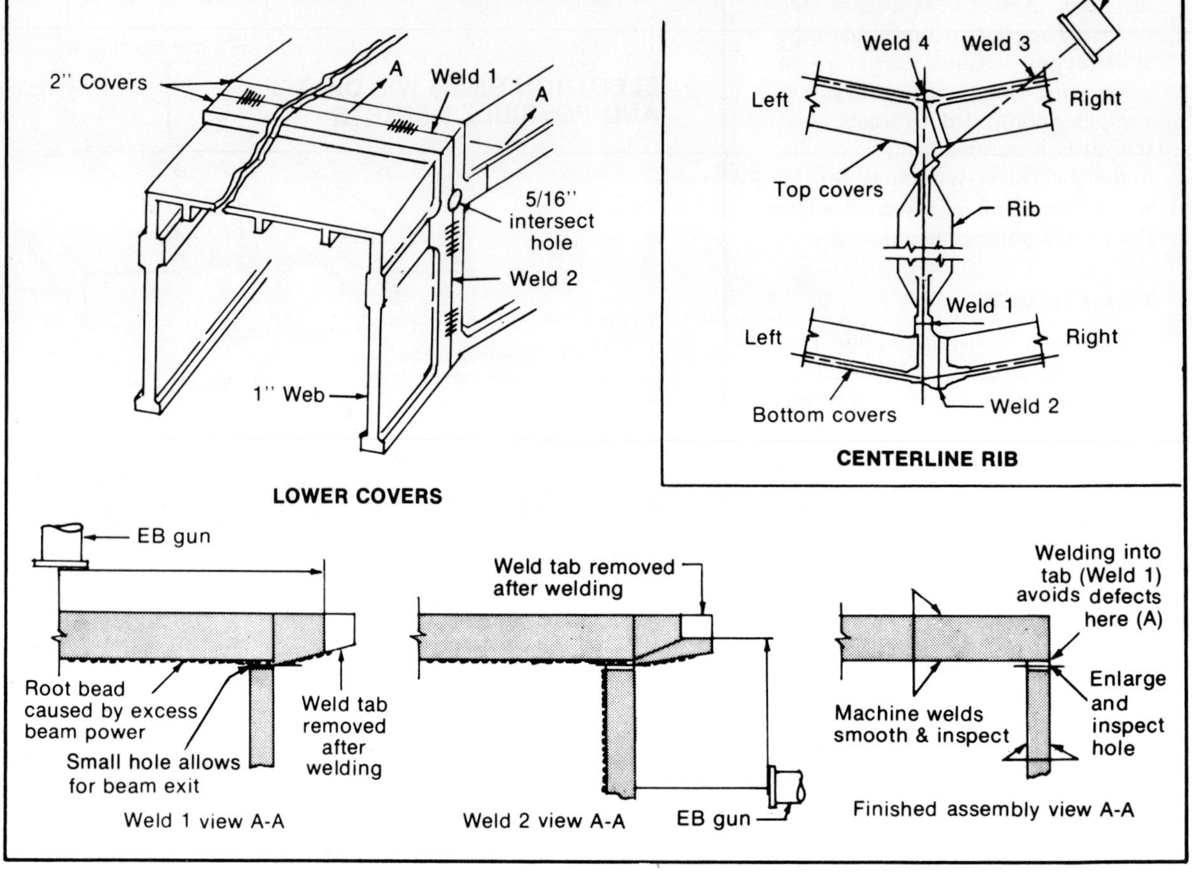

ribs (see drawing). The ribs must be located so that the load lines intersect at a common point. Also, the weld plane that joins cover to cover must be located as close as possible to the rib in order to save weight. Yet the offset must be large enough for the electron beam to exit, for machining, and for inspection.

The ribs carry the compression and tension loads generated by the cover back faces. Therefore, the ribs themselves are stiffened to withstand axial loads. To avoid thick rib welds, the stiffeners are 2 inches deep and tapered to 0.40 inch at the weld joint.

Avoiding flaws

Some of the critical flaws that afflict EB welding have been designed out of the F-14 wing box.

Solidification defects (associated with starting and stopping) are avoided by the use of external tabs—integral or welded in place—and start-stop plugs for rib-intersection holes. The tabs and plugs are machined off after weld inspection acceptance.

Spiking or cold shut (caused by partial penetration, solidification, and contraction) leads to minute voids and cracks, which are hard to detect. Grumman's welding sequence, shown in drawing of lower covers, minimizes spiking.

A 5/16-inch-diameter intersect hole is drilled to provide electron beam exit for the first, cover-to-cover weld. The second and third welds are then made by starting at the free end of each web and ending in a stop tab. This prevents defects from occurring in the lower cover at point A in drawing.

Using X-rays of the intersect hole and the inside surface of the structure, the Using X-rays of the intersect hole and the inside surface of the structure, the hole is enlarged to a specified diameter and the external tab is machined off. To counteract the stress concentration at the hole, the design calls for a symmetrical weld land of total thickness roughly twice the thickness of the lower cover cap. An A-286 steel interference-fit pin is then driven into the whole to bring the effective stress concentration from 3.0 for the hole to 1.6.

Material control

The lower covers and all pivot platters are of premium grade annealed Ti-6Al-4V. It is bought to a specification that controls everything from raw materials going into the ingot and all melting and forging. It has a guaranteed fracture toughness of 60 k.s.i. This figure is called the plane strain-stress intensity factor and it indicates the lowest load that will significantly extend a crack. It measures ability to sustain load in the presence of a flaw.

Early experience in producing thick welds indicated that hydrogen in the titanium out-gassed during the welding (melting) and the bubbles

Grumman fatigue tests a complete wing center section in this intricate setup.

caused porosity. Cure: vacuum degas all parts to reduce the hydrogen level from 130 ppm to 75 ppm.

To insure that production is the equal of development, Grumman imposed on itself a fracture control program. Grumman financed a research program at the Battell Laboratory to determine the growth characteristics of flaws in Eb-welded T1-6A1-4V sheet and plate. Battelle's data plots the size of flaws that can be tolerated at various stress levels. Conservativly, Grumman inspection control at least halves these values.

On the first production run inspectors cut fracture toughness specimens from the parent material of the most critical machined part when the material arrived. The lab tested one set of specimens as-received and another set accompanied the parent part through all processing, including EB welding, and then was tested. These tests verified design data and proved that production processing did not harm the design properties.

Nondestructive testing

In the course of development, Grumman discovered that nondestructive tests with ultrasonics, X-ray, and penetrants were unreliable by themselves. Grumman devised a different, combination scheme. It starts immediately after welding with visual inspection for underfill, root suck-back, penetration and witness lines for possible missed seams. Machining follows to remove surface irregulatities. Then comes X-ray for internal voids. Next, ultrasonic inspection detects missed seams and internal voids. Finally, after stress relieving and descaling, the wing box is dye-penetrant inspected for surface flaws.

Stress relief

After cooldown, EB welds, and welds of other methods that produce narrow fusion zones, usually have higher residual stresss and sharper changes in microstructure than welds made by other methods.

Residual stresses approaching the yield point have been measured in EB titanium weldments, but they have been stress-relieved with minimum distortion. Published data show that residual stresses of annealed Ti-6A1-4V become negligible when the material is held at 1,200° F for 4 hours. Accordingly, the whole wing box assembly is stress relieved in this way.

The process produces a tenacious oxide that must be removed because it lessens fatigue life. An acid dip does the job, followed by a light shot peening to obtain residual compressive stresses in all surfaces.

The payoff

At this time 100 wing center section boxes have been EB welded, 30 aircraft have been delivered, and 3,500 flight hours have been accumulated.

The F-14 has flown to speeds higher than Mach 2, pulled over 7 g's in maneuvers, and landed at sinking speeds of better than 24 f.p.s.

The crowning touch is that the wing box is 19 lb. lighter than its target weight of 2,065 lb., and has become the second-least expensive structure to produce on the F-14 in terms of dollars per pound. ■

This article is the work of WD&F's Eastern executive editor, Henry Lefer. It is based partly on a paper by Carlos Paez, structural designer, and Thomas Taglarine, structural analyst, of Grumman Aerospace Corp., Bethpage, N.Y., and partly on Lefer's interviews with Grumman people.

SECTION A-V:
Quality and Process Control

Factors Affecting the Quality of Electron Beam (EB) Welding

By Richard G. Wilcox and Walton R. Yerger

ELECTRON BEAM (EB) WELDING is generally selected for precision assemblies requiring high-quality welds. For this reason, users and potential users should have an understanding of the factors that affect the quality to help them develop in-house welding procedures or evaluate the performance of jobbers.

Factors include raw material inspection and preparation, welding techniques and equipment, personnel, postweld thermal and mechanical processes, and preliminary and final weld inspections. They determine weld precision, soundness, and mechanical and metallurgical properties.

Example: EB welding of high-speed steel to a lower cost alloy steel to produce continuous bimetallic strip for metal cutting bandsaws is used here as an example of the high quality required. In most cases, teeth are subsequently cut into the strip so that the tips are high-speed steel and gullets and saw backing are alloy steel. The weld must withstand the maximum stress put on the teeth because it joins the cutting edge of each tooth to its base. Weld performance is critical — breaking the weld of a single tooth can lead to a chain reaction in which the other teeth are ripped off.

Wallace Barnes Steel has supplied bimetallic strip to bandsaw manufacturers for 12 years. The company also supplies EB welded strip for other applications, including hacksaws and hole saws, but none is as demanding as the bandsaw blade.

Raw Material Inspection and Preparation: Key to Quality

Inclusion content of incoming material is critical — the high temperature of the electron beam combined with the vacuum environment may cause these impurities to gasify and expand at the beam impingement point, forming pores inside the weld or rising to the surface to erupt and leave cavities or blowholes. The extent to which a weld is weakened by internal or surface pores depends on their number, size, and location.

Inclusion content is determined by examining samples under the microscope. If an initial observation indicates that purity is questionable, the inclusions can be counted and the large ones measured. The material should be rejected if inclusion number and size exceed a predetermined standard.

Purity standards can be established on the basis of product performance. If a weld performs well on the job, the purity of its component metals is adequate or better. If the weld fails because of porosity, purity may need improving. Standards can be estimated at first, then refined via experimentation and experience.

Bandsaw stock at Wallace Barnes consists of D6A alloy steel strip for the saw backing and molybdenum high-speed steel (AISI M-series) wire for cutting edges. D6A is a modification of SAE 6150 containing 1% Mo and 0.5% Ni. The air melted, electric furnace remelted grade has the toughness and fatigue strength a backing needs to withstand flexing over pulleys. The high-speed steel (several grades are used) has the hardness the teeth need to resist wear.

Wallace Barnes' purity standard for both materials is a maximum inclusion size of 0.0003 in. (8μm) thick and a maximum inclusion count of 20 000/in.^{2} (31/mm^2), based on an in-house-developed method of calculating inclusion quantity. One out of every ten coils from each lot of material received is inspected.

Residual gases may also lead to weld porosity. Gas content is minimized by proper selection of melting practice by the material supplier.

Dimensions: The welded strip is supplied to bandsaw manufacturers in several thicknesses, ranging from 0.025 to 0.050 in. (0.63 to 1.3 mm). Widths range from ¼ to 1¼ in. (6.4 to 32 mm).

Wallace Barnes purchases the D6A backing steel in hot rolled coils measuring up to 13 in. (330 mm) wide. Coil stock is repeatedly annealed and cold rolled to reduce

An electron beam joins high-speed to D6A alloy steel to produce bandsaw strip at the Bristol, Conn., plant of Wallace Barnes Steel. The 0.005 in. (0.13 mm) in diameter beam generates a temperature of approximately 4000 F (2200 C) at its impingement point. Photo was taken through the EB welder's viewscope.

Photomicrograph 2 (100X) is a transverse section of a sound, annealed EB weld in bimetallic bandsaw strip. The high-speed steel edging is to the left of the weld, the D6A alloy steel backing is to the right. Micros 1 and 3 (1000X) show the weld interfaces with the edging and backing, respectively. The high-speed steel's structure consists of complex carbides in a softer matrix. They include molybdenum, chromium, and vanadium carbides and Mo-Cr-V carbide. Because of its response to the nital etchant used, the alloy steel's structure is not revealed as well as that of the HSS and weld. Some details can, however, be seen: ferrite (white), grain boundaries (dark lines), and pearlite and carbides (shapes).

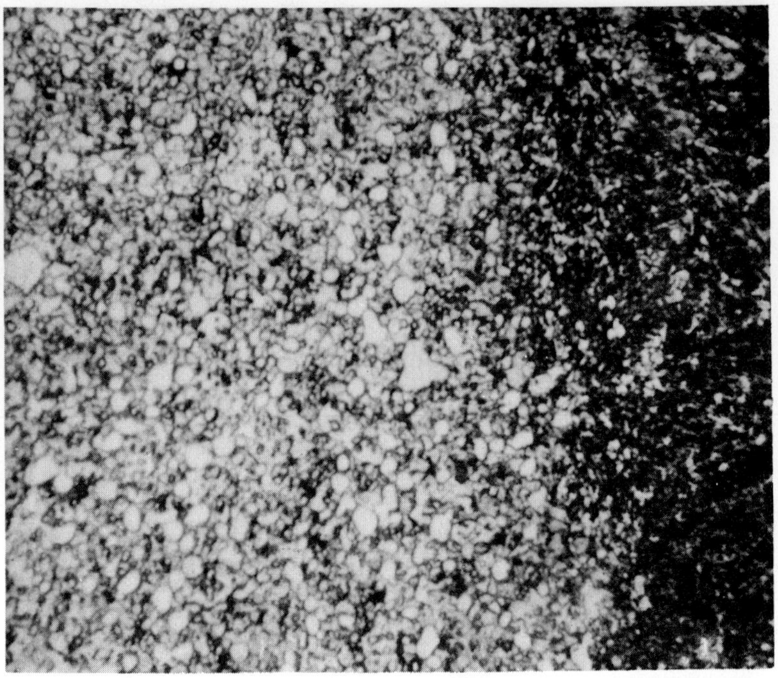

1

thickness to specification. It is then slit to width. The high-speed steel is generally purchased as annealed wire which is subsequently cold rolled to required dimensions.

Edge Preparation: Mating surfaces of components to be welded should be chemically clean. Surface impurities such as grease or dirt particles may — like inclusions — cause the formation of internal or external weld cavities. Surfaces can be adequately cleaned in an alkaline detergent bath followed by a hot water rinse. Cleanliness can be inspected by wiping the metal with a white tissue or cloth or simply by a superficial visual examination. Dirt or grease that could cause significant porosity or blowholes is not difficult to spot.

Mating surfaces should also be mechanically prepared so that they will be in intimate contact over the entire joint area during welding. If the fit is not precise, various weld defects, such as undercut and excessive dropthrough, can be introduced. Undercut develops when surface irregularities or rounded edges leave spaces between the mating surfaces. The solidified weld is lower than the base-metal surfaces because not enough molten metal was produced to fill the gap. If bumps or other defects create large gaps, molten metal runs out, or "drops through," leaving very little weld or, if the gap is as wide as the electron beam, no weld at all.

Mating surfaces should also be parallel to the electron beam. If surfaces fit together on a slant, like the sides of a parallelogram, the beam will not penetrate the entire thickness of the joint, resulting in a partial penetration weld. Poor fitup can also produce this defect.

For the continuous welding of bimetal strip, mating edges must be straight, uniform, and square. Straight and square edges can be prepared by cold rolling, grinding, filing, another mechanical process, or by a combination of processes. Strip received with "fully prepared edges" should be checked for straightness (camber) and squareness as part of the incoming material inspection.

Camber, or deviation from a straight line, can be determined by placing a straight edge against a length of strip and measuring the maximum gap between the two. Standard camber values are based on an 8 ft (2.4 m) straight edge, but measurements based on other straight-edge lengths can be converted to the standard equivalent. Wallace Barnes' standard for edge straightness is ¼ in. (6 mm) or less deviation in 8 ft (2.4 m) of strip.

Squareness can be determined with an optical comparator. At Wallace Barnes, edges must be absolutely square with a 0.003 in. (0.08 mm) maximum corner radius.

Welding Gun's Electron Optics are Important

Electron beam welds can be made in a hard vacuum, in a semivacuum, or out of vacuum, depending on weld width.

3

For a wide weld, a vacuum may be unnecessary but, for a narrow weld, a hard vacuum may be needed to eliminate nitrogen, oxygen, and other molecules in the atmosphere that tend to disperse and interfere with the electron beam.

A hard vacuum or semivacuum is required for bimetal strip welding. Two continuous systems are commonly used.

In the air-to-vacuum-to-air system, the two coils unwind simultaneously, feeding into a sealed vacuum chamber where the strips are brought together under the electron beam by a fixed guidance system. The welded strip emerges from the opposite end of the vacuum chamber and is wound onto a takeup reel.

In the all-vacuum system, coils and takeup reel are also included in the sealed chamber, which is opened after each welding cycle for coil changing.

Wallace Barnes has two air-vacuum-air systems, each with its own hard vacuum chamber and electron gun. Strip travels under the electron beam at speeds ranging from 25 to 40 ft/min (125 to 200 mm/s).

Guidance: The precision of the fixed guidance system that aligns the two strips under the electron beam is critical in any continuous welding operation. The strips must move through the system's jaws so that they're both in intimate contact and horizontally aligned at the point of beam impingement. If they're not in contact, an undercut (or no weld at all) will be produced. If they're not horizontally aligned, one component will be higher than the other, a defect called a mismatch.

Guidance systems are one-of-a-kind designs, depending on the particular manufacturer's equipment, methods, and products. One rule of thumb can, however, be stated: minimize the number of rotating parts that might seize in the high-temperature vacuum environment.

Electron Optics: EB gun quality directly affects weld quality — the better the gun's electron optics, the more precise control it affords over beam energy, diameter, and location, and the more rapid its response to correction of beam focus or alignment.

Beam dimensions and location can be observed directly through a viewscope. The decision as to their correctness is made indirectly, by examining the welded strip as it exits the vacuum chamber.

An electron gun that features nearly instantaneous response to operator commands can pay off in terms of increased product yield and, conversely, a decreased scrap rate by allowing the operator to correct defect-causing, improper focus and misalignment conditions as soon as they're observed.

Wallace Barnes' EB welders, the first of which came on stream 12 years ago, use a current or 7.5 mA. Beam diameters and energies range from 0.0003 to 0.0006 in. (8 to 15μm) and 1 to 1.5 kW, respectively.

People: Even the best equipment is only as good as its operators. Alert, capable, well-trained personnel are essential for a high-quality, EB-welded product. They must be able to detect a problem early, analyze it, and immediately take corrective action.

For example, an operator spotting undercut in welded strip emerging from the vacuum chamber might make either electronic or mechanical corrections. If the undercut is severe because of defective raw material or edge preparation, a segment might be cut out of one of the coils, or one or both of the coils might be replaced. But, if the undercut is due to a wandering electron beam caused by magnetism in the backing steel, use of the demagnetizer would be called for. A misaligned electron beam — another potential source of trouble — would be corrected via an adjustment to the welder's dc beam deflection control.

With experience, good operators develop the ability to detect by eye or hand very slight deviations from quality standards — some can even feel a weld mismatch of only 0.001 in. (0.03 mm).

In-house training can generally provide the needed technical background; the rest depends on the operator's intelligence and motivation. It's important, however, to have at least one engineer with an in-depth knowledge of EB welding.

Preliminary Weld Inspection Spots Gross Defects

Continuously welded strip should be inspected frequently after it exits the welder and before it undergoes thermal or mechanical treatment. This is done to discover defects early on, and eliminate or minimize them in the remainder of the welded coil.

Visual examination under a magnifying glass and strong light can immediately detect gross external defects. A mirror, preferably at the bottom of a lightbox, reveals the underside of the weld. Samples should also be cut from the coil for on-the-floor microscopic examination for smaller flaws.

External defects that might show up under the glass or microscope include pores, mismatch, undercut, partial penetration, or stitching. Stitching is a series of undercuts alternating with segments of sound weld. It can be caused by imperfections in the mating surfaces, jerky movement of the strip through the welder, or insufficient beam energy.

A high-quality weld should have a smooth, continuous, slightly raised bead along the entire length of the strip.

Defects Inevitable: Internal examinations of the weld are conducted by taking a sample, "tearing" it apart at the weld, and inspecting the joint under the microscope for pores and cracks. While gross external defects can often be eliminated after they're discovered, internal microscopic defects can only be minimized — the inclusions present in even the highest quality strip and the inherent brittleness of the weld make porosity inevitable to an extent.

At Wallace Barnes, bandsaw strip is visually examined every 20 ft (6 m) under the magnifying glass. For external and internal microscopic examinations, sections are cut from the beginning and end of every welded coil. This initial microscopic examination ensures that the weld meets the required standard. (Note: a final weld examination, performed before the bandsaw strip is shipped, also provides the basis for quality ratings within that standard. In both cases, examinations are conducted in the laboratory as well as at the welding station.)

Other microscopic samples are annealed and ground in the laboratory and then examined to determine how the parent coil will be affected by these postweld processes.

Postweld Thermal and Mechanical Processing

Annealing is needed to make the EB weld ductile for subsequent operations such as forming, rolling, and milling. Hardening, tempering, freezing, or other thermal treatments may also be required after fabrication to obtain desired metallurgical and mechanical properties. Regardless of the specific treatment, close control of temperature and time is critical.

When a coil of welded strip is thermally treated, thermocouples can be buried in the center of the coil to ensure that the entire mass reaches the required temperature. As a double check, heat treaters often place sample blanks in the coil which are removed after processing and then hardness tested.

In bimetallic bandsaw strip, the anneal is used to soften the weld sufficiently for tooth cutting by the saw manufacturer, and to remove the work hardening produced by cold rolling in the edging and backing steels. The goal is to make all three strip components hard enough to machine cleanly (without burrs) but soft enough to ensure long cutting tool life.

Two-Step Process: Because the critical temperatures of the weld and edging steel are much higher than that of the alloy backing steel, Wallace Barnes uses a two-step annealing process. Coils are placed in a gas-fired, batch anneal furnace and heated to 1450 F (790 C) to anneal the weld and remove the work hardening from the edging steel. This

D6A strip and high-speed steel wire are fed from coils into the EB welder's vacuum chamber (bottom).

temperature is subcritical for these materials, but exceeds the critical for the backing steel. Consequently, furnace temperature is dropped to 1225 F (665 C) to reanneal the backing steel as soon as weld hardness has been sufficiently reduced. Temperature and time are carefully controlled to avoid hardening the backing steel while the weld is being annealed. The process takes three days.

After tooth cutting, the bandsaw manufacturer will harden and temper teeth and backing in a single operation.

Straightening: If component metals have different coefficients of thermal expansion, asymmetric heating and cooling during welding or heat treating can distort them, causing camber, or edgebow, in the welded strip. In bandsaw strip, the high-speed steel edging tends to shrink at a rate sufficiently different from that of the alloy steel backing to produce some camber — the cutting edge is "shorter" than the backing. The amount of camber is enough to require straightening to meet bandsaw manufacturer specifications.

Wallace Barnes uses a thermomechanical straightening technique to stretch the edging steel or shrink the backing. Part of the process is a low-temperature furnace treatment performed immediately after welding. A side benefit of the furnace treatment: the weld is moderately stress relieved, making it less susceptible to cracking before annealing.

Decarburization: Grinding can be used to remove the weld bead or eliminate undercut. In the former, the weld is ground down even with the component metal surfaces. In the latter, component surfaces are ground even with the weld. Grinding also removes decarburization.

Because grinding can reduce strip thickness, purchased strip should be thicker than end-product strip in proportion to the amount of grinding anticipated. As much as 0.003 in. (0.08 mm), for example, may be removed from the surface of bandsaw strip to remedy undercut.

Weld Acceptance Hinges on Ultrasonic Test Results

A continuous nondestructive test is required to ensure that the entire coil of welded, straightened, heat treated, and ground strip is defect free. Only two standard test methods are suitable: ultrasonic and magnetic particle inspection. X-ray inspection can't meet the continuous criterion and eddy current testing can't detect internal defects. Wallace Barnes uses ultrasonic immersion testing, a more costly, but faster and more sensitive technique than magnetic particle inspection.

In Wallace Barnes' setup, the strip runs continuously through an open tank of water. Sound waves are passed through the backing steel, weld, and high-speed steel and then bounced off the edge of the high-speed steel. The system features preset "go/no-go" limits — an alarm bell rings if a void, such as a surface cut, pore, or internal crack, exceeding a certain size is detected.

As soon as a defect is indicated, the operator removes the strip from the tank and examines it under a magnifying glass. If no external blemishes or defects are visible, the weld is pressure tested for internal integrity. The suspect area of the strip is marked if either the visual or pressure test is failed.

The operator also makes a note of the defect's location in terms of its distance from the end of the coil. These notes constitute the "ultrasonic history" compiled for each coil.

Testing will not proceed until the cause of the defect indication has been determined, and a decision has been made as to whether the flaw is significant. Minor grinding cuts are examples of flaws that do not impair weld performance.

Flawed areas marked as significant are cut out during the post-testing recoiling operation. Strip ends are then resistance welded together and the joint is annealed.

Typical defects in EB-welded bimetallic bandsaw strip (100X, Nital).
1. Transverse section shows a blow hole in annealed strip. A pipe in the D6A backing (at left) gasified and erupted during welding.
2. Transverse section shows weld porosity caused by the gasification of nonmetallic inclusions in the backing (left).
3. Transverse section shows metal expelled from the joint by excessive gasification of inclusion stringers in the backing (right).
4. Transverse section shows a partial penetration EB weld.
5. Longitudinal section shows cracks formed in the brittle weld prior to annealing.
6. Transverse sections show mismatch between edging (left) and backing (right).

Other final inspection operations include measurements of thickness, width, and straightness, and a microscopic inspection for decarburization and weld defects.

Standards: Quality standards for determining whether a defect is acceptable or unacceptable vary with the intended application and the nature of the defect. Extreme undercut or partial penetration are nearly always unacceptable.

A relatively large amount of porosity, on the other hand, may be acceptable if little or no stress will be placed on the weld in service. Bimetal strip for appliance thermometers is an example. Metal-cutting bandsaws, however, must withstand very large stresses and standards are proportionately higher. Wallace Barnes' limit on porosity in bandsaw strip, for instance, is no more than three holes larger than half the width of the weld, or approximately 0.004 in. (0.10 mm) in diameter per 6 in. (150 mm) of welded strip. This represents a probability of well below 1% that a defect could cause ripping of a tooth.

Ultimately, appropriate quality standards result from joint welder-user efforts involving the balancing of user quality requirements against the added cost of quality control.

1

2

3

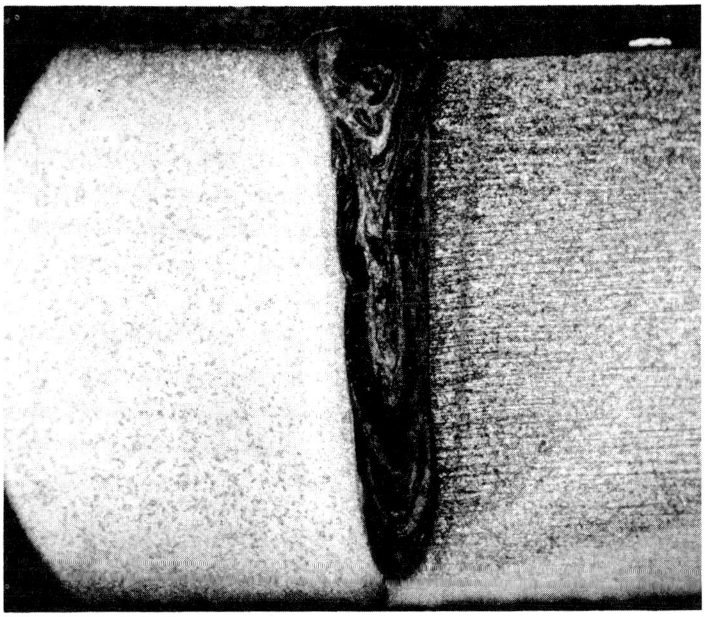

4

5

6

IN PROCESS MONITORING OF THE INTEGRITY OF ELECTRON BEAM WELDS*

S.R. Doctor and F.L. Becker
Nondestructive Testing Section
Battelle, Pacific Northwest Laboratories, Richland, WA

Electron beam welds are being used in many applications when conventional welding techniques are unsuitable. The nature and geometry of some of these welds are such that reliable inspection by conventional nondestructive testing techniques is impossible. Therefore, a research and development program* was pursued to design, develop and demonstrate an improved technique for assuring adequate depth of penetration of electron beam (EB) welds. The technique selected monitors the depth of penetration as the weld is being made and has the advantage that defective areas are detected so that rewelds are made immediately with the part in place.

SYSTEM DESCRIPTION

The instrument, shown in Figure 1, contains from left to right at the top, a Metrotek MP203 pulser, a Metrotek MR101 receiver, a Metrotek MG701 gate, a Lack-of-Penetration monitor module, a Tektronix SC-502 oscilloscope, a Hewlett-Packard 7100B strip chart recorder and in the foreground a shielded transducer with a wire acoustic waveguide. The top trace on the scope illustrates the pulser drive signal and the relfection from the end of the waveguide. The lower scope trace shows the video detected signal.

Figure 2 shows conceptually the monitoring technique. The wire acts as a waveguide and guides ultrasonic energy from the transducer to the wire end. As the electron beam weld is made, a solid/liguid interface that reflects ultrasonic energy is formed in the wire and this interface will advance with the welding. Therefore, as the weld is made, the wire waveguide is shortened. By measuring the time-of-flight the interface position or wire end is determined. When lack-of-penetration occurs, the interface stops moving and the time-of-flight becomes a constant. Hence, the good weld areas will have a slope (plot of time-of-flight versus time) different from zero and bad welds will have a slope equal to zero. The chart paper in Figure 1 shows a good weld with one defect area (zero slope).

A block diagram of the system is shown in Figure 3, containing a pulser to drive the transducer, a receiver to extract any reflected ultrasonic echoes, a gate to restrict the time frame over which valid signals will be accepted, a lack-of-penetration module to generate the sync pulse for the pulser and to provide a threshold control to discriminate against noise, an adjustable video detector, a time-of-flight digital-to-analog converter output and finally, an adjustable time marker output; the scope is used to set up the gate, threshold and video adjustment; the strip chart recorder displays (see Figure 1) along the left-hand margin a series of "ticks" to mark time intervals and the signal running across the chart paper is the time-of-fligh analog output with the jog (zero slope) being an area where there was lack of electron beam penetration.

*Work sponsored by Rockwell International, Rocketdyne Division

Wire Waveguide

The goal is to make the wire diameter as small as possible because the electron beam position resolution would, thereby, be increased. However, laboratory testing demonstrated that if the wire size was decreased, the signal-to-noise ratio was reduced. The reason for this is that the thin rod ultrasonic propagation mode has energy distributed between surface waves and bulk waves. As the wire diameter is reduced, the ratio of bulk wave energy ($\sim\pi r^2$, proportional to cross sectional area) to surface wave energy ($\sim 2\pi r$, proportional to circumference) is rapidly reduced. The principal mode for energy loss is through the surface. Smaller diameter wires have more energy in the surface waves (for constant total energy into the wire). Therefore, the smaller wires will have greater losses and a much smaller signal which reduces the overall signal-to-noise ratio. With the present system, wires in the 0.020 and 0.025 inch range appear to be the smallest that produce an acceptable signal for electronic processing.

Transducer

The feasibility was established using shielded 0.100 inch by 0.100 inch, 1 MHz PZT-5 transducer epoxy bonded to the wire waveguide. The primary problem with the piezoelectric transducer is that the echo contains ringing. The ringing results from the transducer being undamped. Damping obviously reduces the ringing but also reduces the signal-to-noise ratio.

Electronic Packages

The system design was based on using off-the-shelf modules when possible. All modules were off-the-self except for the lack-of-penetration module. This module was designed to use the time-of-flight for classifying the quality of the weld.

The echo signal required massaging before an accurate time-of-flight measurement could be made. Specifically, the echo had to be passed through a video detector so that a reference level could be used for triggering the time measurement circuitry. This adjustable video detector was included in the lack-of-penetration module.

The electronics for measuring true time-of-flight were designed, fabricated and tested. The digital time-of-flight is converted with a DAC (digital-to-analog converter) to an analog signal that is used to drive the strip chart recorder. Figure 4 shows a simplified block diagram for the lack-of-penetration time-of-flight module.

Testing Program

EB welds were made to evaluate the system performance. Stainless steel samples were 3 inches wide by 8 inches long and 0.75 inch thick. The weld root is at 0.5 inch and, therefore, has a 0.25 inch backing. The wire waveguide is inserted into a slot that has been cut into the test specimen with a saw angled at 45 degrees. Scabs of various lengths were placed over the weld centerline to simulate EB lack-of-penetration.

Discussion of Results

The system was demonstrated and found to predict the status of the weld as the weld was being fabricated. Specimen A did represent the type of record expected for a good weld and was verified by micrographs. Test specimen B demonstrated resolution capability with lack-of-penetration of 1/4 inch being clearly resolved.

Test specimen E created some anguish because the beam parameters were the same as for the previous tests. Examination of the micrographs showed that the negative excursions of the strip chart record corresponded to partial fusion of the waveguide. The reason the strip chart record deflects as it does is that for partial fusion the video detected signal is too small to trigger the time-of-flight electronics. Consequently, the end of the gate region or maximum time-of-flight is read out to the strip chart record. Note, the same thing happened just at the interface to the one inch scab on specimen B.

In evaluating the effectiveness of the technique, we have three different conditions as represented by the results on samples A, B and E. In the first condition, (represented by sample A), the ramp is uniform and exhibits no zero slope areas greater than 1.4 inch. For this strip chart record we have very high confidence that the weld is acceptable. In the case of sample B, the ramp is smooth with zero slope areas corresponding to the programmed lack-of-penetration areas. Again, we have high confidence that the results are conclusive; i.e., areas indicated as acceptable are acceptable and zero slope areas are not acceptable.

For sample E, the confidence level is reduced. The programmed lack-of-penetrations are easily detectable and even in a production environment there would be little doubt that indications of this type represent defective welds, provided that the position of the wire relative to the joint is adequately controlled. The uncertainties in evaluating the "goodness" of the weld lie in the areas where partial fusion of the wire occurs. The areas of partial fusion result in negative excursions on the strip chart record. These areas on sample E were acceptable welds. However, there is a possibility that indications (negative excursions) of this type could represent defective weld areas particularly if the positioning of the wire is not closely controlled.

CONCLUSIONS

The EB weld monitor was demonstrated and evaluated on stainless steel test samples. The system is quite easy to set up and can be used to establish the welding parameters faster, easier and more economical than sectional weld samples.

The present monitoring system does make a permanent record of the weld. From this record the conditions of complete, partial or lack-of-penetration can be assessed. Then if rewelding is required, another pass over the suspect areas can be made. This in-situ capability with the permanent record of the first EB pass makes the system extremely useful in determining the quality of EB welds. However, an extensive testing program has not been conducted on partial penetration conditions. These conditions backed up with in depth metallurgical studies need to be made to further quantify the EB weld quality.

Modification of the present system could be made to make the system less operator dependent by using automated decision procedures based on pattern recognition techniques and supported by metallurgical studies of the welds. In conclusion, we believe that this technique can, with additional development, substantially improve the confidence in the acceptability of electron beam welds.

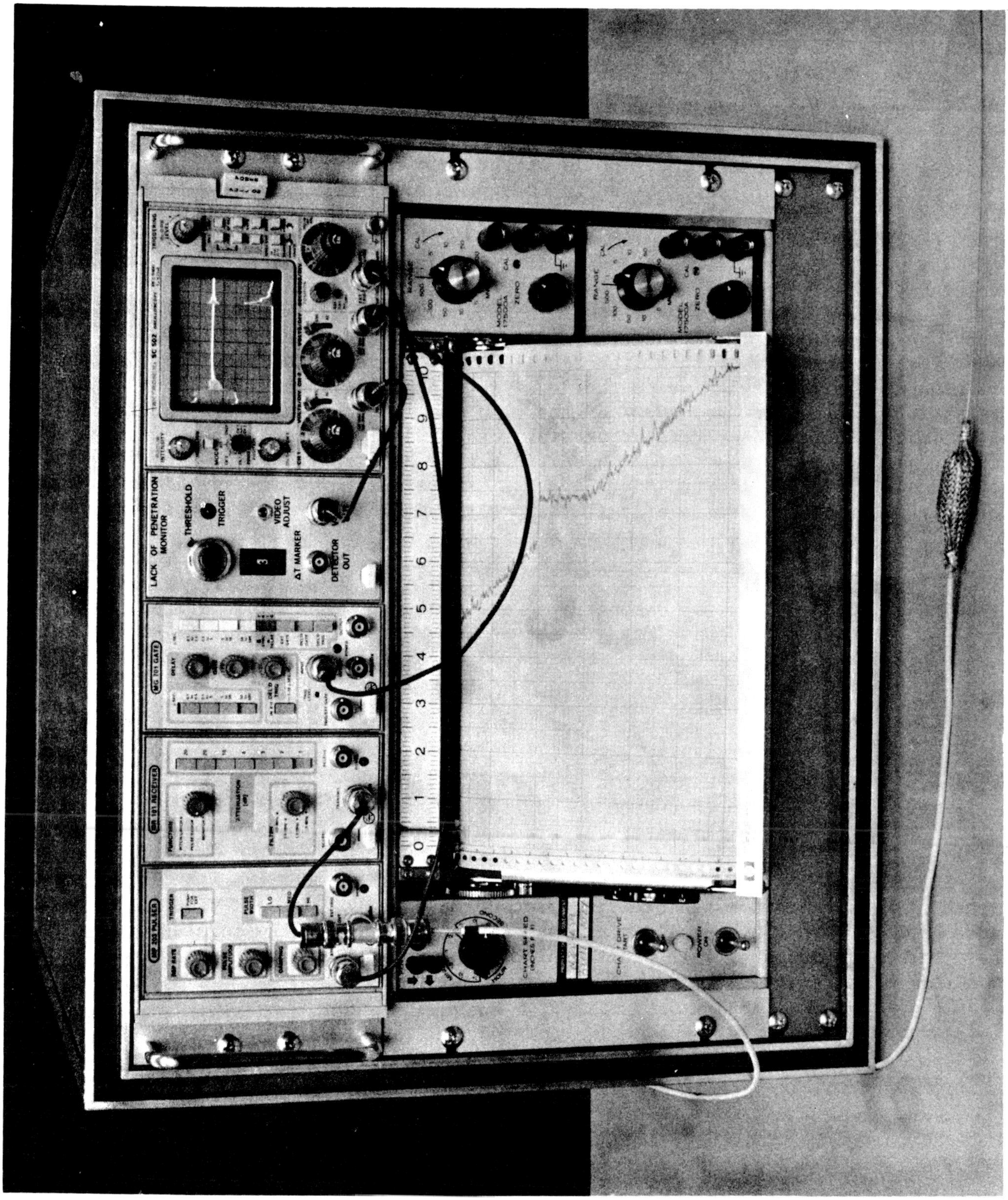

FIGURE 1--Electron beam weld monitor.

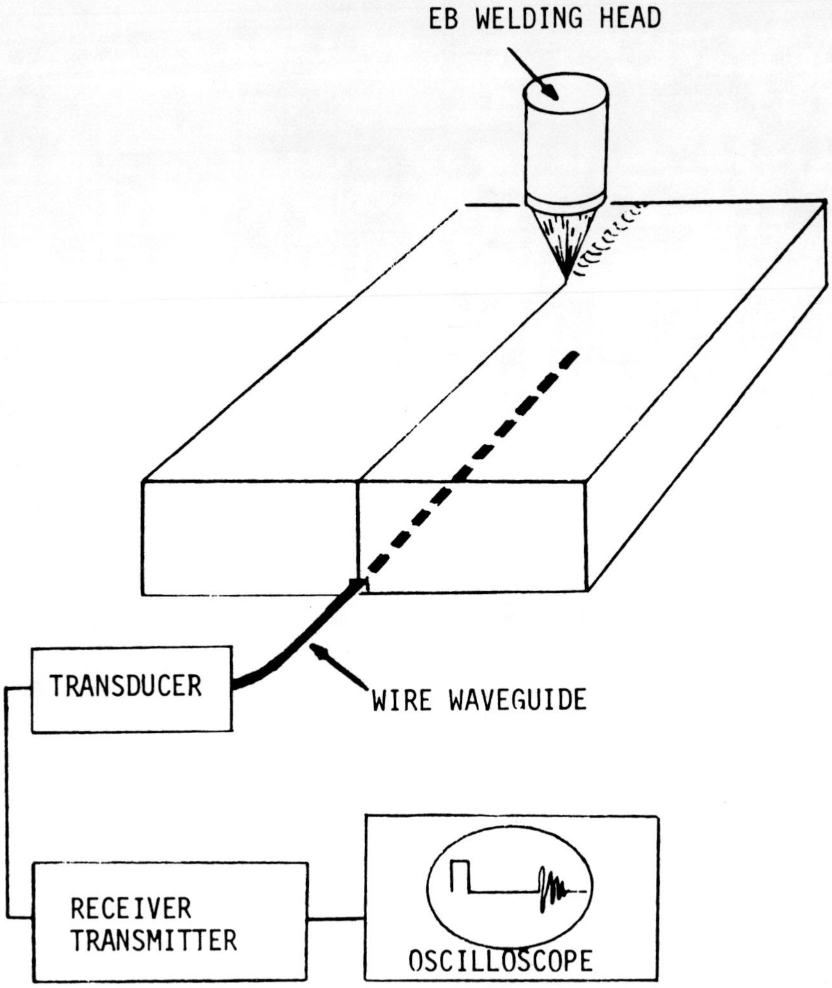

Figure 2. Electron Beam Weld, a simplistic diagram of the monitoring technique and a typical oscilloscope trace showing the transducer excitation pulse and an echo from the waveguide end.

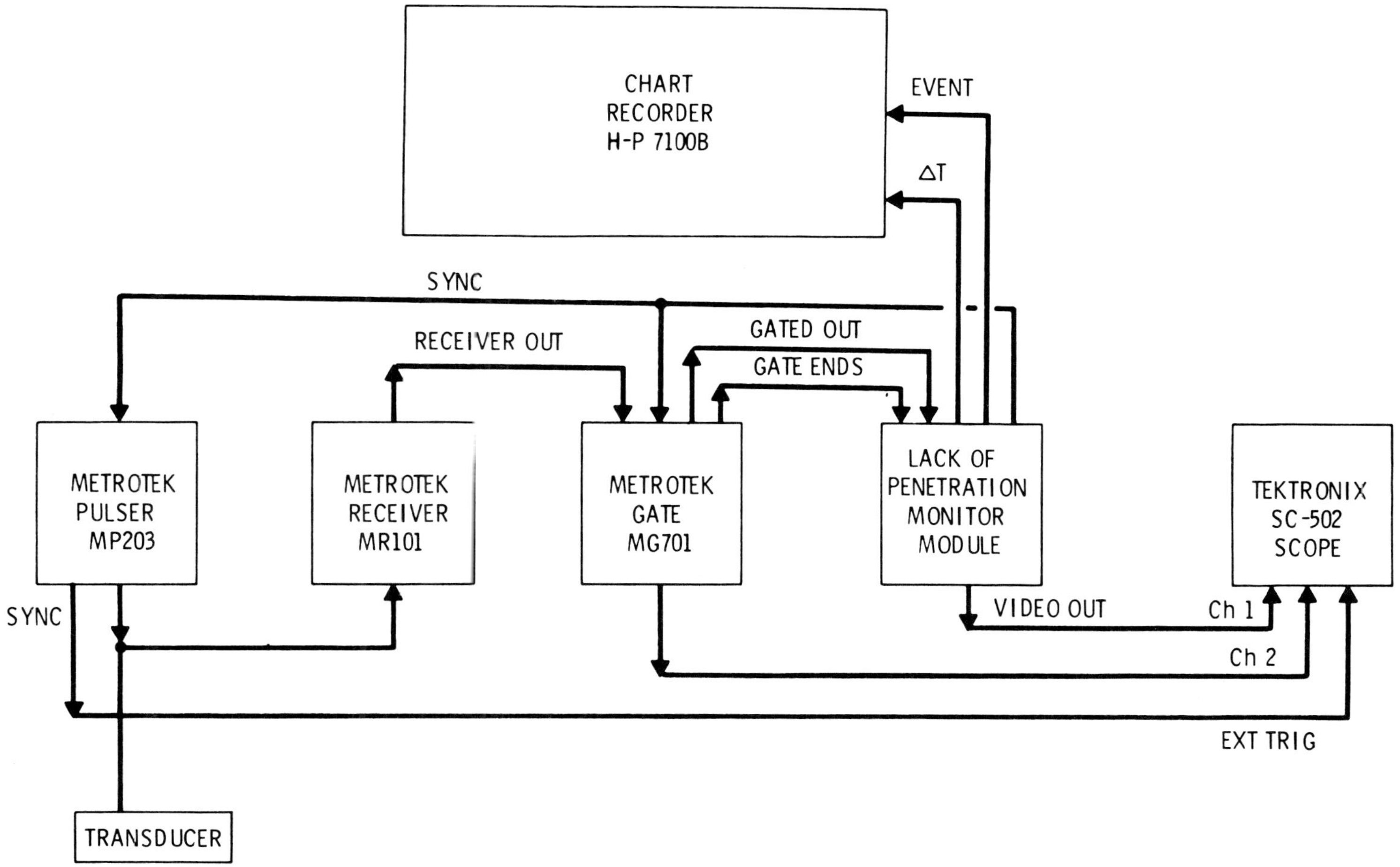

FIGURE 3--Block diagram of the electron beam welding monitor.

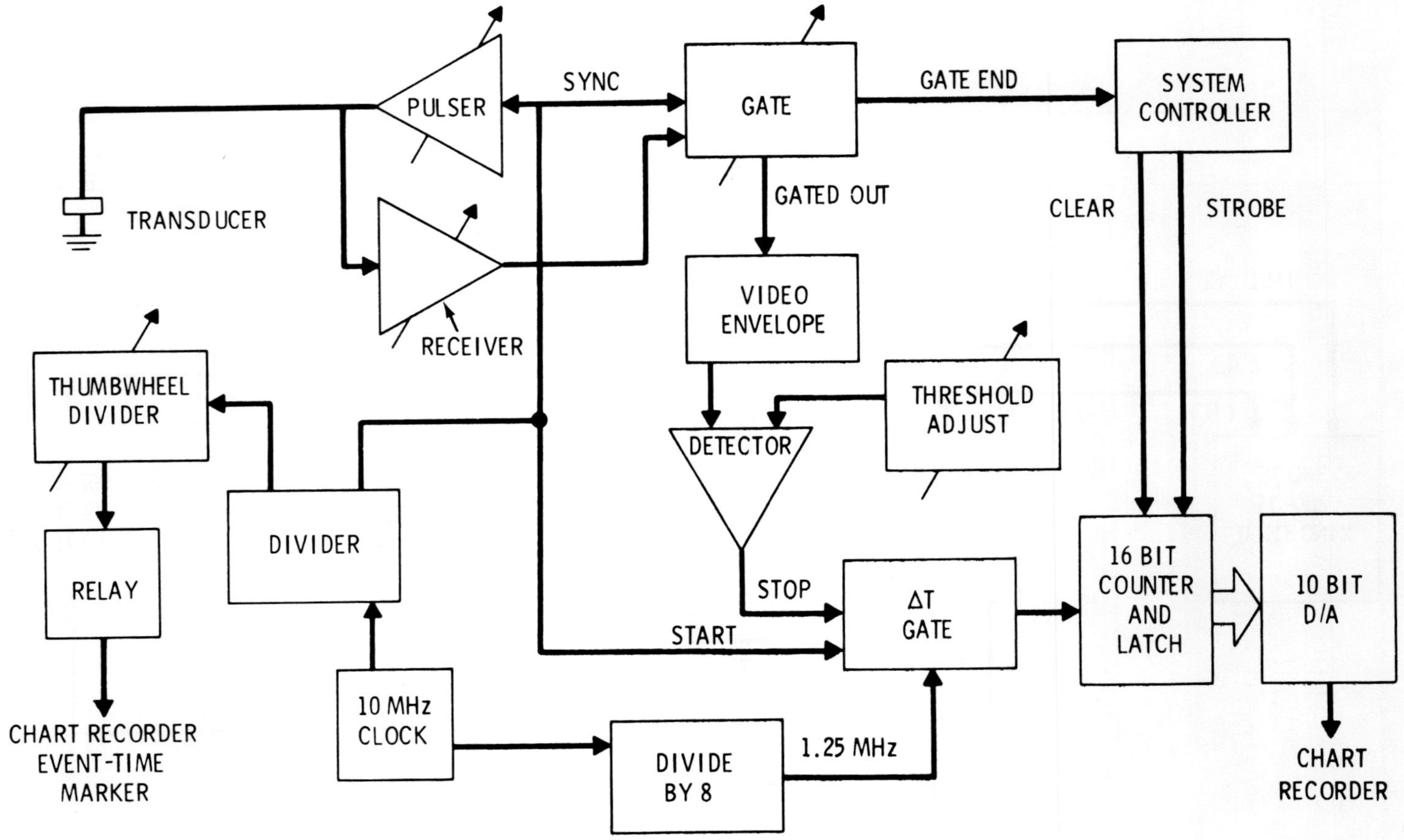

FIGURE 4--Block diagram of the lack of penetration module.

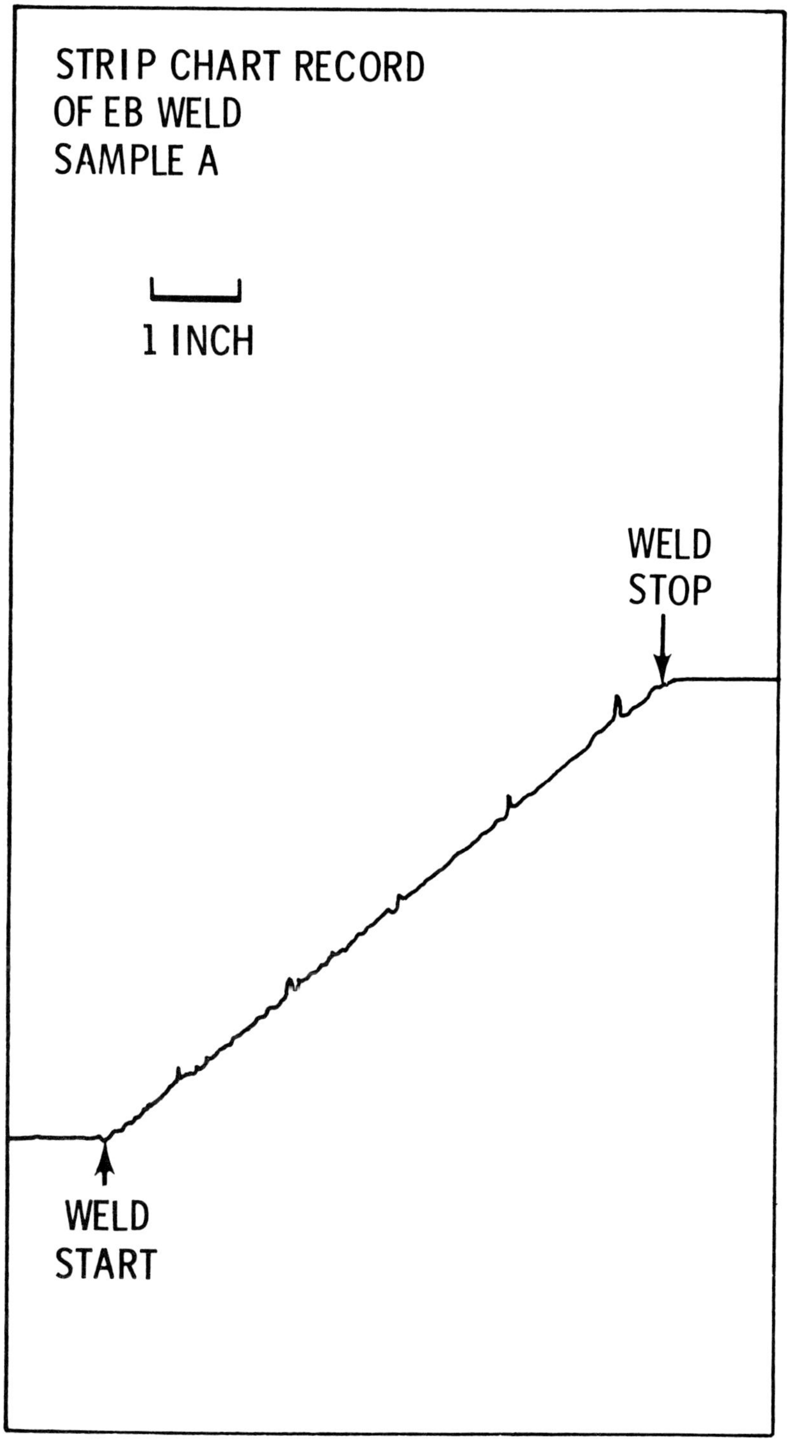

FIGURE 5--Strip chart record of the weld on specimen A.

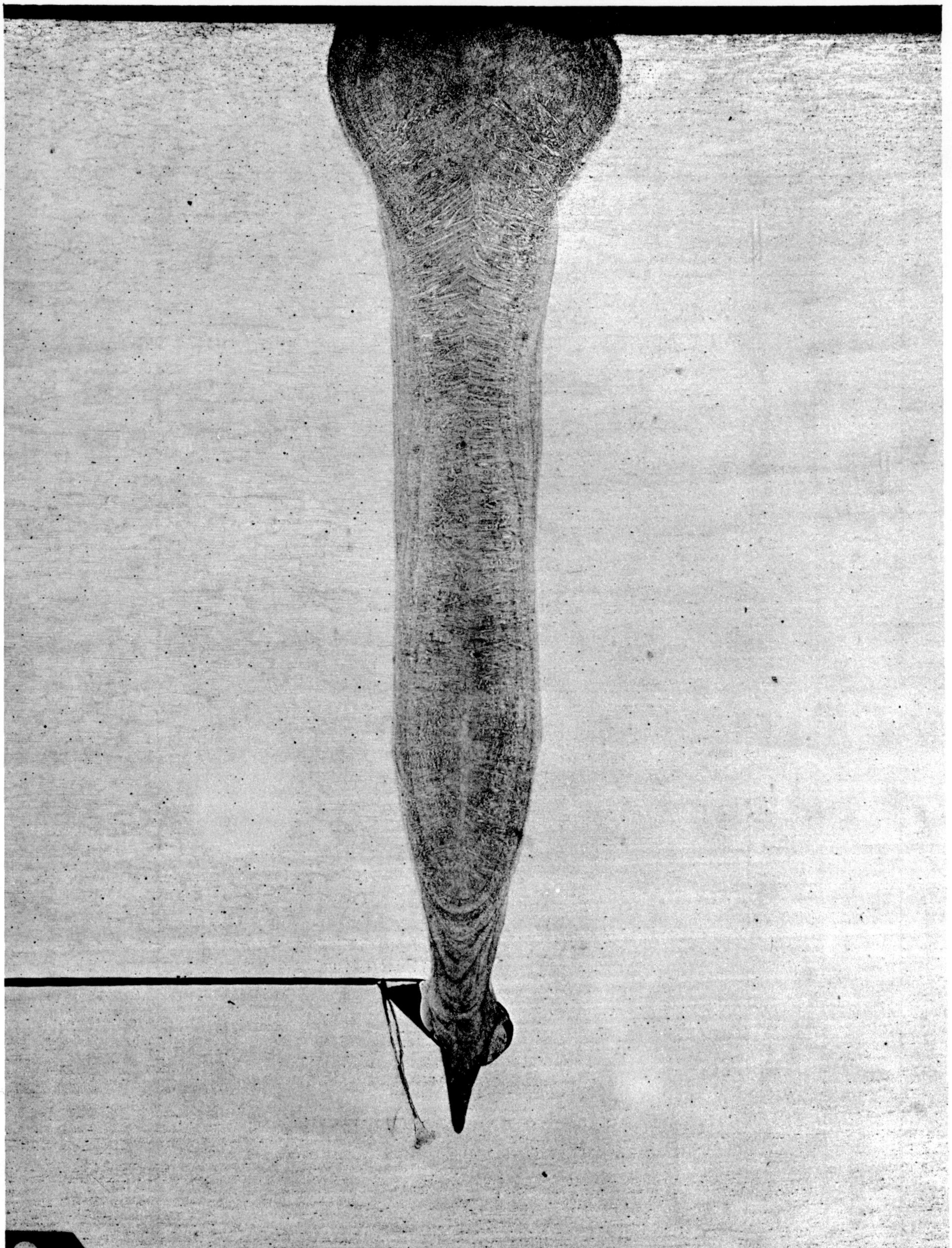

FIGURE 6--Showing a typical cross section of test specimen A. (x6)

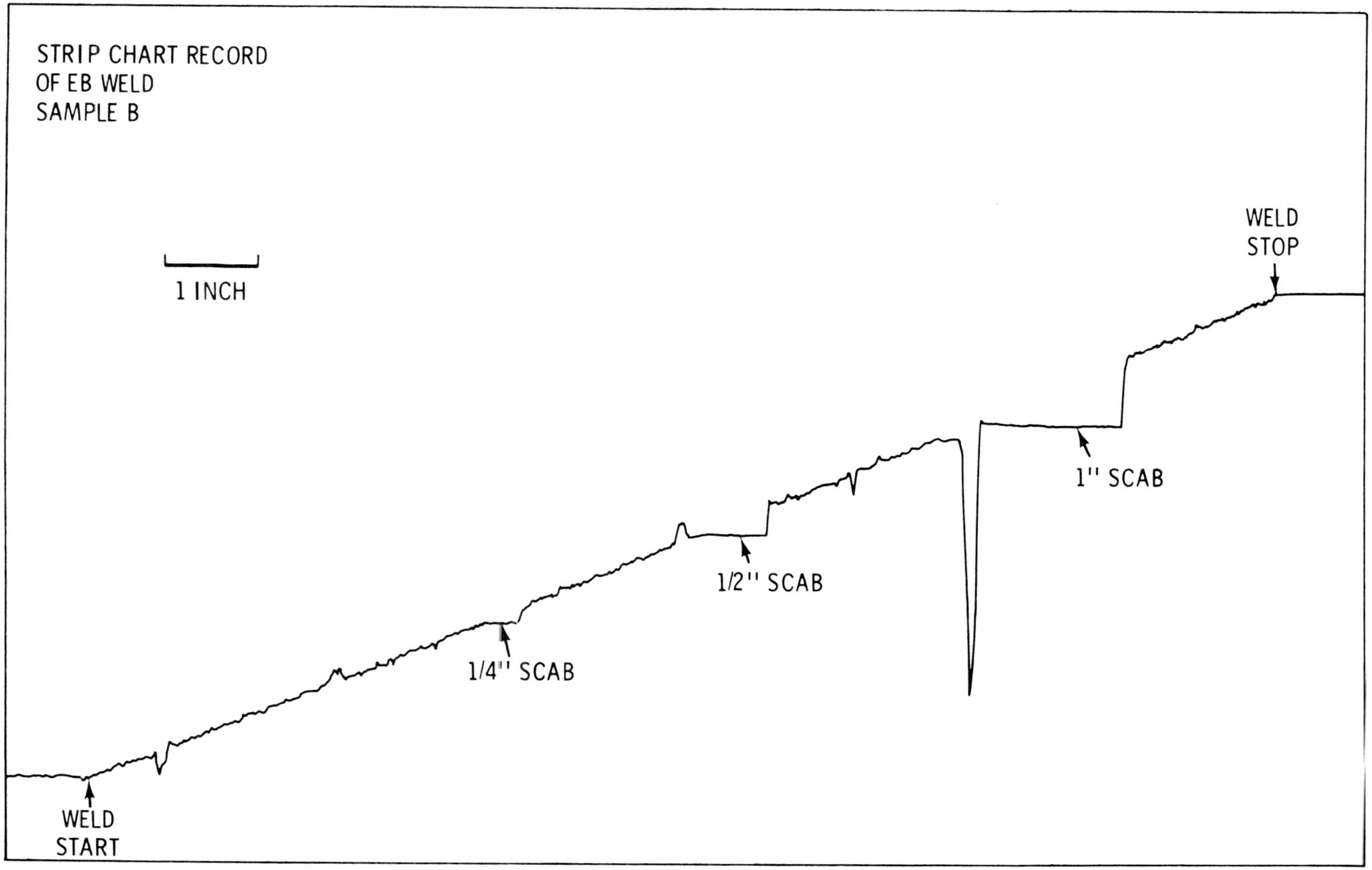

FIGURE 7--Strip chart record for weld specimen B.

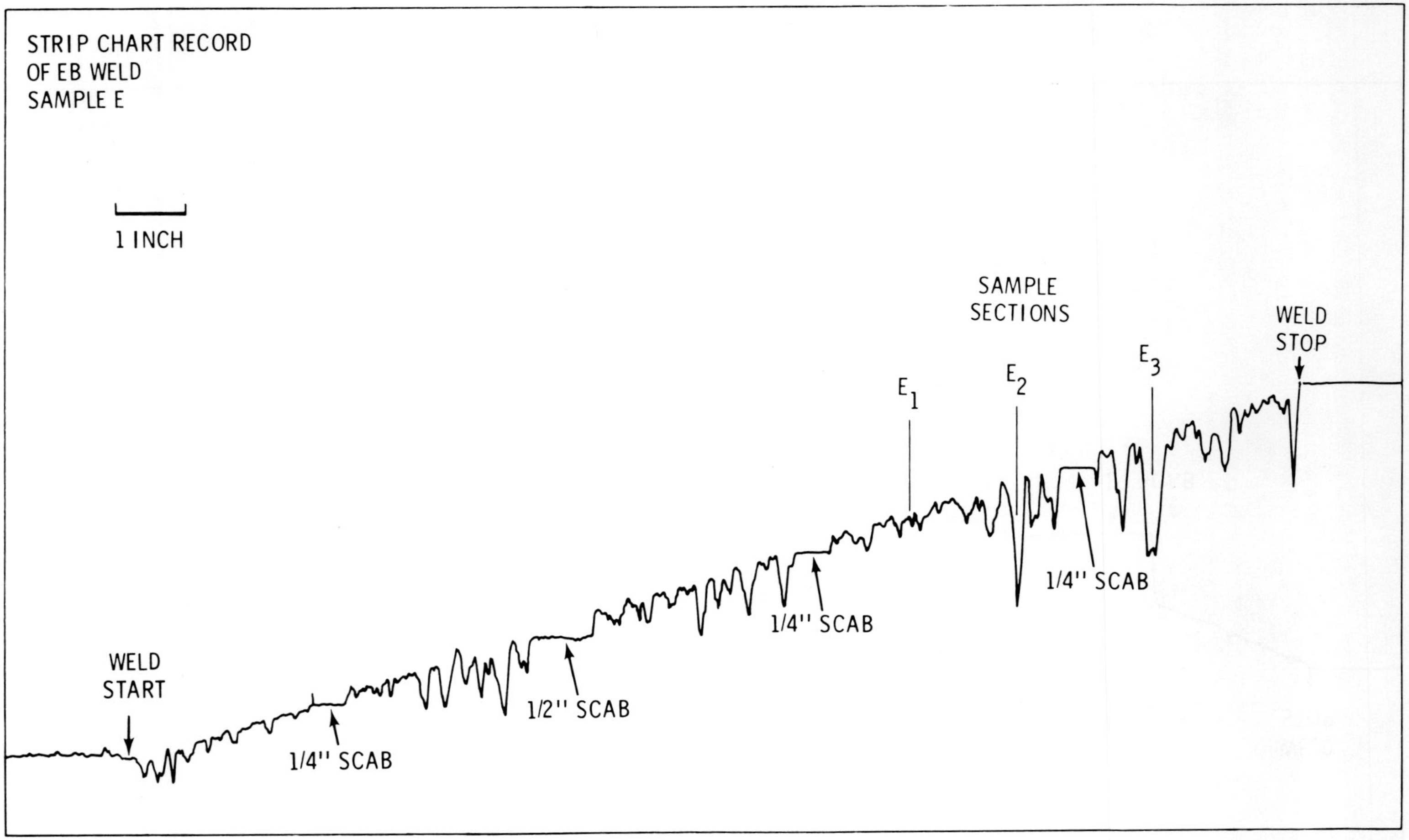

FIGURE 8--Strip chart record of specimen E and the location of the three sectioned samples.

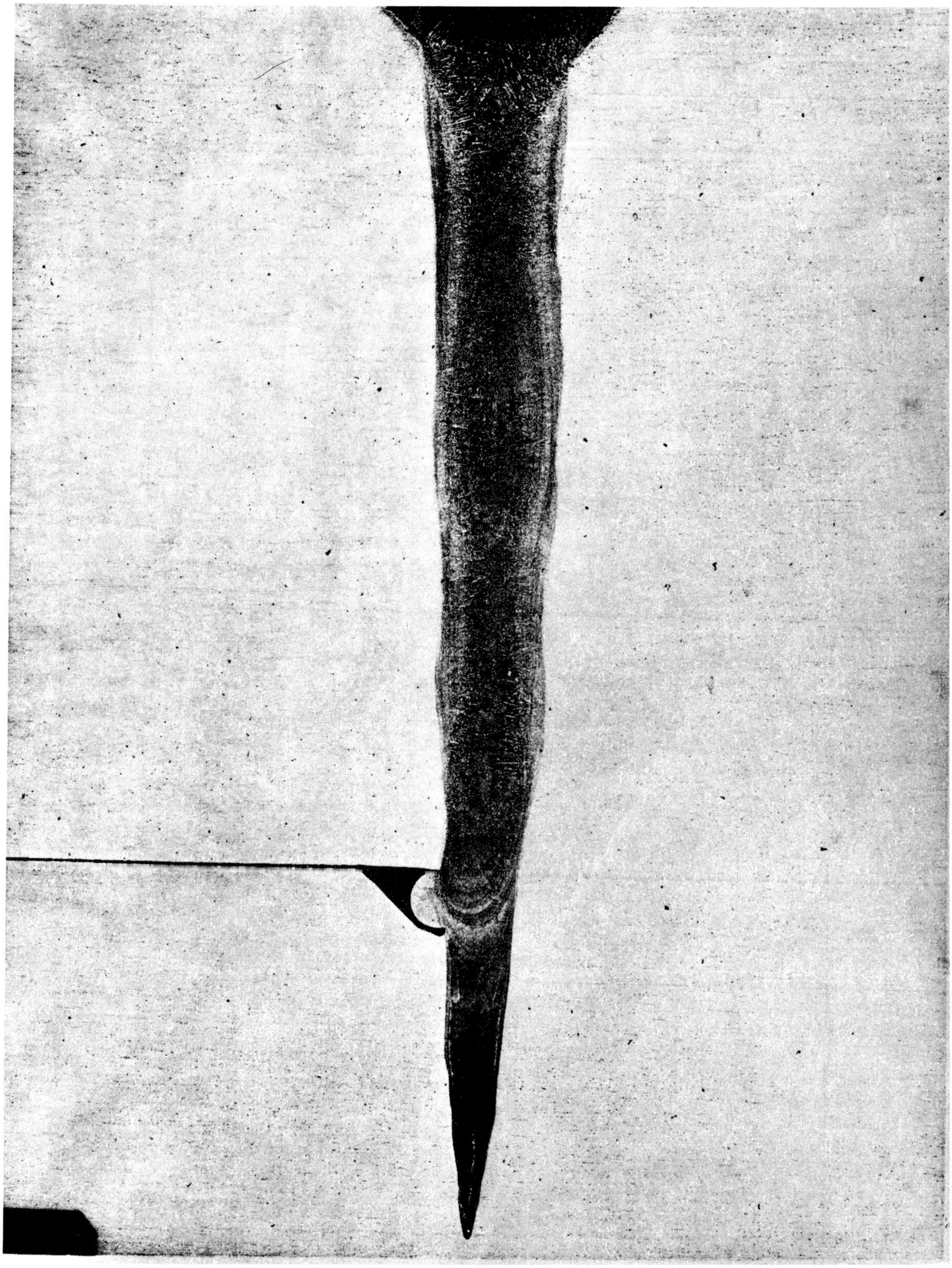

FIGURE 9--Micrography of sectioned sample E_3. (x6)

USE OF A MICROCOMPUTER-BASED DATA ACQUISITION SYSTEM OF IMPROVE PROCESS CONTROL IN AN ELECTRON BEAM WELDING MACHINE

By

Hugh Casey, Fred Flick

and

Victor Vigil
University of California
Los Alamos Scientific Laboratory

ABSTRACT

As part of a program to upgrade electron beam welding equipment, a microcomputer-based data acquisition system has been built and interfaced with a Hamilton Standard W-2 machine. This presentation will review the design and performance of the system. The practical application of this and similar systems for improved process control will be discussed. Data will be presented relating to the significance of process parameter variations (as monitored by the data acquisition system) on weld geometry and the occurrence of weld defects.

INTRODUCTION

There is an increasing use of computers and computer-related devices in control systems for sophisticated automatic welding equipment . In many cases the function of the computer is to control the switching mechanisms of the power supplies in the welding machine. The computer can thus be programmed to execute a particular weld cycle with a high degree of reliability and repeatability. This type of system lends itself to fully-automated mass-production weld processing.

By contrast, the system which is described in this presentation is designed to improve the performance and enhance the capabilities of an electron beam welding machine which is used for welding (and braze-welding) of prototype components for research and development activities. The system was designed with the following objectives:

1. improve resolution and accuracy of instrumentation
2. provide high sample rate monitoring of parameters and indicate deviations from predetermined tolerance bands
3. provide a storage medium for weld data for quality assurance and research and development activities
4. introduce the potential for feedback control of the parameters

To provide a better understanding of these objectives, it is appropriate to describe applications typical of LASL hardware:

The tubular assembly, shown in Fig. 1, is part of an experimental laser cavity. The cavity is ~ 40-mm-long and is made by roll forming and electron beam lap-welding ~ 0.025-mm-thick Inconel 600 foil. The joint is required to be vacuum tight and ~ 100% weld efficiency is desirable.

Fig. 2 shows a honeycomb structure made by electron beam lap welding ~ 0.0255-mm-thick aluminum foil. These foil welds are made with beam currents typically < 1 mA which is outside the range of resolution of the conventional instrumentation.

Fig. 3 shows the various components which comprise one design concept for an artificial heart nuclear heat source. Plutonium fuel is to be contained within the welded assembly which is composed of tantalum, platinum, and stainless steel alloy components. For obvious reasons, the quality control requirements are exceptionally stringent. The weld shown in Fig. 4 is the capillary venting tube attachment on this assembly. The orifice is 0.125-mm-diam and the integrity of this weld is crucial to the performance of the design.

A helium vent assembly for a similar application is shown in Fig. 5a and the transverse weld profile is shown in Fig. 5b. The base plate is Pt-30 Rh and the screen mesh is Pt-30 wt% Rh-8 wt% W alloy. The electron beam weld operation must not alter the leak rate through the mesh.

A longitudinal weld profile from a high strength steel pressure vessel is shown in Fig. 6. The wall thickness of this pressure vessel is ~ 5 mm and the weld is required to penetrate 85% of the wall, but must not penetrate the interior surface.

The photomicrograph in Fig. 7 shows an electron beam weld 'spike' of over penetration which resulted from a voltage transient. This section was removed from a diaphragm gas injector which failed as a result of this defect. The spike could not be detected by non-destructive testing techniques.

It should be apparent, from this brief hardware description, that LASL weld processing encompasses a wide variety of welded prototype components and an unusually large range of metals and metal alloy systems. In addition, since most of the hardware is in some way associated with nuclear devices, the quality assurance requirements are exceptionally demanding.

Unlike our production agency counterparts, we have difficulty justifying sophisticated processing accessories on the basis of a specific assembly. The emphasis with regard to the current computerized system design is, therefore, placed on flexibility and adaptability to miscellaneous operational modes.

SYSTEM DESIGN

Figure 8 shows the general layout of the experimental electron beam welding machine with the computer and accessory equipment in position. The Hamilton Standard W-2 machine is ~ 12 years old, but has been somewhat upgraded with a motor generator set to a 7.5-kW-maximum-capacity. Minor modifications include a vernier switch control on the beam current and a LASL-modified optical viewing system. Tooling is LASL designed and built and all systems are driven by Inland motors with tachometer feedback control. The systems are driven through an Electrocraft motor controller unit.

Although all initial testing is to be carried out on this machine, a basic requirement of the computerized system is that it can be interfaced with other LASL high voltage electron beam welding machines, including a new Leybold Hereaus system due for installation this year. This requirement can be met with only minor modifications to the internal wiring of the basic electron beam welding equipment.

The microcomputer package is shown in detail in Fig. 9. The system was designed and built by LASL's Mini-microcomputer Systems Group, E-5, and utilizes an Intel 8080 Microprocessor and necessary support devices to allow data to be monitored and stored in a volatile 32 k random access memory (R.A.M.). The program code and data base is stored in a 14 k programmable read-only memory (R.O.M.). A Texas Instruments 733 ASR Data Terminal is used for data communication. The signals acquired from the welding machine are conditioned and isolated by means of Neff signal-conditioner-amplifier filters and a set of low-pass filters. The data can be displayed on the digital instrumentation panel, recorded onto tape, or stored into memory under operator control.

SIGNAL CONDITIONING

The signal channels and display scale are indicated in Fig. 10. The parameters to be monitored are typical of most electron beam welding operations and the system has provision for eight additional channels for future development.

The sources of these signals on the Hamilton Standard welder are not particularly suitable for monitoring; the selection of sources and/or addition of transducers to obtain the signals is a compromise between the least possible alteration of the beam welder circuits and the best signal. Some signal sources are 'floating' above ground and have common-mode noise, principally 60 Hz and 120 Hz. Guarded inputs to the instrumentation amplifiers are used with excellent input to output isolation and good common mode rejection characteristics. Additional 60 Hz and 120 Hz components must be removed by filters before measurements are made in the analog-to-digital converter in the system. At this stage of the program the instrumentation package is designed to detect changes in weld parameters in an attempt to correlate the variations with weld zone inconsistencies. The data sampling rate is, therefore, much less than that required for closed loop control. The overall system bandwidth is 10 Hz which will give a 0.03 sec. rise time response to a step input and will adequately remove the 60 Hz and 120 Hz noise.

OPERATION

The system is designed with five operational modes; these are described as follows:

<u>Mode #1</u>. This is the setup mode. The tooling package is loaded and the machine prepared for operation in a normal manner. The operator communicates with the computer using the teletype to produce a log sheet describing the work piece and the desired weld parameters. The system is designed primarily for rotary welds at this time, and the computer uses the information on part diameter and tooling gear ratio to compute the linear weld travel speed.

The computer will also request details on parameters including acceptable tolerance bands for each parameter (as determined, for example by prior weld testing). The parameters to be monitored are selected, and it is possible to monitor any individual parameter or any combination of parameters. For standard operation, the accelerating voltage, beam current, and travel speed are sampled at ten times the rate of the remaining channels. Maximum sample rate is limited to about 300 samples per second. On completion of the setup, the log sheet can be used as a documentary record of the particular weld. A typical setup log is shown in Fig. 11.

Mode #2. This is the display mode. The electron beam is positioned and focused on a suitable refractory surface (as in normal operation) and the weld parameters are displayed on the digital instrumentation. At this stage, the resolution of the instrumentation and the sample rate can be altered from the keyboard.

Mode #3. This is a 'wait' mode. The keyboard is deactivated and the machine can be left unattended without fear of interference with the setup.

Mode #4. This is the welding mode. During this mode the computer monitors the machine parameters. The computer 'looks' for a predetermined level of beam current (usually 10% of the prescribed welding current) and on sensing this threshold level, begins to store the parameter data in the R.A.M. The display panel is not updated during the weld cycle and displays only the setup parameters established in Mode #2.

Mode #5. This is the data retrieval mode. The parameter data is stored in the R.A.M. in the octal numerical form. The data can now be played onto magnetic tape and/or printed out on the teletype as indicated in Fig. 12. The right hand digit of each octal number indicates the channel number and, if a particular reading is out of the designated tolerance (as defined in Mode #1), the excursion will be noted with an asterisk thus giving an immediate indication of a potential problem. Obviously, the data at this time is unsuited for further analysis and some data reduction is required to make the information more useful. It should be emphasized, however, that at this point, the system provides a positive or negative indication reflecting the performance of the machine during the weld cycle. At this time, the programmable R.O.M. does not have sufficient capacity to allow further reduction of the data. However, the system is tied directly to LASL's Central Computer Facility (C.C.F.) and the data contained on the magnetic tape can now be transferred to C.C.F. over a telecommunications linkage and processed using one of several available codes. Using this technique, data plots of the type show in Fig. 13 can be produced.

APPLICATIONS

The application of this system allows weld parameters to be prescribed and monitored during welding with a high degree of resolution. The first phase of our test program (currently underway) is to characterize the performance of the welding machine and define just exactly what constitutes a significant parameter variation. For example, the data to date has shown that the operational stability varies significantly with the accelerating voltage/ beam current combination, i.e., it was found that a 120 kV accelerating voltage produces a more stable current than a 100 kV accelerating voltage for beam currents in the range 0-5 mA. This is presumable a characteristic of this machine and may be associated with other extraneous variables. The important feature is that a stable operating range can be defined. In this context it should again be emphasized that LASL processing often calls for operating at the extremes of the operating range of the equipment where the stability is understandably questionable.

In determining the significance of parameter variations it is essential that the application and weld performance requirements be defined in detail. Consider, for example, the extreme case of incipient arcing in the electron gun, often experienced when welding high vapor pressure metals, particularly aluminum and magnesium alloys. Many EB machines are designed with over-voltage protective circuits which switch off the beam when a voltage transient is experienced. The effect of an incipient arc may be insignificant in terms of the weld performance, but the cavity left by the abrupt termination of the beam can be difficult to eliminate by rewelding. The sensitivity of the overload circuitry should be adjusted to afford protection to vulnerable components in the power supplies but an overreaction to minor transients should be avoided. It is intended to use the data acquisition system to make an extensive study of the effects of arcing and establish direct correlations with the weld bead itself.

Of more immediate interest to LASL applications, is the effect of parameter variations on penetration control. It is an established fact that parameter variations do occur, e.g., day to day changes in focus settings result from filament deterioration, overheating of lens coils, etc...; changing guns or replacing filaments can introduce substantial differences in weld parameter settings; the effect of line voltage fluctuations induced by other equipment operating in the vicinity are possibly significant.

In a qualitative sense, these effects and numerous other equipment related factors have been recognized and substantial improvements have been made by the manufacturers of commercially available equipment to reduce their impact on the reliability and repeatability of EB weld processing. Our system will be used to generate quantitative data on these effects. This data can be used to identify equipment problems, establish upgrading procedures, and define practical weld tolerance values for application in welded component design.

The application of the system as a quality assurance tool is obvious. Many of our assemblies are difficult, if not impossible, to inspect using standard nondestructive testing techniques. By characterizing and certifying the weld cycle parameters, equipment related defects can be positively identified. Detailed records of the welds can be filed on tape, microfilm, etc. for future reference.

FUTURE SYSTEM DEVELOPMENT

In its present form, the system operates simply as a data acquisition package. Additional channels will increase this particular capability to include, for example, thermal history of a welded component, strain gauge data to monitor weld stresses, and possibly vacuum gauge transducers to compare vacuum in the region of the weld pool with the standard port gauges. Positional encoders are to be added to the rotary tooling packages to allow better correlation of parameter variables and weld quality.

For the purpose of developing closed loop feedback control, it is accepted that this system is not fast enough to handle beam current/high voltage control. There are commercially available systems which can handle these parameters more efficiently. Some control functions, are, however, possible with this system. As an example, a high speed photographic recording system is being built as part of a program to study the weld metal flow characteristics of certain alloys. It is planned to use the microcomputer to trigger the camera when parameter transients are detected in order to photograph the effects on weld pool behavior. Also, there are applications involving nonlinear weld seams, for which it should be fairly simple to program a beam deflection path and use the computer to control the electromagnetic lens signal which positions the beam.

Data acquisition is, however, the primary function of this package, i.e., the objective is to monitor the control systems. One proposed application of this function relates to the use of modulated wave form electron beams. It is postulated that the efficiency of the EBW process can be improved by pulse mode operation and that pulsed mode welding may allow some increased control over the weld pool solidification modes and grain size. It is proposed to study this phenomena using the D.A.S. to monitor pulsed wave forms used for electron beam welding of molybdenum and molybdenum alloys.

CONCLUSIONS

A microcomputer-based multichannel data acquisition system has been used to extend the capability and improve quality assurance procedures with an electron beam welding system.

The system is designed to be used for quality control in low production electron beam weld processing and also as a research tool for analysis of weld parameter effects on weld performance.

Additional capability is provided by the potential for closed loop control of some of the operating parameters.

ACKNOWLEDGEMENTS

The authors wish to acknowledge the valuable assistance of the following Los Alamos Scientific Laboratory personnel: Alphonse Criscuolo, Group E-5; W. C. Erickson, Group CMB-6; James D. Gallagher, Group CNC-11; Stephen E. Newfield, Group CMB-6; and William M. Seifert, Group E-5.

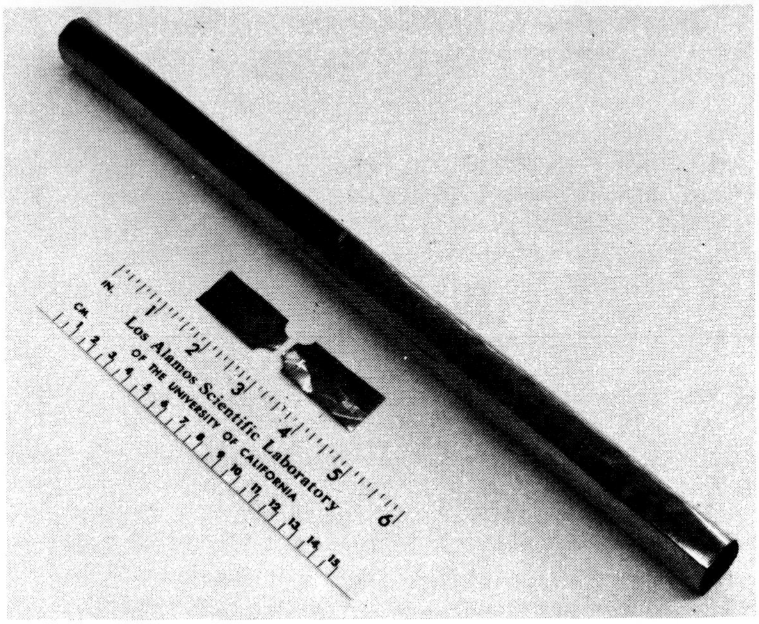

E. B. Welded Inconel 600 Foil (.025-mm-thick) Laser Cavity - Tensile Sample Fractures Outside Weld Zone

Fig. 1

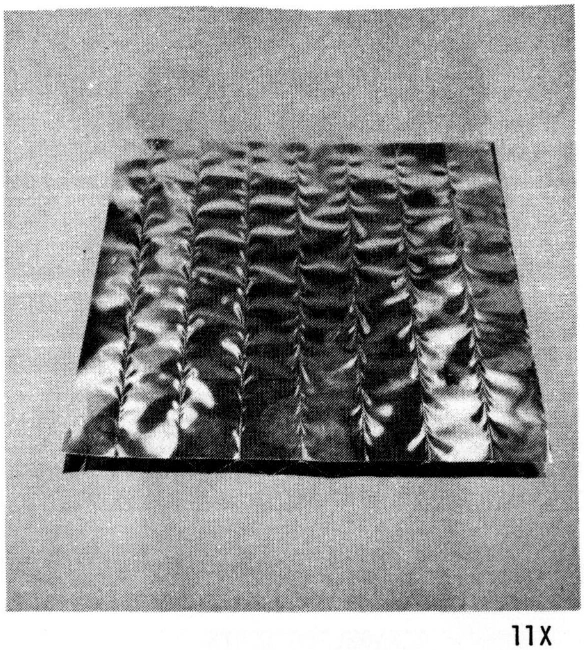

11X

E.B. Welded Aluminum Foil (.025-mm-thick) Honeycomb Panel for Conductivity Tests

Fig. 2

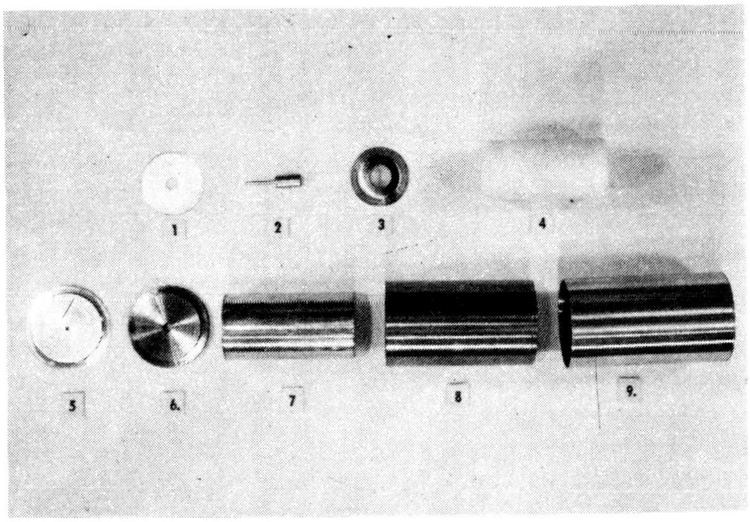

1. Pt-20 wt% Rh liner top cap
2. PRD (pressure relief device)
3. Spacer
4. Fuel (mockup)
5. Pt-20 wt% Rh alloy cladding top cap
6. T-111 alloy strength member top cap
7. Pt-20 wt% Rh alloy liner
8. Tantalum strength member
9. Pt-20 wt% Rh alloy cladding

Exploded Diagram Showing Various Components

Fig. 3

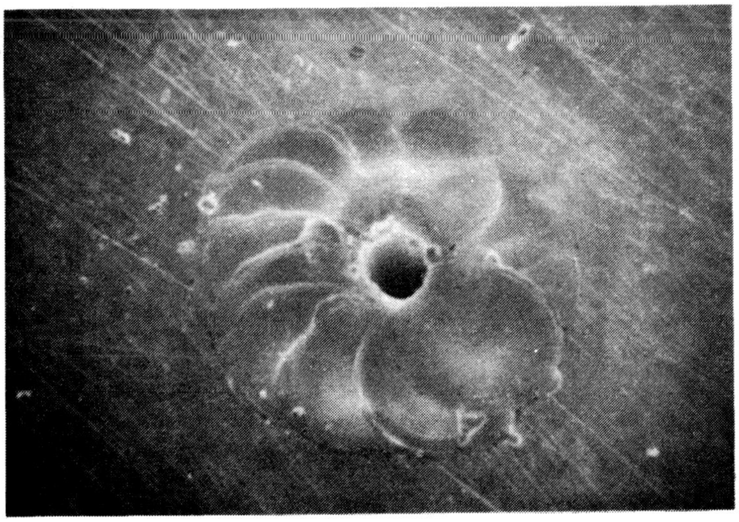

50X

Electron Beam Weld of Capillary Tube to Outer Lid

Fig. 4

A. Top view of vent "as welded" 11X

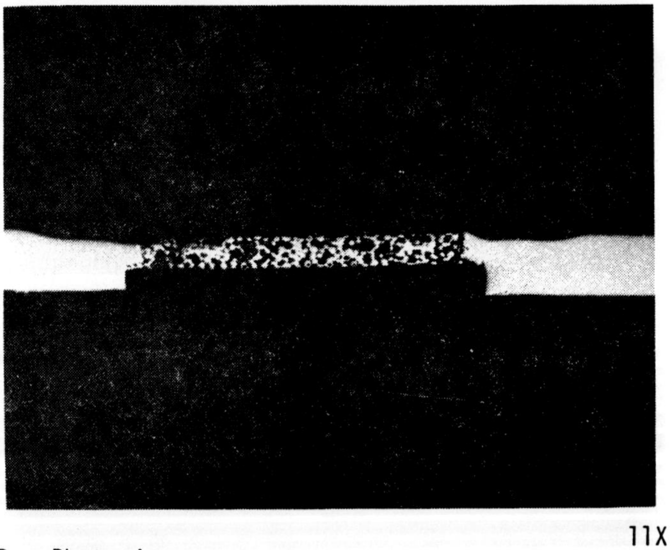

11X
B. Photomicrograph of cross-section through vent as welded to the platinum holder

Figure 5

HELIUM VENT - ELECTRON BEAM WELD DEVEOPMENT PIECES

15X

Photomicrograph Shows Longitudinal Section Through Partial Penetration Weld Zone in 21-6-9 Stainless Steel. Weld Penetration is Nominally 85% of Wall Thickness.

Fig. 6

100X

Photomicrograph Shows Longitudinal E.B. Weld Section In Stainless Steel Illustrating Area Of Overpenetration Due To A Voltage Transient.

Fig. 7

General Layout of Experimental E.B.W. Area.
Shows Computer and Accessory Equipment

Fig. 8

Microcomputer Package and Signal
Conditioning and Display Panels

Fig. 9

PARAMETER	AMPLIFIER	GAIN	DISPLAY
BEAM CURRENT	1	100	5.000 mA
	1	1000	50.00 mA
HIGH VOLTAGE	2	100	150.0 kV
TACHOMETER	3	100	500.0 RPM
FILA. CURRENT	4	100	1.500 A
X DEFLECTION	5	100	40.00 mA
Y DEFLECTION	6	100	40.00 mA
FOCUS	7	100	1.200 A
VACUUM	8	500	5000 (ARBITRARY)

FIG. 10 SIGNAL CHANNELS AND DISPLAY SCALE

```
◆◆◆◆◆◆◆◆◆◆◆◆◆◆◆◆◆◆◆◆◆◆◆◆◆◆◆◆◆◆◆◆◆◆◆◆◆◆◆◆◆◆◆◆◆
◆                                           ◆
◆               C M B - 6                   ◆
◆             E L E C T R O N               ◆
◆           B E A M   W E L D E R           ◆
◆         DATA ACQUISITION SYSTEM           ◆
◆               VERSION 1.0                 ◆
◆                12/10/76                   ◆
◆◆◆◆◆◆◆◆◆◆◆◆◆◆◆◆◆◆◆◆◆◆◆◆◆◆◆◆◆◆◆◆◆◆◆◆◆◆◆◆◆◆◆◆◆


>ENTER THE TIME OF DAY : 11:16:00
>ENTER DATE : 02:11:77
>ENTER PROGRAM CODE :D548
>PART DESCRIPTION :FUEL CLADDING
>MATERIAL :TA-10%W
>SERIAL NO. :AHS-004
>JOINT TYPE :BUTT W STEP
>PENETRATION :4.0MM-85%

>MODE:1
>INITIALIZE MODE - HIT RETURN TO PROCEED
>ENTER THE WELD PIECE GEAR RATIO : 12:1
>ENTER DESIRED UNITS - MM OR IN : MM
>ENTER WELD PIECE DIAMETER : 23.4
>ENTER WELD RATE (IN/MIN OR MM/SEC) : 12.5
>DO YOU DESIRE AUTO SAMPLING? NO
>ENTER HIGH SAMPLE RATE CHANNELS : 0:2:5:
>ENTER LOW SAMPLE RATE CHANNELS : 1:6:
>ENTER DESIRED BEAM CURRENT (MA) : 5.4
>ENTER CHANNEL TOLERANCES :5:10:10:10:10

>INSERT TAPE CASSETTE 1
>SET ALL FUNCTIONS EXCEPT ◆PLAYBACK◆ 1J ◆LINE◆
>INITIALIZE MODE COMPLETED
>MODE:2
>DISPLAY MODE - HIT RETURN TO PROCEED
.MODE:3
>DISPLAY MODE COMPLETED
>WAIT MODE - HIT ESC TO PROCEED
>MODE:4
>WELD MODE READY - HIT RETURN TO PROCEED
.MODE:5
>WELD MODE COMPLETED
>RECORD DATA MODE - HIT RETURN TO PROCEED
>SWITCH TERMINAL SPEED TO ◆HI◆, THEN HIT RETURN TO PROCEED
     NOTE: AFTER ALL DATA IS RECORDED, ACKNOWLEDGE WITH ◆RETURN◆
```

Fig. 11

```
********************************************************************
*                                                                  *
*              WELD DATA FOR H. CASEY & CMB-6                      *
*                                                                  *
*        DATE: 02:11:77          TIME WELD BEGUN: 11:16:00         *
*                                                                  *
********************************************************************
    000000 177542 177745 000000 177542 177745 000000 177542 177745 000000
    177542 177745 000000 177542 177745 000000 177542 177745 000000 177542
    177745 177541 177746 000000 177542 177745 000000 177542 177745 000000
    177542 177745 000000 177542 177745 000000 177542 177745 000000 177542
    177745 000000 177542 177745 000000 177542 177745 000000 177542 177745
    000000 177542 177745 177541 177746 000000 177542 177745 000000 177542
    177745 000000 177542 177745 000000 177542 177745 000000 177542 177745
    000000 177542 177745 000000 177542 177745 000000 177542 177745 000000
    177542 177745 000000 177542 177745 177541 177746 000000 177542 177745
    000000 177542 177745 000000 177542 177745 000000 177542 177745 000000
    177542 177745 000000 177542 177745 000000 177542 177745 000000 177542
    177745 000000 177542 177745 000000 177542 177745 177541 177746 000000
    177542 177745 000000 177542 177745 000000 177542 177745 000000 177542
    177745 000000 177542 177745 000000 177542 177745 000000 177542 177745
    000000 177542 177745 000000 177542 177745 000000 177542 177745 177541
    177746 000000 177542 177745 000000 177542 177745 000000 177542 177745
    000000 177542 177745 000000 177542 177745 000000 177542 177745 000000
    177542 177745 000000 177542 177745 000000 177542 177745 000000 177542
    177745 177541 177746 000000 177542 177745 000000 177542 177745 000000
    177542 177745 000000 177542 177745 000000 177542 177745 000000 177542
    177745 000000 177542 177745 000000 177542 177745 000000 177542 177745
    000000 177542 177745 177541 177746 000000 177542 177745 000000 177542
    177745 000000 177542 177745 000000 177542 177745 000000 177542 177745
    000000 177542 177745 000000 177542 177745 000000 177542 177745 000000
    177542 177745 000000 177542 177745 177541 177746 000000 177542 177745
    000000 177542 177745 000000 177542 177745 000000 177542 177745 000000
    177542 177745 000000 177542 177745 000000 177542 177745 000000 177542
    177745 000000 177542 177745 000000 177542 177745 177541 177746 000000
    177542 177745 000000 177542 177745 000000 177542 177745 000000 177542
    177745 000000 177542 177745 000000 177542 177745 000000 177542 177745
    000000 177542 177745 000000 177542 177745 000000 177542 177745 177541
    177746 000000 177542 177745 000000 177542 177745 000000 177542 177745
    000000 177542 177745 000000 177542 177745 000000 177542 177745 000000
    177542 177745 000000 177542 177745 000000 177542 177745 000000 177542
    177745 177541 177746 000000 177542 177745 000000 177542 177745 000000
    177542 177745 000000 177542 177745 000000 177542 177745 000000 177542
    177745 000000 177542 177745 000000 177542 177745 000000 177542 177745
    000000 177542 177745 177541 177746 000000 177542 177745 000000 177542
    177745 000000 177542 177745 000000 177542 177745 000000 177542 177745
    000000 177542 177745 000000 177542 177745 000000 177542 177745 000000
    177542 177745 000000 177542 177745 177541 177746 000000 177542 177745
    000000 177542 177745 000000 177542 177745 000000 177542 177745 000000
    177542 177745 000000 177542 177745 000000 177542 177745 000000 177542
    177745 000000 177542 177745 000000 177542 177745 177541 177746 00

********************************************************************
WAIT MODE - HIT ESC TO PROCEED
```

Fig. 12

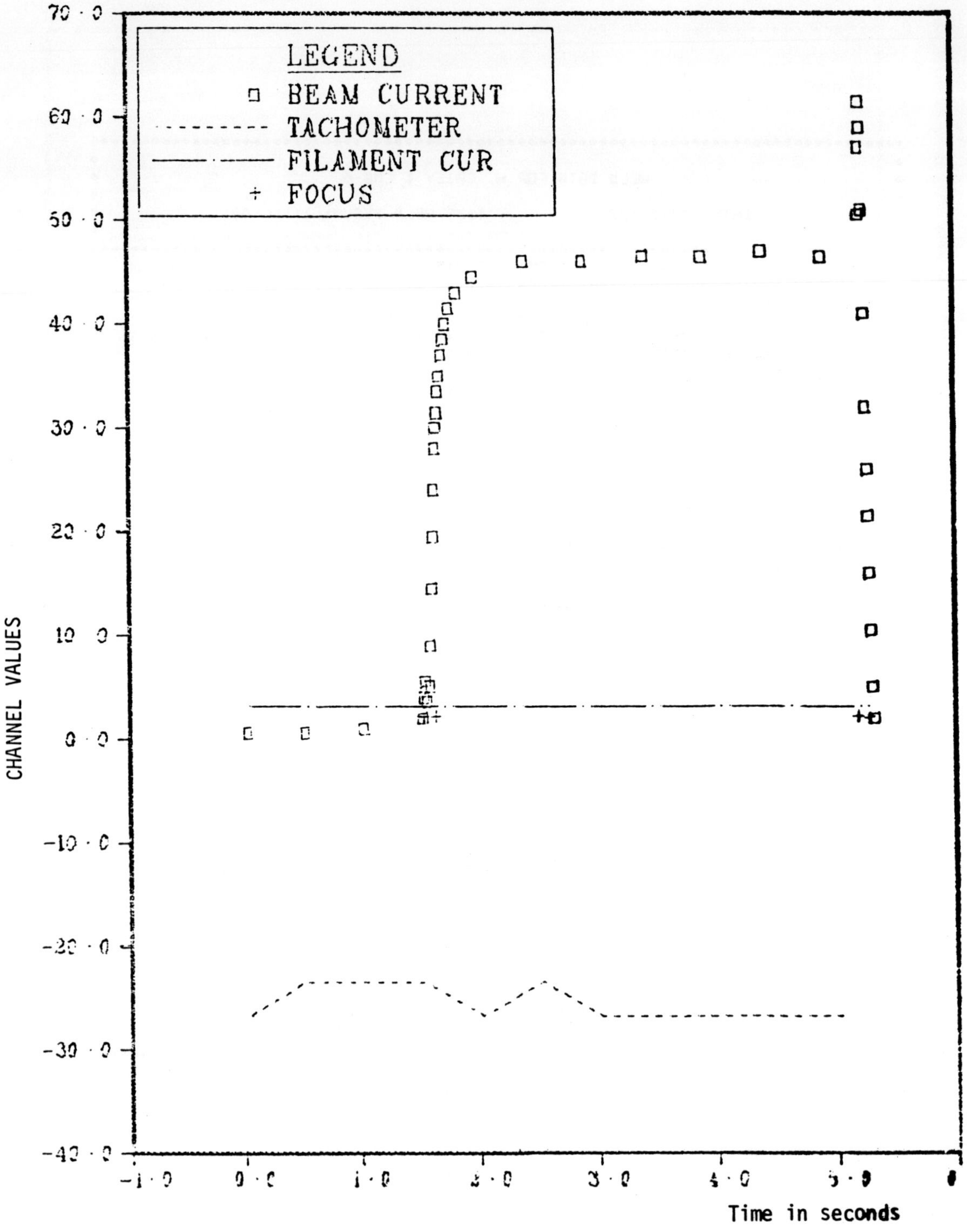

Fig. 13a Data Plot For Four Channels

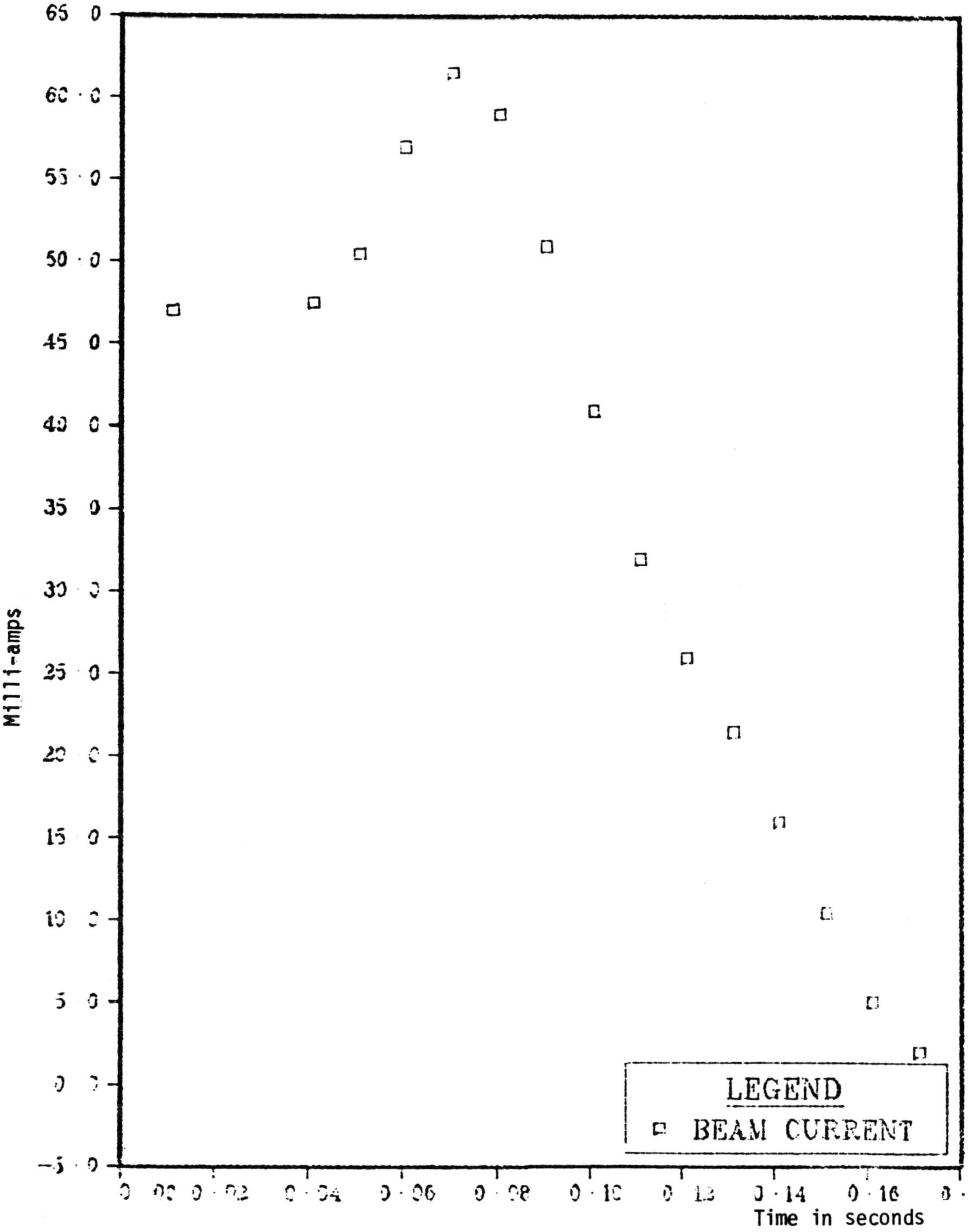

Fig. 13b Beam Current Detail in Region of "Overcurrent"

Ceramic X-ray tube's small focal spot size is key to short film-focal distance

Microfocus gets in close without sacrificing radiograph quality in inspecting electron beam welds on turbofan main engine mount.

In general, the rules for getting sharp X-rays are to keep the X-ray source far from the workpiece and the film or recording surface close to the workpiece. To disregard these rules is to invite fuzzy and distorted images. The size of the X-ray spot also influences sharpness. There is a definite relationship between the so-called focal spot size of the X-ray ube and the definition in the radiograph. A large-focus tube, although capable of withstanding high energies, does not permit the delineation of as much detail as a small-focus tube. Long source-film distances (film-focal distance of f.f.d.) will aid in showing detail when a large-focus tube is used, but it is generally advantageous to use the smallest focal spot permissible.

Additionally, the central axis of the X-ray beam should be perpindicular, and the area of greatest interest (weld joint, or whatever) should be parallel, to the film plane.

The AiResearch Manufacturing Company of Phoenix, X-Ray inspects electron beam (EB) welds on the main engine mount of its TFE 731 turbofan jet engine. Each mount has 64 EB welds, some only 0.3-mm (0.01 inch) wide.

Good radiographic quality was hard to come by, because an ordinary X-ray tube could not always be positioned well. Furthermore, the focal spot of an ordinary X-ray tube was too large to produce an image with high enough graphic resolution.

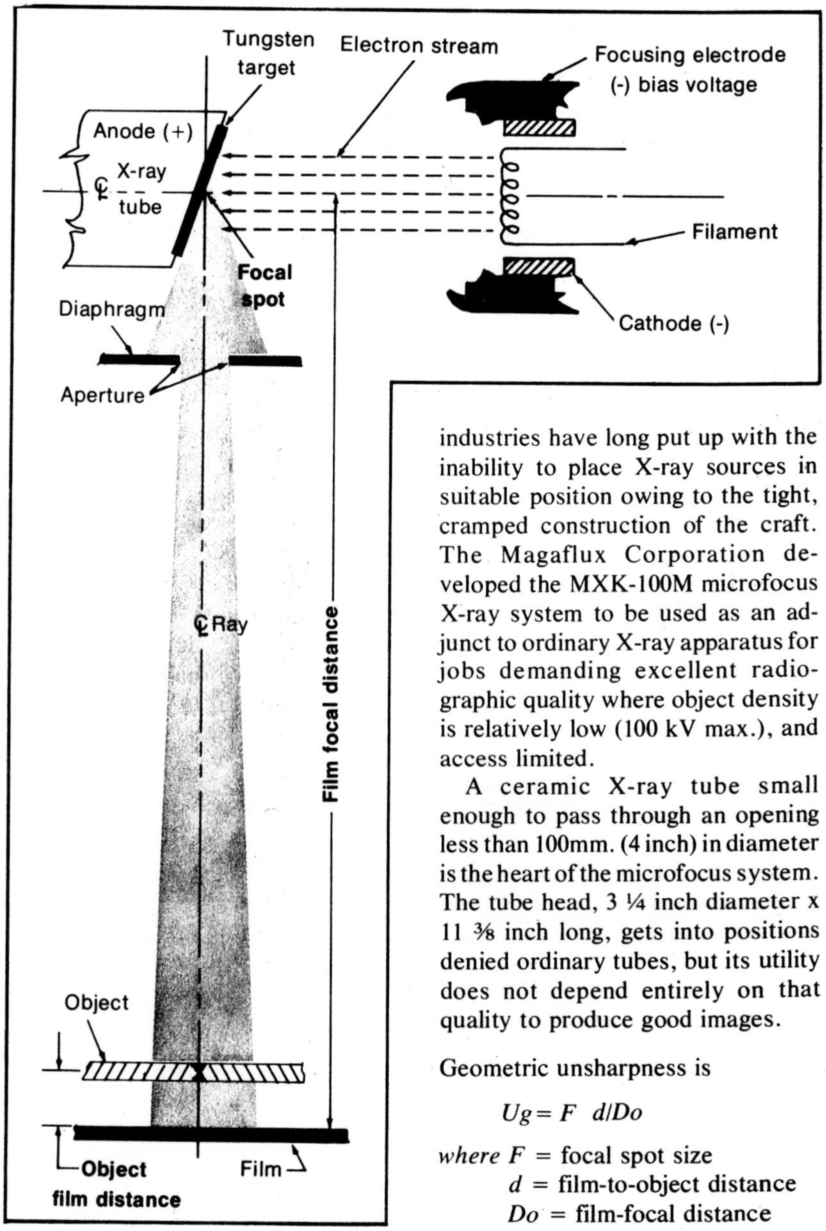

The aircraft and aerospace industries have long put up with the inability to place X-ray sources in suitable position owing to the tight, cramped construction of the craft. The Magaflux Corporation developed the MXK-100M microfocus X-ray system to be used as an adjunct to ordinary X-ray apparatus for jobs demanding excellent radiographic quality where object density is relatively low (100 kV max.), and access limited.

A ceramic X-ray tube small enough to pass through an opening less than 100mm. (4 inch) in diameter is the heart of the microfocus system. The tube head, 3 ¼ inch diameter x 11 ⅜ inch long, gets into positions denied ordinary tubes, but its utility does not depend entirely on that quality to produce good images.

Geometric unsharpness is

$$Ug = F \; d/Do$$

where F = focal spot size
d = film-to-object distance
Do = film-focal distance

If d increases or Do decreases, and F is constant, then geometric unsharpness Ug increases. If the focal spot size F is decreased to a point source, however, geometric unsharpness becomes negligible.

The size of the focal spot of an X-ray tube is determined by the filament and the beam forming structure. The tube of the MXK-100M system can produce a focal spot as small as 0.05 mm (0.002 inch). The device electrostatically focuses, by applying a negative bias voltage to a focusing electrode, the electrons streaming from the cathode to a tungsten target. Varying the bias voltage from 0 to 120 volts, produces a range of focal spots from 0.5 to 0.05 mm (0.02 to 0.002 inch). Tube voltage through this range runs from 20 kV to 100 kV. Current rating is 1 mA maximum.

The maximum allowable current for a small focal spot is relatively low, because most of the energy applied to an X-ray tube is transformed into heat at the focal spot. The smaller the spot, the less energy the target material and anode will be able to withstand without damage. So small is the focal spot that even the maximum current rating of 1 mA for the MXK-100M tube exceeds the ordinary target loading factor of 600 watt/mm^2. Target thickness far exceeds the diameter of the focal spot, so the heat generated is dissipated axially and radially.

The low current rating of the microfocus tube is not in itself a disadvantage, especially when two other points of X-ray geometry are considered. Like the laws of light. the intensity of radiation falling on an area, is inversely proportional to the square of the film-focal distance; that is, if I_1 and I_2 are the intensities at distances d_1 and d_2 respectively, then

$$\frac{I_1}{I_2} = \frac{(d_2)^2}{(d_1)^2}$$

and the milliamperage distance relation, the milliamperage required for a given exposure is directly proportional to the square of the film-focal distance. Put another way, where M_1 and M_2 are the exposure times (mA/sec) at distances d_1 and d_2 respectively, then

$$\frac{M_1}{M_2} = \frac{(d_1)^2}{(d_2)^2}$$

Suppose the tube current is constant and the X-ray beam travels a distance $d_1 = 24$ inches and covers 16 square inches on the recording surface. Then reducing the film-focal distance to $d_2 = 12$ inches, the X-rays will cover only 4 square inches on the recording surface — an area 4 times smaller. Then since,

$$\frac{I_1}{I_2} = \frac{(12)^2}{(24)^2}$$

$$I_2 = 4I_1$$

When distance is d_2 the intensity will be 4 times greater than at d_1. An exposure adequate with distance d_1 need only ¼ as long as distance d_2 to produce a radiograph of equal den-

The microfocus X-ray tubehead in position to radiograph electron beam welds on the main engine mount of the TFE-731 turbofan jet engine.

Microfocus X-ray reveals chain like porosity defect in electron beam weld of engine mount. Taken at 75 kV, 0.8 mA, 10-inch film-focal distance, 90 volts negative bias, 100 micron focal spot size, 30° angle to perpendicular.

sity. In practice the milliamperage could be decreased. The reduction of radiation intensity, when working in close to the object, also eases the task of protecting operators against radiation hazards.

If a radiograph is made at a film-focal distance d_1 of 100 cm with an exposure M_1 of 300 mA/sec., and should the ceramic tube be moved to d_2 10 cm from the object, then

$$M_2 = \frac{M_1(d_2)^2}{(d_1)^2}$$

$$= \frac{300 \text{ mA/sec } (10\text{cm})^2}{(100\text{cm})^2}$$

$$M_2 = 3 \text{ mA/sec}$$

With a tube current of 0.2 mA the exposure time would be 15 seconds.

The examples prove that the intensity of radiation and the milliamperage required can be decreased considerably because of the short-film focal distance of the MXK-100M system. The ceramic tube is capable of making an equivalent, or often better, radiograph at 10 cm with an exposure time of 3 mA/sec.

Often where 1-to-1 radiographs are inadequate because complex configurations and/or tiny internal parts require blow-ups for clarity, to measure distances, to make comparisons, and to detect joint misalignment better, AiResearch finds it advantageous to use direct X-ray magnification. The microfocus tube can directly magnify the object while retaining sharpness because of the small focal spot. If in the second example a X2-diameter magnification was required, the film-object distance could be made equivalent to the object-focal distance. This would result in a total film-focal distance of 20 cm and would change the exposure value to 12 mA/sec., or 0.4 mA for 30 seconds.

The alloy used for fabrication of the TFE-731 main engine mount is common grade titanium (64L4V). It is welded with a Sciaky electron beam welder. High and low penetration welds are made depending on the cross-sectional thickness of the parts being joined. Weld chamber vacuum is 5 X 10^{-5} Torr.

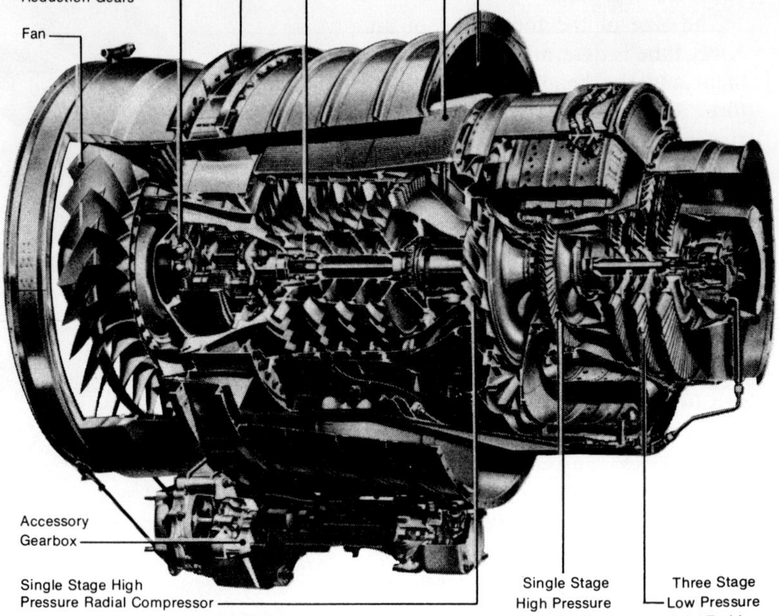

Cutaway view of the TFE-731 turbojet engine showing the location of the main engine mount. F.A.A. certified in August of '74, the engine is rated at 3,500 lb. takeoff thrust. Its 52.8 x 32.2 x 39.5 inch dimensions house 710 lb. of dry weight or 750 lb. with optional cooler and oil tanks. The engine powers the Learjet 35 & 36 models, Lockheed's Jet Star II, and the French Dassault 10.

Microfocus X-ray showing transverse and longitudinal cracks in a Tig weldment.

Engine mount components are machined to close tolerance and carefully assembled. A number of process variables can spoil an electron beam weld. Slight angular variations of the electron beam can lesson penetration. Bad welds, however, generally stem from improper fitup. Imperfections stemming from metallurgical flaws such as porosity are few.

An average weld time of 8 hours is required to complete the assembly. Eight seperate weld setups are used. The titanium is acid cleaned before welding.

X-ray quality control procedure

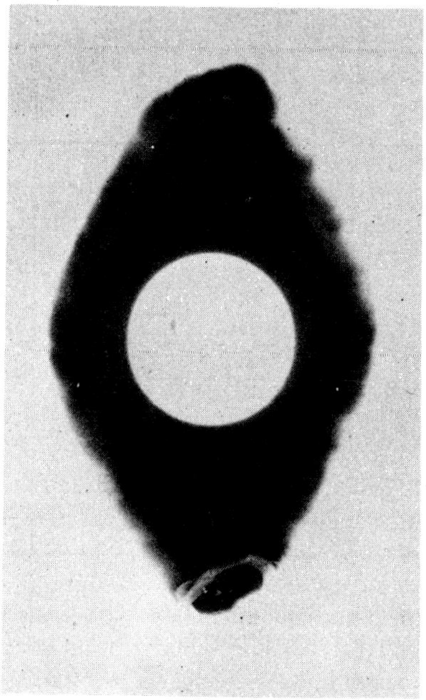

Microfocus radiographs reveal (above) a gas hole with a linear tail in a Tig weld, and (below) gas holes in several areas of an electron beam circumferential weld.

covers three distinct production stages of engine mount assembly: the raw material, intermediate, and finished stage.

In the raw material stage, X-rays are shot at 100kV, 0.8 mA. Exposure time is 60 to 100 seconds at a film-focal distance range of 4 to 12 inches. The thickness of the titanium in the raw material stage is ⅜ inch.

In the intermediate stage, X-rays are taken at 70 to 80 kV, 0.8 mA. Exposure time is 45 to 70 seconds at a f.f.d. range of 4 to 12 inches. At the intermediate stage the material has been reduced to one-half the original cross-sectional thickness. At this stage the material is stress relieved.

Final stage X-rays are taken at 50kV, 0.5 mA. Exposure time is 25 seconds at a f.f.d. range of 4 to 12 inches. By the time the engine mount reaches the final stage, all machining has been completed, and all welds have been dressed down. With milling and drilling operations completed, the structural attachments have been made and the engine mount has been finished to tolerance.

Four to 5 hours of X-ray inspection time is required on each mount. Each of the 64 welds are radiographed at each stage. A minimum of 192 X-rays are taken of each engine mount. If a weld must be repaired, additional radiographs are required.

Kodak Industrial X-ray Film, Type M, 70 mm. (lead pack) is used. Type M is an extra fine grain, high contrast film for light metals, where critical radiography is required. AiResearch has used fixture angles varying from 20 to 40 degrees using this film in conjunction with the MXK-100M system. Magnifications greater than X20 have been made without significant loss in geometric sharpness.

AiResearch also uses the microfocus X-ray equipment to check the position of carbon ring seals inside steel or titanium bodies and inspect air-cooled, high temperature turbine blades and vanes, nozzles, turbine ducts, and valve bodies. ■

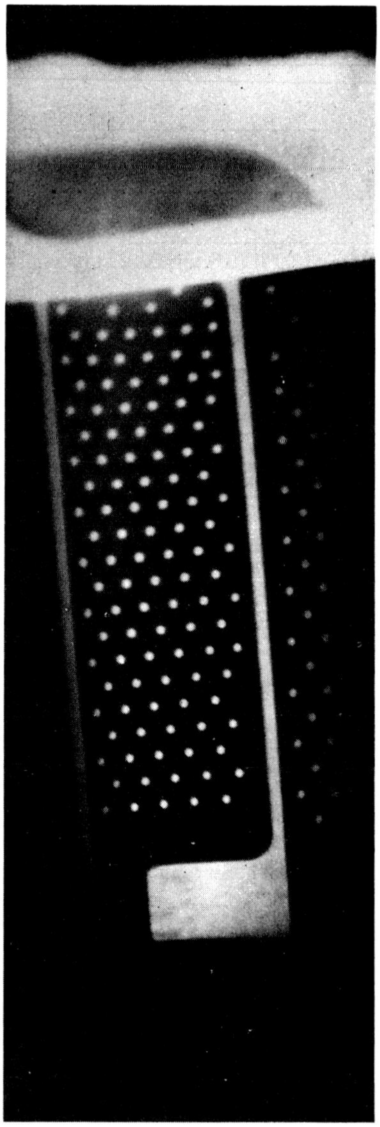

Micrfocus X-ray of core stiffener, magnified to four times diameter. The radiographs are used to measure the spacing of cores.

Based on an article by William H. Spaulding, NDT engineer, manufacturing quality engineering, AiResearch Manufacturing, Company, Arizona. Technical information on the microfocus X-ray system, based on an article by Henry J. Ridder, regional applications engineer, Magnaflux Corporation.

PART B:
Laser Welding

LASER WELDING-THE PRESENT STATE-OF-THE-ART

E. M. BREINAN and C. M. BANAS, United Technologies Research Center
M. A. GREENFIELD, Air Force Wright Aeronautical Laboratories

SECTION I

INTRODUCTION

Laser system development has been extremely rapid since the first successful operation of a ruby laser in 1960 (1)*. Ruby as a lasing medium was followed quickly by neodynium-glass and subsequently by helium-neon and argon-ion gas laser systems (2,3). Further developments in solid-state lasers involved the addition of neodymium to yttrium aluminum garnet (YAG) and to yttrium aluminate (YALO) which provided the solid-state laser with a capability for continuous-wave (cw) output (4). Intense activity has led to the further development of a wide range of gas, liquid, and solid-state lasers having characteristic wavelengths ranging from the ultraviolet to the far infrared (2). In pulsed units, peak power levels of the order of 10^{16} watts (5) have been attained. Because pulsed systems are not capable of generating continuous, deep-penetration welds, however, pulsed systems are not generally applicable to most welding applications.

From the standpoint of welding applications, the solid-state YAG and the carbon dioxide gas laser have been the most significant developments. The first carbon dioxide laser was operated by Patel (6) in 1964 and exhibited an initial power output of several watts. By the addition of nitrogen and helium to the carbon dioxide in this system (7), the performance was improved to a level of approximately 100 watts per meter of length. This technology was utilized to construct a 790 feet (240 meter) long, folded, CO_2 laser (8) which, in 1967, produced a continuous power output in excess of 8.0 kilowatts. Despite the power level at-

*Numbers in ( ) denote references listed at end of text.

tained, the complex optics and multimode output characteristics of this device did not lend themselves to welding applications.

Since the laser is a relatively low efficiency energy conversion device, one of the principal requirements for a high-power system is the effective removal of waste heat energy. Thus, development of convectively-cooled systems (9) led to marked improvement in CO_2 laser performance. The first of the convectively-cooled systems, the mixing laser (9), provided high output power per unit volume but was not applicable to industrial welding because the laser gases were expelled to the atmosphere and operating costs were high. Closed-cycle, convectively-cooled units (10) eliminated the primary operating costs of the mixing unit and are currently the basis for high-power, production-rated equipment to 10 kilowatts output.

Carbon dioxide lasers suitable for production applications depend upon excitation from a high-voltage, abnormal glow discharge. Instability of such discharges at high current levels is one problem associated with construction of compact, efficient, high-power systems. In DC discharges, the electrical conductivity of the gas mixture is intimately associated with the current which passes through it. Consequently, the discharge operates on a voltage-current characteristic determined by the gas composition and operating pressure. One means for stabilizing DC discharges to higher current levels is to provide an auxiliary means for ionizing the gas medium and thereby uncoupling operating voltage from the DC current. Possible auxiliary ionization techniques include ultraviolet light, high-voltage electron beams, high-frequency electrical energy, radioactive particle ionization, and preionization by a secondary DC discharge. Although all of these techniques have been investigated in prototype systems, only the electron-beam ionization technique is currently offered in a commercial package.

Requirements for waste energy removal also limit the operating performance of solid-state, Nd-YAG systems. In this case, continuous operation is achieved by excitation of the YAG rod by illumination from a rare gas discharge lamp. Xenon or krypton discharge lamps are currently used and electrical conversion efficiency is substantially lower than that for the electrically-excited, CO_2 laser. A single, liquid-cooled laser rod typically provides an output power of approximately 100 watts. In order to increase output, several rods are arranged in series; a maximum continuous output of 1100 watts (11) has been attained.

More recently, substantial effort has been devoted to the development of carbon monoxide lasers (12) as well as to various forms of chemical lasers (13). In view of the fundamental operating nature of the latter, i.e., laser excitation occurs as the result of a chemical reaction, chemical laser systems do not lend themselves to closed-cycle operation and hence do not, at present, appear to be well suited to industrial applications. The carbon monoxide laser, on the other hand, does offer the promise of future potential because its maximum theoretical operating efficiency can approach 100% as compared to 42% for CO_2 laser systems. This follows from the fact that the lasing transition in CO is from the excited energy level to the ground state, whereas in CO_2 the lower laser level is substantially above ground state energy. Lower power CO systems (12) have demonstrated operating efficiencies of over

50%. Unfortunately, current CO systems require cryogenic temperature operation to maintain high operating efficiency and high power units suitable for industrial use have not yet been developed. The potential for high efficiency and the somewhat shorter output wavelength of the CO laser system (which should slightly improve coupling to metals) make it a promising candidate for future systems. At the present state of development, however, optically-pumped, continuous Nd-YAG systems with maximum power levels to 1 kw and electrically-excited, CO_2 systems with output power levels to 10 kw represent the only high-power systems available for current application to industrial processes.

SECTION II

HISTORY OF LASER MATERIALS PROCESSING

As a result of the pulsed output characteristics of the first laser systems, initial utilization of laser energy for materials processing was directed toward hole drilling. For several years, the processing capability of such pulsed laser systems was gaged by the number of razor blades which a focused beam could pierce. In 1965, Western Electric adapted a pulsed system for production drilling of diamond wire-drawing dies (2). In this application the processing time for drilling was reduced from 24 hours by conventional techniques to less than 1 hour by means of the laser. Other successful commercial applications of pulsed laser systems have included piercing of baby bottle nipples, trimming of resistors, scribing of microelectronic substrates, pulsed welding of microelectronic components, and drilling of cooling passages in gas turbine parts (2,3). In each production application, the unique features of the laser which permit delivery of a precisely controlled quantity of energy to a highly localized region have been responsible for process adaptation and improvement.

By using a technique of overlapping spots, the welding capability of the pulsed laser was extended to seam welding. Since the maximum practical pulse length for current laser systems is of the order of ten milliseconds, however, the time for conduction of thermal energy into the materials is severely restricted and the maximum attainable penetration in metals is limited to approximately 0.060 in. (1.5 mm). This limit is imposed by the requirement that melting take place over the entire material thickness prior to vaporization at the surface. Since the thermal diffusivity controls the rate at which thermal energy moves into the material, maximum penetrations in the pulsed mode are attained in high diffusivity materials. This penetration limitation, coupled with the extremely low welding rate which can be attained by the overlapping spot technique, has restricted application of pulsed welding to small, thin-section assemblies requiring precise control of welding energy. One such case is that of microelectronic components. In this application the laser provides an added advantage over electron beams since X-rays, which might damage some of the solid-state components, are not generated on interaction of the laser beam with the material as they are when the electron beam is used.

In general, industrial laser welding applications require continuous-wave (cw) laser operation. Using carbon dioxide lasers having output powers of less than 1000 watts, initial experiments were conducted which demonstrated maximum penetration capability to approximately 0.120 in. (3 mm) in the stainless steel (14,15,16). It was found (17) that the CO_2 laser operating at several hundreds of watts output was more suitable for cutting than for welding. Numerous applications including the slitting of steel rule dies (18), sheet metal pattern cutting (19), cutting of plastic and vinyls for use in the automotive industry (20), and the computer-controlled cutting of fabric for the garment industry (21) were developed. Significant welding applications were to follow the development of multikilowatt systems.

In 1971, results were reported for laser welding with a multikilowatt unit (22). It was shown that the high-power laser was capable of producing deep-penetration welds similar to those produced by electron beams. In the deep-penetration mode, energy is delivered to the workpiece surface more rapidly than it can be removed by conduction and radiation from the incident spot. Under these conditions, which typically involve an incident power density of the order of 10^7 w/in.^{2} (10^6 w/cm^2), a hole is drilled into the material such that energy is deposited through the material depth. With appropriate motion of the workpiece relative to the focused beam, this cavity can be translated through the material to produce a high depth-to-width ratio weld whose characteristics can be predicted by a moving, cylindrical heat source model (23).

With the development of higher-power, CO_2 laser systems (24), the maximum single-pass penetration in steel has been increased to approximately 0.80 in. (20 mm) at a laser power level of 20 kw (25). At the other end of the spectrum, thin gage materials have been welded at speeds of over 2360 in./min (100 cm/s). Typical welding performance in stainless steel involves welding speeds of 59 in./min (2.5 cm/s) in 0.24 in. (6 mm) thick material at 5 kw and a similar welding speed in 0.47 in. (12 mm) thick material at 10 kw (25,26,27). More importantly, the high welding speeds and low specific energy input have provided some unique weld properties (28,29); these characteristics will be described in more detail in the following sections.

SECTION III

HIGH-POWER LASER EQUIPMENT

In general, most solid-state laser systems available for materials processing are of the pulsed output type. A notable exception is the continuous-wave, Nd-YAG unit produced by the Holobeam Corporation (11). The Holobeam unit is compactly packaged and can readily be integrated with a tooling package to provide flexible process capability. As may be noted in Figure 1, the Holobeam unit employs a modular construction with each module containing a YAG element in an elliptical cavity. Krypton arc lamp excitation is utilized and a maximum output of approximately 125 watts per module is obtained. For higher output power levels, the number of modules used in the assembly is increased; a 1000 watt version of this model is offered by the manufacturer. At this power level the unit is capable of welding 0.12 in. (3 mm) thick stainless steel at 10 ipm (0.42 cm/sec) and thinner gage material at comparably higher welding speeds (30). The equipment lends itself to precision cutting and welding of thin gage components.

Several organizations offer low-mass-flow, conventional, carbon dioxide laser systems. Among these are the Coherent Radiation Laboratories, Ferranti Ltd, and Photon Sources. Coherent Radiation offers folded, multitube units for operation to maximum outputs of 300 and 600 watts in the fundamental, TEM_{00} mode. Discharge tubes are mounted on an I beam structure to insure optical stability.

In the Ferranti system (19), a series of discharge tubes is mounted on the frustrum of a cone in order to provide a compact configuration.

(PHOTO COURTESY OF HOLOBEAM,INC.)

Figure 1. 800 Watt Industrial Laser Welder

Output of the unit is of the order of 500 watts in the TEM_{00} mode. Primary adaptations of the unit have been for the oxygen-assisted cutting of metals.

Photon Sources provides laser systems with output power levels of 375, 500 and 1250 watts (31). The units are designed for industrial applications and utilize a monolithic granite beam to provide optical stability. The largest unit has a folded, double-Z tube arrangement to minimize the number of mirrors required to produce a compact package. Output at 1000 watts is in the TEM_{00} mode; the unit can also be operated in pulsed mode with a repetition frequency to 1 kHz.

In the area of convectively-cooled, CO_2 lasers, GTE-Sylvania produces a closed-cycle unit (Figure 2) which is capable of a maximum power output of 2 kw. The system is being marketed by the Linde Division of Union Carbide together with a universal, computer-controlled, work-handling table (Figure 2). Output in the TEM_{00} mode is obtained to power levels of approximately 1200 watts. Since this is a convectively-cooled system it should be readily amenable to adaptation to higher power levels, and the manufacturer has indicated that higher power level units are being designed.

Two suppliers exist for multikilowatt, cw, CO_2 laser systems. The Hamilton Standard Division (HSD) of United Technologies Corporation offers nominal 3 kw and 6 kw systems based on fast axial flow technology. As may be seen in Figure 3, one of the 6 kw units has been equipped with extensive tooling and computer control for an automotive underbody welding application. A 3 kw system is currently being integrated into a

(PHOTO COURTESY OF UNION CARBIDE CO.)

Figure 2. 2 kw Industrial Laser Welding System with Work Station and Computer Control

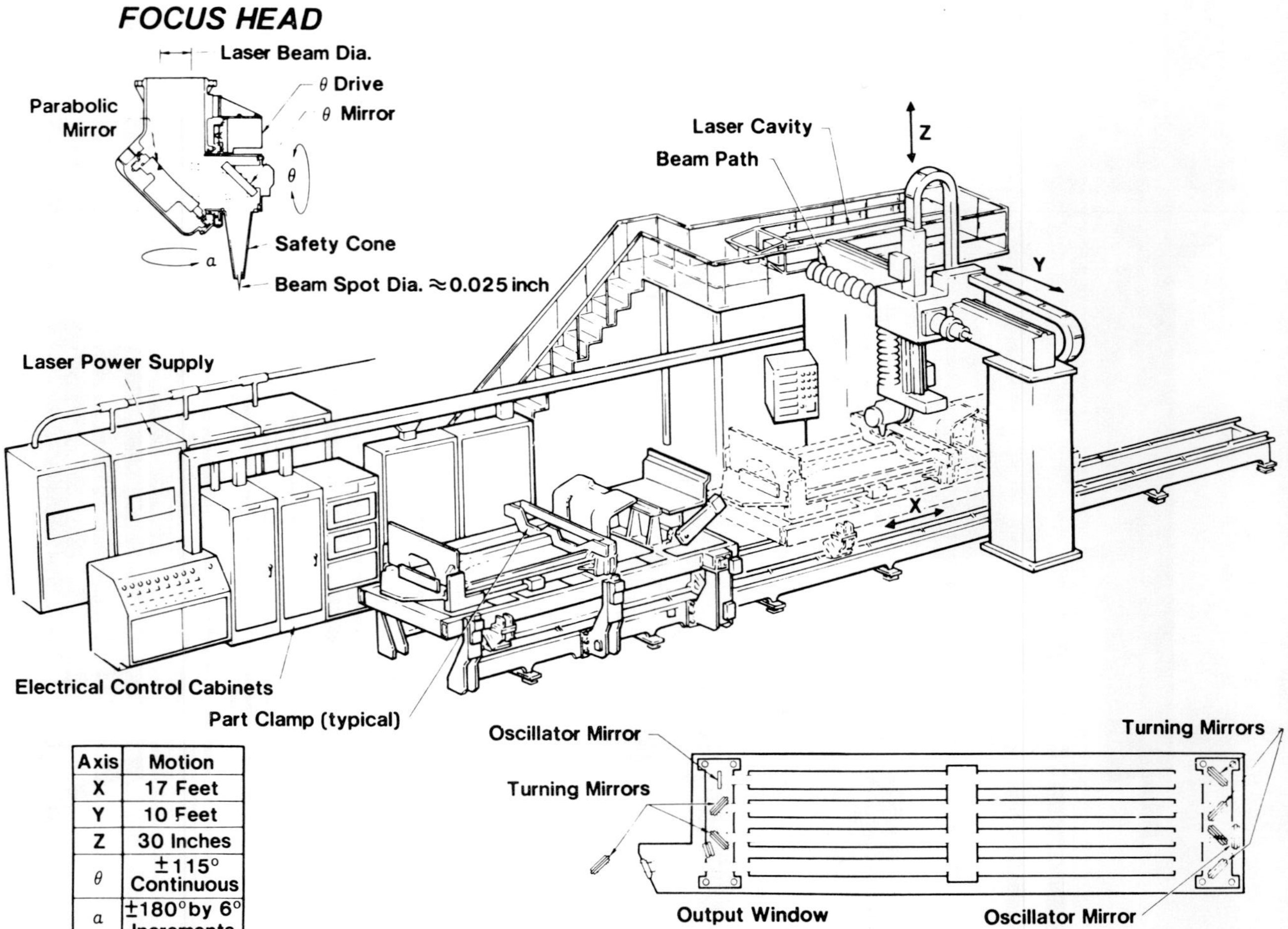

Axis	Motion
X	17 Feet
Y	10 Feet
Z	30 Inches
θ	±115° Continuous
α	±180° by 6° Increments

(COURTESY OF UNITED TECHNOLOGIES CORPORATION)

Figure 3. 6 kw Multi-Axis Welding System

production welding system for lead-acid storage batteries (Figure 4). In these units excitation is provided by a direct current, high-voltage discharge and the laser cavity is equipped with unstable resonator optics which provide an annular output beam with a magnification ratio (ratio of outer-to-inner beam diameter) of 2.0. Zinc selenide output optics are utilized.

The AVCO Corporation produces a nominal 10 kw, CO_2 laser system (HPL-10) which utilizes electron beam discharge stabilization to provide high power output in a compact system. The HPL-10 unit (Figure 5) is marketed for metalworking applications by the Sciaky Corporation. High power output is taken from the laser cavity through an aerodynamic window in order to avoid materials problems associated with solid windows at this wavelength and power level. Output of the AVCO unit is provided by unstable resonator optics which provide an annular output beam with a magnification ratio of approximately 1.7. The unit incorporates a beam scanner to provide capability for surface heat treating and a fast response power control to permit sloping out of deep-penetration welds. Both the HSD and AVCO units are conservatively designed and provide maximum output power levels approximately 50% above the nominal rating at less than 100% duty cycle or at 100% duty cycle with only slight reduction in optical quality.

In addition to the currently available commercial systems, several organizations have made significant strides in the development of convectively-cooled, multikilowatt laser systems which may be the basis for future commercial developments. The British Welding Institute has developed a 2 kw, fast axial flow system (32) which is compact in design

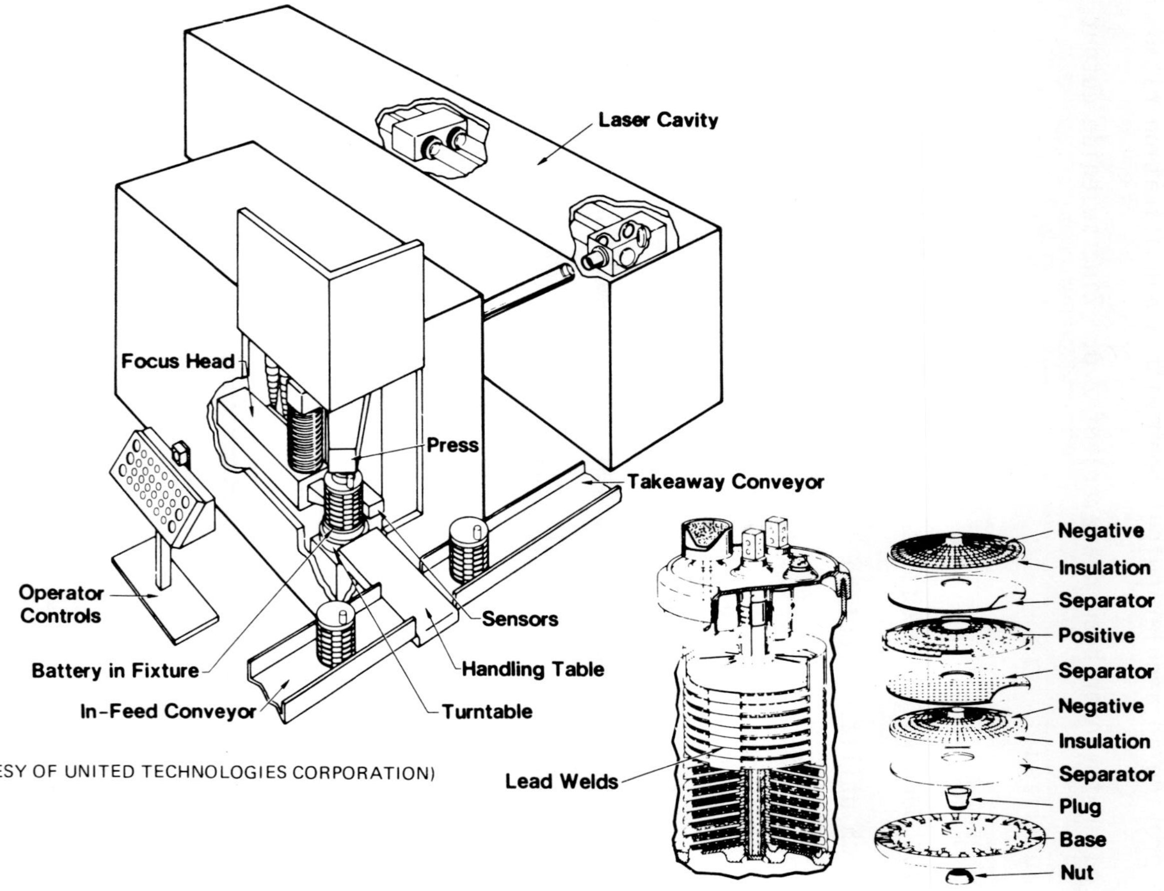

(COURTESY OF UNITED TECHNOLOGIES CORPORATION)

Figure 4. 3 kw Battery Welding System

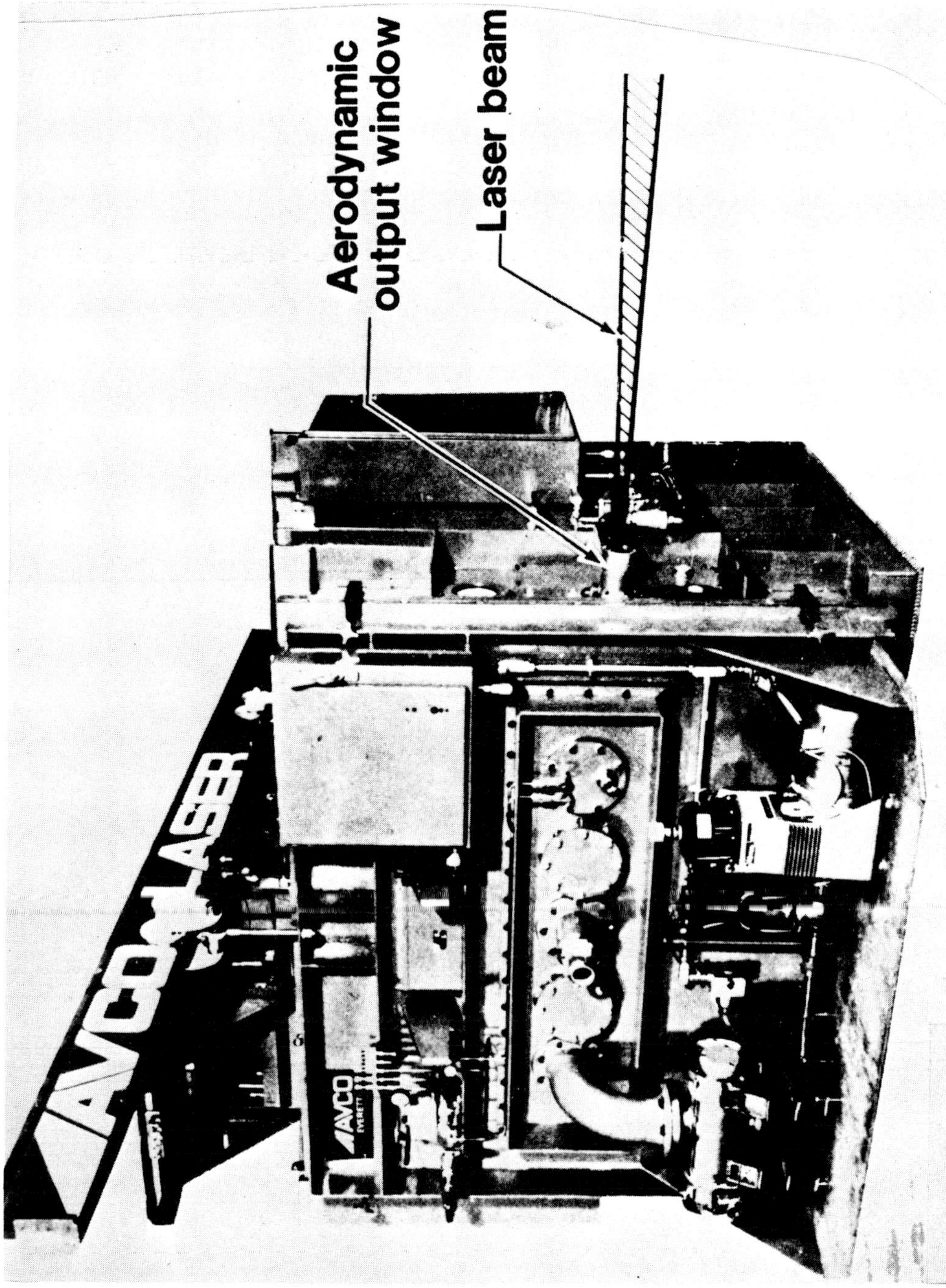

(PHOTO COURTESY OF SCIAKY BROTHERS, INC.)

Figure 5. 10 kw Industrial Laser

and is shown in Figure 6. Westinghouse Research Laboratories (33) have operated a transversely-excited, CO_2 laser at the 5 kw level which utilizes DC preionization in order to stabilize the discharge. It has also been reported that Culham Laboratories in England has developed a 10 kw unit (34), although the information available on this unit has been sparse to date. It is obvious that the potential user of high-power, industrial laser equipment may soon have a broad range of units from which to select the most appropriate for his application.

(PHOTO COURTESY OF THE BRITISH WELDING INSTITUTE)

Figure 6. 2 kw Industrial Laser Welding System

SECTION IV

LASER WELDING PERFORMANCE

Since the first reported demonstration of deep-penetration laser welding (22), numerous investigations have been undertaken to examine the welding capabilities of the laser (25-29). It has been shown that the high depth-to-width ratio welds can be formed in a variety of materials with a specific energy input per unit length of weld substantially below that for conventional arc processes. As may be noted in Figure 7, weld penetration in several materials has been shown to increase approximately with the 0.7th power of the laser power level, to a maximum of approximately 0.80 in. (20 mm) in stainless steel at the 20 kw level. Typical penetrations in stainless steel at 15 kw are shown in Figure 8. By contrast it is noted, with reference to Figure 7, that 0.80 in. (20 mm) penetration can be obtained with only about 1.5 kw using an electron beam in hard vacuum. On the other hand, the maximum penetration per unit power level for non-vacuum electron beams appears to be of the same order of magnitude as that for the CO_2 laser.

The maximum penetration which can be attained with the laser is related to the beam power, power density, and focusing characteristics. As noted in (35) the laser compares quite favorably in performance level with the electron beam at higher operating speeds. As speed is reduced in order to increase penetration, the deep-penetration cavity collapses due to the large mass of molten metal surrounding it and typical fusion welding characteristics pertain. The speed at which this transition occurs is substantially higher than in the case of the electron beam

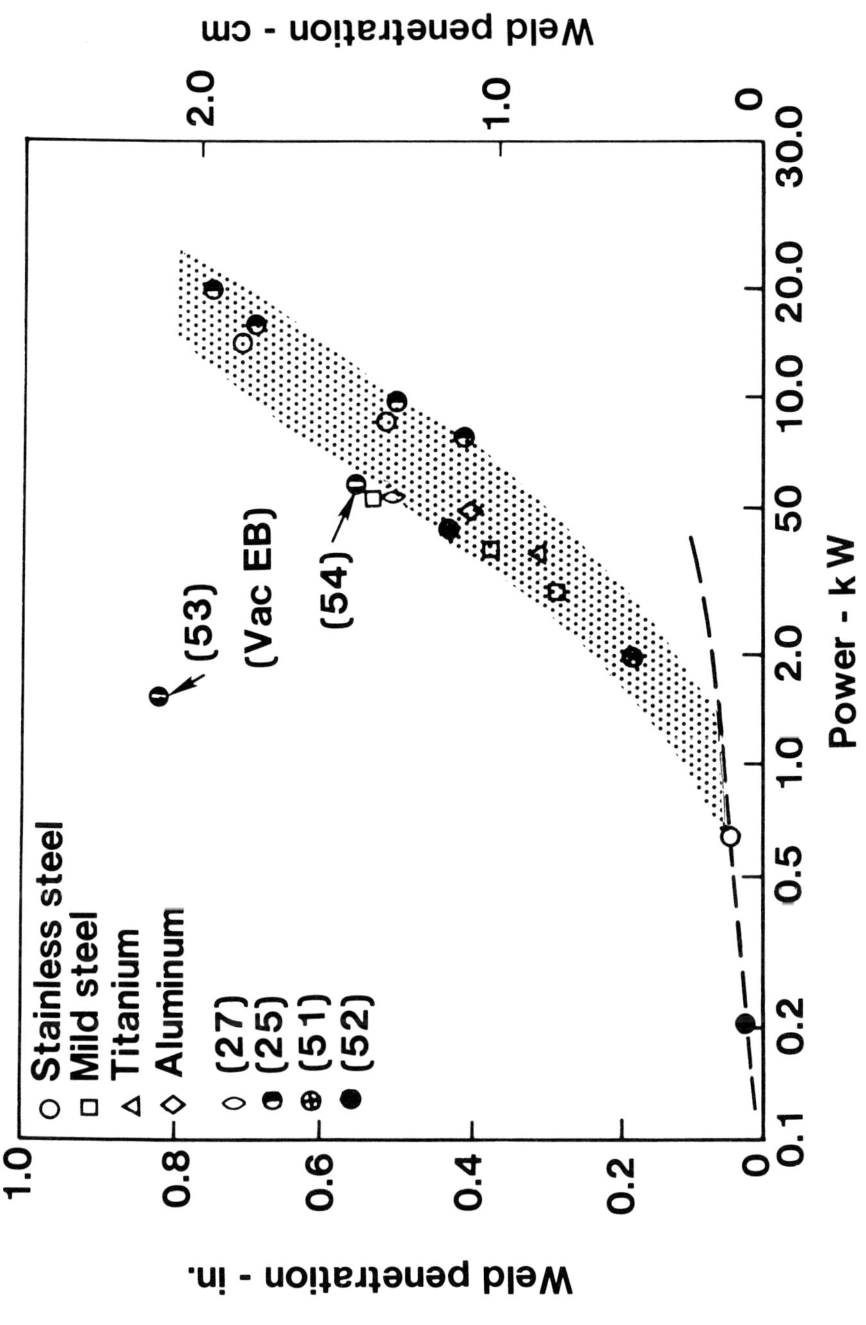

Figure 7. Effect of Power on Maximum Weld Penetration

Bead-on-plate tests
Laser power : 15 kW
Material : 304 SS
Plate thickness : 1 inch (2.54 cm)

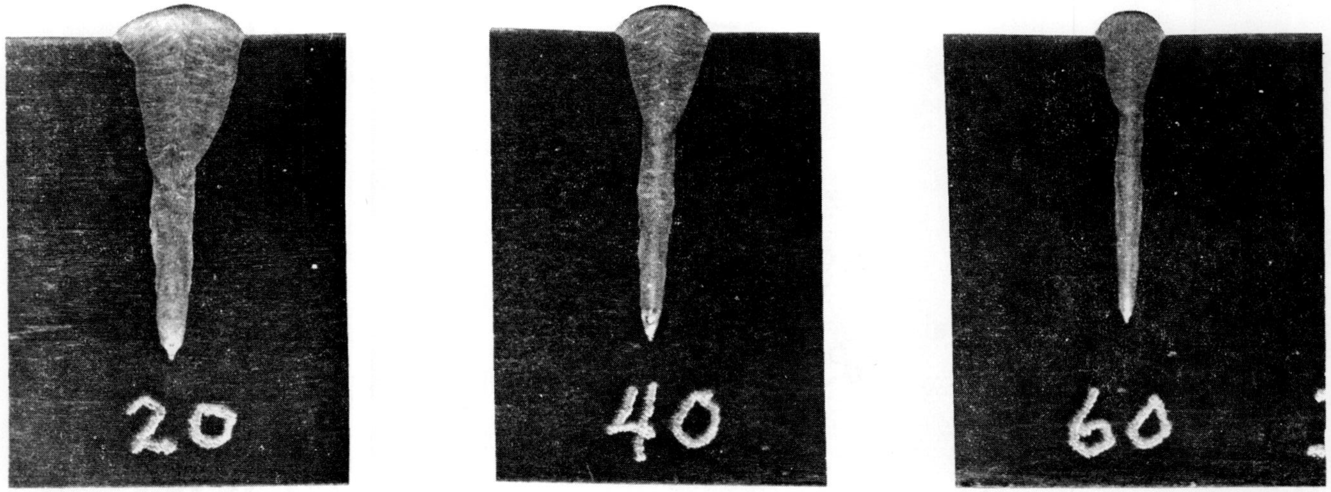

20 ipm (0.85 cm/sec) | 40 ipm (1.70 cm/sec) | 60 ipm (2.54 cm/sec)

Figure 8. Laser Weld Penetrations

because of the shorter depth of field and plasma attenuation problems encountered with the laser; a smaller maximum penetration is therefore obtained.

Another means of comparing laser and electron beam performance is in terms of a nondimensional correlation which was initially developed for electron beam welding by Hablanian (36). Using dimensional analysis he assumed that the depth of penetration, h, can be expressed as a function of laser beam parameters (power, P and focused spot diameter, d); material properties (thermal conductivity, $\bar{k}$, thermal diffusivity, $\bar{\alpha}$, and adjusted melting temperature, $\bar{T}_M$) and a process parameter (welding speed, V). Existing electron beam performance data was correlated by Hablanian (36) using the following nondimensional combinations of variables

$$\frac{h \bar{k} \bar{T}_M}{P} = f \left[\frac{Vd}{\bar{\alpha}} \right]$$

This correlation is shown in Figure 9 together with laser welding performance data for a broad range of materials exhibiting widely varying physical properties. It may be seen from Figure 9 that laser welding performance is comparable to that of the electron beam provided that operation occurs well within the deep-penetration mode.

Another means for establishing the welding performance of the laser is to examine the efficiency with which beam energy is utilized for materials fusion in the weld zone. Since the initial reflectivity of metals is generally quite high for the 10.6 micron infrared wavelength

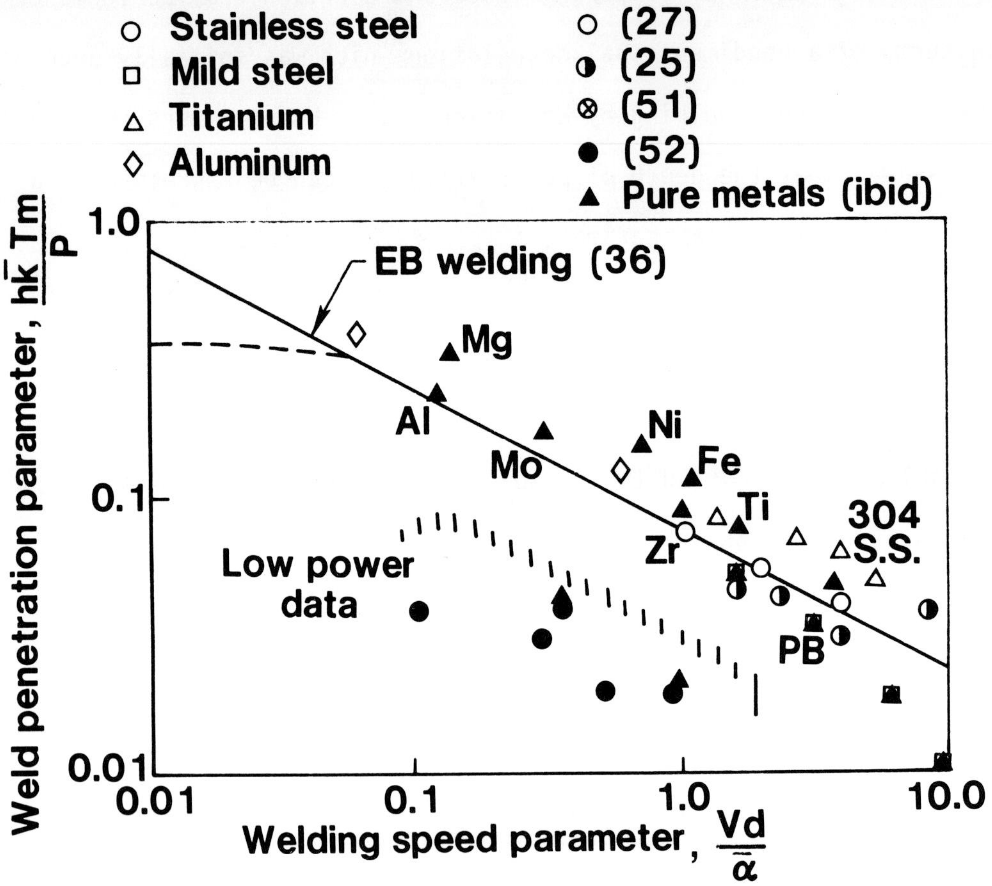

Figure 9. Comparison of Laser and Electron Beam Welding Performance

of the CO_2 laser, it is important to determine what fraction of the beam energy is actually absorbed in the welding process and how effectively it is utilized. Calorimetric measurements of the absorbed energy have been made which show that beam absorption generally approaches 90% provided that deep-penetration occurs. The melting efficiency has been computed by dividing the energy required to melt the material in the fusion zone by the absorbed laser energy. Welding penetrations on which these computations were based were formed at a laser power level of 5 kw and a welding speed of 40 in./min (1.69 cm/sec).

As may be noted in Figure 10, the melting efficiency of the laser welding process at the speed investigated approaches values of over 70% for materials of low thermal diffusivity. This level is substantially above that estimated in the approximate analysis of Swifthook and Gick (37) who proposed a maximum efficiency in the high speed limit of approximately 48%. If their relationship is modified to include the heat of fusion, H_f, a new limiting efficiency expression is obtained

$$\eta_{melt} = \frac{0.483}{1 - H_f/H}$$

which, in view of the fact that the heat of fusion (H_f) of most metals is about 20-30% of the total energy difference between the room temperature state and the liquid state at the melting temperature (H), provides maximum values in the 70% range. It should also be noted that welding at higher speed, or more correctly at higher values of $V/\bar{\alpha}$, should provide higher melting efficiencies for the higher diffusivity materials as well. In any event it is quite apparent that the CO_2 laser is a highly

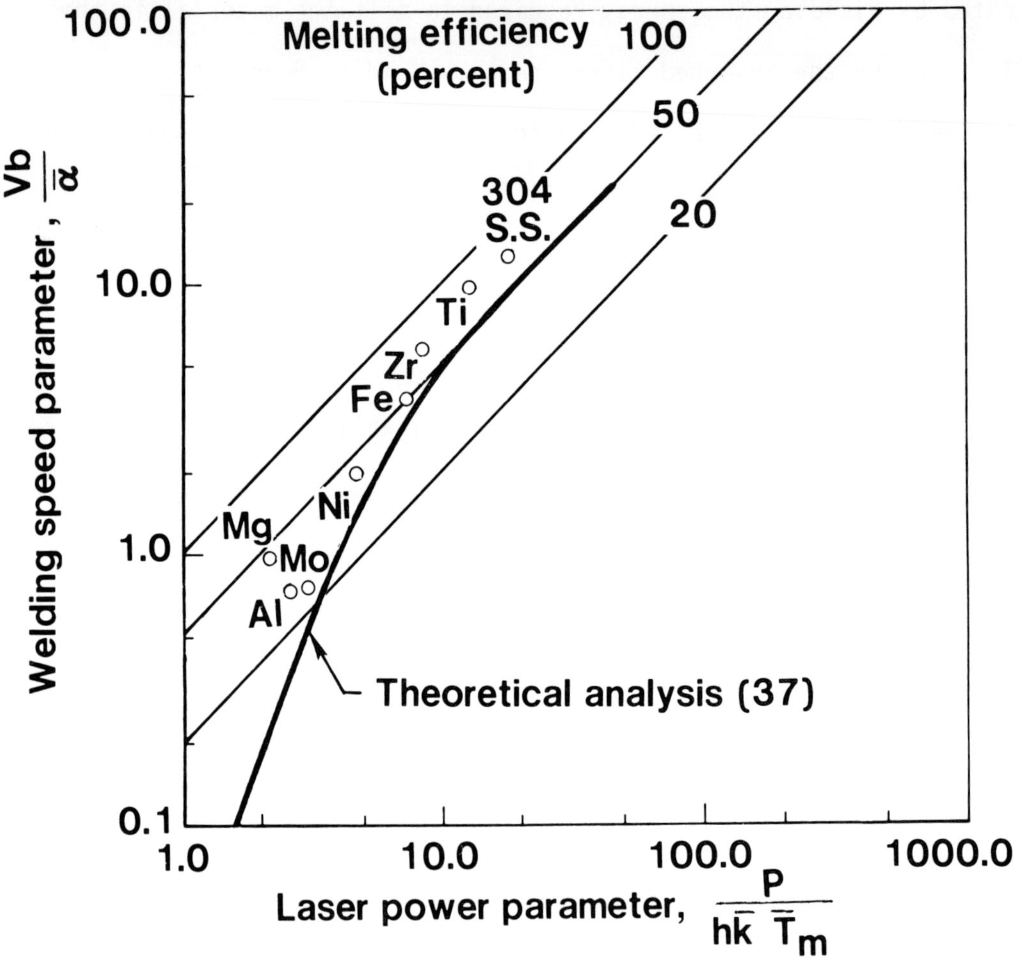

Figure 10. Laser Welding Energy Correlation

efficient welding tool which provides the capability for forming welded joints with a minimum of excess energy expenditure.

SECTION V

LASER WELD CHARACTERISTICS AND PROPERTIES

1. FERROUS ALLOYS

Although full penetration 1/4 in. (0.64 cm) stainless steel welds were first formed in 1969 (16), little, if any, formal characterization work on these earliest welds was conducted. In 1971, Brown and Banas (22) reported examination of a deep-penetration laser weld in 1/8 in. (0.32 cm) thick 304 stainless steel which was found to be acceptable under X-ray evaluation. This weld specimen failed in the base metal when tensile tested. It was also noted at this time that the laser welds examined showed less tendency for underbead spiking than electron beam welds. In that same year, Banas and Lessard (38) evaluated the capability for laser welding of rimmed steel sheet and noted that, although the welds possessed some gas porosity, the characteristics of these welds appeared to be suitable for automatic sheet metal joining applications. Results in HY-alloy steels were described by Banas (26) in 1972. It was stated that initial welds in HY-130 exhibited embrittlement and low fracture toughness as compared to base metal; however, it was also noted that welds having tensile and Charpy impact strengths equivalent to or greater than base metal had been formed in several low carbon and low alloy steels. It was anticipated that problems in the HY-steels could probably be solved by improved gas shielding, with possible use of preheating, postheating, or filler metal addition.

More recently, detailed analysis of the mechanical properties of laser welds in several ferrous base alloys has been made. These include tanker construction steels (39), HY-130 alloy (29), and X-80 Arctic pipeline steel (28). In the study of the applicability of laser welding to merchant ship construction, grades A, B, and C ship steel were welded in thicknesses ranging from 3/8 in. (.95 cm) to 1/2 in. (1.27 cm) for grade A; 1/2 in. (1.27 cm) to 3/4 in. (1.90 cm) for grade B; and 1.0 in. (2.54 cm) to 1.125 in. (2.86 cm) for grade C. These steels are all in the 0.23 percent max carbon, 0.60-1.03 percent Mn class, and increase in deoxidation from grades A through C. Welding was accomplished with a cw, convectively-cooled, high-power CO_2 laser at power levels from 5.5 to 12.8 kw. At these power levels, it was necessary to weld the 3/4 and 1.0 in. (1.90 and 2.54 cm) thick materials in two passes, one from each side. Typical weld cross sections obtained are shown in Figures 11 and 12 for a 1.0 in. (2.54 cm) dual pass weld and a tee joint of 3/8 in. (0.95 cm) to 1/2 in. (1.27 cm) thick material. The welding of the tee section was accomplished by impinging the laser beam on the seam at a low angle of incidence (6 to 9 degrees) to the base plate.

Material properties of the welds obtained in this study are summarized in Table I. All welds made within the thickness capability of the equipment passed X-ray inspection, but porosity appeared at the root of deeper welds which were somewhat marginal in penetration. The generation of root porosity was deemed to be speed/power related. Root porosity in dual-pass welds was thought to be related to gas evolution without sufficient time for removal from the melt and was reduced by appropriate selection of welding parameters.

Tensile properties of welds in grades A, B, and C steel were shown to be excellent, with fractures occurring in the base plate rather than

Thickness: 1.0 in.
Laser power: 12.0 kW
Welding speed: 30 ipm (dual pass)

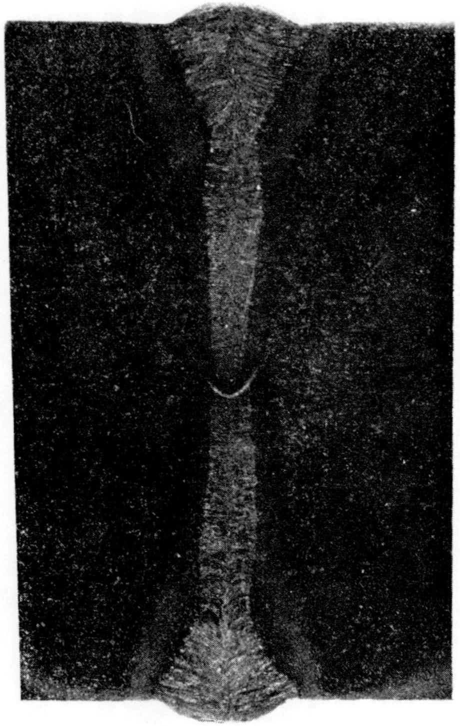

Bead cross section

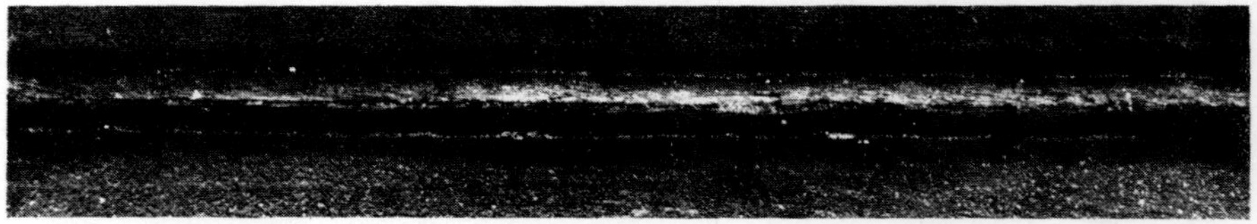

Surface bead

Radiograph

Figure 11. Laser Weld Characteristics in Grade B Ship Steel

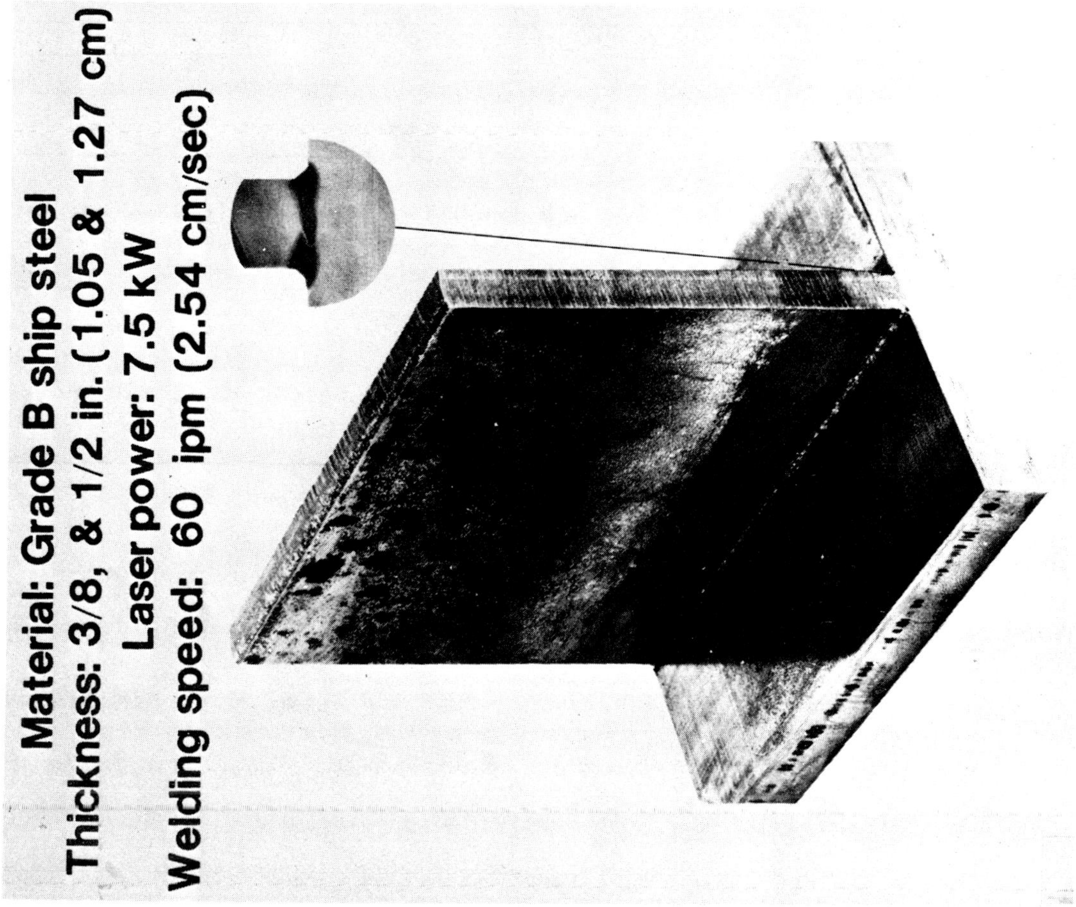

Figure 12. Laser Tee Weld

Table I

Weld Properties of Tanker Construction Steels (39)

No. & Grade	Thickness in.	Thickness (cm)	X-Ray	Tensile	Side Bend	Charpy Impact Energy at 0°F ft-lb (Joule) Weld	Charpy Impact Energy at 0°F ft-lb (Joule) HAZ
1-A	3/8	(.95)	OK	Failure in Base Plate	OK	14.3(19.4)	12.7(17.2)
2-A	3/8	(.95)	"	"	"	8.0(10.8)	10.3(14.0)
3-A	1/2	(1.27)	"	"	"	9.7(13.1)	7.3(9.9)
4-A	1/2	(1.27)	"	"	"	11.7(15.8)	11.7(15.8)
5-B	5/8	(1.59)	"	"	"	36.0(48.8)	14.0(19.0)
6-B	3/4	(1.90)	"	"	"	10.7(14.5)	5.0(6.8)
7-C	1.0	(2.54)	Aligned Root Porosity	"	"	12.7(17.1)	11.7(15.8)
8-C	1.0	(2.54)	"	"	"	26.7(36.2)	15.3(20.8)
9-C	1 1/8	(2.86)	Large Scale Porosity	NO MECHANICAL TESTS			
12-C	1.0	(2.54)	OK	Failure in Base Plate	OK	8.8(11.9)	5.3(7.1)
13-C	1.0	(2.54)	"	"	"	8.0(10.8)	11.5(15.6)

in the welds. Adequate weld ductility was demonstrated with side bend tests. Charpy impact tests were also conducted as a qualitative demonstration of weld soundness in this material which is not impact qualified. It was concluded (39) that high power laser systems offer significant potential for shipyard applications. Autogenous, deep-penetration laser welds exhibited mechanical properties which compared favorably with conventional arc welds using filler metal. Since the weld energy per unit length is relatively low, narrow fusion and heat-affected zones were obtained and thermal distortion was reduced to a minimum. Further evidence (to be discussed in detail below) pointed to some material purification during inert gas-shielded laser welding, along with some restructuring and redistribution of inclusions.

In reporting the detailed mechanical properties of laser welds in ferrous base alloys (29), it was shown that the tensile strength and hardness of laser welds was equal to or greater than that of the base metal, providing that good bead geometry (no serious undercut) and freedom from major porosity were achieved. This was expected since a higher cooling rate was characteristic of the laser welds than was experienced at any time during the manufacture of the base metal. Ductility of these relatively low hardenability steels, as qualitatively measured in bend tests, was observed to be satisfactory and so increased attention was directed to the toughness characteristics of these welds as reported in (28).

Single- and dual-pass laser welds in 0.520 in. (1.32 cm) thick X-80 steel as fabricated by Breinan and Banas (28) are shown in Figure 13. X-80 is an experimental (80,000 psi yeild strength) pipeline alloy with

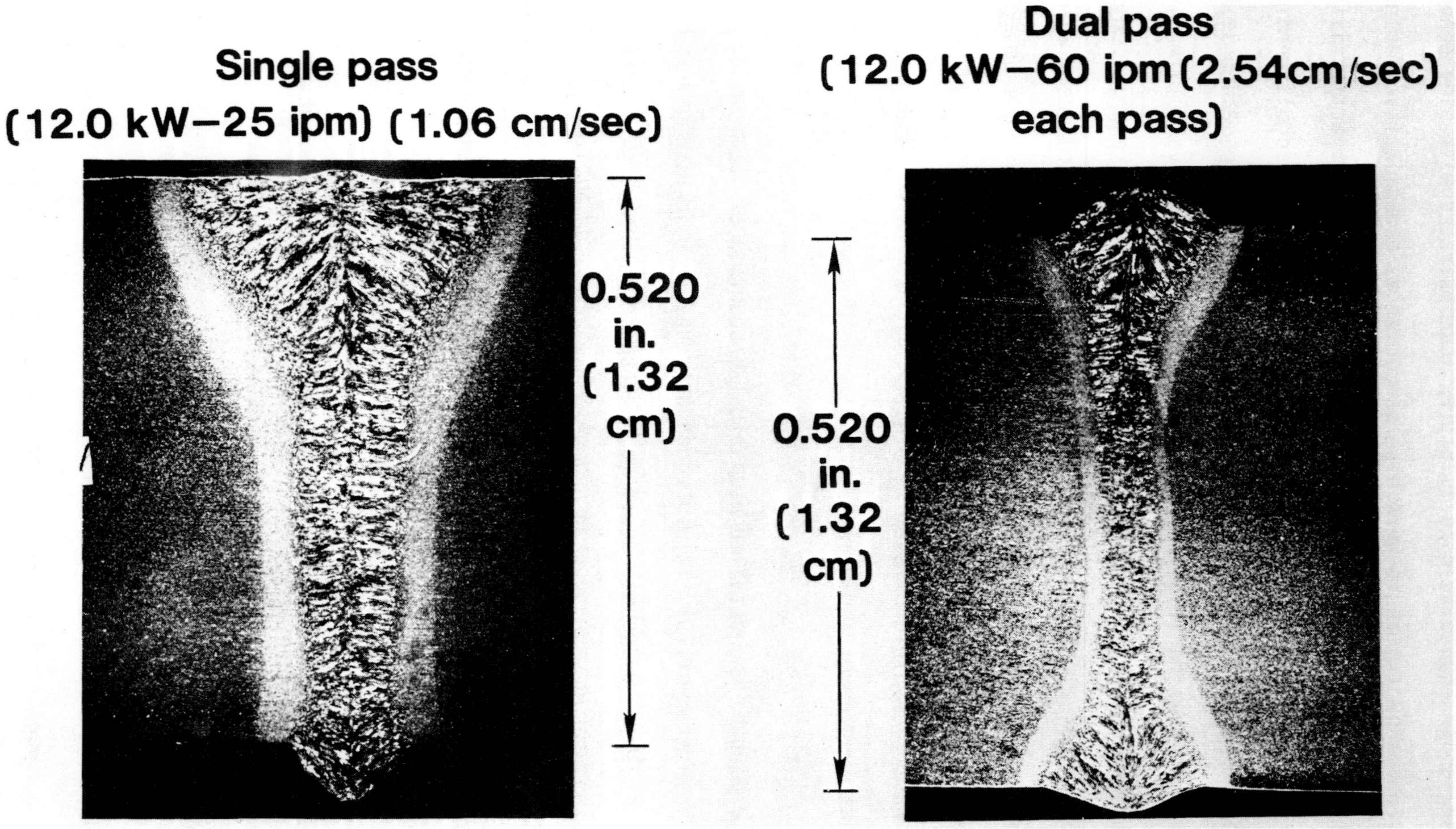

Figure 13. Laser Welds in X-80 Arctic Pipeline Steel

0.06 wt % C, 1.35% Mn, 0.6% Si, 0.065% Al, 0.10% Cb, 0.002% S, 0.011% N and 0.025% Ce. The laser welds in this alloy, made under the conditions specified in the figure, exhibited characteristic columnar grains growing toward the cold base plate in the direction of maximum thermal gradient. It is noted that all cross-weld tensile tests conducted on this material resulted in failures in the base metal. The welds also exhibited excellent ductility when subjected to root, face, side and transverse bends. Magnaflux inspection of the bent specimens did not reveal any traces of cracking. Impact test results for base metal, submerged-arc welds, and single- and dual-pass laser welds are summarized in Figure 14. It was observed that, in contrast to the base metal which had a 70°F (12°C) impact strength of 181 ft-lb (245 J), both the single- and dual-pass laser welds exhibited an upper Charpy shelf at this temperature which was in excess of the 264 ft-lb (358 J) capacity of the impact tester. The base metal, however, showed a gradual, linear decrease in Charpy energy with temperature down to a value of 90 ft-lb (122 J) at -100°F (-73°C), while the single-pass laser welds exhibited a sharp ductile-to-brittle transition ot a lower Charpy shelf energy of approximately 10 ft-lb (13.6 J) in the temperature range centering around 15°F (-9°C). Metallographic inspection of these welds revealed a decrease in the inclusion content which was felt to be responsible for the increase in Charpy shelf energy, and a relatively coarse grain size (ASTM No. 3 in the weld zone as compared to ASTM No. 8 in the base plate) which was felt to be responsible for the high transition temperature. Subsequently, dual-pass welds were fabricated in order to reduce the energy input per unit length of weld on each pass, thereby increasing the cooling rate and producing a finer grain size. A grain size of ASTM No. 4 was achieved, and the ductile-to-brittle transition temperature of these

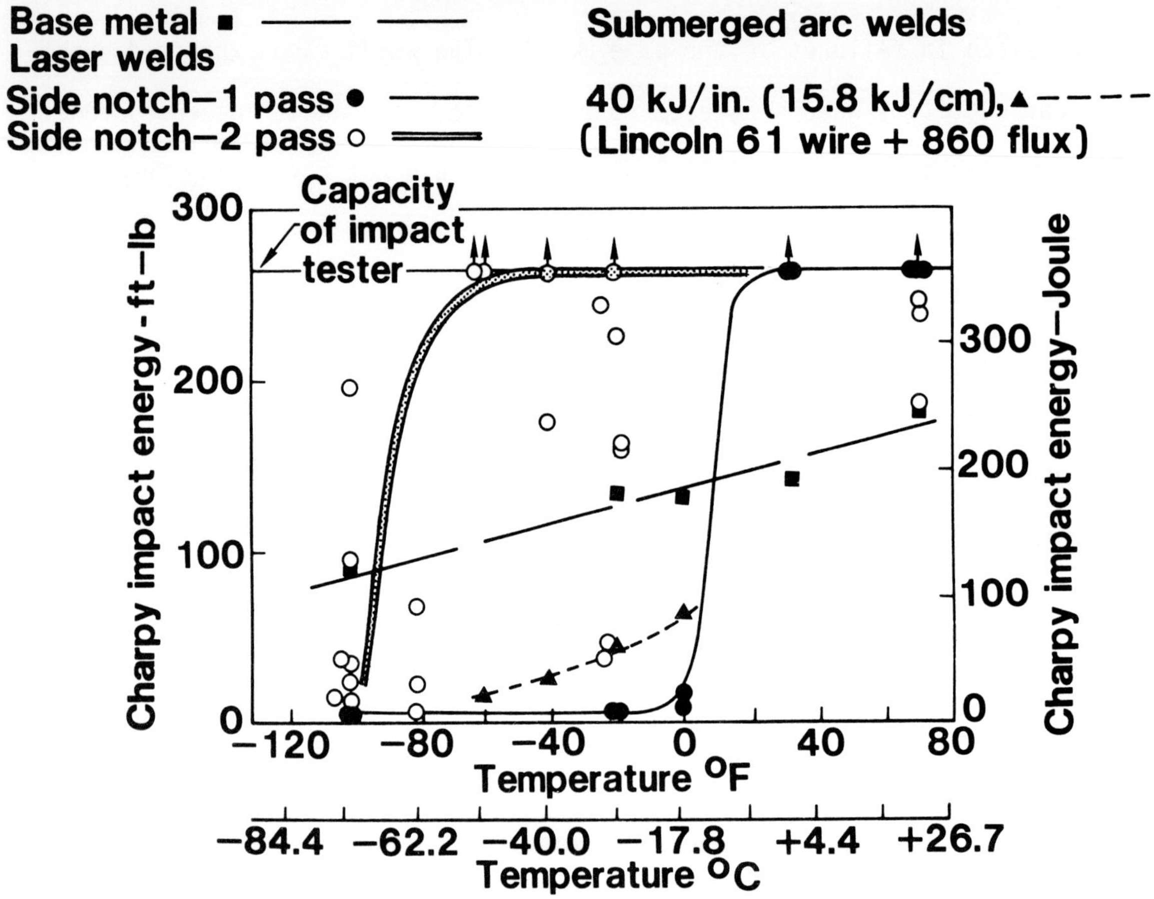

Figure 14. Impact Test Results for Laser and Submerged Arc Welds

welds was reduced to below -60°F (-51°C). Also included in Figure 14 is comparison data for filler metal submerged-arc welds using Lincoln 61 wire and No. 860 flux. These welds, although the best of a large series of developmental welds, could not approach base metal properties.

Additional detailed evaluation of laser weld impact properties and microstructure was presented in (29). In this reference, a thorough study of laser welds in HY-130 steel was conducted. Characteristic bead shapes at a variety of welding speeds in HY-130 are shown in Figure 15. The welds were reported to have tensile strengths and hardnesses in excess of base metal properties and, when tested, also exhibited impact resistance in excess of the base metal. Typical impact values at a nearly-optimum welding speed together with characteristics of the fracture surfaces are shown in Figure 16. A change was noted in the aspect ratio of fracture surface inclusions as well as an apparent decrease in the volume fraction of inclusions on the fracture surface. Chemical analyses of the welds and adjacent base metal revealed significant decreases in the oxygen and nitrogen contents of the welds. Carbon, sulfur, and hydrogen content was essentially unaffected. The decrease in oxygen was cited as further evidence of the reduction of manganese silicate type inclusions in the weld fusion zone area.

The observation of fusion zone purification during laser welding of several different ferrous base alloys indicated that this may be a unique characteristic of deep-penetration, autogenous laser welding. A mechanism for the reduction of inclusion content based on the energy gradients in the vicinity of the fusion zone has been proposed (29) and is under further investigation.

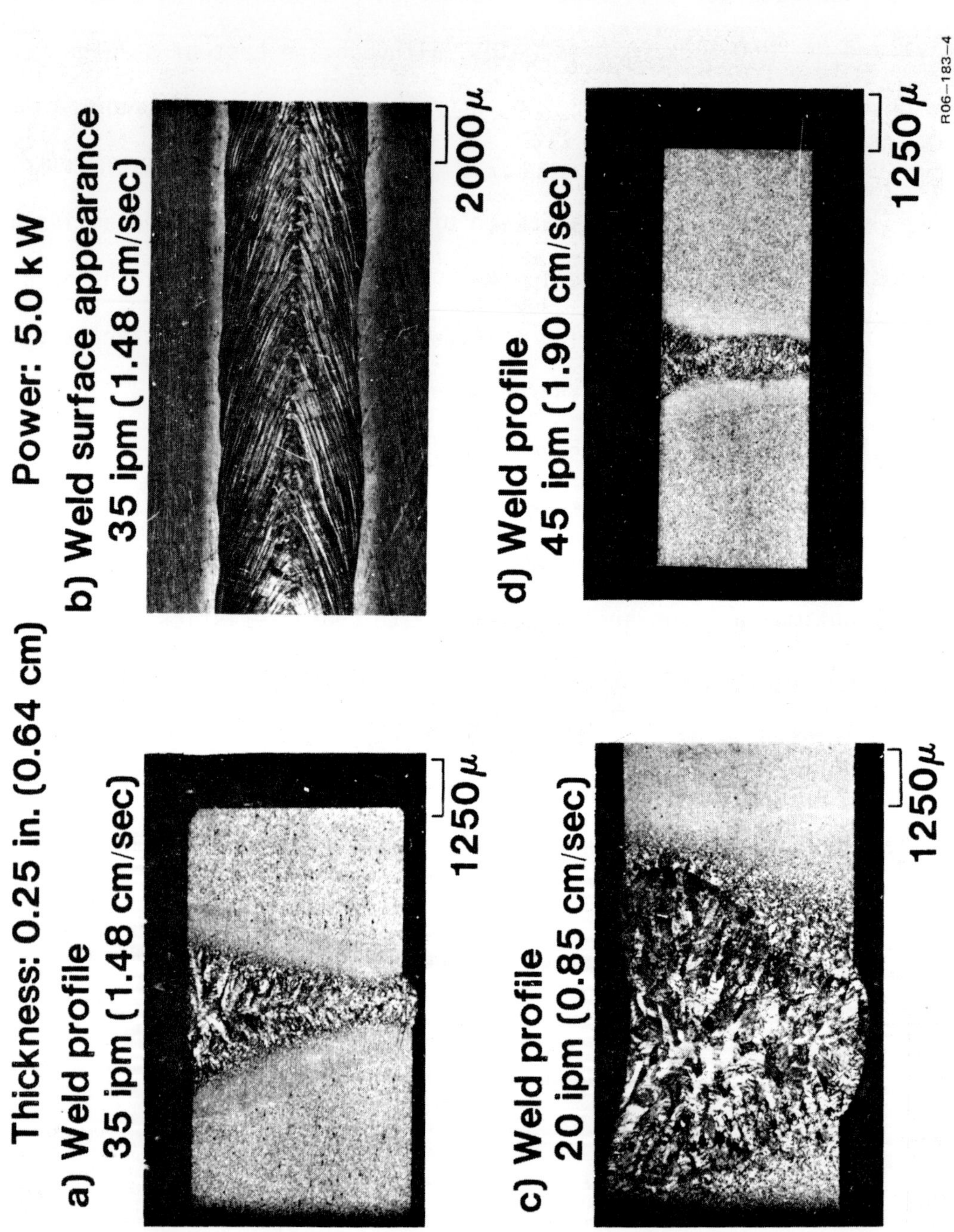

Figure 15. HY-130 Weld Characteristics

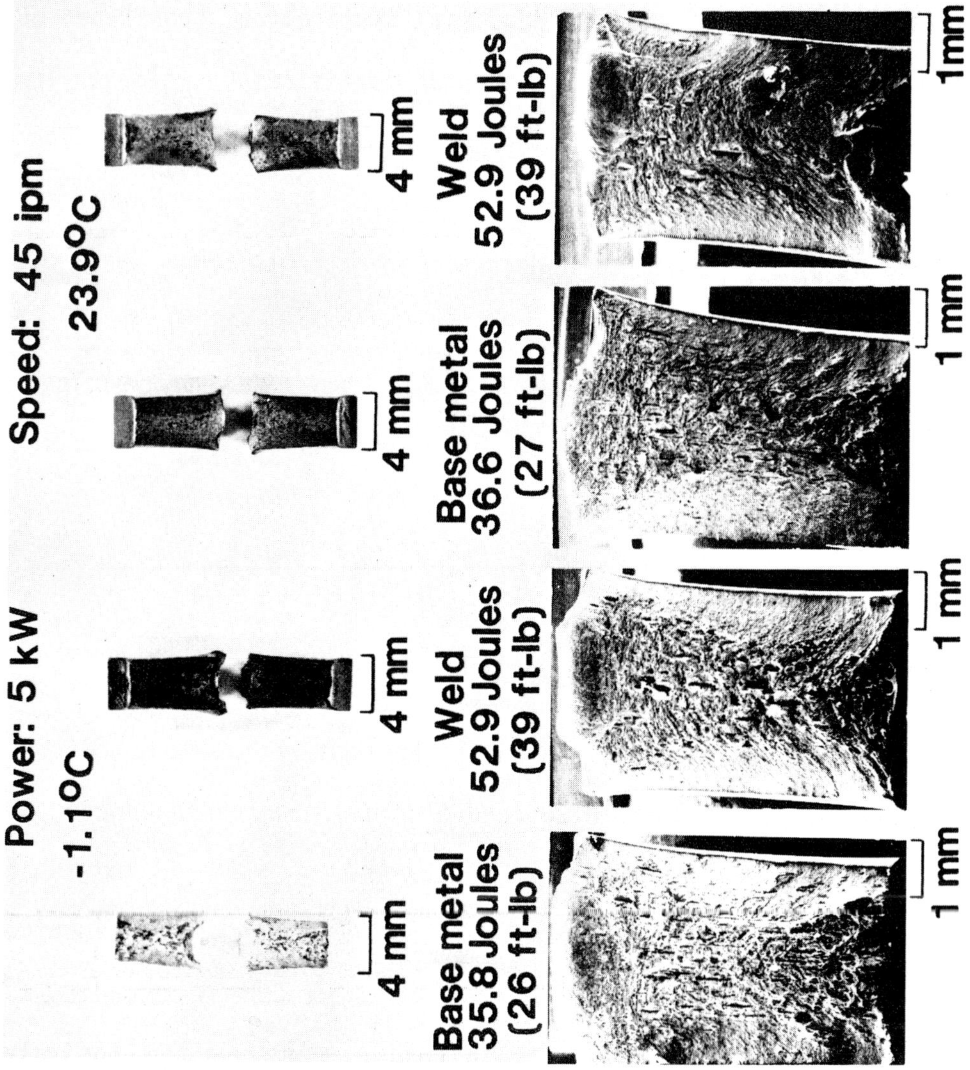

Figure 16. HY-130 Laser Weld Impact Fractures

With reference to the mechanical and metallurgical properties of ferrous base alloys in actual industrial application, Yessik and Schmatz (40) have indicated the ability to obtain satisfactory welds in low carbon and HSLA grade steels. They cited properties equivalent to base metals in HSLA steel welds, whereas the higher heat input necessary for arc welding often results in a decrease in strength below base metal values in these alloys. Figure 17 shows a penetration laser weld in a laminated heater motor yoke assembly. The feasibility of welding underbody sections and other relatively thin-section welding applications have also been demonstrated.

2. TITANIUM ALLOYS

A comparison study (41) of electron beam, laser, and plasma/gas tungsten-arc welds in Ti-6Al-4V in thicknesses of 0.040, 0.060, 0.080, and 0.250 in. (0.10, 0.15, 0.20, and 0.64 cm) has been conducted in which butt weld specimens of each thickness were prepared and subjected to metallographic analysis, and both nondestructive and destructive mechanical tests. Radiographically sound welds were produced by all three techniques and all welds exhibited tensile strengths equivalent to or exceeding base metal values following a stress relief of 2 hours at 1000°F (538°C).

Since electron beam and laser welds were made with substantially lower specific weld energies than arc welds, smaller grain size and higher fusion zone hardness were characteristic of these welds. Interestingly, the finest grain sizes were achieved in the laser welds, even though the specific energies were slightly higher than the electron

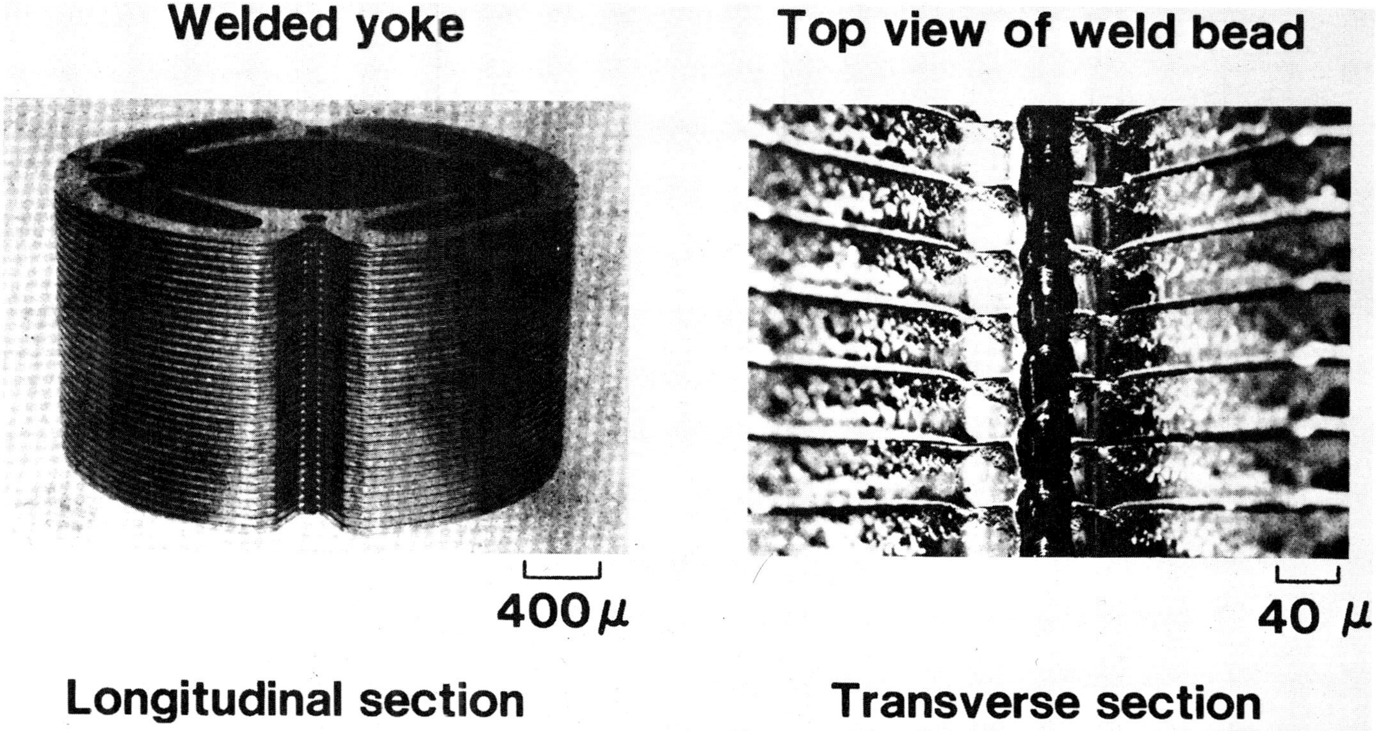

Figure 17. Laser-Welded Heater Motor Yoke Assembly

beam welds, indicating that these welds experienced higher cooling rates. The higher cooling rates of the laser welds were explained by the fact that these welds were made in air under gas shielding whereas the electron beam welds were formed in hard vacuum. Of the three weld processes studied, the coarse grained plasma-arc welds exhibited the highest fracture toughness. This is primarily due to the better energy absorbing characteristics of the coarse, Widmanstätten structure in the fusion zone of the more rapidly cooled welds. It was concluded that both electron beam and laser welds would require additional post-weld processing for applications requiring high fracture toughness.

The observed small grains and fine microstructures of the laser and electron beam welds were expected to be more resistant to fatigue crack initiation than welds exhibiting coarser microstructures. The fatigue properties of laser butt-welds in 0.230 in. (0.580 cm) and 0.140 in. (0.360 cm) Ti-6Al-4V sheet have been determined (42). Typical welds in these materials are shown for a variety of speeds in Figure 18. A wide variety of welding conditions was investigated, as well as the mechanical properties and tendency for porosity formation in welds made with two types of lasers - an oscillator-amplifier (Gaussian output beam) and an unstable resonator (hollow beam). All tensile specimens were reported to have failed in the base metal, and preliminary screening for oxygen contamination in 3-point bend tests indicated that the specimens were not contaminated when properly shielded. Welds were X-rayed using a high resolution technique capable of resolving pores down to 0.006 in. (0.015 cm) diameter. No noticeable porosity was observed in the final test welds, with the exception of one specimen. Smaller pores were impossible to detect radiographically; however, their presence in fatigue

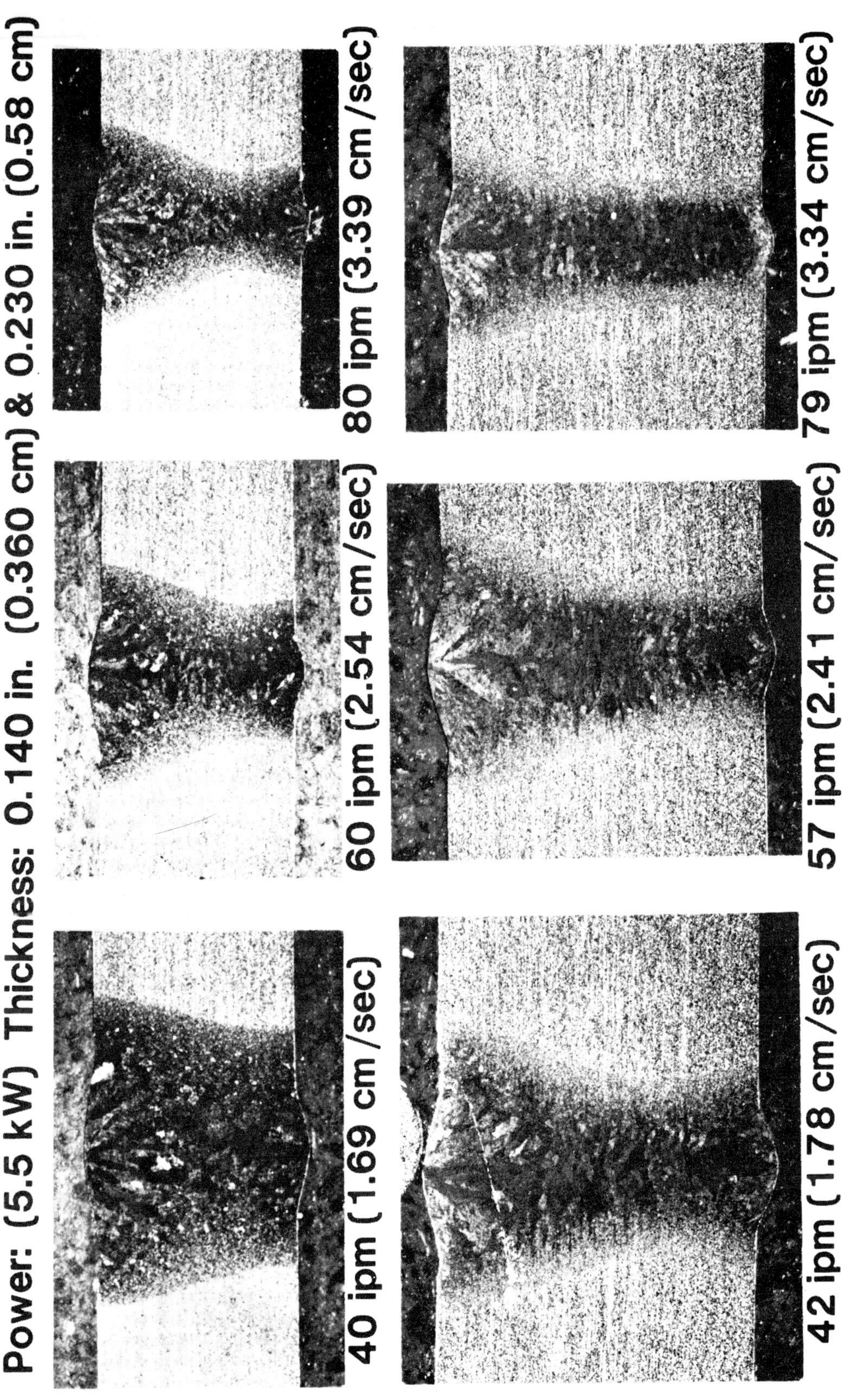

Figure 18. Macrographs of Laser-Welded Ti-6Al-4V

tested samples was noted on the fatigue fracture surfaces. The fatigue test results on all laser weld specimens are compared with the best previously-made plasma-arc welds and the base metal in Figure 19. Two series of 0.140 in. (0.360 cm) and one series of 0.230 in. (0.580 cm) specimens were tested; an unstable resonator was used for the initial 0.140 in. welds and an oscillator-amplifier was used for all other welds tested. While weld beads appeared similar, more tendency for porosity and cold shuts was apparent in the unstable resonator welds. The fatigue properties of the welds, as shown in Figure 19, show that, under proper welding conditions, laser welds can be made in Ti-6Al-4V which exhibit base metal fatigue characteristics. The best laser weld failures initiated at sites in the base metal, whereas other weld failures originated at undetected small pores. Where failures initiated in the base metal, it was concluded that no porosity or weld defects of sufficient size to preferentially initiate fatigue fractures was present. The technique of laser welding was concluded to be promising for welding of thin to medium sections of Ti-6Al-4V.

3. ALUMINUM ALLOYS

No published literature on the mechanical properties of high-power laser-welded aluminum alloys is available. Some preliminary research on aluminum welding has been conducted (43) and the results are summarized herein. Aluminum alloys have proved the most difficult to laser weld of the commonly used structural alloys due to their high initial surface reflectivity for CO_2 laser radiation. The main difficulty in the formation of laser welds in aluminum alloys has been the generation of porosity, which is also the area of greatest difficulty in arc welding. Some success has, however, been achieved in laser welding of 2219 (Figure 20)

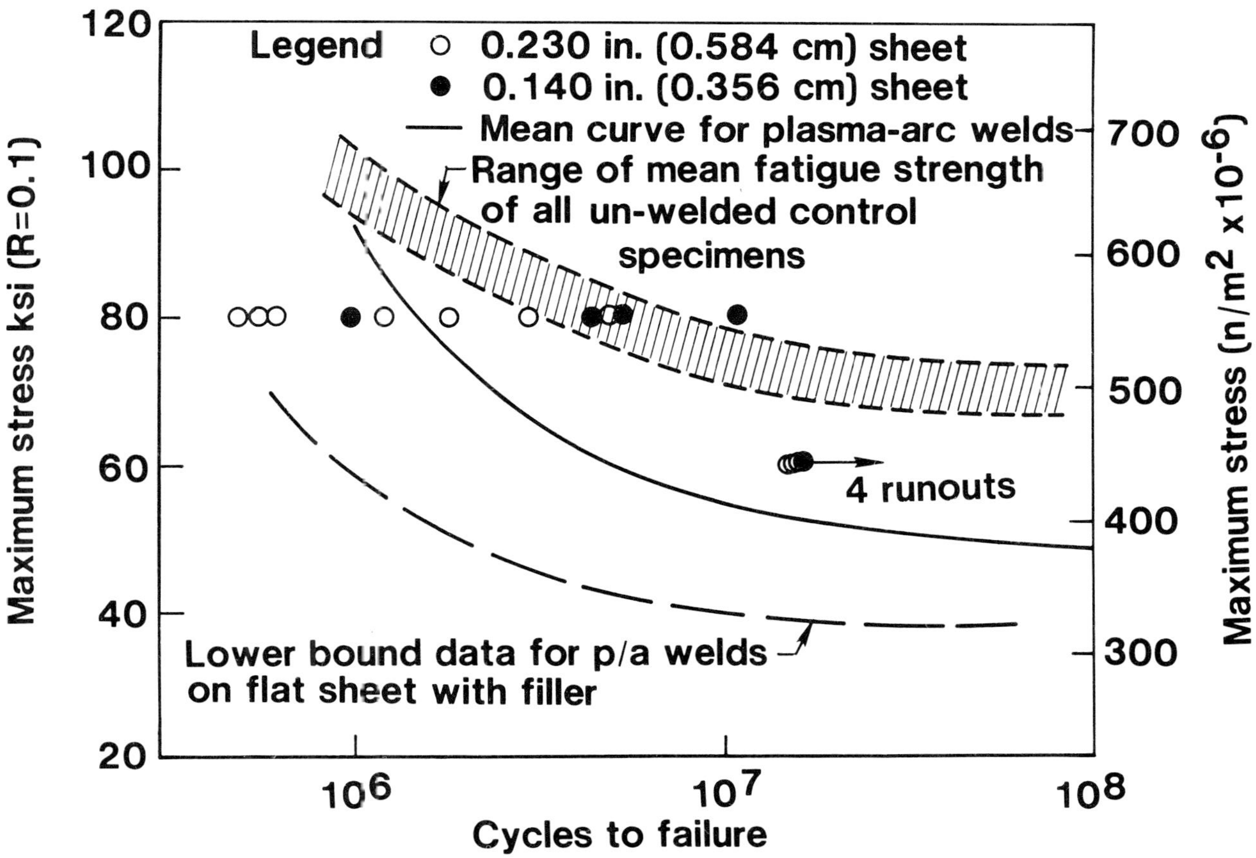

Figure 19. Fatigue Test Results on Laser-Welded Ti-6Al-4V Sheet

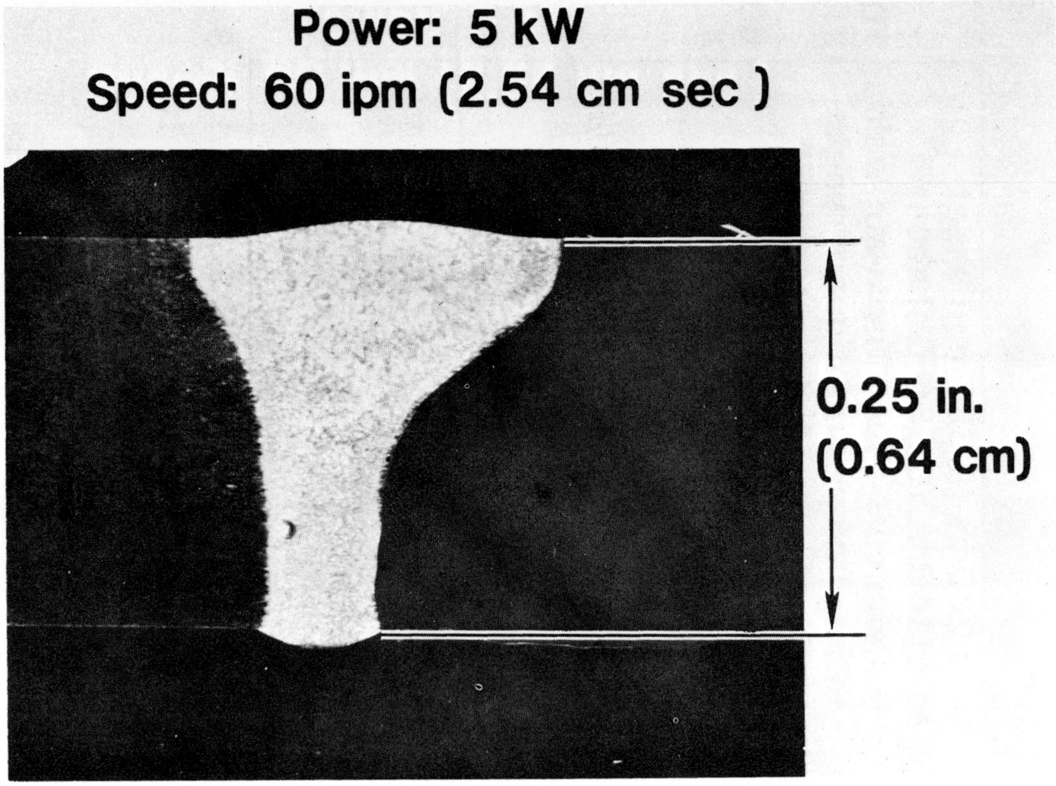

Figure 20. Laser Weld in 2219 Aluminum Alloy

and 5456 (Figure 21) alloys. Both of these materials have been welded in thicknesses up to 0.375 in. (0.95 cm) with acceptable bead geometry and microstructure. It has not yet been possible to completely eliminate porosity and a typical, large pore is shown in the weld in Figure 20. A weld cross section and a bend test specimen in 0.110 in. (0.28 cm) thick 5456 aluminum are shown in Figure 21. Tensile tests at 70°F (21°C) on these welds resulted in failure by diagonal shear through the welds at an average strength of 49,700 psi (343 N/mm^2) vs a parent metal UTS of 50,000 psi (345 N/mm^2). Thus, the welds failed at very near parent metal strength, with the fractures appearing to be located within the welds primarily as a result of slight undercuts and related stress concentrations present at the weld edges. Due to the increasing use of heavy section aluminum construction, much emphasis on aluminum laser welding is anticipated in the next year. It appears that the laser welding process can be made suitable for aluminum welding, although even closer control of process parameters appears to be necessary than has been the case with the ferrous and titanium alloys which have been successfully welded to date.

4. HIGH TEMPERATURE ALLOYS

The most extensive research on laser welding of high temperature alloys has been conducted at the British Welding Institute (44) where a fast axial flow CO_2 laser with an output of up to 2.0 kw has been used to weld nickel base alloys such as PK 33, N 75, and C 263 in thicknesses of 0.039 and 0.079 in. (1.0 and 2.0 mm). A high temperature ferrous base alloy, Jethete M 152, was also studied. Compositions of these alloys are summarized in Table II. All materials were successfully welded, with the 0.079 in. (2.0 mm) thick Jethete welds being optimum at

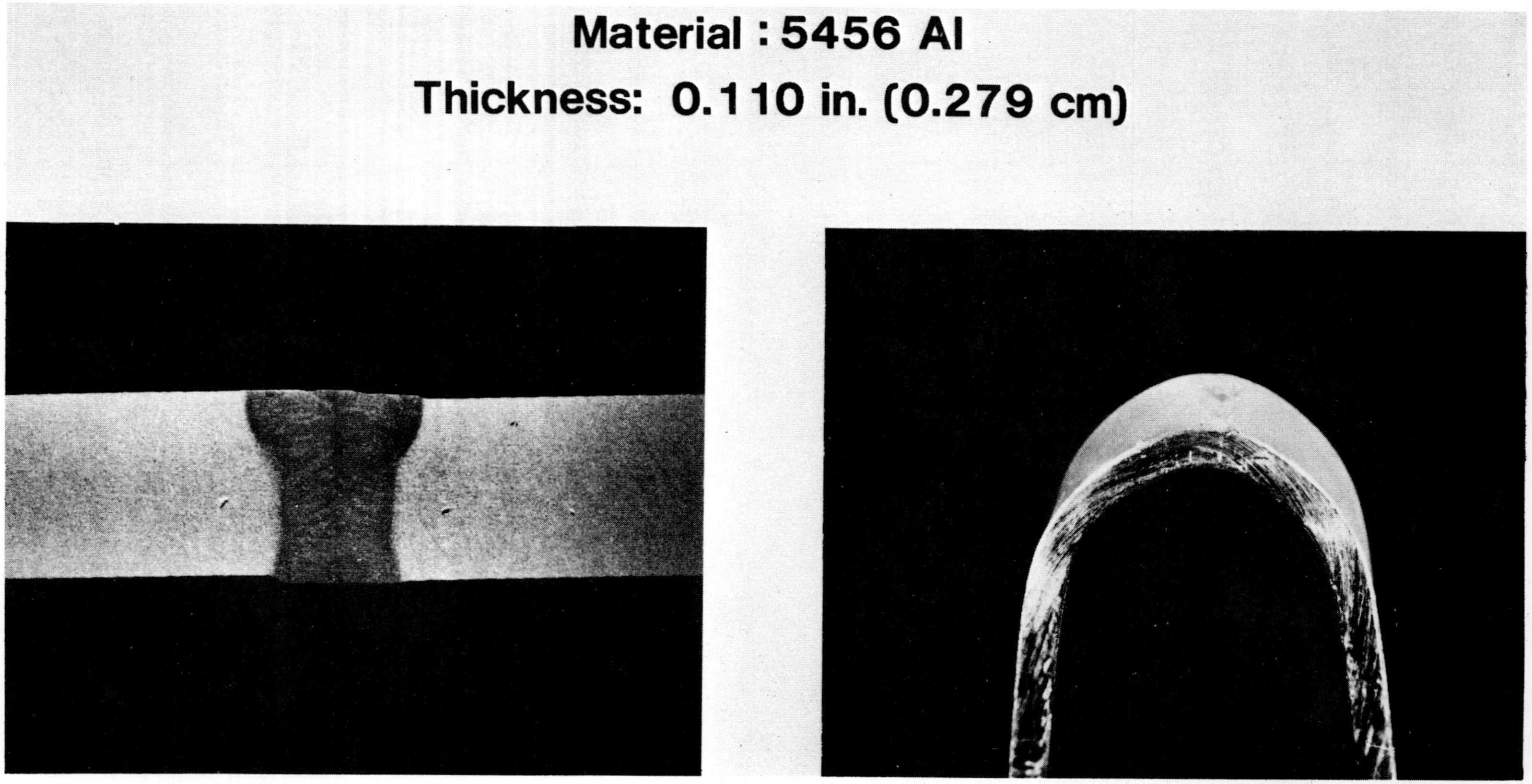

Figure 21. Laser Weld and Bend Test Specimen

Table II

Compositions of Laser-Welded Ni and Fe Base High Temperature Alloys (44)

Element	Alloy Designation			
	PK-33	N-75	C-263	M-152
Ni	Bal.	Bal.	Bal.	2.34
Fe	1.0	5.0	0.7	Bal.
C	.07	.08-.15	.04-.08	0.1
Si	0.5	1.0	0.1-0.4	0.27
Cu	0.2	0.5	0.2	-
Mn	0.5	1.0	0.2-0.6	0.55
Cr	16-20	18-21	19-21	11.9
Ti	1.5-3	0.2-0.6	1.9-2.1	-
Al	1.7-2.5	-	0.3-0.6	-
Co	12-16	-	19-21	-
Mo	5-9	-	5.6-6.1	1.6
P	-	-	-	0.02

19.6 in./min (0.83 cm/s), the 0.079 in. (2.0 mm) nickel at 14.6 in./min (0.62 cm/s), and all 0.039 in. (1.0 mm) thick materials at 80.3 in./min (3.40 cm/s). A typical weld shape in 0.039 in. (1.0 mm) Jethete is shown in Figure 22. In this thin sheet material, the depth-to-width ratio is nearly unity as the result of parameter selection to promote a slower cooling rate in the weld zone to prevent cracking. It was reported that 14 of 18 tested tensile specimens failed in the base metal, away from the weld zone. The Nimonic 75 specimens, however, failed preferentially at the edge of the fusion zone at about 90% of parent metal strength. Ductility in these specimens was 12-13% as compared to 33-35% in the parent material. Jethete welds, as illustrated in Figure 22, were observed to have a cellular dendritic structure which was martensitic. Nimonic PK 33 also showed cellular dendritic structure in the welds. The hardness of Jethete M 152 welds was 480 (Leitz microhardness tester, 0.220 lb (100 g) load) as compared to 300 for the parent material, confirming the presence of martensite in these welds. The Nimonic 75 weld, which failed at 101.5 ksi (78 N/mm^2) exhibited a ductile cup and cone fracture as shown in Figure 23. The tensile test data for the laser welded ferrous and nickel-base alloys are summarized in Table III.

For welding of Ni and Fe base superalloys, it was found that an energy density threshold existed below which the reflectivity of the workpiece was high, preventing useful welding. This has also been found to be true for other metals and alloys. For the superalloys, it was possible to exceed this threshold only with a low order mode laser. Above this threshold, reflectivity was low and the laser energy could be used for welding efficiently. It was concluded that a 2 kw, low-order-

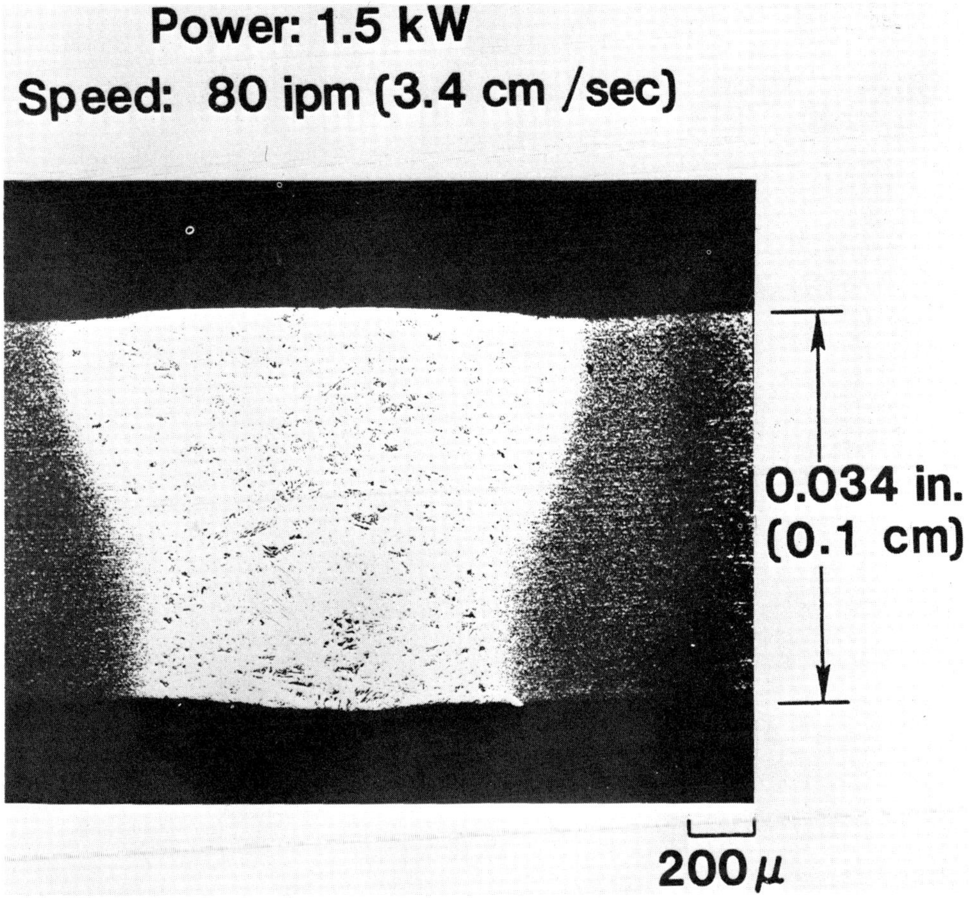

Figure 22. Cross Section of Laser Weld in Jethete M152 Alloy

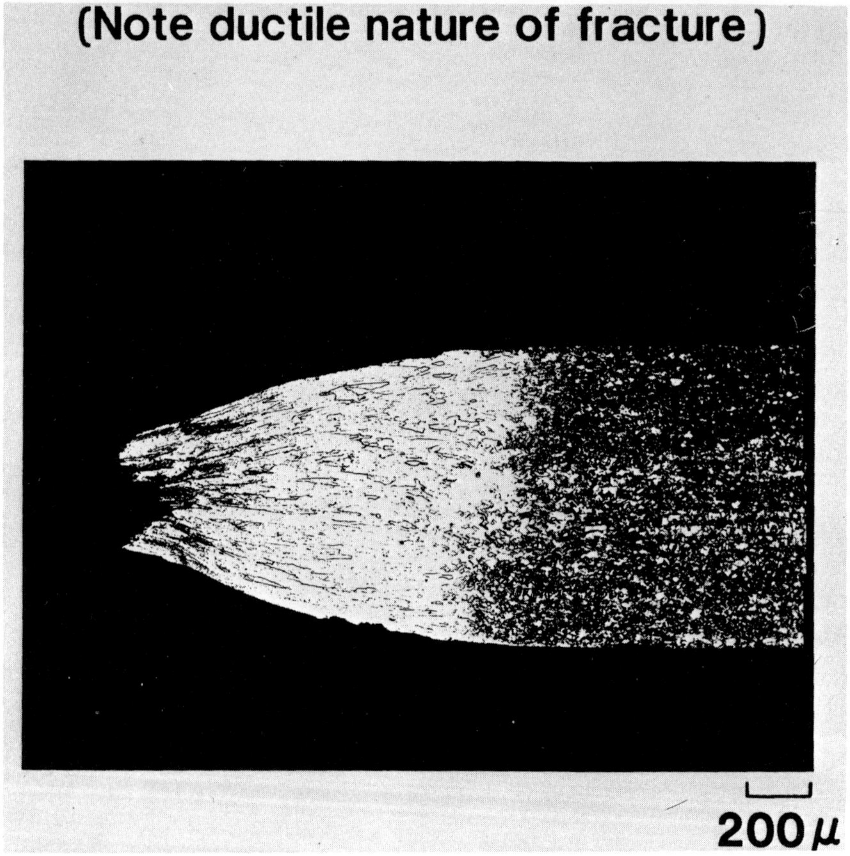

Figure 23. Nimonic 75 Laser Weld Tensile Specimen

Table III

Tensile Properties of Laser Welds in Nickel Base and Iron Base High Temperature Alloys(44)

Alloy & Weld or Parent Material	Nominal Thickness		Average UTS		Elongation in 1.98 in. (50 mm) %	Fracture Location
	in.	(mm)	ksi	(N/mm^2)		
PK33-W	0.039	(1.0)	142	(982)	20.0	Weld
PK33-P	0.039	(1.0)	147	(1017)	23.0	-
PK33-W	0.079	(2.0)	155	(1067)	26.0	Parent
PK33-P	0.079	(2.0)	151	(1042)	27.0	-
C263-W	0.039	(1.0)	145	(997)	31.0	Parent
C263-P	0.039	(1.0)	141	(973)	30.0	-
C263-W	0.079	(2.0)	141	(972)	23.0	Weld
C263-P	0.079	(2.0)	143	(988)	35.0	-
N75-W	0.039	(1.0)	104	(714)	13.0	Weld
N75-P	0.039	(1.0)	117	(807)	34.0	-
N75-W	0.079	(2.0)	106	(729)	18.0	Weld
N75-P	0.079	(2.0)	117	(804)	32.0	-
JM152-W	0.039	(1.0)	129	(887)	15.0	Parent
JM152-P	0.039	(1.0)	133	(919)	16.0	-
JM152-W	0.079	(2.0)	126	(872)	11.0	Parent
JM152-P	0.079	(2.0)	126	(871)	16.0	-

mode, CO_2 laser could weld a significant amount of material used in aircraft engine fabrication. The nickel and iron base alloys appeared to demonstrate good response to laser welding and exhibited acceptable mechanical properties in the thickness range investigated.

5. ADDITION OF FILLER METAL TO LASER WELDS

Preliminary investigation of filler metal addition to welds in 300 series stainless steel has been performed at United Technologies Research Center (45). These experiments have demonstrated that no special difficulties exist in filler metal addition and it should therefore be possible to add filler metal when desirable to change fusion zone composition to prevent undercut, etc.

The effects of filler metal addition to laser welds are shown in Figure 24. As may be noted in Figure 24, the weld surfaces were initially machined to provide a half-plate-thickness, 30 degree, vee groove at the weld interface. A section of the weld was formed as a deep-penetration butt weld without added filler, and displays a concave upper bead surface and undercut (Figure 24). When filler was added, bead reinforcement was achieved. It is noted that the weld bead was narrower when filler was added. This is due in part to the fact that energy is expended in melting the filler which would otherwise be available for melting of the base metal. It is further noted that the welds shown in Figure 24 exhibit good melt-back of filler into the base metal. Although some porosity has been found in these initial welds, it is expected that it can be controlled or eliminated by optimizing process parameters.

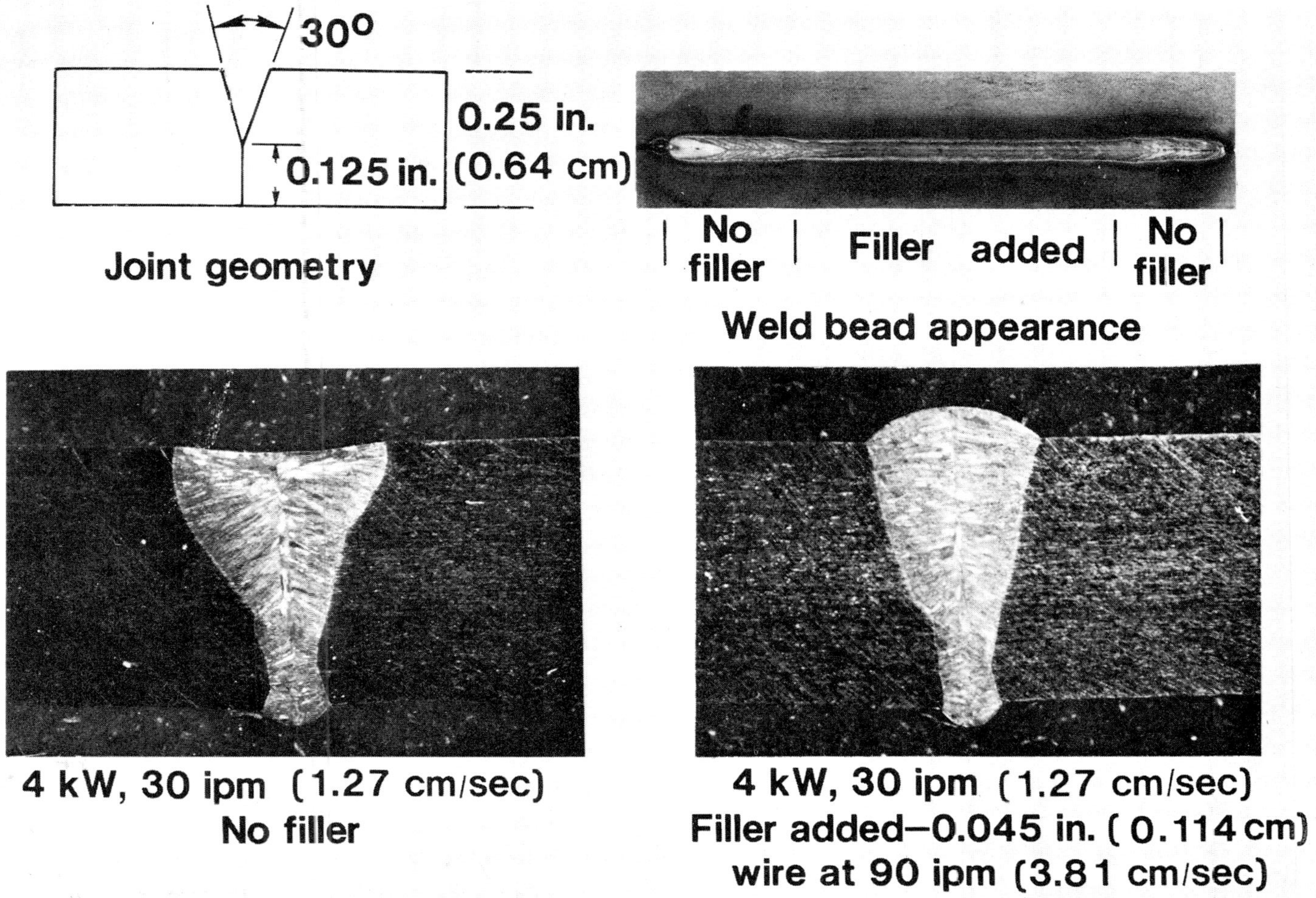

Figure 24. Laser Weld Filler Metal Addition

SECTION VI

CURRENT INDUSTRIAL LASER WELDING APPLICATIONS

The unique capabilities of the laser, as noted in preceding sections, have led to a number of important laser processing applications. Currently, most industrial applications involve average pulsed or cw power levels up to approximately 1 kw. A second category of applications, involving laser power levels from 1.5 to 10 kw, is under advanced development and several production applications are imminent. Finally, research is being conducted on a number of potentially important applications at power levels above 10 kw. It is anticipated that this work will contribute significantly to the development of future industrial processes. A discussion of applications within the three noted power ranges follows.

1. LOW POWER APPLICATIONS (<1 kw)

An example of current industrial laser welding applications which utilize the precision and control that laser welding affords to weld very small parts is the welding of wire leads on small contact wires for wire-wound relays (46). In this application, 6 joule pulses from a ruby laser are used to make extremely small spot welds in dissimilar materials (Figure 25a). The large difference in electrical resistivity of these materials (Paliney 7* welded to a phosphor bronze spring) makes resistance welding difficult, but does not adversely influence the laser process. Because of the intense energy concentration available, the laser produces a controlled, high quality weld with thorough intermixing of the two different materials. A magnification of the bond, demonstrating this intermixing, is shown in Figure 25b. The weld bead is approximately 0.030 in. (0.8 mm) in diameter and precise control of the weld extent is achieved by controlling the beam spot size.

A second example of a current low-power industrial laser welding application is the formation of a continuous weld in a small impingement tube for use in a gas turbine engine (30). In this application, as shown in Figure 26, a cap is welded to the vane with a cw neodymium-YAG laser operating at up to 400 watts. Again in this application, the ability to precisely control the level and location of energy input leads to significant process advantages. A clean, strong joint is formed with minimal thermal distortion and production is facilitated by full tape-control of the welding process.

The above-noted are representative of a wide range of present-day industrial laser welding applications utilizing both cw and pulsed lasers. Materials being processed include low carbon steels, stainless steels, low alloy steels, nuclear reactor materials, superalloys, and dissimilar materials. Further examples of such current applications are in the fabrication of nuclear reactor fuel rods (47), relay housings (48) and small magnesium parts (49).

2. MEDIUM POWER APPLICATIONS (1.5 to 10 kw)

Several laser welding applications have recently evolved in this power range. Because the laser provides a basis for cost-effective, high volume production, strong interest has centered around applications

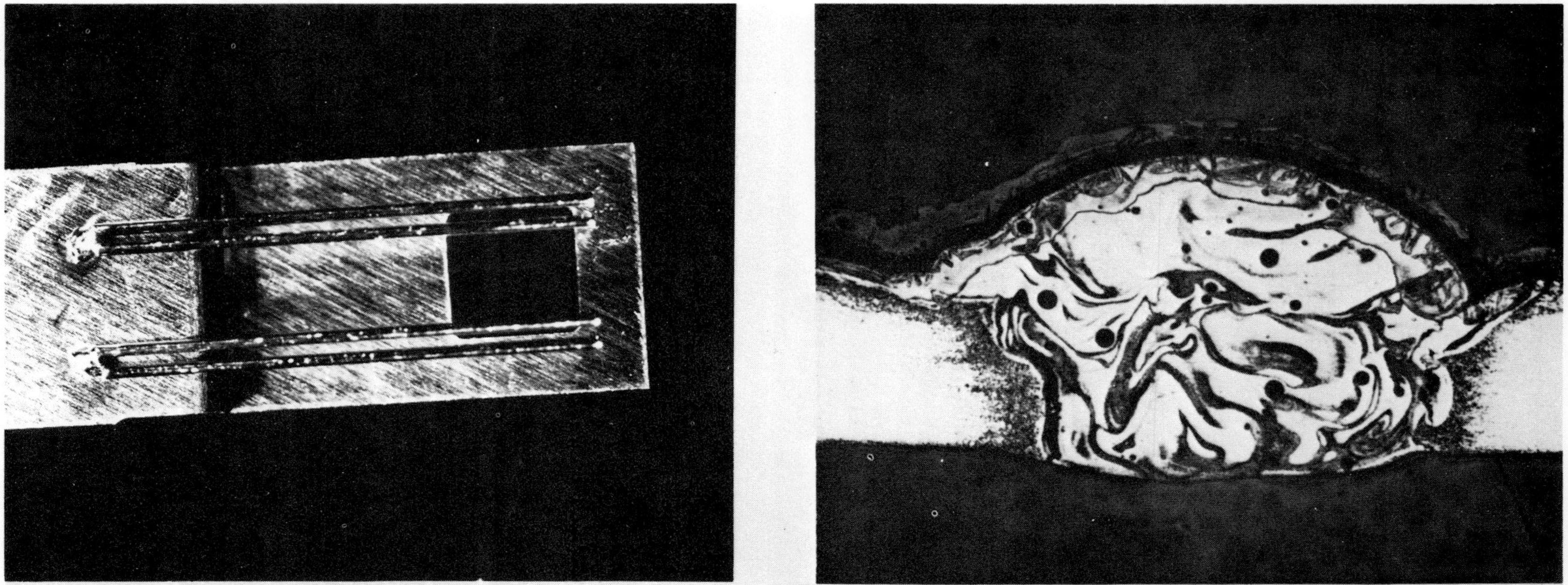

a) Macro photograph
Weld bead diameter ≈ 0.8 mm

b) Transverse section of welded wire pair

(PHOTOS COURTESY OF WESTERN ELECTRIC CO)

Figure 25. Laser Spring-Wire Contact Welding

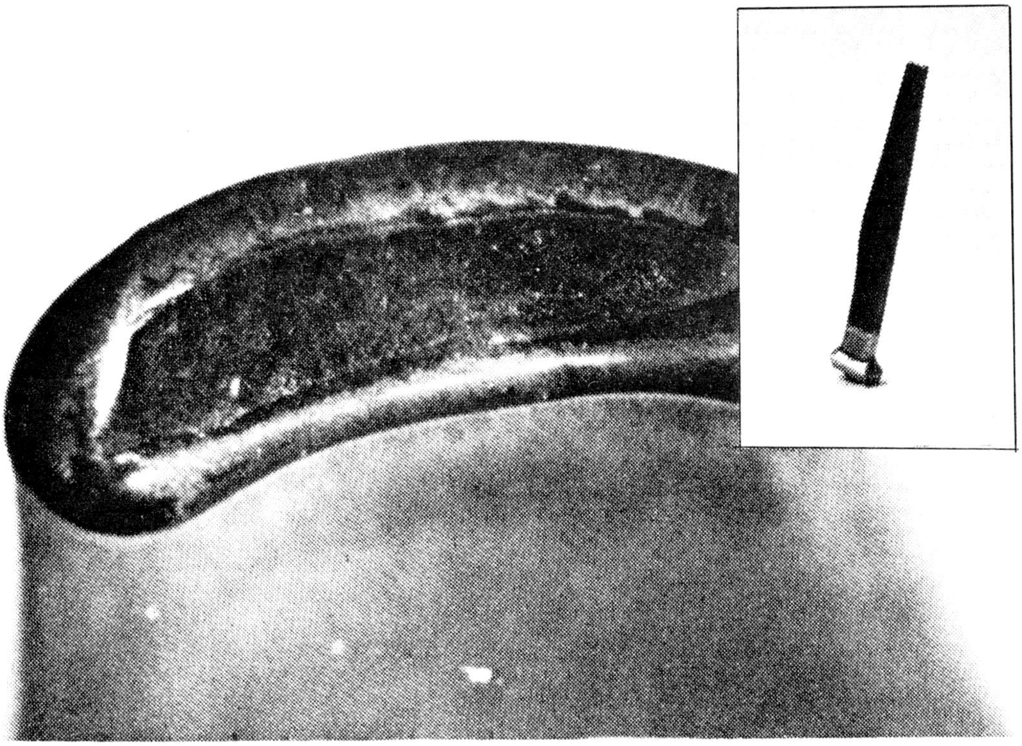

(PHOTO COURTESY OF HOLOBEAM INC.)

Figure 26. Laser Welded End Plug

in the automotive industry. An example of the advantages of laser welding for automotive production is provided by a description of the automatic underbody welding system developed for the Ford Motor Company (27).

A schematic diagram of the underbody welding system is shown in Figure 3. As may be noted in the figure, the laser remains stationary and motion of the output beam is provided in four degrees of freedom by a combination of moving mirrors. Beam translation along horizontal and vertical axes is achieved by moving a turning mirror and the focus head along horizontal and vertical tracks. Two rotational degrees of freedom for the beam are provided by a unique focusing system employing a 90°-off-axis, parabolic mirror. The ease with which the beam may be directed with this optical arrangement, coupled with a longitudinal degree of freedom obtained by moving the automotive underbody, facilitates welding over the complex contours associated with the various humps and ridges of the underbody.

With this system, penetrations of 0.1 in. (0.25 cm) have been achieved at speeds greater than 200 ipm (8.47 cm/sec), and penetrations of 0.060 in. (0.15 cm) have been achieved at speeds near 500 ipm (21.1 cm/sec). This welding performance allows up to 60 underbodies to be welded per hour on an assembly line basis. A lap weld is formed which provides a strong, continuous joint and further seals the joint area such that a secondary sealing process (required for spot welding) is eliminated. The flexibility of the computer-controlled system allows welding parameters for a number of different underbody designs to be stored in the central computer memory and selected as required. Further, since weld parameters may readily be programmed for other designs, the cost of retooling the system to accept new shapes is dependent only on requirements for new clamping structures. The underbody welding offers striking proof of the ability of the laser to produce high-speed welds in large, complex parts.

Another example of a production laser welding application in the medium power category is a lead-welding process which is discussed in (46). As shown in Figure 27, smooth welds can be made with the laser beam at speeds fo 200 ipm (8.47 cm/sec) at a power of 2 kw. This high welding speed is unmatched by other processes as noted in (46). It is further noted that this application requires a precisely-controlled, partial penetration which is readily facilitated by the close control of energy input provided by the laser process. The high welding speed and minimal specific weld energy also promote rapid cooling such that metal does not slump out of the weld zone during processing. A unique capability for high-speed welding of lead is demonstrated by this application.

Other sample applications in the medium power range include welding of transmission gear clusters and box beams for the automotive industry (49) and welding of thin sheet titanium and aluminum for the construction of high-speed surface vessels (50).

3. HIGH-POWER WELDING APPLICATIONS (>10 kw)

As available laser power increases, more attention is being directed toward heavy industrial welding, where the laser can provide economic advantage over conventional techniques. Relatively little

(PHOTO COURTESY OF WESTERN ELECTRIC CO)

Figure 27. Laser Butt Weld in Pure Lead

detailed information on heavy section welding has appeared in the literature, although applications have been considered in the areas of thick wall pipeline welding (28), shipbuilding (39), drill bit manufacturing, nuclear plant fabrication, farm machinery manufacturing, etc. Heavy section welding has been demonstrated in thicknesses up to 1 in. (2.54 cm) by welding from both sides (Figure 11).

Basic laser advantages in heavy section welding stem from the lack of filler metal requirement (which can dominate the welding cost for certain closely-controlled, high-strength, low alloy steel types such as used in gas line construction), from high welding speed and from reduced thermal distortion. Further, for quality-critical applications such as the welding of gas line for use in Arctic environments, unique weld characteristics have been demonstrated with laser welding (28).

As commercially available laser power continues to increase and new welding techniques are developed, it is anticipated that many new applications in the area of heavy section welding will be developed with substantial fabrication cost savings to industry.

SECTION VII

THE FUTURE OF LASER WELDING

Industrial laser development has progressed rapidly within the last decade and several commercial systems are now offered in the medium power (1.0 kw - 10.0 kw) regime. The increased availability of such systems has justified more and more industrial welding applications and still others require only system cost reductions to justify economic implementation. It is readily apparent that the introduction of new systems with potentially higher efficiencies and/or improvements in the design of current systems is to be anticipated and that the resultant lower capital costs will further promote industrial laser welding applications.

From a technical standpoint, it has been demonstrated that very high quality laser welds can be formed in ferrous base materials, titanium alloys, and nickel base alloys up to an inch in thickness. Although comparable success has not been attained in aluminum, results generated indicate that at least some aluminum alloys will prove amenable to laser welding.

Apart from weld quality and cost considerations, more widespread application of laser welding awaits improvement in penetration capability; such improvement may stem, in fact, from an increase in laser power, but will also require additional refinement in plasma suppression techniques. Plasma formation can occur in the atmosphere above the workpiece as well as in the metal vapor within the deep-penetration cavity.

Such plasmas absorb laser energy and influence both the amount and the distribution of energy transfer to the workpiece. The requirements for deeper penetration, i.e., increased power and power density, generally tend to aggravate the plasma formation problem. Barring unforeseen obstacles, however, it is not unreasonable to assume that single-pass laser weld penetrations of over an inch in ferrous-base materials can be anticipated in the near future.

Increasing knowledge of the laser welding process, the expanded variety and quality of available laser equipment, and the success of the initial industrial laser applications will undoubtedly contribute to rapid expansion of the industrial use of laser welding. The unparalleled adaptability of the laser process to automation will be a major factor in speeding its acceptance by industry. The growing number of potential users and suppliers of commercial laser welding equipment attests an expanding future for this newest of welding techniques.

REFERENCES

1. Maiman, T. H., "Stimulated Optical Radiation in Ruby Lasers", Nature, Vol. 187, 1969, p 493.

2. Charschan, S. S., Editor, Lasers in Industry, Van Nostrand Reinhold Co., New York, 1972.

3. Ready, John F., Effects of High-Power Laser Radiation, Academic Press, New York, 1971.

4. Geusic, J. E. et al, "Laser Oscillations in Nd-Doped Yttrium Aluminum, Yttrium Gallium and Gadolinium Garnets", Applied Physics Letters, Vol. 4, 1964, p 182.

5. "Four Kilojoule Pulses from Sandia's Chemical Laser Obtained with E-Beam Initiation", Laser Focus, Vol. 11, No. 6, June 1975, p 24.

6. Patel, C. K. N., "Continuous-Wave Laser Action on Vibrational-Rotational Transitions of CO_2", Physical Review, Vol. 136, 1964, p A1187.

7. Patel, C. K. N., P. K. Tien and J. H. McFee, "CW, High-Power CO_2-N_2-He Laser", Applied Physics Letters, Vol. 7, No. 11, Dec. 1965, pp 290-292.

8. Harrigan, F. A., R. I. Rudko and D. T. Wilson, "High-Power Gas Laser Research", Raytheon Report S-1037 on Contract DAAJ01-67-C-1589, Jan. 10, 1968.

9. Brown, C. O., "High Power, CO_2, Electric-Discharge Mixing Laser," Applied Physics Letters, Vol. 17, No. 9, Nov. 1970, pp 388-391.

10. Davis, J. W. and C. O. Brown, "Electric Discharge Convection Lasers," Paper 72-722 presented at the AIAA 5th Flux and Plasma Dynamics Conference, Boston, MA, June 1972.

11. Holobeam Laser Inc. Sales Brochure, Nov. 1974.

12. Bhaumik, M. L., W. Lacina and M. M. Mann, "Enhancement of CO Laser Efficiency by Addition of Xenon", IEEE Journal of Quantum Electronics, Vol. QE-6, No. 9, Sept. 1970, pp 575-576.

13. Hinchen, J. J. and C. M. Banas, "CW HF Electric-Discharge Mixing Laser", Applied Physics Letters, Vol. 17, No. 9, Nov. 1970, pp 386-388.

14. Alwang, W. G., L. A. Cavanaugh, and E. Sammartino, "Continuous Butt Welding Using a Carbon Dioxide Laser", Welding Research Supplement, Mar. 1969, pp 110s-115s.

15. "CO_2 Laser Welding Joins the Parade", Welding Engineer, Aug. 1970, pp 42-44.

16. Banas, C. M., "The Role of the Laser in Materials Processing", Presented at the Canadian Welding and Materials Technology Conference, Toronto, Canada, Sept. 29 to Oct. 1, 1969.

17. Sullivan, A. B. J. and P. T. Houldcroft, "Gas-Jet Laser Cutting", British Welding Journal, Aug. 1967.

18. Miller, S. Ray Jr., "Laser Diemaking Machines in Production", SME Paper MR74-966, 1974.

19. Belforte, David A., "CO_2 Laser Cuts Metals - and Costs", Electro-Optical Systems Design, May 1975.

20. Yessik, M. and D. J. Schmatz, "Laser Processing at Ford", Metal Progress, May 1975, pp 61-66.

21. Wick, D. W., "Lasercutting - Current Applications, Future Potentials", SME Paper MR74-965, 1974.

22. Brown, C. O. and C. M. Banas, "Deep-Penetration Laser Welding", Paper presented at the AWS 52nd Annual Meeting, San Francisco, CA, April 26-29, 1971.

23. Tong, H. and W. H. Giedt, "Depth of Penetration During Electron Beam Welding", ASME Paper No. 70-WA/HT-2.

24. Burwell, W. G., "Review of High-Power Laser Technology", Presented at the 3rd RPI Workship on Laser Interactions and Related Plasma Phenomena, Troy, NY, August 13-17, 1973.

25. Locke, E. V., E. D. Hoag and R. A. Hella, "Deep-Penetration Welding with High-Power, CO_2 Lasers", IEEE Journal of Quantum Electronics, Vol. QE-8, No. 2, Feb. 1972.

26. Banas, C. M., "Laser Welding Developments", Proceedings of the CEGB International Conference on Welding Research Related to Power Plant, Southampton, England, Sept. 17-21, 1972.

27. Ball, W. C. and C. M. Banas, "Welding with a High-Power, CO_2 Laser", SAE Paper No. 740863, Oct. 1974.

28. Breinan, E. M. and C. M. Banas, "Preliminary Evaluation of Laser Welding of X-80 Arctic Pipeline Steel", WRC Bulletin No. 201, Dec. 1974.

29. Breinan, E. M. and C. M. Banas, "Fusion Zone Purification During Welding with High Power CO_2 Lasers", To be presented at the Second International Symposium of the Japan Welding Society, Osaka, Japan, Aug. 25-29, 1975.

30. Staff Report, "State of the Art in Continuous Nd-YAG Laser Welding", Metal Progress, No. 1970.

31. Barber, R. H., "Laser Welding Applications", SME Paper No. AD74-405, 1974.

32. "British Welding Institute Works with 2 kw, CO_2 System", Lasersphere, Vol. 3, No. 1, Jan. 15, 1973, p 3.

33. Wutzke, S. A. et al, "Discharge Scaling Studies for a Continuously-Operating, Fast Flow, Electrically-Excited (COFFEE) Laser", Presented at 1975 IEEE/OSA Conference on Laser Engineering and Applications, Washington, DC, May 28-30, 1975.

34. News Release, "New 10-kw Laser to be Available Shortly", Culham Laboratories, Oxford, England, 1973.

35. Locke, E. V. and F. D. Seamon, "Materials Processing with a Multi-kilowatt CO_2 Laser", Paper presented at the SAE National Aerospace Engineering and Manufacturing Meeting, San Diego, CA, Oct. 1-3, 1974.

36. Hablanian, M. H., "A Correlation of Welding Variables", Proceedings of the 5th Sumposium on Electron Beam Technology, 1963.

37. Swifthook, D. and A. E. F. Gick, "Penetration Welding with the Laser", AWS Journal, Nov. 1973.

38. Banas, C. M. and R. D. Lessard, "Evaluation of CO_2 Laser Welding Capability in Rimmed Sheet Steel", United Technologies Research Center Report UAR-K158, 1971.

39. Banas, C. M. and G. T. Peters, "Study of the Feasibility of Laser Welding in Merchant Ship Construction", Final Report to Bethlehem Steel Corp. in Support of Bethlehem's Contract 2-36214 with the U.S. Dept. of Commerce, Aug. 1974.

40. Yessik, M. and D. J. Schmatz, "Laser Processing in the Automotive Industry", SME Paper MR74-962, 1974.

41. Banas, C. M., D. M. Royster, D. Rotz and D. Anderson, "Comparison of Electron Beam, Laser Beam and Plasma-Arc Welding of Ti-6Al-4V", Presented at the 56th Annual Meeting of the American Welding Society, Cleveland, Ohio, Apr. 21-25, 1975.

42. Breinan, E. M. and C. M. Banas, "Fatigue of Laser-Welded Ti-6Al-4V Alloy", Presented at the ASM/WRC Conference on Joining of Titanium for Aerospace Applications, Beverly Hills, CA, Feb. 5-6, 1975.

43. Breinan, E. M. and C. M. Banas, "High Power Laser Welding of Aluminum Alloys", Unpublished, United Technologies Research Center, E. Hartford, CT, 1974.

44. Adams, M. J., "CO_2 Laser Welding of Aero-Engine Materials", Report No. 3335/3/73, British Welding Institute, Cambridge, England, 1973.

45. Banas, C. M. and E. M. Breinan, "Filler Metal Addition to Laser Welds in Stainless Steel", Unpublished, United Technologies Research Center, E. Hartford, CT, 1975.

46. Gagliano, F. P., "Interaction of Laser Radiation with Metals to Produce Welds", SME Paper MR74-954, 1974.

47. Palmer, J. B., "Laser Fabrication Techniques for Use in the Nuclear Fuel Rod Industry", Holobeam, Inc. Technical Report No. 171-2.

48. Mason, J. H. and J. W. Wasko, "Spot and Continuous Welding with Solid-State Lasers", SME Paper No. MR74-955, 1974.

49. Seaman, F. D., "Welding with a Multikilowatt, CO_2 Laser", SME Paper No. MR74-957.

50. Banas, C. M., "Laser Welding of Navy Ship Construction Materials", Final Report to Naval Ship Systems Command under Contract N00024-72-C-5585, Aug. 1973.

51. Baardsen, E. L., D. J. Schmatz and R. E. Bisaro, "High Speed Welding of Sheet Steel with a CO_2 Laser", Welding Journal, April 1973.

52. Webster, J. W., "Welding at High Speed with the CO_2 Laser", Metal Process, Nov. 1970.

53. Meier, J. W., "High Power Density Electron Beam Welding of Several Materials", Paper presented at the Second International Vacuum Congress, Washington, DC, Oct. 10-16, 1961.

54. Meier, J. W., "Recent Developments in Nonvacuum Electron Beam Welding", Presented at the International Conference on Electron and Ion Beam Science and Technology, Toronto, Canada, May 6, 1974.

Laser welding simplified

Confused about lasers? Here is the straight dope from a machine builder and user, centered on a typical laser welder and showing how, where, and why it works

Starting as a laboratory curiosity about six years ago, laser welding is rapidly becoming a practical industrial tool, strongly competing with conventional and electron-beam welding, both on performance and cost.

Laser welding is especially good on small, delicate assemblies; it can weld dissimilar metals; and it does not require a vacuum. It can make spot welds, and it can weld seams by making a rapid series of overlapping spots.

In addition, it can weld in inaccessible places, and the beam can even pass through glass or other transparent material to make a weld on the other side. A 20-joule machine, such as the Holobeam Series 660, can make weld spots from 1 to 100 mils (0.001 to 0.100 in.) in diameter and can join wires up to about 80 mils in diameter.

Nevertheless, there is still much confusion on the part of the potential user—the engineer responsible for doing welding in a shop. His main problem is that laser welding is a new approach to joining, and it involves specifications and terminology that seem almost like a new language to him. However, once he understands the principles, he shouldn't have any serious difficulty in putting them into practice.

Let's take a look at a typical 20-joule laser welder, its mechanism of welding, the interpretation of laser parameters, a typical setup procedure, fixturing, and a few welding examples.

A typical 20-joule machine has a business end consisting of the laser head, an optical focusing system, work table, power supply, and control console (see sketch). The worktable can be driven in X and Y directions by stepping motors controlled by the operator, by standard NC punched tape, or by computer. With manual, dial-in control, the operator can maneuver the table to line up the workpiece spot to be welded under the crosshairs of

By Lee Benson, operations manager, **Joseph Nussenbaum,** and **Dennis Mulvaney,** senior staff engineers, Holobeam, Inc, Paramus, NJ

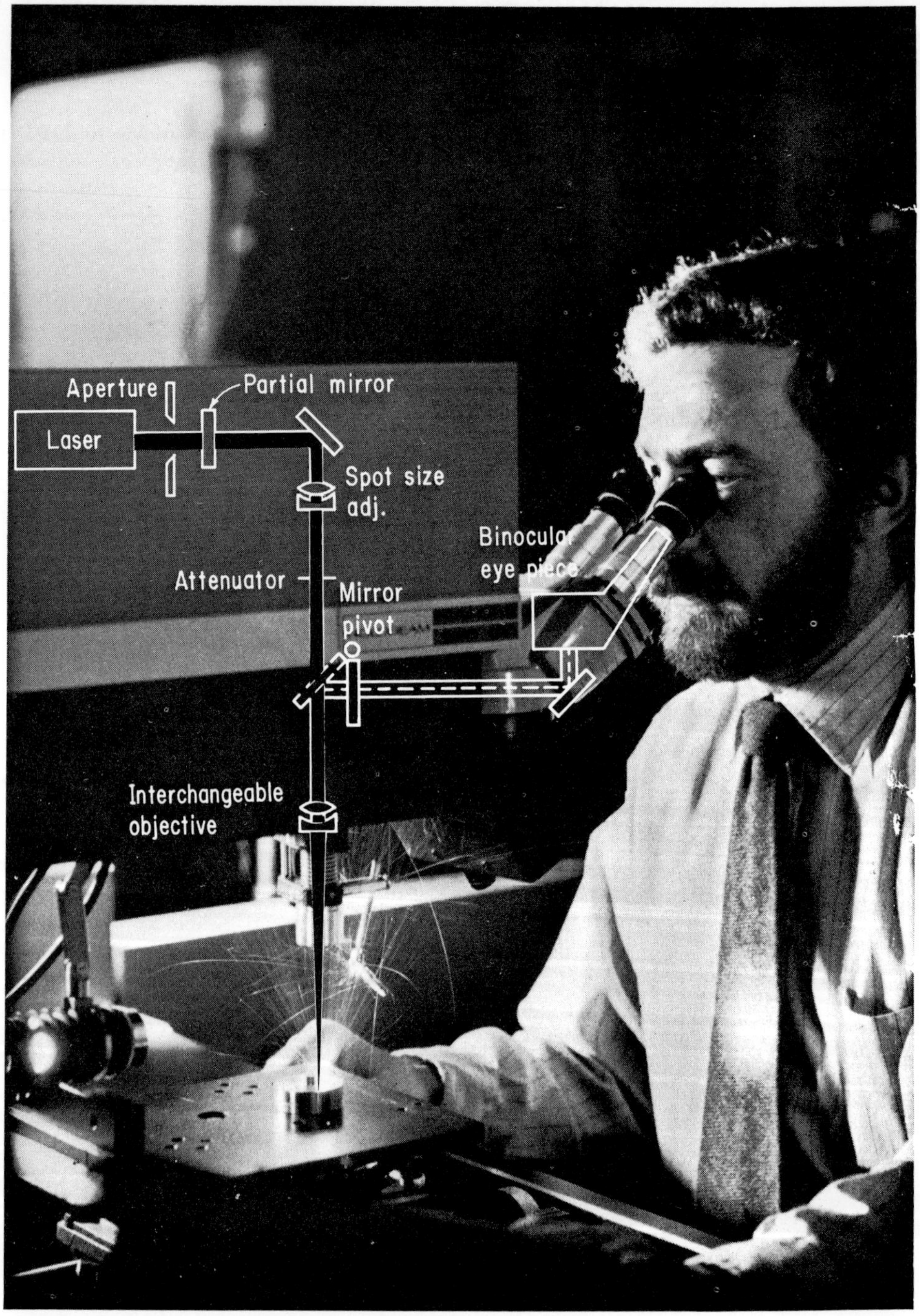

Laser welding-drilling machine operates through quick automatic cycle when 'fire' button is pressed, including motion of safety mirror, charging of capacitor bank, and electrical discharge through flashlamp. In addition, a television or film camera can be mounted on view unit to make a record

a 100-X microscope. There is a manual 'joy stick' control for final positioning and a pushbutton for firing the laser. Present controls include laser-beam focus, spot size, and other adjustments.

When the button is pushed, the power-supply capacitors charge up in a fast cycle to a preset level at which the machine fires. For safety, this is done with an electrical signal that flips a moveable mirror to protect the operator's eyes from the intense light. Then the mirror trips a precision switch that triggers the power supply to discharge the capacitors and fire the laser. All this takes less than a second, and the machine will repeat at a rate of 60 per minute.

Thus, firing is very quick, and speed of welding depends on operator speed and how quickly he can locate the work under the laser. A new operator can be trained to produce 30 precision welds per minute, and if the table is positioned automatically, the production rate is 60 welds/min.

How the weld is made

When the electrical energy in the capacitors is discharged into the flashlamp, the laser rod emits a very fast pulse of light (typically 1 to 10 milliseconds). This light is focused by the optical system into a small spot on the work. Some is absorbed, and some is reflected.

As the absorbed energy converts into heat in the work, some of the heat will be conducted away through the workpiece. If the amount and rate of heat buildup is greater than the rate of conduction, the spot will heat to the melting point. Then, as the metal melts, the spot will start to absorb more energy and reflect less, so the rate of heating will increase.

This must not be allowed to continue unchecked because the metal may start to boil, and to evaporate, which will produce a hole rather than a weld. This is the reason for a weld schedule (or pulse shape) that reduces energy input at a specified point in time.

From these fundamentals, it follows that the capability and efficiency of a laser welder are determined by these factors:

- Laser output energy and its focus.
- Wavelength of the emitted light.
- Duration of the energy pulse.
- Control of pulse shape (time).

Laser output is usually specified in joules (watts x seconds), which represents total energy supplied. This must be combined with a time factor and

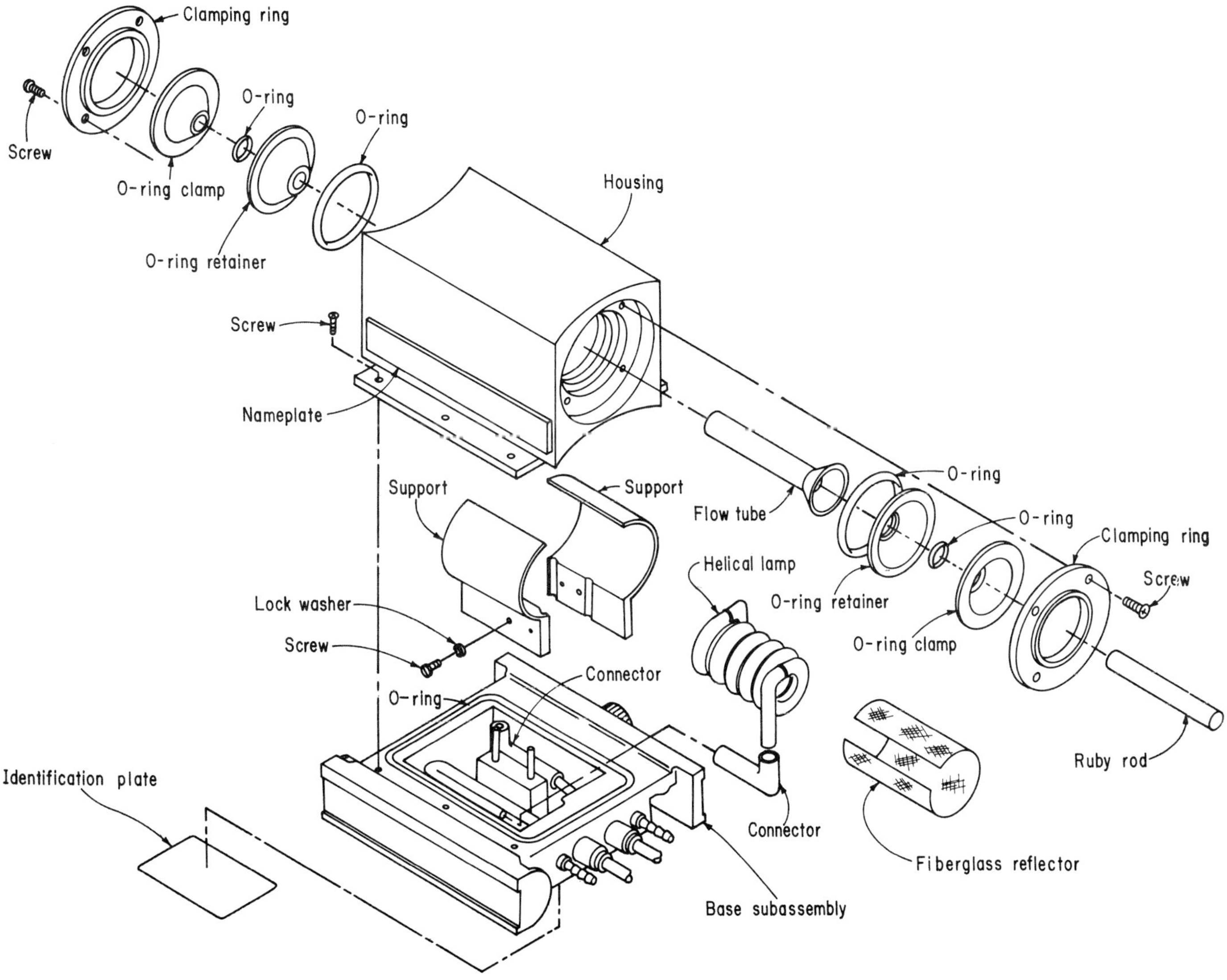

Laser mount includes the ruby or YAG rod, a helical flashlamp, and the reflectors and mechanical fittings to support the two together in alignment. Laser rod is about 6 in. long

spot size to evaluate the energy rate.

An output of 20 joules in 1 millisecond is equivalent to 19,000 BTU/sec. But this is only the figure at the source; the reflectivity of the work material and its heat conductivity must be figured in to calculate the welding effect. For reference, a Nd:YAG (neodymium doped yttrium aluminum garnet) laser rod emits light with a wavelength of 1.06 microns, which is 20% absorbed by a flat polished titanium surface or 5% absorbed by polished gold.

Laser (YAG) output, W/cm²x10⁶ (at 1.06-micron wavelength)

Work material	Effective output
Gold	120
Copper	54
Tantalum	18
Nickel	13
Chromium	9
Iron	8.5

Certain other machines use CO_2 lasers, which have a wavelength of 10.6 microns, but this is less efficient in terms of reflectivity than ruby or Nd:YAG lasers. Even though a CO_2 laser has greater electricity-to-light conversion efficiency, the total welding efficiency is often much lower.

Welding schedules

Since absorption increases as the metal melts, the best power-output curve must be some kind of drooping curve, as illustrated in an accompanying sketch. This takes into account the lower power required to maintain melting temperature in the work, once the melting point is reached.

Exact calculations for laser output are complex, and they usually require the use of a computer. But in many cases a good weld schedule can be established by making a few trial welds, based on an understanding of the principles and familiarity with a simple set of guidelines.

Working out the settings

The first step in planning the setup is the spot size, since the size of the welded area is usually known in advance. After that, the energy level and pulse width (length of time) can be figured. For a material with high thermal conductivity, the energy level should be high and the pulse short—say 10 joules and milliseconds.

For a low-conductivity material, the best approach would be a longer pulse and lower energy level, perhaps 3 to 5 joules at 4 milliseconds. The reason for this is that heat will penetrate the work slowly, and too much energy at

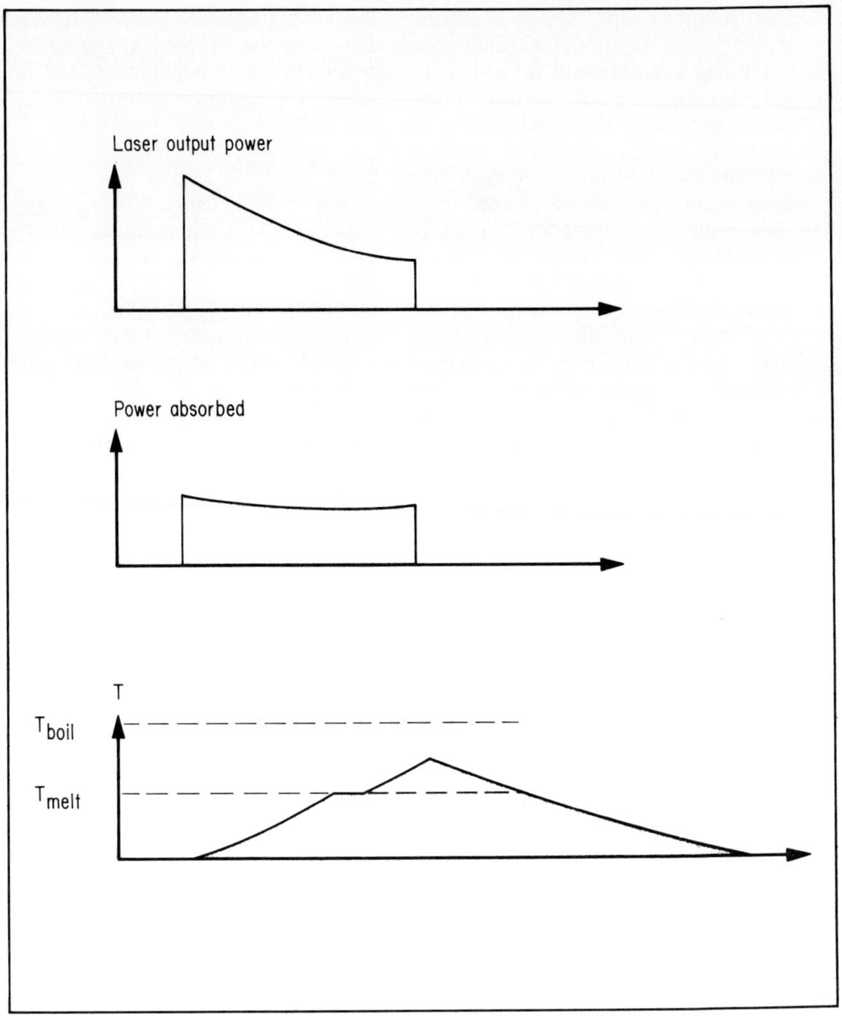

Power output of the laser itself, power absorbed by the work, and the resulting temperature change on the surface of the work are plotted

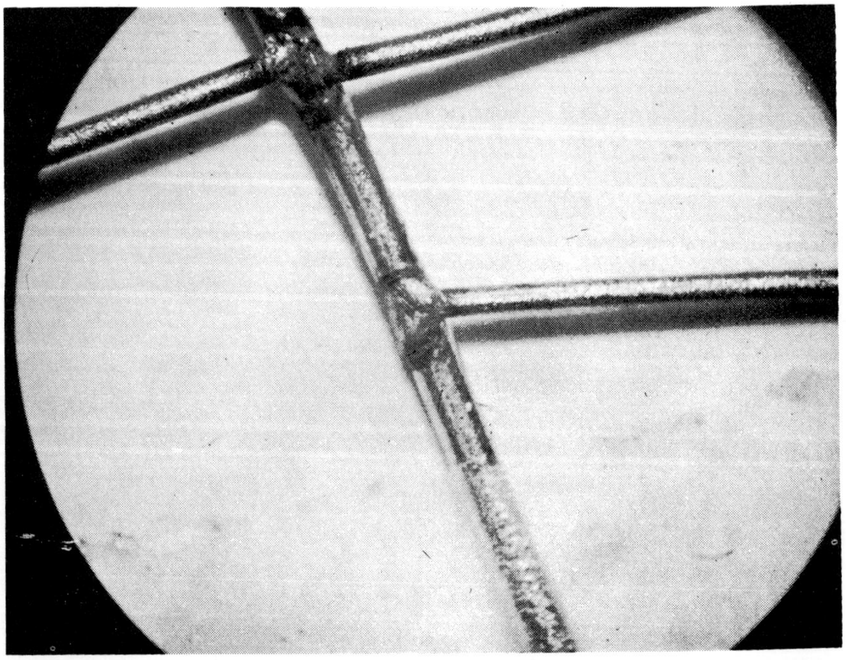

Typical crossed-wire weld indicates simplicity of working with a laser. The wires merely have to be within half a spot diameter of each other

Special field problem involved wires inserted in connectors, and then sealed in a potted plastic enclosure. Some wires were not making proper contact. Laser was employed to split the wires, making them spread to the sides of the connectors, then to weld them at reduced power

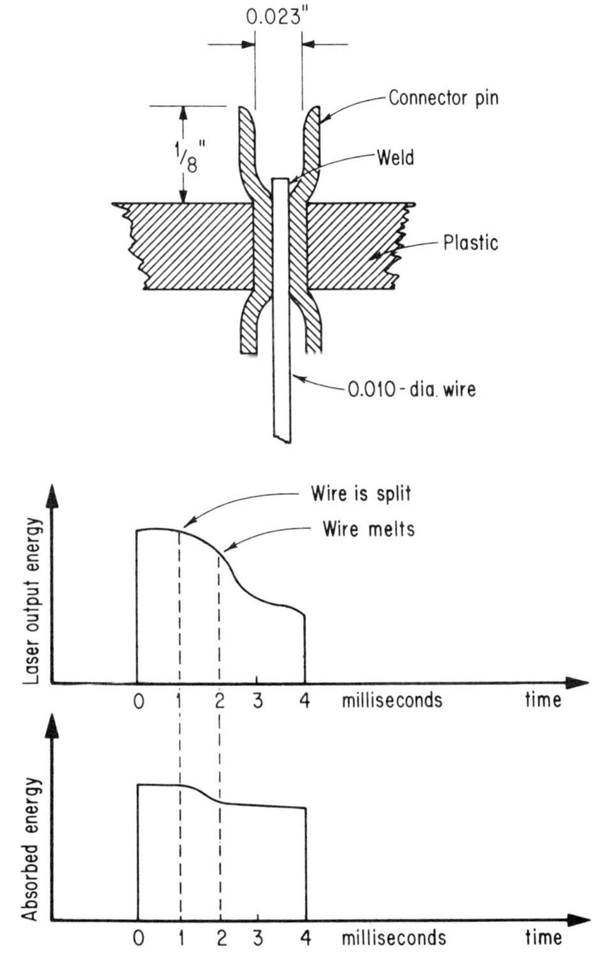

the surface would produce local boiling and evaporation.

For those who like to see the calculations, here is a typical example:

$$T(x, t) = (2q \sqrt{at/k}) \, [ierfc \, (x/4at)]$$

Where T(x,t) is temperature at distance x below the work surface at time t after start of constant heat input with intensity q.

a is thermal diffusicity in cm^2 sec.

k is thermal conductivity in calories/-cm-sec-C°.

ierfc is indeterminate error function.

x is desired penetration depth, cm.

t is pulse duration in seconds.

q is absorbed heat input, watts/cm^2.

For a job where copper is being welded to copper:

T(melt) = 1083 C.
a = 0.95 cm^2/sec.
k = 0.9 cal/cm-sec-C°.

Assuming that desired weld penetration is 0.025 cm, a 4 millisecond pulse with a constant absorbed heat input of 8.8×10^4 watts/cm^2 yields the desired depth of penetration without raising the surface temperature above the vaporization point.

The temperature equation given above, which can be filled in with data from a number of engineering handbooks, gives an approximate value for the required laser energy. A couple of trial welds will quickly point out how much the settings should be changed.

In general, penetration and spot size depend on the span between the melting and vaporizing temperatures of the work; the greater the span, the deeper and larger the weld will be. Temperature at the center of the laser spot is highest, and vaporization at this point must be avoided. In the case of copper, the span in temperatures is large (1083 C and 2300 C), and deep welds can be made.

Simple fixturing

In laser welding, parts that are merely in loose contact or in close proximity (up to half the diameter of the spot) can be joined without any holding fixtures. Two wires have even been welded together without stripping their insulation.

In cases where the workpieces must be held together, the fixture must not interfere with the laser beam. However, there's a note of flexibility in the fact that the laser can 'see' through transparent material, so the holding device can actually cover the weld area if it's made of glass or thin Teflon.

Typical jobs

The 20-joule Holobeam welder has joined 0.031-in.-diameter titanium wire to 0.016-in.-diameter gold wire without any hold down, using a spot size of 7 mills and 4-joule output. And an iron-constantan termocouple was made by welding together wires of each material, lying free but in contact, with a 12-mil spot and 4 joules.

Seam welds can be made by overlapping spot welds. For example, two pieces of stainless steel tube (½-in. OD, 0.034-in. wall) were butt-welded with 25-mil spots and an energy level of 7 joules. The tubes were rotated, and, at the maximum rate of 60 pulses per minute, the weld was made in under three minutes.

Welds can also be made in inaccessible locations, as well as on parts sealed in glass and other transparent envelopes. On one job, 0.010-in. wires were welded inside 0.023-in.-ID connector pins after than had been potted. With initial high energy, the wire split (by drilling) to pin it against the wall of the connector, and then a lower energy level in the same pulse did the welding.

Welding costs

In the past year or so, the cost of a laser weld has dropped to about half a cent per weld for total costs including machine amortization, assuming full-time utilization. This dramatic reduction is due partly to improvements in the machines, related to the life and cost of components, and partly to higher welding speeds.

The flashlamp is one of the most important and expensive parts that must be replaced regularly, but the cost for the Holobeam 20-joule unit, for example, has stabilized at $180, and this is good for 200,000 welds before replacement. Development work now going on may yield even longer life.

Solid-state electronic circuitry has been developing along with the flashlamp materials, and close control of the lamp current eliminates damage due to current surging. It has also permitted designing a number of different pulse shapes to produce specific effects. A helical flashlamp provides a long current path, which means high energy input to a laser rod of given length at minimum lamp loading. All this improves expected life.

Laser-head cooling has also become important, and the 20-joule Series 660 machine includes a sealed recirculating system with deionized and chemically treated water that cannot build up deposits.

Another contributor to lower cost is the laser rod, which may be either ruby or Nd:YAG. The rod length has been optimized and a special holder developed to free the rod ends and eliminate the usual rod-end deterioration.■

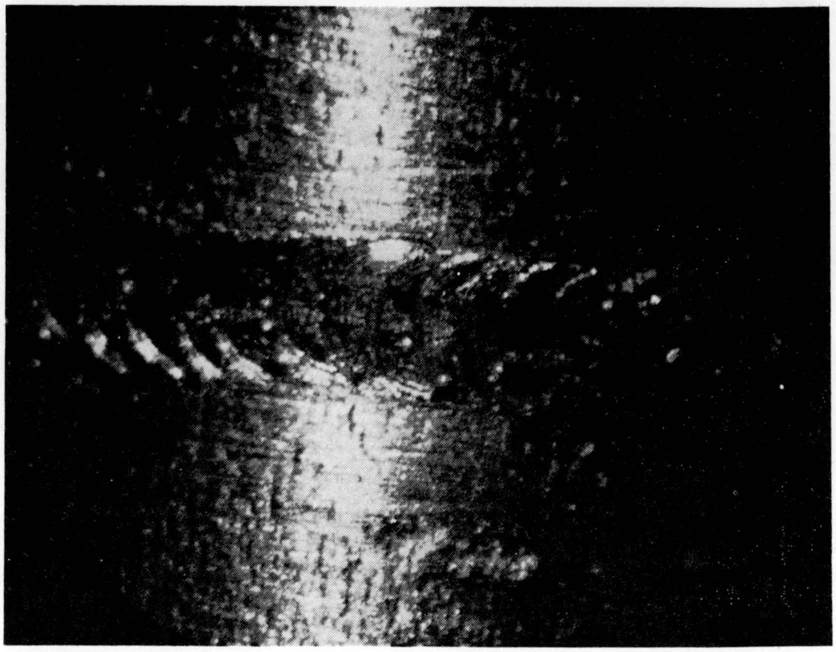

Seam weld is produced by overlapping weld pulses to produce a continuous joint. This seam was made in ½-in. stainless steel tubing (0.034-in. wall)

Typical laser welder-driller specifications:

Laser rod	Ruby or Nd:YAG	Nd:YAG operates efficiently at higher repeat rate & lower power
Wavelength	1.06 microns (YAG) 0.69 (ruby)	Wavelength determines absorption efficiency by various materials
Pulse duration	1 to 6 millisec, or to 10 millisec	Determines rate at which heat is supplied (for given energy level)
Laser output	Up to 20 joules	Maximum energy limit of machine
Beam divergence	3 to 5 milli-radians	Beam spread limits depth in narrow holes or slots. Smaller angles are available
Repitition	30 pulses/min (60 available)	Determines max speed of welding
Spot size	1 to 100 mils	Also adjusts heat intensity
Working distance	1 to 4 in.	Proportional to spot size
Viewer magnification	100X	Viewer also will accept TV or camera
Size	36 x 98 x 36 in.	
Weight	1500 lb	
Cost	$30,000	$180 for replacement flashlamp

Typical cost breakdown for 20-joule welder-driller

Maintenance:		
Flashlamp replacement	$180	
Additional	40	Total: 0.11¢
Labor:		
Burden, per hr	$5	
Welds per hr	1800	
Operator inefficiency factor	20%	Total: 0.33¢
Amortization machine cost, per weld		Total: 0.30¢

Based on machine cost of $30,000 and 10 million welds/yr

Design and Tooling

Design for laser beam welding

LBW offers the advantages of EBW with no need for a vacuum

by JOHN R. SHEWELL, *Lasertool Consultants, Glenview, Ill.*

Automotive, consumer products, controls, aerospace, and electronics industries — all use laser beam welding (LBW) to join the gamut of materials from precious metals and alloys to the more common ones like copper and its alloys, carbon steels, high alloy steels, stainless steels, and lead.

The two kinds of industrial laser welding processes, conduction-limited and deep-penetration, are autogenous. They use only the parent metal with no added filler. Conduction-limited laser welding is the more common method. In conduction-limited LBW, the metal absorbs the laser beam at the work surface. The sub-surface region is heated entirely by thermal conduction. The term laser welding normally refers to this method, which uses solid state and moderate power CO_2 lasers.

The second type, deep penetration welding, requires a high power CO_2 laser. Thermal conduction does not limit penetration. Laser beam energy is delivered to the through depth of the weld, not just to the top surface. Deep-penetration laser welding is very similar to high power electron beam welding in vacuum.

The boundary between deep penetration and conduction-limited laser welding is not well defined. Some low kilowatt continuous power welders (see chart, *Industrial Lasers for Welding*) make welds that have the properties of deep penetration welds. Future developments are expected to decrease the power per depth required for deep penetration. The drill and fill technique uses lower power lasers to achieve deep penetration welds. Despite this growing overlap of welder capability, for practical design purposes it is reasonable to draw distinctions between them on the basis of the two LBW techniques. Ordinary laser welding handles gages from foil to 0.100 inch (2.54 mm); deep penetration welding joins sections to half an inch and more, depending on the material.

Laser welds may be either continuous or spot. Seam welds made by overlapping spot welds yield high quality hermetic seals in thin sections. The first industrial laser welding applications used pulsed solid state lasers for spot welds and for seam welds using overlapping spots. These lasers will continue to have industrial application in welding, even though total pulse time is limited. If welding time is greater than about 10 ms (1 ms = 0.001 second), the right laser is a continuous one, either CO_2 or YAG.

Both kinds of laser welding, like EBW, produce joints with lower energy input than standard welding methods. The result is less distortion and a small heat-affected zone (HAZ).

Gas shielding

Laser welding normally uses an inert gas shield, typically argon. For welding with a pulsed laser or a moderate power continuous laser the shielding gas feeds to the weld along the laser beam axis. Production welding with a high power CO_2 laser requires a shielding fixture that will completely cover the weld.

To save on operating expense, pulsed solid state laser spot welding can be done without cover gas at a slight sacrifice in mechanical prop-

INDUSTRIAL LASERS FOR WELDING

Type	Designation	Output, kW	Maximum Pulse Rate, pulses/sec	Characteristics
Moderate Power	CO_2	To 1.5	1000	Continuous or pulsed from 0.2 ms up
	YAG[a] (Solid State)	To 2	—	Continuous
High Power	CO_2	1.5 to 20	—	Continuous
Pulsed Solid	YAG (Solid State)	Maximum J/ pulse 100	200	Pulse 0.1 ms to 8 ms; 200 W maximum average power
	Nd: Glass[b] (Solid State)	100	4 (per minute)	Pulse 0.6 ms to 8 ms
	Ruby (Solid State)	100	1	Pulse 1 ms to 5 ms; 12 W maximum average power

a. Yttrium-aluminum-garnet *b.* Glass doped with neodymium.

Source: *Welding Design & Fabrication*, June 1977

erties of the weld. Single-pulse solid state laser welding is possible without a cover gas, because the weld puddle is molten for only a very short time. The laser pulse itself lasts for 1-10 ms. The puddle life is even shorter. Seam welding by using overlapping spot welds is a different situation. Here, a cover gas is recommended.

Joint preparation and fit-up

For general laser welding, joint preparation and tooling suitable for autogenous automatic GTAW is a good starting point. If you need highest quality joints, use preparation and fixturing for vacuum EBW. These are guidelines only. Production welding has yielded an acceptable butt joint with a 0.040-inch gap (1.02 mm).

For continuous laser welding of ferrous and nickel alloys the gap in butt joints should not exceed 0.15 times the material thickness. Vertical misalignment should not exceed 0.25 times the material thickness. Both tolerances refer to the thinnest gage in the joint. Reduce these tolerances for materials with higher thermal conductivity.

Pulsed laser spot welding of lap joints demands tighter tolerances. The maximum gap for ferrous and nickel alloys is 0.15 the material thickness, less for higher conductivity materials.

Joint design is very similar to that for EBW (see sketches). The corner weld, *I*, gives a very rigid joint between sheet metal parts. With enough power to melt the flanges, puddle flow and surface tension form a bead (grey area). Rigid fixturing is important. Chill plates may be needed. Similar beads can be formed on flange and edge joints.

J, called the kissing weld because the puddle forms where the two pieces just kiss, offers a small angle between the two parts that traps most of the energy of the laser beam. Very little pressure, if any, is required for welding, but the faying surfaces must fit well. A gap between the sheets will allow radiation to escape. This joint successfully joins thin foils in cases where spot welding burns through. It should find use for thicker sections also.

Beam trapping as in the kissing

JOINT DESIGN FOR LASER BEAM WELDS — SHEET METAL
(Arrow shows direction of laser beam)

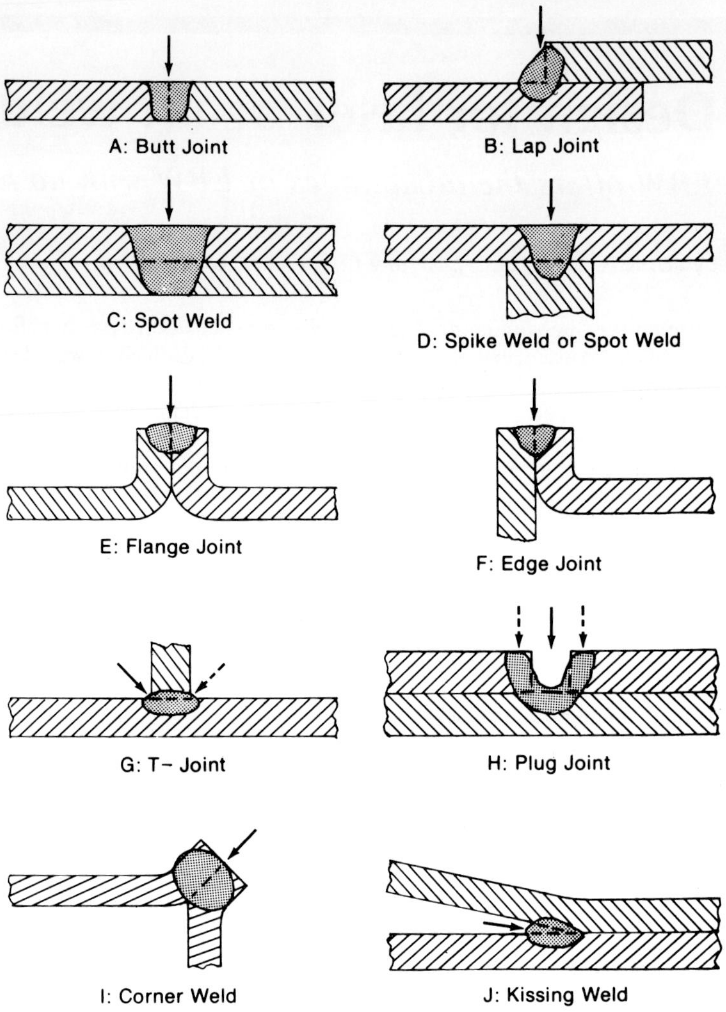

JOINT DESIGN FOR LASER BEAM WELDS — WIRE
(Arrow shows direction of laser beam)

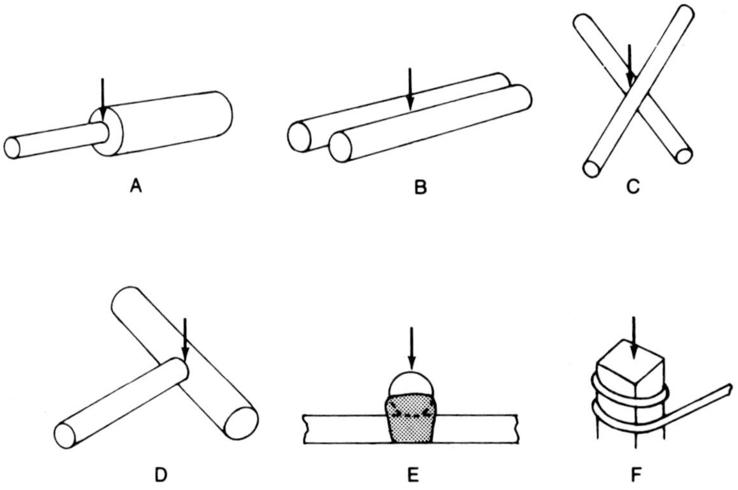

weld allows full penetration welding of a T-joint from one side only, an advantage when the far side of the T is inaccessible. Beam trapping aids deep penetration of the laser beam for welding.

Joint configurations for wires were developed by the electronics industry. These welds most often use a pulsed solid state laser, though that is not a requirement. For wire-to-wire joints the two wires must share the incident laser energy. In cross joint, *C*, for example, direct the laser beam at the intersection of the wires so that both wires get the direct energy of the beam.

The terminal or lug weld, *F*, is best made when the thermal mass of the lug equals the thermal mass of the wrapped wire. Directing the laser beam at the lug helps to avoid burn-through of the wire. Laser welds for electrical connections can be made without stripping of the wire. The laser beam vaporizes the insulation.

The laser beam — how small?

The laser beam is a beam of light. Of the industrial welding lasers, only the ruby emits visible light; the others emit light in the infrared range. The light from the laser is delivered to the work through an enclosed optical system of shutters, mirrors, and focusing elements. The focusing elements concentrate the light at the work, putting enough power into the weld joint to melt the base metal.

Light cannot focus to a point. A lens or mirror concentrates the laser beam into a cylinder of diameter $2d$ and height $2h$. Beam size can be adjusted. General welding with pulsed lasers and moderate power CO_2 lasers offers radii 0.002 cm to 0.1 cm. The range of radii for high power lasers starts at 0.015 cm. Special microwelding pulsed lasers have beam radii as small as 0.0005 cm.

The depth of the beam, $2h$, determines the tolerance on the work-to-lens distance. The tolerance is proportional to the square of the beam diameter. For a 10 percent loss in intensity at the focal plane, the depth for different lasers is:

Ruby: $h = 1.508 \times 10^4\ d^2$
Nd: $h = 9.879 \times 10^3\ d^2$
CO_2: $h = 9.879 \times 10^2\ d^2$

Use d in centimeters to get h in

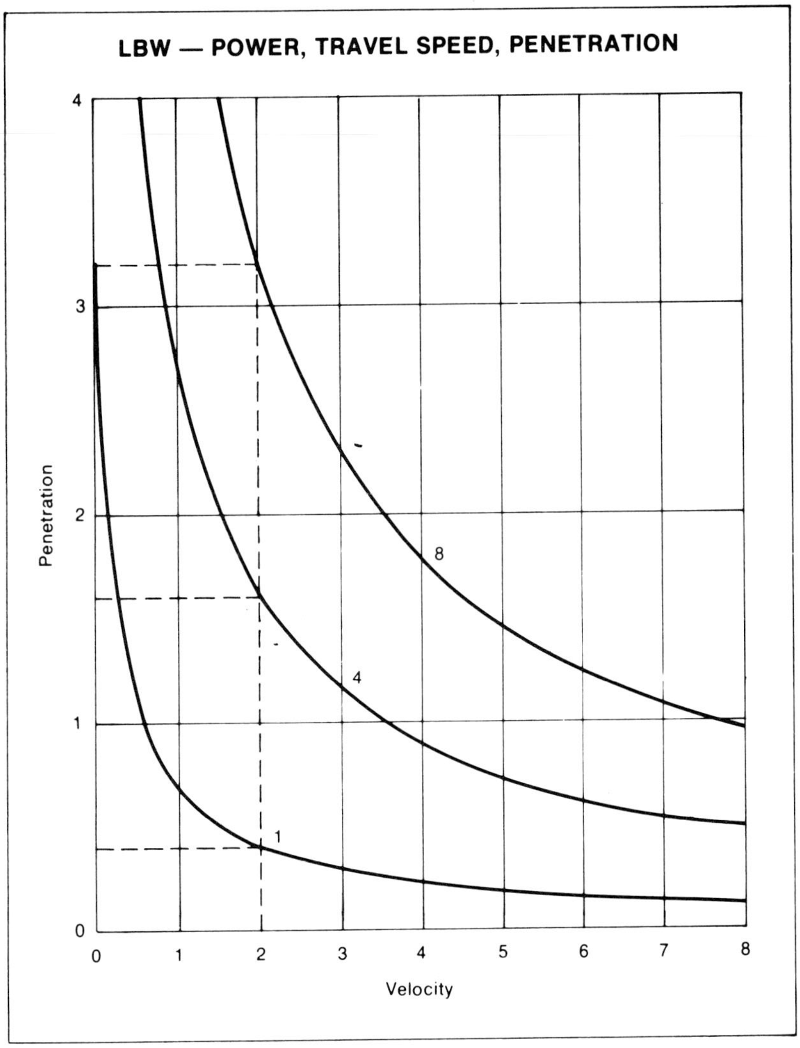

Constant power curves show changes in penetration with changes in welding velocity for continuous welding. At twice unit velocity, increasing power input by four times increases penetration four times; upping the power level to eight times the original increases the amount of penetration eight times. At low velocity and high power the curves become vertical, reflecting weld burn-through.

centimeters. The effective depth varies also with the base material.

Laser beam absorption

Laser welding depends on absorption of light energy by the workpiece. Metals are poor absorbers of light. A typical absorption coefficient for a low intensity YAG laser beam is 10 percent. For a low intensity CO_2 laser beam it is 1 percent. But at a sufficiently high intensity, the threshold intensity, the absorption coefficient increases dramatically. The threshold intensity for carbon steel is about 10^4 W/cm^2; 10^5 W/cm^2 for aluminum and copper; 10^6 W/cm^2 for tungsten. At the threshold, absorption is 50 percent no matter what the wavelength of the light. In the absence of data for a specific metal, absorption above threshold intensity is conservatively estimated as 0.5. At threshold intensity surface melting is instantaneous.

Temperatures in LBW

Conduction-limited laser welding depends upon the thermal properties of the metal: thermal conductivity, thermal diffusivity, melting point (T_m), latent heat of fusion (L_m), specific heat (C), boiling point (T_b), and the latent heat of vaporization (L_v).

In deep penetration welding and welding of light gage material an isothermal surface at temperature T_c extends through the center of the melt. For partial penetration welds and heavier gages T_c is an average temperature through the molten zone.

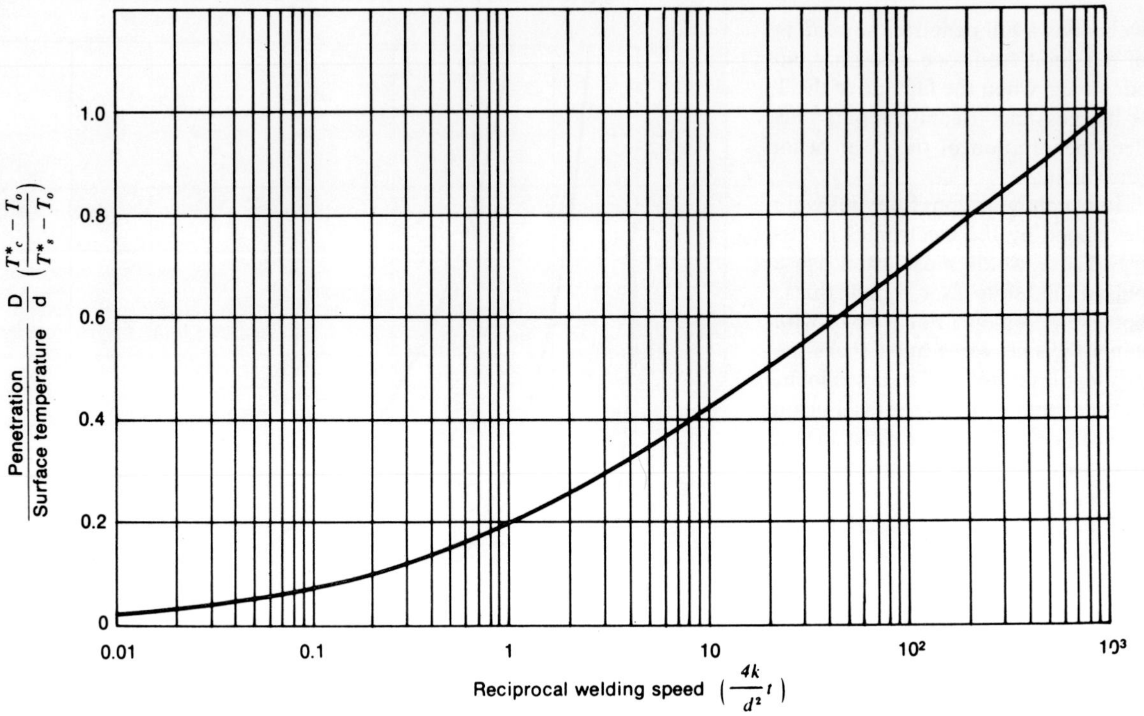

HOW TO SELECT LBW PARAMETERS

To design for continuous welding or spot welding, choose a modified surface temperature, T_s*, that is within the limits of required weld quality, and choose a beam radius. Use the required weld penetration and the required welding rate with the first graph to determine the modified center temperature, T_c*. This number will be large. In all welding, the center temperature is considerably above the melting point. Arc welding shows an increase in T_c over T_m of 30 percent. The temperatures in LBW, as in EBW, run even higher.

Use the modified central temperature with the second graph to find the absorbed power, $\propto P$, that is needed. P is the power needed from the laser welder; $\propto$ is the absorption coefficient. These calculations yield the welder power and the beam radius necessary to give the required penetration at the required speed. k is the thermal diffusivity of the base metal, T_o its starting temperature.

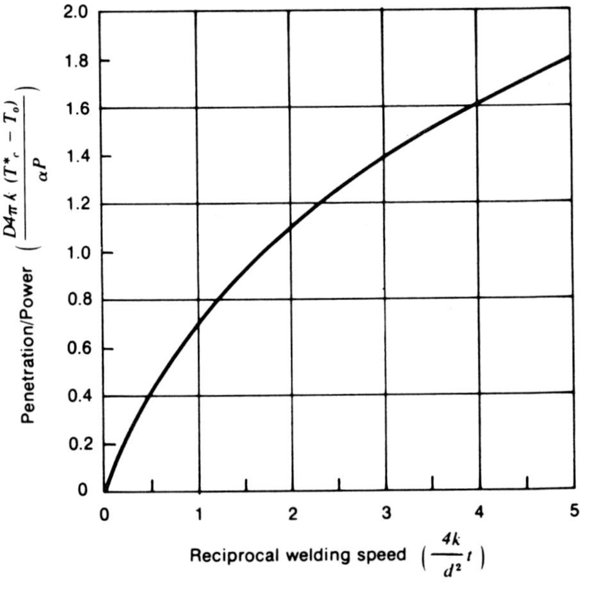

The temperature at the center of the weld, T_c, is higher than the melting point. Absorption of the laser beam takes place at the surface, so the temperature at the surface, T_s, is even higher than T_c. Energy input to the weld depends on the heat content. Modified temperatures (*) account for heat content.

$T_m{}^* = T_m + L_m/C$
$T_c{}^* = T_c + L/C$
$T_s{}^* = T_s + L/C.$

L is L_m or the sum of L_m and L_v, depending on whether melting or melting and vaporization is occurring. The latent heat of fusion, L_m, introduces only a modest correction in most cases. But the latent heat of vaporization, which for metals is from ten to thirty times L_m, is significant.

Limits on surface temperatures

As with other welding processes, LB welding temperatures are not directly controlled. The operating conditions are varied to obtain an acceptable weld. Since the laser beam is absorbed right at the surface of the metal, increasing the power increases the surface temperature. Increasing travel speed reduces surface temperature, because less energy goes into any spot on the surface. Decreasing beam size increases surface temperature, because the power is more concentrated. For a given speed and beam radius, penetration increases with increasing surface temperature. Maximum penetration for any speed and beam radius requires the highest power consistent with weld quality.

If material loss or alloy depletion through vaporization cannot be tolerated, then the surface temperature should not exceed the boiling point of the base material. Sometimes there are ways to get around this restriction. In the case of LB welding of AZ31B magnesium alloy, a good mechanical joint and a hermetic seal require two-pass welding. The second pass eliminates porosity resulting from zinc vaporization during the first pass. For joints in which material loss can be tolerated, a conservative upper limit for temperature is

$$T_s^* = T_b + (L_m + L_v)/C$$

Laser welding with this surface temperature is more like drill-and-fill than conduction limited.

Penetration, speed, power

Penetration in LBW depends on the absorbed power and the weld rate. Increase power to increase penetration. In continuous welding, reduce speed to increase penetration. In spot welding with a continuous laser welder at constant power, increase time to increase penetration. In spot LBW with a pulsed laser welder at constant energy, decrease welding time to increase penetration, because for this kind of welder, reduced welding time means higher power.

Practical continuous welding speeds for powers up to the kilowatt region range from a few ipm to 1000 ipm (25 m). Practical welding times for spot welding range from 1 ms to several hundred ms. The graph shows how welding rate, penetration, and power are related. ■

TOOLING UP FOR LASER WELDING

By

Simon L. Engel
Development Engineer
GTE Slyvania Inc.

ABSTRACT

This paper presents laser metal welding information aimed to acquaint manufacturing and tooling engineers with this technology. Laser related parameters such as power density, mode and power stability are related to the welding process. Expected weld performance is presented in terms of weld speeds, weld geometries and tolerances. Power programming (ramping) and N/C welding are dealt with in detail. The paper concludes with a review of some of the most common laser welding problems in the field.

INTRODUCTION

The substantial effort that went into the development of high power lasers, then into laser welding techniques coupled with the vast number of technical publications is showing its rewards, at last. The companies that used to "watch closely" the progress of laser technology now assign manufacturing and tooling engineers to qualify laser equipment for both quality and profitability of performance.

Most physics aside, the laser is merely a heat generator that requires utilities, space and maintenance, just like other conventional machinery. Its performance is judged by its ability to turn out a product with consistent quality. The tooling and controls are, for the most part, of conventional design as well, that are adapted to suit the laser welding process. It is the purpose of this article to present the characteristics of laser welding: what laser parameters are of significance; what are the typical weld parameters, geometries and fit-up tolerances; the need for power programming and the selection of the right numerical controller.

The data presented were generated by using GTE Sylvania's Model 971 continuous CO_2 Gas Transport Laser.

Review of Laser Characteristics

The interaction of laser energy (infrared, at 10.6μm wavelength for CO_2 lasers) with metal surfaces is well identified, see Reference 1. Fortunately for the metal working industry, metal surfaces become good absorbers of laser energy when they are in the molten state. One of the requirements is, however, that a certain power density be present at the surface to produce spontaneous melting. For ferrous metals the power density required is 10^{+5} watts/in^2; for good conductors such as aluminum and copper it is 10^{+6} watts/in^2, or higher.

The focusability of a laser beam is function of its wavelength and its divergence (spreading angle). The divergence is function of the transverse excitation mode (TEM) of the laser. The wavelength of the laser is determined by the lasing medium, or gas, and is normally a very steady parameter. However, changes in the mode of the laser will cause quantum changes in the focusability of the beam, i.e. the spot size and power density will change substantially. Such changes in power density may mean the difference between a good weld and no weld at all. Hence mode stability is the most important laser parameter. Second most important parameter is the short and long term power output level stability. Generally, a $\pm5\%$ power output stability is adequate for welding purposes.

To summarize, the type of metals to be welded will determine the power density requirements, that in turn dictate the required mode quality and stability of a laser to be selected. (Productivity, i.e. rate of welding, dictate the total output power requirements.)

The Physical Parameters for Laser Welds

In the power range in question (up to 1500 watts) focusing

of laser beams is accomplished using ZnSe lenses, coated on both sides for maximum transmission. Typical focal lengths are 2.5 to 10 inches. It has been found, Ref. 1, that shorter focal length lenses give deeper penetration than longer ones for a given laser power and weld rate. As may be expected, the shorter focal length lenses also require more critical focusing, sometimes to the point of impossibly tight tolerances. Hence, if corrugated or out-of-flat material is to be welded, longer focal length lenses should be considered, say 5 to 10 inch f.l.

Figure 1 shows a typical inverse relationship between welding rate and penetration. Because of the substantial gain in welding rates in thinner gage material, it is well worth while for the engineer to review the gauge specifications for the materials to be welded. When very deep welds are tried at very slow speeds, due to severe conduction the weld bead starts to spread "sideways" rather than into the material. The heat-affected zone increases substantially and weld porosity may occur because of the partial vaporization of the metal caused by the excess laser power absorbed. This particular problem is likely to be encountered in N/C controlled contour welding of small radii corners. The solution lies in reducing the laser power for that portion of the weld cycle.

Weld Geometries and Tolerances

Most weld geometries used in the traditional thermal welding processes are suitable for laser welding. The most frequently used laser weld geometries are the butt and lap type. Figures 2 and 3 indicate the fit-up requirements in terms of material thickness. Note, that although distortion during laser welding is minimal, some clamping is still recommended, to retain the pieces in position while being welded.

The tolerances shown are applicable to ferrous and nickel

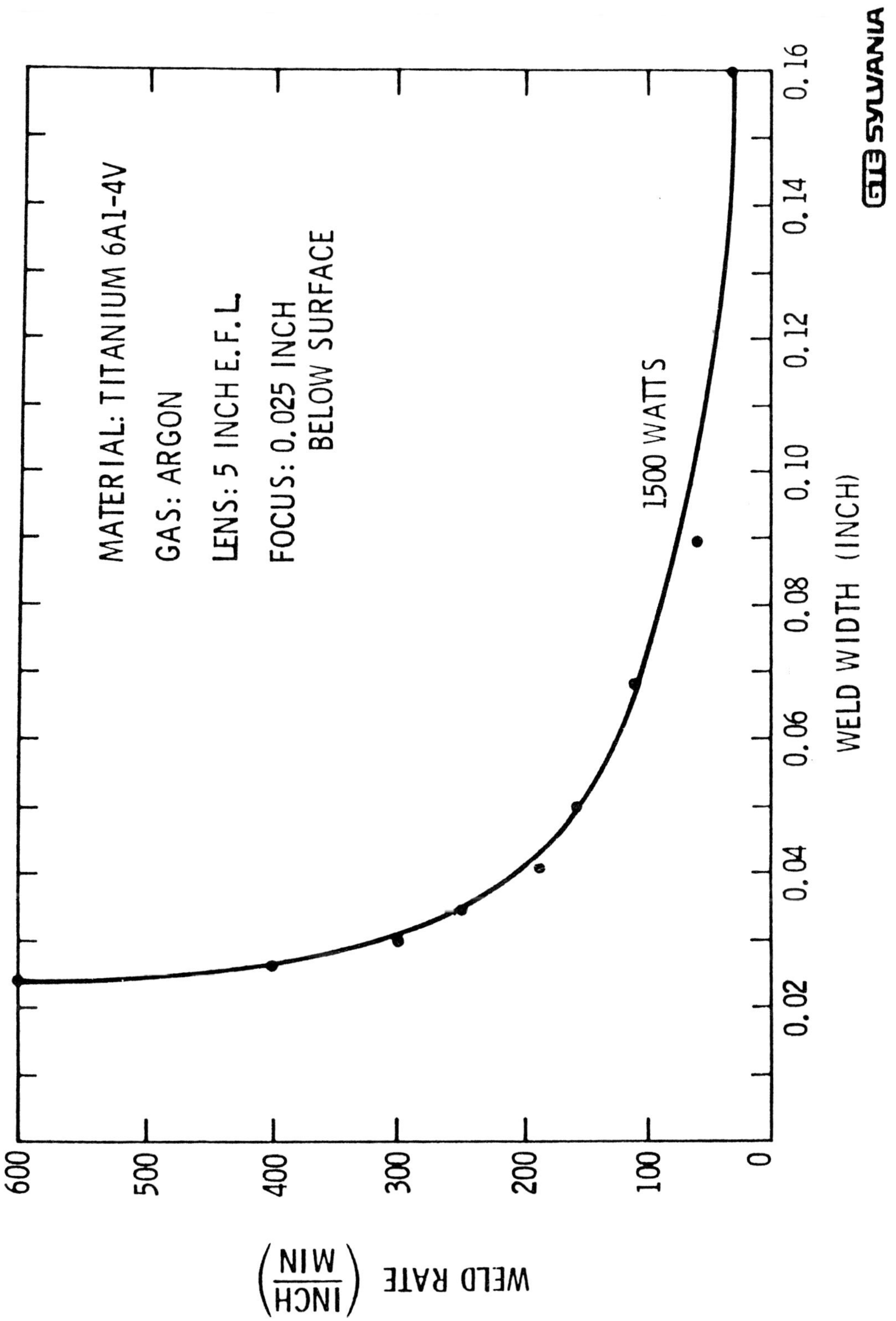
MATERIAL: TITANIUM 6A1-4V
GAS: ARGON
LENS: 5 INCH E.F.L.
FOCUS: 0.025 INCH
BELOW SURFACE
1500 WATTS
WELD RATE (INCH/MIN)
WELD WIDTH (INCH)
600
500
400
300
200
100
0
0.02
0.04
0.06
0.08
0.10
0.12
0.14
0.16
GTE SYLVANIA

FIGURE 1

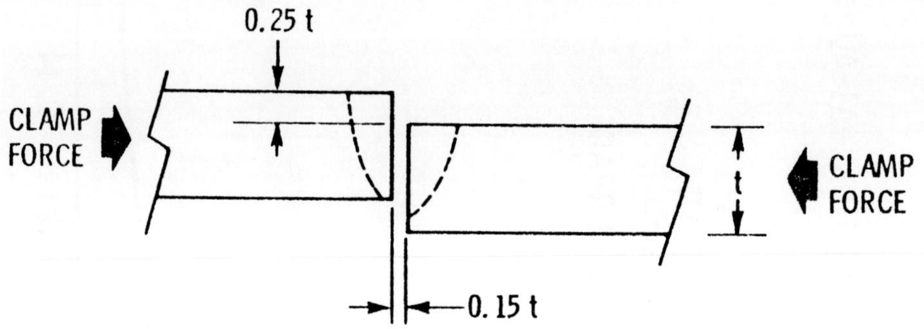

Figure 2 - Butt-joint tolerances

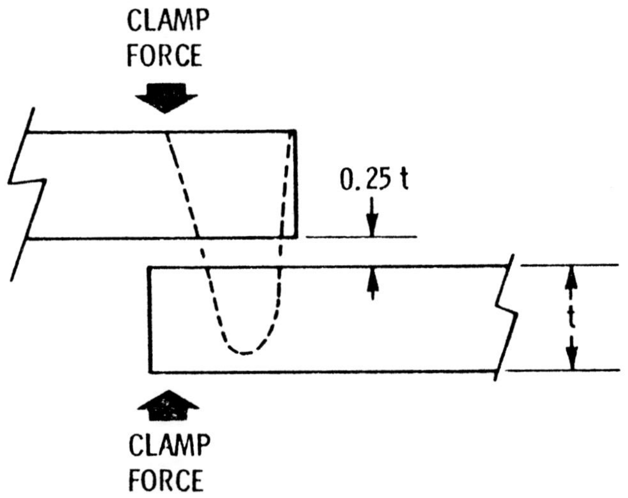

Figure 3 - Lap-joint tolerances

alloys. Somewhat tighter tolerances are required for the good conductors such as aluminum and copper alloys. The amount of gap in butt welds will also determine the degree of concavity of the weld bead,since no filler metals are normally used. In lap welds the 25% material thickness tolerance refers to the thinner of the two metals. It is preferable to weld the thinner metal to the heavier one. Excessive air gap will cause the top piece to burn thru rather than to weld to the lower piece.

In recent times interest in the welding of sheet metal components has resurfaced both in the automotive and the utensil industries. Figure 4 shows an interesting laser welded corner seam, where the bent edges are melted down to form a rod-like weld bead that acts as a stiffener. With the proper weld parameters the metal can be made to fill in the gap between the two sheets completely, forming an inside corner that is easy to clean and keep hygenic. When compared with conventional double-lock seams, this joint offers material and time savings, lower die costs and allows the use of metals that have limited plastic flow. This weld is made possible by the precise power control and smooth turn-on and turn-off of the laser beam. Figure 5 shows a laser welded single lock seam. The welding process provides a hermetic seal; however, the lap joint tolerances of Figure 3 must be observed especially between the lowest two of the three plates to be joined. Minimum or no heat sinking is required for this joint, while good heat sinking is required for the joint shown in Figure 4.

Laser Power Programming (Ramping)

Laser welding is not totally unlike electron beam welding. Even in shallow welds (say < 0.060 inch) some "key hole" effect is present in the welding process. When the laser power is turned off at the end of the weld there will be typically a "dimple" at the end of the weld bead. Further, it

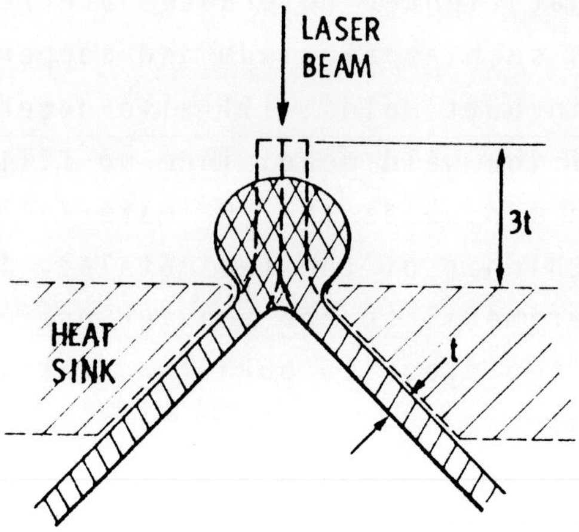

Figure 4 - Laser formed corner seam.
Note the recommended heat sinking
and material dimensions to achieve
the rod-like corner support.

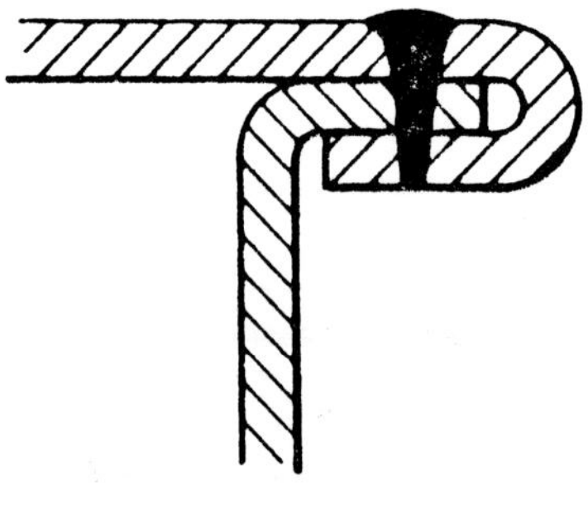

Figure 5 - Laser welded single lock seam.

has been noted that when the laser weld over laps a previously welded section, there seems to be a greater absorbtivity of the laser beam, resulting in the possible overheating of the components, or the possible generation of porosity. Therefore, for critical linear, or all overlapping hermetric welds it is desirable to program the laser power to take care of the above mentioned problems. A typical power programmer, or ramp generator, will provide adjustable ramp-up time (during which time the laser output power is electronically raised from 0 to a preset power level), adjustable weld time and adjustable ramp-down time. Resolution of 0.1 second during ramping has been found adequate. Total time range of 10 seconds for the ramping and 10 minutes for welding has been found to satisfy most applications. The power programmer should have facilities to be triggered externally by a micro-switch or Numerical Controller. From the safety point of view, there should be means of aborting the cycle in progress any time, as well as safety guards, such as electronic line filters to prevent accidental triggering due to EMI effects. Figure 6 shows a weld test pattern where correct power programming was applied to blend the start and finish of the weld.

Numerically Controlled (N/C) Welding

With the exception of continuous welding or welding of very simple geometries, numerically controlled (N/C) equipment is replacing the traditional hard-tooling and cam operated equipment. In view of the nature of laser welding, the selection of the proper N/C and certain of its options must be made with care.

There are basically two types of N/C systems. In the point-to-point system the table or optics are moved from point-to-point as per programming instructions on the tape. To generate a straight or curved line between two points, each point must be programmed along the path, and the table will zig-zag its way

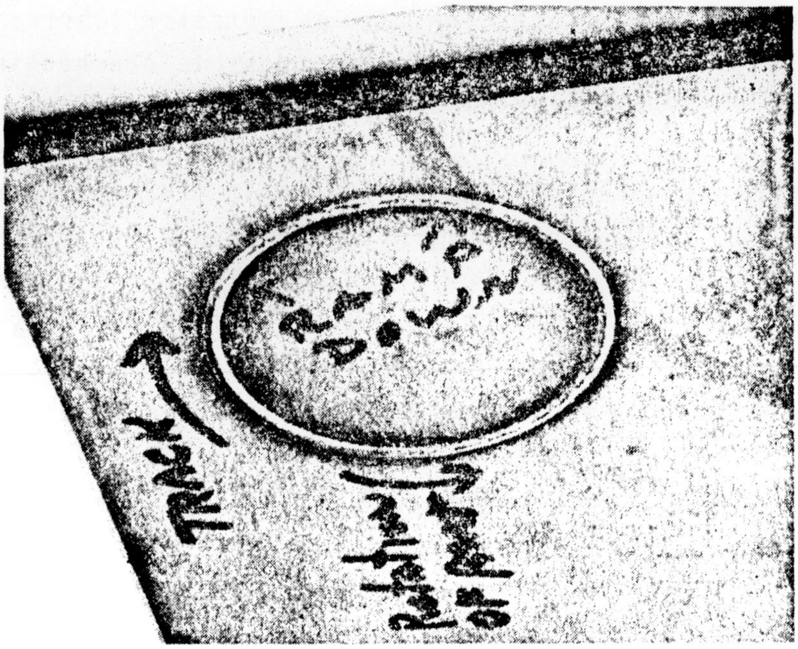

Figure 6: Bead-on-plate test of Power Programmer. Laser power was ramped down at the overlap to produce a smooth, porous free circumferential weld.

to the end of the line. The stepping motors that drive the table start and stop for each point. Such a system is not suitable for laser welding, because during the dwell time between the steps (i.e., non-uniform motion) excessive laser energy is absorbed and an uneven or porous weld bead will be generated.

In the contouring N/C system, internal electronics generate a signal that will simultaneously power both (or all three) motors to generate the desired straight ("linear interpolation") or curved ("circular interpolation") line between two points. The motors turn continuously rather than step by step. Such a controller is a good basic equipment. Programmably adjustable speed control up to about 300 inch/minute in 0.1 inch/minute increments is desirable. Because of the mass of the table or components to be moved, there is certain acceleration and deceleration involved in executing a programmed motion. This situation occurs for each new block of informamation to be executed, or when corners of less than 1 inch radius are generated. As noted in Figure 1, changes in traverse rate result in changes in weld penetration. It is desirable to have a "deceleration inhibit" option with the controller, to achieve uniform weld performance. This option is usable only when lines are tangent to each other, however, and therefore the problem of acceleration/deceleration still exist in non-tangent situations such as square corners. The inability of the conventional N/C to read a new command while it's executing a previous one makes the triggering of an external power programmer, for instance, impossible. Indeed what is needed is a parallel processor that would adjust power levels or perform other switching actions at the right moment. Such a processor would receive its instructions both from the programmed tape and the detectors that detect table position and velocity. Parallel processing, automatic height sensing, automatic correction for fit-up variations and automatic correction for misalignment may be expected to be all part of the future N/C system dedicated to laser welding.

The Most Common Welding Problems in the Field

Perhaps because of the novelty of the laser as a manufacturing tool, a lot of engineers who are just getting acquainted with laser welding are apprehensive about the metallurgical and physical qualities of the welds. They are also concerned about the effects of the welding process on the laser itself.

It must be remembered that the laser beam is only a heat source used to melt the metal and that all metals that can be welded by traditional methods and all metals that are metallurgically compatible, can also be laser welded. Companies that manufacture and market industrial lasers have application labs where the ability of the laser to satisfactorily weld a customer supplied sample is quite quickly established. Normally, by the time the equipment is purchased and is installed, the metallurgical and physical tests have all been completed. Therefore, such problems seldom crop up in production, unless unexpected changes in the received material itself occur. Further, assuming proper welding set-up, the laser itself receives no feedback about the welding process, and it operates for extended periods of time in production without any noticeable deterioration. Note, however, the qualifier in the previous statement, "assuming proper set-up."

The most common welding problems in the field have been found to be related to misaligned optical components external to the laser and to misaligned tooling. Several hundred watts of misaligned laser power is sufficient to heat up mirror and lens holders causing physical distortion that will break the optical elements. Such catastrophic events may also cause thermal damage and destruction of the optical elements. A misaligned gas shield nozzle, for example, may "clip" the laser beam and besides heating up, it may scatter the beam to the extent that it will no longer focus properly on the metal. If the required power density to melt the surface of the metal is not present, for whatever reason, the metal will reflect substantial percentage of the laser beam causing thermal damage to lenses, hoses, etc. The damage occurs in a fraction of a second, much too fast for the operator to react and prevent it. Yet, for some unexplained reasons, lot of people have shown tendency to adjust the laser cavity optics to correct the above problems.

Some welding problems have been found to be related to the

way shield gases are applied. In practice the gas flow is applied coaxially with the focused laser beam, thru a cone shaped nozzle that has a relatively large, say, 0.250 inch jet diameter. Flow rate is set at 10 - 15 s.c.f.h. The shield gas serves two purposes, one is to protect the lens and second is to shield the molten metal from the atmosphere. Increasing the flow rate too much will cause the molten metal to splatter, while decreasing the flow rate will allow metal vapors and smoke to deposit on the focusing lens, increasing energy absorption in the lens and thus destroying it. Gas flow switch interlocks and flow gages (rather than pressure gages) will prevent problems related to shield gases.

Other welding problems have been found to be related to the material handling means. The feed rollers and turn-tables used for TIG or torch welding are normally not accurate enough for laser welding. In addition to the fit-up tolerances shown in Figures 2 and 3, the positioning accuracy of the fit-up must be within ±0.003 inch (for 1.5 KW laser power focused by a 5 inch focal length lens). The accuracy of the fixtures used in laser welding should be similar to the fixtures used in electron beam welding. Of course, there is no need to build these fixtures from stainless steel. The clamping blocks or rollers closest to the weld area should be water cooled, or at least have copper faces.

In closing, it is recommended with great emphasis, that the laser technicians servicing the equipment be trained not only how to set-up and operate the laser, but also how to check and set the alignment of the optical and mechanical components downstream of the laser. They should understand the importance of proper focus, shield gas and fixture alignment.

Conclusion

This paper presented laser metal welding information aimed at manufacturing and tooling engineers. Beside the basic desirable laser parameters, emphasis was placed on the physical weld parameters that relate to tooling and weld quality. The accessories required to achieve quality welds, such as power programmers and Numerical Controllers were also discussed. A special section was devoted to welding problems in the field, and how they may be traced back to the fundamentals of laser welding. It was noted that such problems are not only easily prevented, but should they occur, they are also easily identified and corrected. The large number of laser welders in the field are the testimony, that laser welding technology has successfully met the challanges of the commercial world.

List of References

1. S.L. Engel, "A guide to laser metal welding - to 1500 watts." SME Paper No. MR75-579, Lasers in Manufacturing Conference, Chicago, Illinois 1975

Laser solves spotwelding problems

Fusion welding of dissimilar metals is difficult, and when the weld must be made at the edge of the material, the troubles compound. Laser welding offers a solution for both problems

The idea of using lasers in metalworking usually conjures up thoughts of ultrasophisticated applications and unlimited budgets. In fact, lasers have often been viewed as a solution looking for a problem. But when you get right down to it, the main attraction of a laser beam is that it provides a noncontact means of getting a high concentration of heat at a specific restricted location without affecting the surrounding area.

This property makes the laser beam highly suitable for the reliable noncontact fusion joining of many engineering materials. And industrial laser systems are now available at prices that can compete effectively with the more conventional processes they replace.

A case in point is the automated Raytheon Model SS-380 laser welder recently installed at Foxboro Co. (see cover), where welding specialist Lorin Marshall was faced with the problem of producing reliable, high-quality welds between Ni-Span C and 430 stainless steel. Foxboro joins these materials to form a lever assembly which transmits motion from a metering diaphragm to a force motor—part of a differential pressure-measuring device it manufactures for the process industries.

Eleven-piece assembly

Ni-Span C is used for five separate hinge plates—thin, flexible bridge pieces called flexures—which connect the six stainless-steel elements of the compound lever arrangement (Fig 3). Ni-Span C was chosen for its constant modulus of elasticity, which is important for instrument applications, and because it has a very low coefficient of thermal expansion.

The flexures must be attached close to the edge of each stainless-steel component in the assembly in order to provide the proper flexing characteristics, and a

By George Schaffer, associate editor

1. Operator loads fixture in which 40 spot welds are made on each of eight parts. Safety enclosure of laser welder (rear) is open to receive the fixture

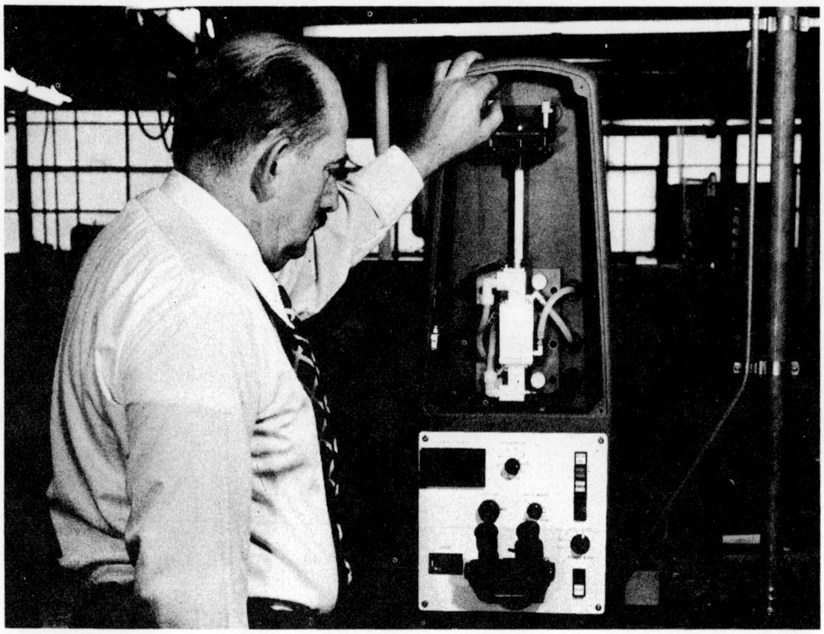

2. Laser is tuned by aligning rear mirror of laser cavity, using two micrometer screws. Other maintenance involves keeping optics clean and replacing flash tubes

3. A total of 52 separate spot welds (close-up above) are used to make up one lever assembly. The flexures must be attached close to the edge of each component

total of 52 separate spot welds are used to make one lever assembly. But fusion-welding of dissimilar metals, such as Ni-Span C and stainless steel, is difficult because the bastard alloy formed in the weld nugget is subject to phase transformations, resulting in precipitation growth of intermetallic compounds during the cooling process—a condition that destroys the strength of the weld.

In addition, the particular configuration of the Foxboro assembly is hard to weld because, with the weld spots close to the edge of the material, heat migration during cooling is unbalanced; there is no material available to absorb the heat that flows toward the edge. The resulting thermal imbalance can destroy the desired mechanical properties of the materials before a good weld is achieved, and there is a good likelihood that cracking will occur in the heat-affected area.

Both these welding obstacles can be overcome by keeping the heat input to the weld at a minimum and by producing a highly localized weld spot. The way to do that is to create very localized high-intensity heat very rapidly. Rapid heating also results in rapid cooling and, in effect, freezes out the precipitation growth of intermetallic compounds. And the small heat-affected area from a small weld spot reduces the heat-migration problem as well.

EB was tried

One way to achieve such rapid localized high-intensity heat input is with electron-beam (EB) welding—and, in the past, that was the only process Marshall found adequate to the task. But that involved subcontracting the work to a job shop, because the relatively high initial cost of the EB-welding equipment could not be justified for the limited applications at Foxboro.

Part of that expense is associated with the enclosures and pumping equipment required for EB welding, which must be done under high vacuum. "It can all add up to a $125,000 package," says Marshall, "and because of the vacuum requirements, you don't have direct access to the part during operation. That means an additional expenditure for a remote positioning system."

Another way to achieve the desired rapid localized heating is with a laser beam. And laser welding requires no vacuum enclosures. In fact, Marshall found that he could get in-house laser-welding capability from Raytheon at about one third of the cost for an EB-welding setup—and that includes a computer-controlled positioning system as well as some initial tooling.

The SS-380 laser welding system at Foxboro consists of a laser head, an optical system, a cooler/deionizer section, a power supply, and the computer-controlled NC (CNC) positioning table—all integrated into a pedestal-mounted work station with an interlocked safety enclosure. The laser head uses a flash-pumped solid-state pulsed laser with a single-crystal rod of yttrium-aluminum-garnet (commonly called YAG) as the lasing medium.

The pulse-repetition rate of the SS-380 YAG laser system at Foxboro is adjustable to 1 or 2 pps, or the laser can be fired manually for single-shot operation. The laser operates at the low end of the infrared spectrum (1.06 microns), and the nominal output energy is 20 joules, with a full-width pulse of 10 milliseconds. The spot size of the beam at the workpiece is about 0.045 in. with the particular optics selected for this application, and the resulting weld nugget has a diameter of about 0.032 in.

The computer (a Computer Automation Alpha LS1-11 mini) controls the movement of a Summit Model 6106F 6 × 6-in. positioning table, and also coordinates the table-positioning motion with the repetition rate of the laser so a shutter is interposed in the optical path of the beam during the table movement between weld locations. When the table is in an appropriate spotwelding location, the shutter is removed, and a laser shot (or several shots, if required) completes the weld.

Some laser shots are wasted with this programming method, because the repetition rate of the laser is faster than the indexing time between weld locations. But if the laser is operated continuously at a constant repetition rate, the temperature of the laser cavity remains constant, and that makes for more-consistent power levels and, therefore, better welds.

Operation is simple

Programs are loaded into the minicomputer via a Teletype input/output terminal, which is also used for setup and initiation. To initiate a welding run, the operator types a homing command, which brings the indexing table to a reference position. The operator then sights through the microscope (the shutter is in place to prevent exposure to stray laser radiation), which has an optical path coincident with that of the laser beam, and uses jog controls to align a reference mark on the part-holding fixture. Another instruction on the keyboard then does the zero-shifting to bring the home position in line with the reference. A final key stroke initiates the welding cycle, which ends with the

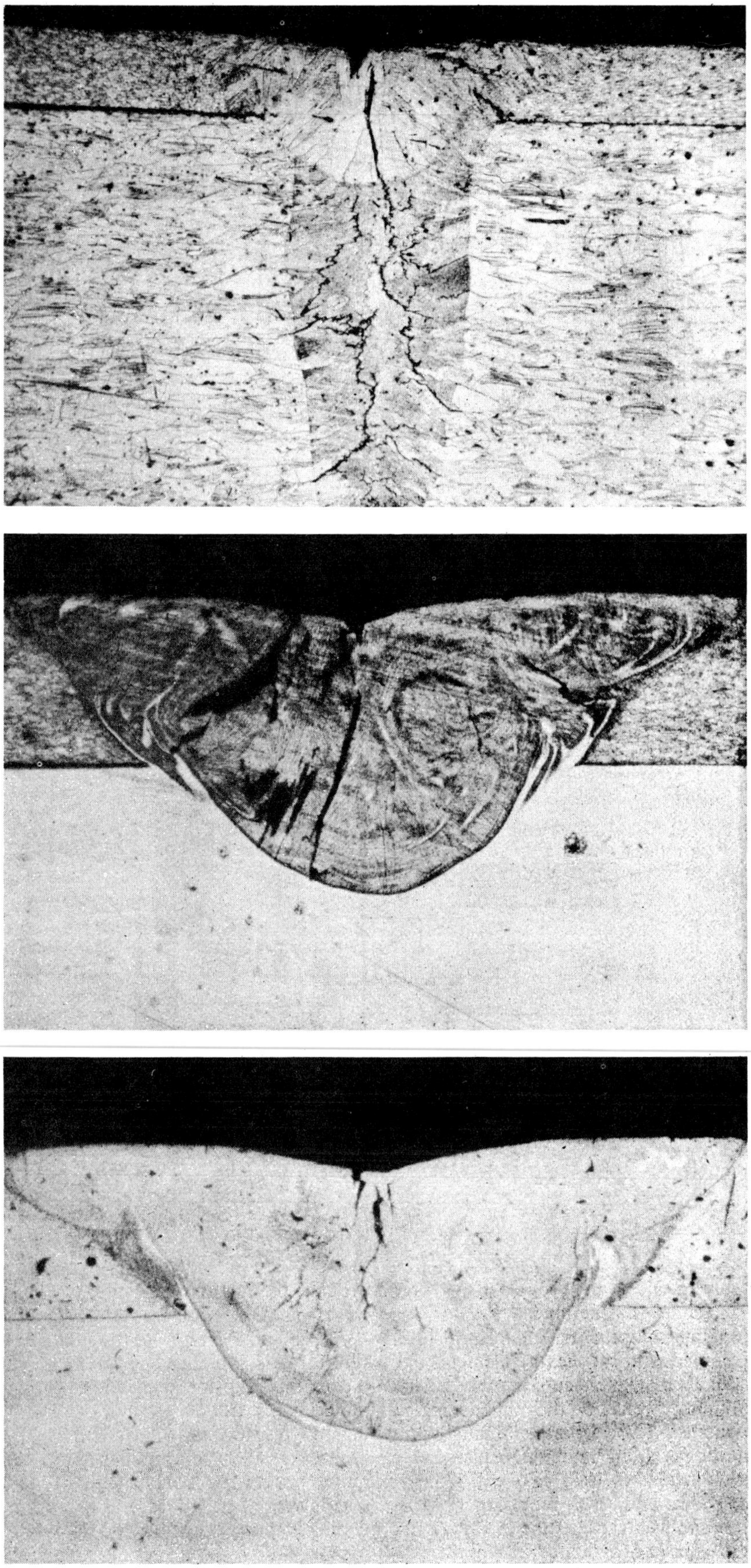

4. Photomicrographs of welds show heat-affected zones. EB weld (top photo) has deep zone, and is unacceptable because of excessive cracks. Laser weld (center) is also a reject. Typical good laser weld (bottom) has no cracks at nugget boundaries

Weld penetration

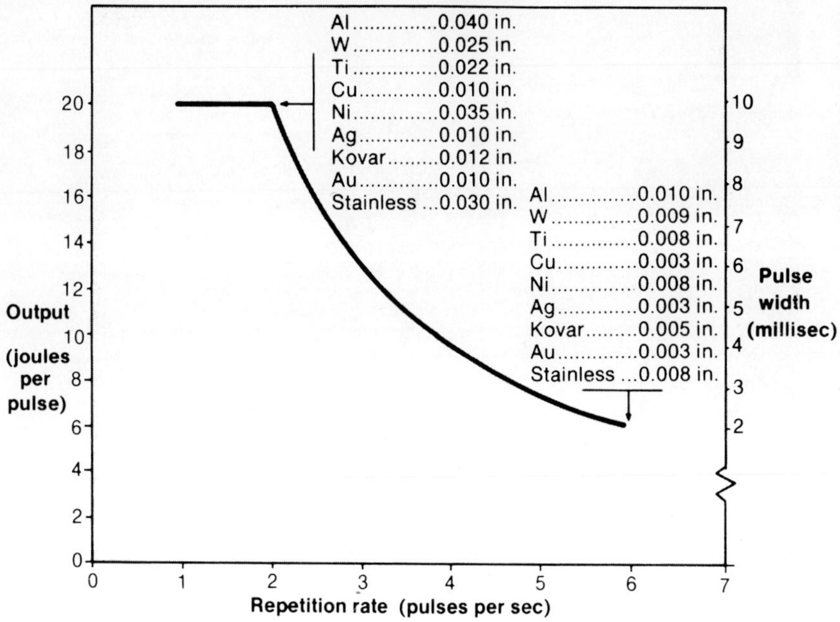

Elements of a laser manufacturing tool

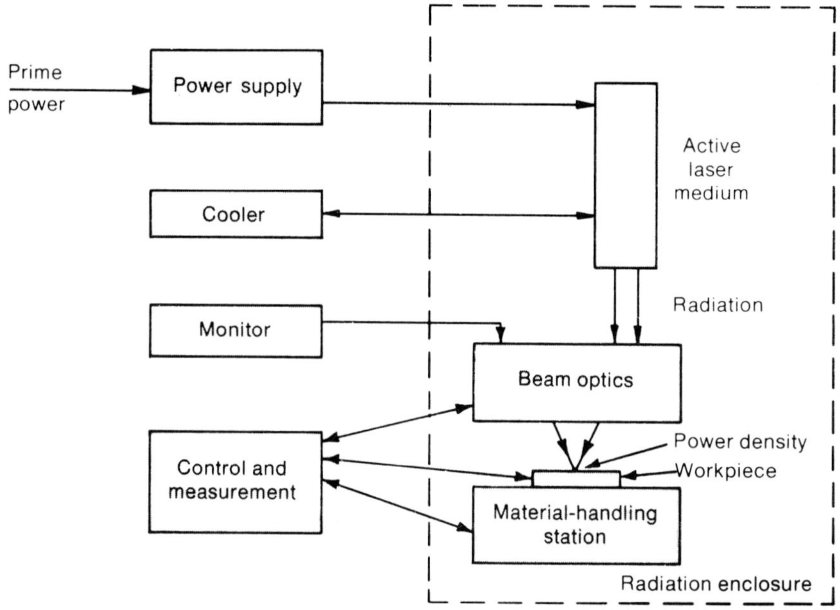

table returning to the unload position.

Each lever assembly is welded in two separate setups. First, the central lever is joined with two flexure elements and a small plate; this subassembly is subsequently welded to the remaining components. Two separate multiple-part nesting welding fixtures are used. One such fixture is used to weld 22 lever subassemblies, each requiring 12 separate spotwelds. All of the 264 spotwelds are executed in about 2½ minutes. A second fixture handles the final welding operation, in which another 40 spotwelds are made on each of eight parts to produce the final lever assembly. These weld runs take about 4 minutes each.

At present, Foxboro has only one fixture for each welding operation, which means that the operator is idle during the weld run. But even under these production restrictions, the system is approaching the 30-parts-per-hour requirement set for the job. Under good conditions, Marshall says, Foxboro is getting 25 parts per hour, and it is planning to build a second set of fixtures, so the operator can load one while the other is working.

The factors that govern weld characteristics are beam-spot size, pulse length, and energy per pulse. Spot size is determined by the optics chosen, and can be varied within limits by an optical adjustment. Pulse length is determined by the discharge time of the flash tube, and can be varied electronically. Energy per laser pulse is controlled either by the length of the pulse duration or by interposing an aperture control in the optical path. Arriving at the appropriate "schedule" is a matter of trial and error, but, once satisfactory conditions are determined, they can be repeated at will.

With a little over seven months of operating experience under his belt, Marshall feels the laser welds are equal to or better than the previous EB-welded joints. And the stiff tolerances on the assembly are easier to hold because there is less heat input to the part. "We have had no rejects for dimensional reasons since the laser welder was put into operation," he says. Tolerances on spacing between assembled components are typically ±0.005 in., and the parts must be held parallel to within 0.005 in. Also, the EB produced a very deep heat-affected zone even with the beam defocused, whereas the laser weld has a more-limited penetration depth. (Fig 4).

Little maintenance needed

Maintenance problems have been minimal. Mainly it has been a matter of flashtube replacement, a normal, expendable item. The flashtubes, which excite the lasing action, are relatively simple to replace because they are straight-line tubes instead of the annular type used in other laser systems. The flash-tube and YAG rod are housed parallel to each other in an elliptical, gold-plated enclosure, with each at one of the foci of the ellipse.

This arrangement provides full flash concentration over the entire lasing medium, and alignment is a simple matter. It is accomplished easily by using two micrometer screws, which align the rear mirror of the laser cavity.

Other maintenance involves keeping the optics clean and removing any accumulation of vaporized weld material, and Marshall has instituted a regular-preventive-maintenance schedule for such housekeeping.

Quality and reliability are very important at Foxboro because its equipment goes into process pipelines, where downtime can easily run to $1000 per hour. For this reason, Marshall has always been on the lookout for improved welding methods. "We like to stay abreast of current development, and we constantly try to improve weld integrity." he says. And it looks like the laser approach is doing just that. ■

SECTION B-III:
Applications

Assessing the laser for power plant welding

R. A. Willgoss, J. H. P. C. Megaw, J. N. Clark

Although laser welders are still expensive items of equipment, their growing availability (at least five manufacturers in the UK) means that the special qualities they offer could find applications in a wider range of practical work. Comparative tests for the CEGB have suggested that the laser could well be used for such jobs as tube-to-tube welding, deep penetration welding and where access is dangerous or difficult.

Commercially available CO_2 lasers are now capable of production welding metal plate up to 10 mm thick in one pass. Because CEGB power plant involves components with 2-10 mm wall thicknesses, the Board initiated a feasibility study of laser welding 6 mm plate in types 316, 310 and Ducol W30 steels—materials typical of those used in CEGB power plant.

After a preliminary assessment of weld profile, 6 mm thick plates were test-welded in the downhand configuration. Tensile, radiographic and hardness tests followed and the results were assessed against the British Standard and parent plate properties.

Laser welding was also compared in terms of capital cost, power requirements, ease of handling and workpiece distortion against TIG, plasma and electron beam welding.

Two lasers, developed and situated at UKAEA Culham Laboratories, were used. Both were operated to give beam powers of 5 kW at the workpiece. Although the workpieces were 6 mm thick, the maximum penetration depth obtainable using the beam focussing optics was also charted (up to 7kW) using 25 mm thick plate.

Laser welding in general

Using 10·6 μm radiation to heat metals is normally an inefficient process: 80—90% of the impinging radiation is reflected from the solid surface at room temperature. However, at a sufficiently high radiation intensity (around 10^{10} W/m^{-2}), the surface temperature can rise rapidly causing increased absorbtion and, with the onset of melting, significant vapourisation and disruption of the liquid surface. Very efficient coupling of radiation (up to 90%) into a workpiece results. The energy is deposited in a thin layer of material at the surface which is intensely heated and then disrupted, eventually forming a "keyhole" within which the beam is almost completely absorbed. For beam powers of a few kilowatts, the keyhole can be several millimetres deep. It is kept open mainly by the recoil vapour pressure from the intensely heated surface (Andrews and Atthey, 1976).

The keyhole phenomenon is employed in laser welding. The beam is made to fully penetrate the plate and is traversed along the line of the workpieces butted together. Weld integrity depends on alignment accuracy and the width of the weld fusion zone produced. It is possible to make parallel sided welds but the transverse section and surface finish can be radically altered by adjusting beam optics and displacing the characteristic plasma plume which appears above the keyhole.

Beam optics focus the power at the workpiece surface. Depth of focus, d, and original beam divergence, a, are related by the expression $d=(2f^2a)/a$ where f is the focal length and a is the unfocused beam diameter. Thus, for a fixed beam divergance the focal spot diameter can only be reduced at the expense of a marked decrease in the geometric depth of focus. Lens aperture is f/a; for a given beam power and beam divergance, there should be a corresponding optimum working lens aperture, observed in some experiments (Locke and Hella, 1974; Breinan and Banas, 1975) to lie between f/5 and f/8. Depth of penetration has been noted as sensitive to the position of the focus with respect to the front face of the workpiece, the optimum position being just inside the workpiece.

A characteristic of laser welding, not presently fully understood, is the plasma plume which often forms just above the laser focus and severely attenuates the beam. The simple theory of the laser keyhole involves the formation of a hot vapour plume (Andrews & Atthey, 1976), but this would contain only non-ionised atoms of the likely thermal equilibrium temperature. Any free electrons formed in the focal region would be rapidly accelerated by inverse bremsstrahlung. The dynamics governing the growth and decay of such an electron population have been discussed by Grey-Morgan, (1975).

The plume can be reduced and displaced by applying a transverse gas jet across the keyhole; this also provides an inert atmosphere at the weld. Failure to remove the plume results in serious beam attenuation (with a subsequent loss of penetration) and oxidation of the workpiece. Helium is normally used to displace the plume, since it produces the least dense ionised plume of all the inert gases. It has also been shown (Rein, et al., 1972) that adding small quantities of hydrogen, sulphur hexaflouride and carbon dioxide to the helium enhances penetration at welding speeds below 40 mm s^{-1}, presumably by increasing the thermal conductivity of the gas shield or by reducing the free electron density by attachment to form negative ions.

Acceptable laser welds

Previously published work on the welding of steels refers mainly to Type 304 stainless steel (Locke & Hella, 1974; Breinan, et al., 1975; Locke, et al., 1972; Breinan & Banas, 1975) at plate thicknesses up to 17 mm and power levels

Willgoss and Clark are with the CEGB Marchwood Engineering Laboratory. Megew is with the UKAEA Culham laboratory.

Table 1. Cast analysis and standards (Percentage additions—max. values inserted for standards).

Element	BS 1501 316-S16	316 Cast	BS 1501 310-S24	310 Cast	BS 1501 304-S15	BS 1501 271 (W30)	Ducol W30 Cast
Carbon	0·07	0·025	0·15	0·06	0·06	0·17	0·12
Silicon	1·00	0·43	1·00	0·56	1·00	0·4	0·29
Manganese	2·00	1·25	2·00	1·5	2·00	1·5	1·49
Nickel	13·0	10·7	22·00	20·1	11·00	0·7	0·24
Chromium	18·5	18·1	26·00	24·8	19·00	0·7	0·46
Molybdenum	3·0	2·6	—	1·1	—	0·28	0·25
Vanadium	—	0·12	—	0·1	—	0·12	0·11
Niobium	—	0·11	—	0·1	—	0·1	0·01
Sulphur	0·03	0·01	0·03	0·02	0·03	} 0·04	0·02
Phosphorous	0·045	0·039	0·045	0·03	0·045		0·026
Copper	—	0·22	—	0·3	—	0·3	0·12
Nitrogen	—	—	—	—	—	0·115	—

approaching 20 kW. For the work on Types 302 and 321 stainless steel plate thicknesses did not exceed 2 mm. Metallurgical assessments have been made on laser welds in thicker section carbon steels (Stoop, 1976; Breinan and Banas, 1974). To our knowledge, no work on laser welding Types 316 and 310 stainless steel and Ducol W30 carbon steel has been published.

Type 316 stainless steel differs from Type 304 in that 3% molybdenum is added to increase strength. Type 310 is a 25% chromium, 20% nickel stainless steel which is resistant to corrosion and oxidation at high temperatures; however it has been known to suffer from solidification cracking when welded autogenously. Cast analyses and standard material compositions are given in Table 1.

Before welding 6 mm thick plate to plate, a series of bead on plate runs were made to investigate changes in profile and surface finish as a function of beam optics, power level and workpiece traverse speed. Additionally some incomplete penetration welds were made on thick section workpieces to investigate penetration depth as a function of beam power. Input parameters used to generate acceptable bead-on-plate welds produced virtually identical profiles in plate-to-plate welding with negligible gap between workpieces.

Previous experience has indicated the need to maintain inert gas cover for several seconds on both workpiece faces after welding to prevent gross oxidation of the surface. Also, front face surface finish is markedly affected by the speed and direction of the inert gas jet which removes the vapour plume from above the keyhole. Initial experiments demonstrated that it was possible to produce acceptable profiles and surface finish with the vapour plume deflected either forward or backward; the optimum depended on the speed and angle of approach of the gas jet. Since it was easier to deflect the plume forward this configuration was adopted for all experiments.

The upper gas shield assembly consists of a trailing shield into which argon or helium flows (Fig. 1b). The sample holder incorporated an undershield to prevent oxidation of the backface of the weld.

Experimental details

The laser and work station optical arrangement is illustrated (Fig. 1a). The electrical discharge takes place in a mixture of helium, nitrogen and carbon dioxide at a total pressure of about 0·1 bar.

Fast recirculation and cooling of the active medium are needed to obtain maximum available power out (Duley, 1976). The discharge current, gas flow and optical axis of the laser beam are made mutually orthogonal. A folded, two-pass discharge with an unstable optical resonator are employed to reduce the overall dimensions and maintain good beam mode control. The annular output beam has an outer diameter of about 90 mm and an inner diameter of about 40 mm.

In Fig. 1b, the focus unit and electrically driven traverse table are shown with the sample clamps removed. No direct measurements of spot size were made in this series of experiments but previous results with the same focus unit gave an effective spot diameter (ie, distance between the 1/e intensity points) of 0·5 mm.

Welds in Types 316 and 310 stainless steel

Position and speed of the gas jet controlling the vapour plume, the traverse speed and the focus position were all adjusted in an attempt to generate acceptable, fully penetrated, bead on plate welds. Laser power at the workpiece was 5 kW. The terms for acceptable welds were defined as no undercut, no surface protrusions greater than 1 mm, smooth fluid flow lines on the weld finish and a front to back weld width of not less than 2 mm. Some acceptable profiles

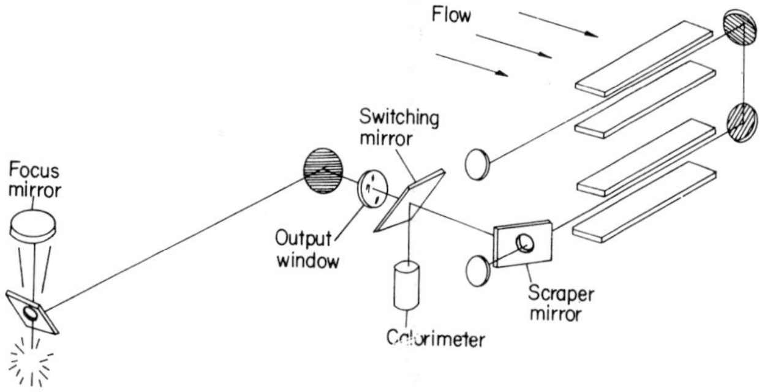

Fig.1 Schematic of output power take-off from laser together with a photograph of the focusing mirrors at the workpiece platform.

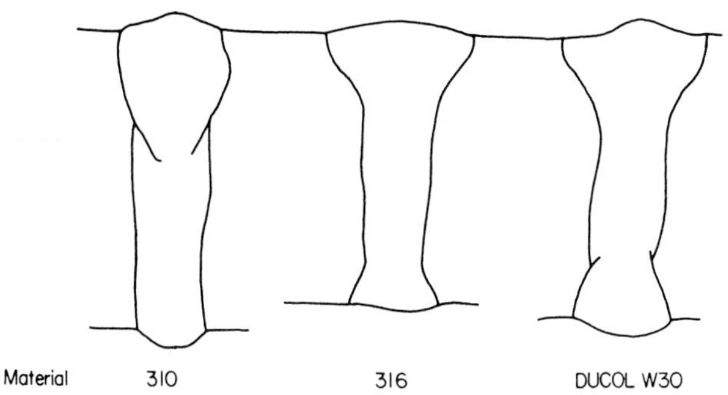

Fig. 2. Acceptable transverse sections of laser welds (Beam power 5 kW, traverse 16 mm s^{-1}).

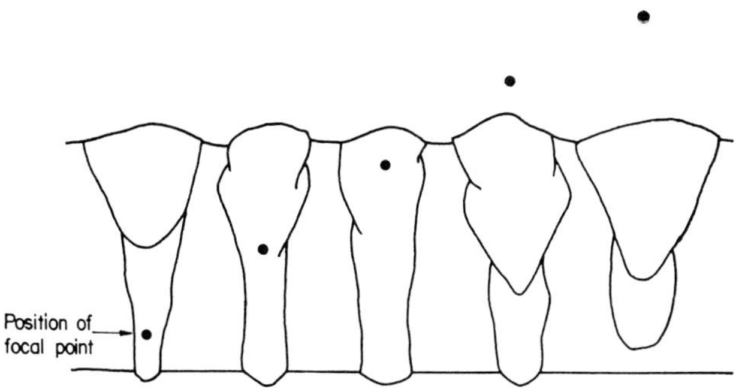

Fig. 3. Transverse profiles as a function of focus position (Material 310, beam power 5kW traverse 16 mm s^{-1}).

are shown in Figure 2.

The optimum position of the focus was checked. Fig. 3 shows transverse profiles generated by moving the focus point in 2·5 mm steps perpendicular to the plane of the workpiece. When the focus point was positioned deep inside the workpiece, a V-shaped weld resulted necessitating more precise alignment than that for a parallel sided weld of the same cross-section. When the beam was focused well above the level of the workpiece front face, a large "nail head" was formed with a consequent loss of penetration.

Optimum focus was approximately 1 mm below the level of the workpiece front face. This produced welds with little or no "nail head" and near parallel sides. Metallographic examination revealed that solidification cracking had occurred in all the welds in type 310 stainless steel. Liquation cracks in the heat affected zone and bands of segregate in the parent plate had acted as initiators for the solidification cracks. Such defects are known to occur in this metal when it is welded autogenously by other processes (eg, TIG.) The results suggest filler wire would be needed and no further welding of type 310 was carried out in this study.

Minimum beam power giving consistent full penetration was used for welds in 6 mm thick type 316 steels. The problems encountered as the power level and corresponding traverse speed were reduced are illustrated (Fig. 4). At 2·5 kW, with full penetration just possible in 6 mm thick plate, the "nail head" became abnormally large, often occupying more than 50% of the weld transverse section area.

Radiographs showed that fully penetrating welds in 6 mm thick type 316 steel, produced with 2·5 kW power were were invariably porous, whereas at 5 kW power they showed no signs of porosity. It was concluded that the power level had to exceed that needed to just penetrate the workpiece so that welding takes place in the regime where a virtually parallel sided keyhole is established. This may be because it is then possible to traverse the workpiece fast enough to prevent unstable sidewall profiles from forming and trapping gas in the body of the weld.

Welds in Ducol W30 ferritic steel

The welding procedure outlined in the previous four paragraphs was adopted. Transverse sections obtained at 2·5 kW and 5 kW (Fig. 5) again demonstrated that the higher the power level, the more symmetrical and parallel sided the welds become. Radiographs showed that porosity occurred more often in Ducol W30 than in the stainless steels, although at 5 kW, porosity-free welds could be made. In welds where the backface width was as wide as the maximum width of the transverse section (excluding the nail head) porosity was not detected, although the parameter bounds for this were not fully investigated.

Penetration depth as a function of beam power

The bead-on-plate welding runs used the keyhole mode in which some of the beam power was not absorbed by the workpiece but lost through the rear of the keyhole. To assess the maximum depth of penetration achievable for a given beam power, bead-on-plate runs were made in thick section material in which penetration was not achieved. The two welding speeds were 16 mm s^{-1} and 5 mm s^{-1}. When welding at 16 mm s^{-1}, only a small nail head was produced and the beam power was most efficiently used to obtain penetration. At 5 mm s^{-1} a much larger nail head was produced, but the speed was slow enough to achieve virtually the maximum penetration possible with a set beam power.

Results of plotting penetration depth as a function of beam power and welding speed for both materials are shown in Fig. 6. Linear regression has been applied to the results at 16 mm s^{-1} welding speed, between the 2·5 and 7 kW power levels, regressing penetration depth on power input. The high correlation coefficients obtained (>0·99 in both cases) indicated that accurate control of depth of penetration by adjusting beam power was possible. Below 2·5 kW, beam power welding took place without a keyhole being formed. Some welding runs exhibited a dual stability. In part of the weld, a small vapour plume would be seen with little surface activity while in other parts of the weld a much larger vapour plume appeared with considerable surface activity. A small vapour plume correlated with a smooth weld surface; a large vapour plume correlated with a ruffled surface,

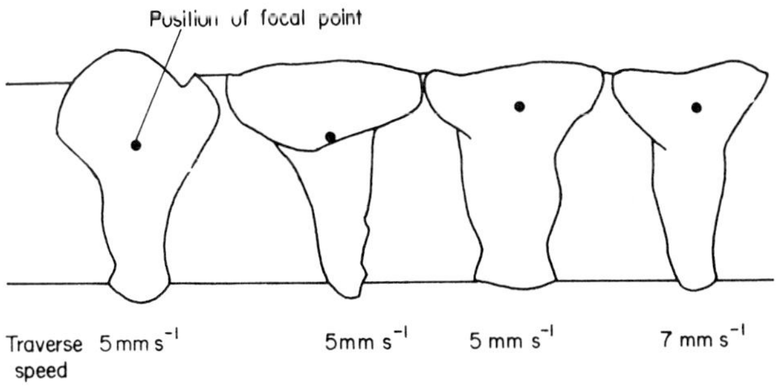

Fig. 4. Transverse profiles in 316 stainless steel at low levels of laser power (Beam power 2·5 kW).

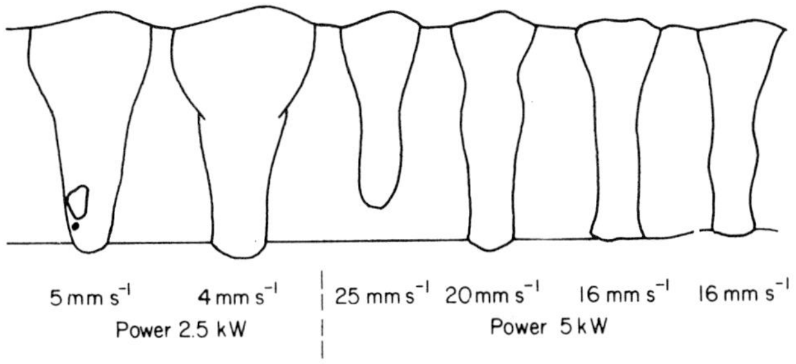

Fig. 5. Transverse profiles in Ducol W30 steel).

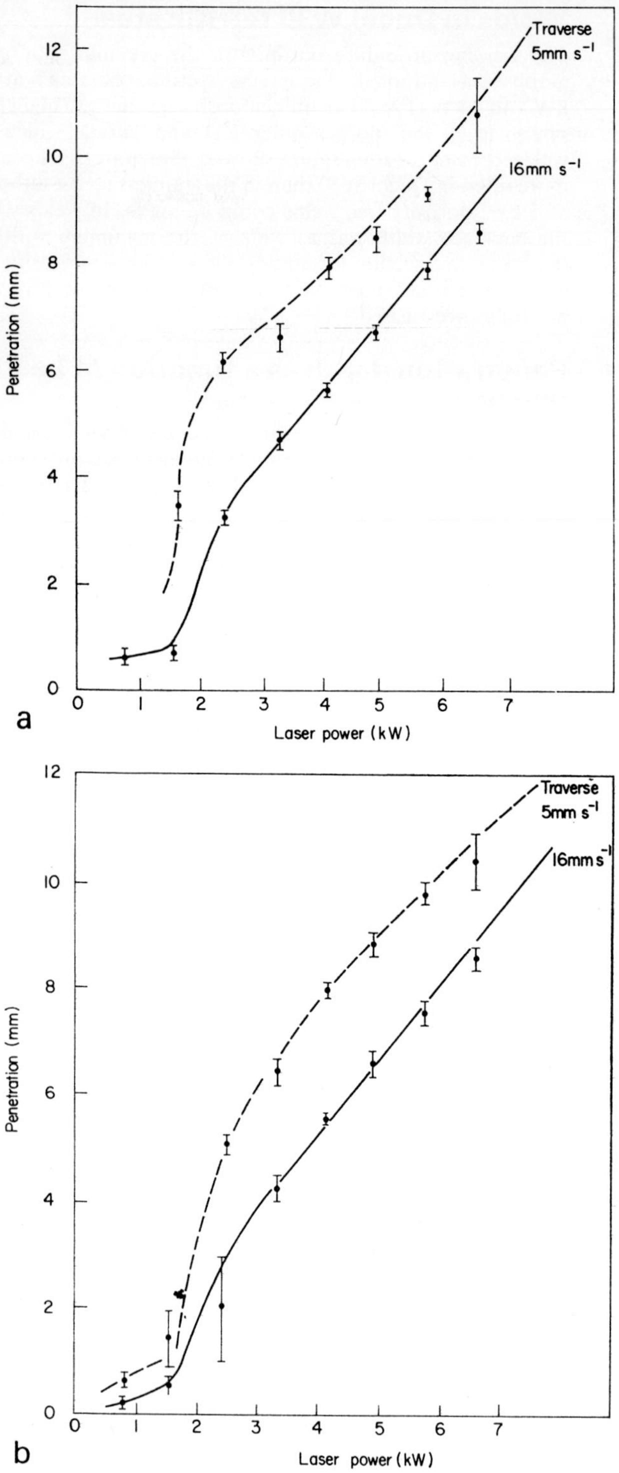

Fig. 6. Penetration depth of laser welds as a function of beam power: (a) type 316 stainless steel, (b) Ducol W30 ferritic steel.

ruffling presumably indicating the presence of the keyhole mode.

Penetration depth as a function of beam power was also plotted in normalized form for comparison with theoretical limits. In Fig. 7 present results are compared with the limits given by Swift-Hook and Gick (1973) and Breinan and Banas (1975). The discrepancy between the predictions is due to the fact that Breinan and Banas allowed for latent heat whereas Swift-Hook and Gick did not. Breinan and Banas (1975) measured the melting ratio (ie, the energy required to raise material to liquid at the melting temperature divided by the energy absorbed by the workpiece) for various metals using calorimetry and obtained values as high as 72% for zirconium, titanium and type 304 stainless steel. Adams (1973) has shown that typically 70% of beam power is absorbed by the workpiece, the rest being reflected. Breinan and Banas (1975) also noted that between 54% and 91% of incident beam energy was being absorbed by the workpiece. The results of Breinan and Banas are recorded such that W, beam power, in the abscissa of Fig. 7 is beam power absorbed. The results of the present experiment have also been adjusted to allow for reflected beam power by taking the value W as 80% of beam power delivered to the workpiece.

The main criteria for assessing the welds in the present experiments related to metallurgical soundness and appearance, but not necessarily an optimum use of beam power. Hence, the results were regarded as concomitant with a satisfactory but non-optimum use of laser power.

Welding procedure

To assess the properties of laser welds in type 316 stainless steel and Ducol W30, downhand butt welds (plate thickness 6 mm) were made using the input parameters which generated acceptable weld profiles for bead-on-plate welding. After the plates were aligned, they were tacked together using laser spot welds at both ends of the plates prior to making the welding traverse. Main welding parameters were

- f/5 focusing system, focused 1 mm below the upper workpiece surface.
- Downhand welding.

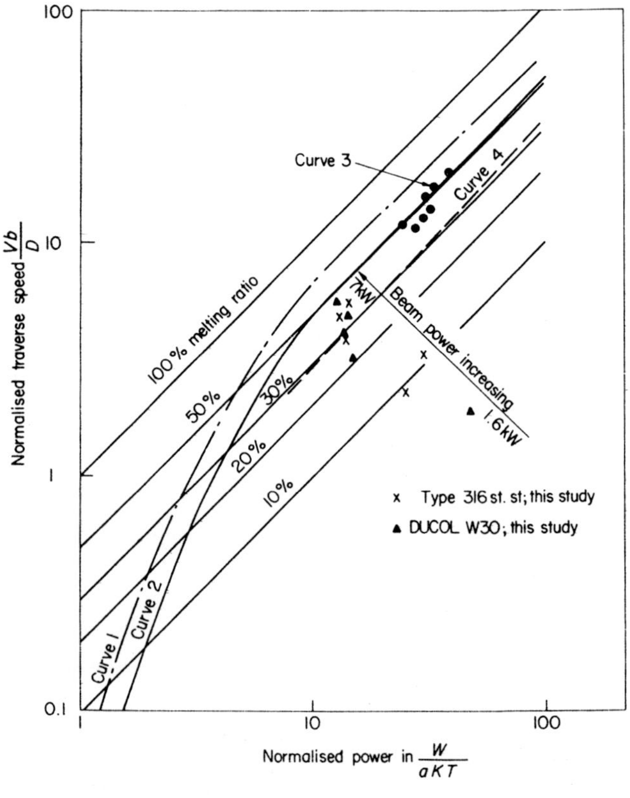

Fig. 7. Laser welding penetration as a function of beam power using normalised axes (V traverse speed, b width of weld, D diffusivity, W beam power, a weld depth, K thermal conductivity, T melting temperature). Curve 1—Breinan and Banas (1975) theoretical limit. Curve 2—Swift-Hook and Gick (1973) theoretical limit. Curve 3—Breinan and Banas (1975) results with 304 stainless steel. Locke and Hella (1974) results.

- Traverse speed: 16 mm s^{-1}.
- Power in beam: 5 kW.
- Length of weld: 120 mm joining two plates $150 \times 50 \times 6$ mm.
- Vapour plume control gas jet: helium at 20 psi.
- Front and backface gas shield helium.

Welds were examined to see that surface finish and profile matched the bead-on-plate welds made previously. No significant differences could be detected. Several more welds were made to check reproducibility of finish and profile both along the length of a weld and from one weld to the next. Reproducibility was good. Transverse sections of the welds were found to be consistent with the previous results for bead-on-plate welds.

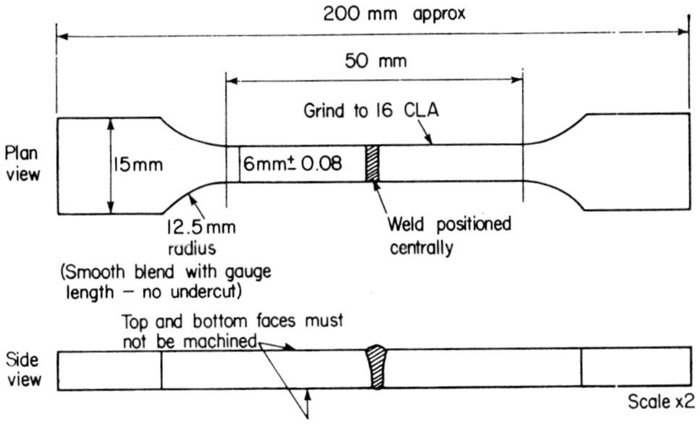

Fig. 8. Tensile test pieces specification.

Radiographic inspection

No signs of porosity nor of any other voids or inclusions were evident in radiographed specimen welds. The limits of radiography were such that voids greater than 5% of plate thickness (0·3 mm) and greater than 0·2 mm in diameter were detectable.

Tensile tests

Tensile tests aimed at determining the mechanical properties of the laser butt welded plates were significantly inferior to those of the parent plate. Flat tensile specimens were manufactured with the plane of the weld running transversely across the gauge length at its centre (Fig. 8). Room temperature uniaxial tensile tests were performend at a crosshead velocity of 1·7 mm s^{-1} corresponding to a strain rate of $2{\cdot}8 \times 10^{-2} s^{-1}$. Failure in each case occurred within the gauge length and away from the shoulders. Specimens tested were: (1) Parent plate (called Parent Plate.), (2) Welds in which the profile and penetration were considered to be satisfactory (called Optimum) (3) Welds not fully penetrated or where the top surface showed undercut or marked rippling (called Non Optimum). These were included to check how sensitive the mechanical properties were to weld quality.

Table 2. Tensile test results.

Material	Welding Conditions	Welding speed (mm s^{-1})	Stress at ultimate tensile strength (MPa)		Strain (%)	
		Power=5kW	Failure in parent (P) or weld (W)	Mean		Mean
316 stainless steel	Optimum	16	P 599		50·0	
		16	P 598	598	50·0	50·0
	Non-optimum	25	W 634		40·4	
		25	W 554	594	51·0	45·7
		16	W 715		57·6	
		16	W 557	636	56·0	56·8
	Parent plate	—	P 592		54·0	
		—	P 590	591	54·0	54·0
310 stainless steel	Non-optimum (but fully penetrated)	25	W 410		16·8	
		25	W 446	428	22·6	19·7
		16	W 506		31·0	
		16	W 510	508	26·4	28·7
Ducol W30 steel	optimum	16	P 766		16·0	
		16	P 769	767	15·0	15·5
	Non-optimum	25	W 666		13·2	
		25	W 754	710	8·0	10·6
		16	P 774		22·8	
		16	P 740	757	21·0	21·9
	Parent plate	—	P 769		18·0	
		—	P 780	775	17·0	17·5
BS 1501 Standard data						
316 (hardness)	Parent plate (143-212)	—	—	at least 520	—	at least 40
310 (hardness)	Parent plate (149-217)	—	—	at least 540	—	at least 40
Ducol W30 (hardness)	Parent plate (not quoted)	—	—	at least 590-700	—	at least 16

Results for type 310 stainless steel are also included in Table 2.

The following observations were made:

●In type 316 stainless steel the elongation and UTS (ultimate tensile strength) values are in excess of the minimum quoted in BS 1501. As the quality of the weld improved, the position of the fracture was displaced to the parent plate. In welded test pieces where failure occurred in the parent plate, necking was also apparent in the weld metal.

●In type 310 steel, both the UTS and elongation values were unacceptably low, failure occurring in each case in the weld. The poor mechanical properties, especially the low elongation was attributed to the presence of solidification cracks in the weld metal, which were evident on the fracture faces.

●The UTS of the Ducol W30 parent plate was slightly in excess of the BS 1051 standard, although close agreement between results from welded test pieces and parent plate was obtained. Elongation exceeded the necessary minimum of the BS 1051 standard except when failure occurred in the weld metal, where, it was believed, localisation of strain had occurred. In contrast to results from the type 316 steel specimens, when tensile failure occurred in the weld metal, no deformation of the weld was evident indicating that brittle fracture due to high hardness had been the likely cause of failure.

Hardness tests

Welds typical of the Optimum tensile specimens were sectioned transversely and subjected to hardness traverses using a microhardness tester (0·2 kg load). Traverses were made from parent material, across the weld and heat affected zones into the parent material on the opposite side of the weld. Indentations were spaced at 0·25 mm intervals. The results are shown in Figs. 9 and 10.

In type 316, the hardness of the weld and heat affected zone closely matched that of the parent plate. In Ducol W30, both the weld and heat affected zone were excessively hard. Hardness was attributed to the high cooling rate depressing the transformation temperature thereby producing hard transformation products. The excessive hardness may be obviated in practice by the use of preheat which reduces cooling rates. The appropriate preheat temperature depends on the plate thickness and joint geometry.

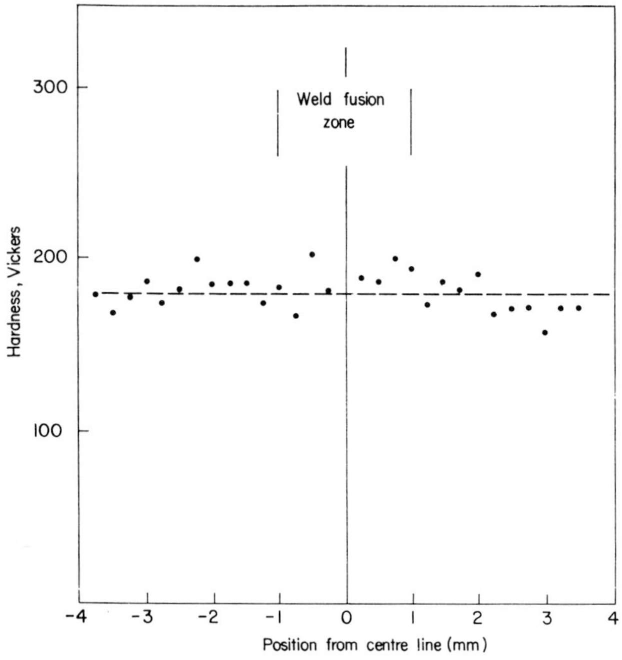

Fig. 9. Hardness test on type 316 stainless steel laser weld (Plate thickness 6 mm, traverse 16 mm s^{-1}, beam power 5 kW).

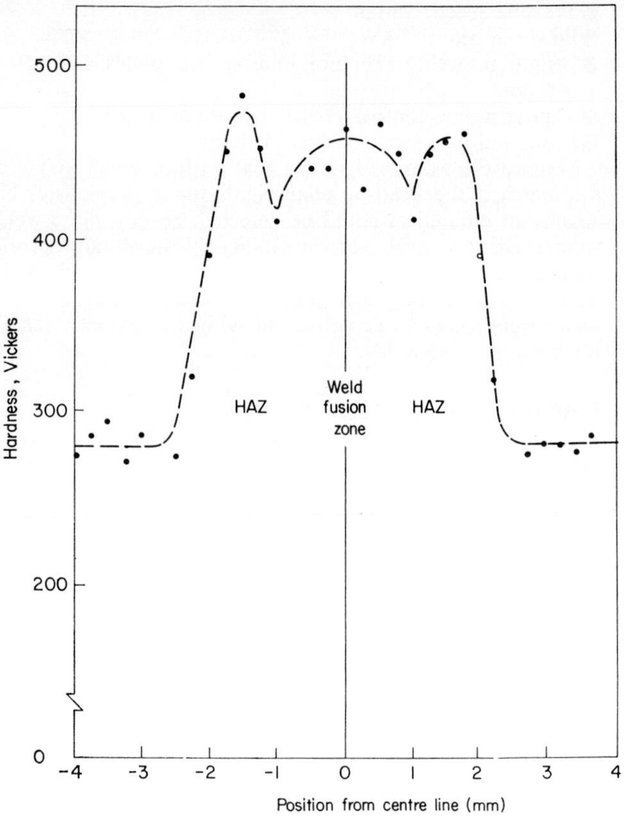

Fig. 10. Hardness test on Ducol W 30 laser weld with no preheat (Plate thickness 6 mm, traverse 16 mm s^{-1}, beam power 5 kW).

Comparison of welding processes

Criteria for comparing laser welds with other welding processes are listed in Table 3. Parameters were chosen as typical of those needed to weld 6 mm thick plate. Extremes of speed and power have been avoided. The parameters for laser, plasma and electron beam could also apply to thicker section plate but those for TIG are set at virtually the practical limit of penetration. Estimates could be in error due to unpublished developments or choice of non optimum conditions but the magnitudes of the figures will not be changed significantly. Transverse weld profiles typical of the processes mentioned, are shown in Fig. 11. The following points deserve consideration:

●The main advantages of laser and EB welds over TIG and plasma are similar (ie, narrow, deep penetration, near-parallel sided welds with small heat affected zones.)

●TIG and plasma welding have relatively low capital investment. TIG, used extensively for many years, offers a large background of experience.

●Laser welding and EB welding have much greater potential for high productivity rates as long as workpiece handling procedures can be significantly improved. EB suffers from the need to work in a vacuum (often made local) but can be used in air giving poorer penetration properties than a laser at the same beam power.

●Distortion from laser and EB processes, which give relatively narrow weld widths and heat affected zones, is minimal.

●EB welding has been used in industry for about 17 years during which time the number of machines and diversity of application has steadily grown. To date, about 400 operational EB welding plants have been installed in the UK. Laser welding is approximately 10 years younger. Similar criteria to those used on EB welding can be applied when assessing laser welding (ie, justification of a high capital investment to attain high productivity, consistent weld profile and penetration, etc).

Application areas

Metallurgical assessment of the laser welds has shown them to be of high quality and reproducibility. Such properties are much in demand for the production of power plant since some welds are, or become, inaccessible or are very costly to repair both in time and money. Against these advantages are the high capital cost of the laser plus its attendant workpiece handling gear. However, applications for laser welding could possibly be found in:

(1) Tube-to-tube welding (up to 20 mm wall thickness with 20 kW beam) using a fixed head rotated tube configuration. It would be possible to fabricate long lengths of the tube of varying diameter which could then be fashioned so that only tube-to-tube plate welds would be necessary to complete internal boiler fabrication.

(2) Platen welding, where long runs of downhand welding in 8 mm wall thickness are required. At present autogenous TIG with a pure argon shield involves high heat losses to the closely butted tubes and a weldpool penetration of only 75% Laser welding would be faster by an order of magnitude and give full penetration.

(3) Deep penetration welding. At present, multipass techniques using TIG, MIG, MMA and SA are used for workpieces thicker than 5 mm. Narrow gap TIG welding reduces the number of passes considerably but still has a

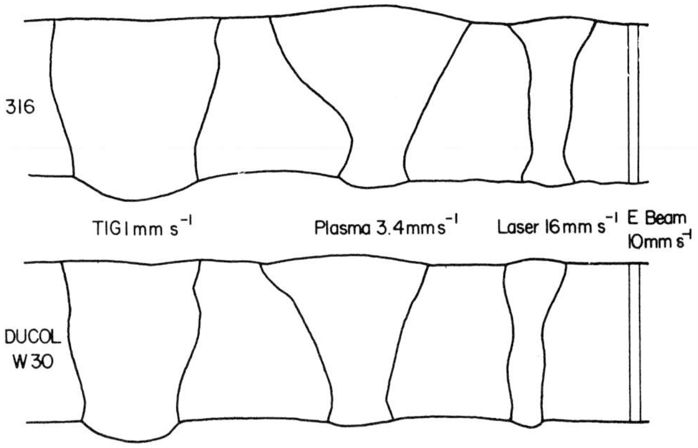

Fig. 11. Comparison of transverse profiles produced by welding processes.

Table 3. Comparison of processes (plate thickness 6 mm).

Process		Laser	TIG	Plasma	Electron beam
Power absorbed by workpiece Total power used		4 kW 50 kW	2 kW 3 kW	4 kW 6 kW	5 kW 6 kW
Typical traverse speed		16 mm s^{-1}	2 mm s^{-1}	6·7 mm s^{-1}	40 mm s^{-1}
Alignment accuracy required		±0·5 mm	±1·0 mm	±1·0 mm	±0·3 mm
Energy input per unit length (absorbed by workpiece)		250 J mm^{-1}	1000 J mm^{-1}	600 J mm^{-1}	125 J mm^{-1}
Capital cost and availability		~£120k for 5 kW output Available on a year's delivery in U. K. and U.S.A.	~£8k per welding set Immediate delivery	~£10k per welding set Immediate delivery	~£50k for 6 kW machine On short delivery
Possibility of all positional welding		YES but requires optics to manipulate the beam. Optimum when moving workpiece	Serious penetration characteristic changes with attitude	Serious penetration characteristic changes with attitude	YES but requires mechanics to move gun. Optimum when moving workpiece.
Distortion	Axial shrinkage	Minimal (small HAZ)	Significant ~1 mm on 5 mm plate	Significant ~1 mm on 5 mm plate	Minimal (small HAZ)
	Angular	Minimal parallel sidedness	Significant V shaped weld	Significant V shaped weld	Minimal parallel sidedness
Surface profile defects		Very fine flow lines	Underside protrusion held in by surface tension	Underside protrusion held in by surface tension	Produces ruffled swarf on backface.
Special requirements for process operation		Safety interlock to guard against misplaced beam reflections	Normal light screening	Normal light screening	Vacuum chamber Local vacuum X-RAY SCREENING
Process END effects	Start	Slight surface protrusion	Smooth	Slight surface protrusion	Slight surface protrusion
	Finish	Smooth		Slight surface protrusion	Slight surface protrusion

slow traverse speed (3 mm s^{-1}). Laser welding has the potential for welding workpieces up to perhaps 30 mm thick in a single pass using 20 kW beam power at speeds in excess of 10 mm s^{-1}. Research is proceeding on increasing the penetration limits of laser welding at a given power but information is invariably proprietary.

(4) Difficult or dangerous access welding. It may be possible to mechanically/optically translate beam power over distances up to 50 m, for 'in reactor' repair welding and similar environments.

Conclusions

- Laser welds in 6 mm thick type 316 stainless steel and Ducol W/30 carbon steel were acceptable in terms of surface finish, weld pool profile, absence of defects and tensile strength.
- Laser welds in 6 mm thick type 310 stainless steel suffered from solidification cracking. This problem can also occur with other autogenous welding processes.

Acknowledgements

The work was carried out at Culham laboratory and Marchwood Engineering Laboratories and is published by permission of the Central Electricity Generating Board and the United Kingdom Atomic Energy Authority.

Thanks are due to D. F. Jephcott and A. S. Kaye for active interest in the project and to E. Hanley and F. Fabling for technical assistance with the welding and metallurgy.

References

Andrews, J. G., Atthey, D. R. (1976). CEGB Laboratory Note No. R/M/N852.

Adams, M. J. (1973). WI Report No. 3335/3/73.

Breinan, E. M., Banas, C. M. & Greenfield, M. A. (1975). United Tech. Res. Centre Report R75-111087-3.

Breinan, E. M., Banas, C. M. (1975). United Tech. Res. Centre Report R76-911989-4.

Crafer, R. C., Pauley, J. T., Bishop, D. I., Hyland, P. J. (1976). WI Report No. 29/1976/P.

Duley, W. W. (1976). CO_2 Lasers, Academic Press, London.

Grey Morgan, C. (1975). Rep. Prog. Phys. *38*, 621-665.

Locke, E. V., Hoag, E. D., Hella, R. A. (1972). IEEE JQE, *8*, 132.

Locke, E. V., Hella, R. A. (1974). IEEE JQE *10*, 179.

Swift-Hook, D. T., Gick, A. E. F. (1973). Weld. J. *52*, 492s-499s.

Rein, R. M. et al. (1972). Patent 1448740. The Patent Office.

Stoop, J., Rep. NRL Prog. Feb. 1976 4-7.

LASER PROCESSING IN THE AUTOMOTIVE INDUSTRY

By

Michael Yessick
Research Scientist
Ford Motor Company

and

Duane J. Schmatz
Principal Staff Engineer
Ford Motor Company

INTRODUCTION

The types of material processing of interest to the automotive industry can be considered in three categories: cutting and drilling, hardening, and welding. The physical mechanisms for these processes are basically thermal in origin and the laser, for our purposes, can be considered simply as a heat source. The main distinction between the laser and most other heat sources lies in its ability to obtain very high power densities of the order of 10^7 w/in^2 ($\sim 10^6$ w/cm^2) or greater. Only the electron beam devices can rival the laser in obtaining high power densities. The advantage that the laser has over all other heat sources is the ease with which the beam can be controlled, both in size and in position. The laser beam is focused using a lens or a focusing mirror and the focused spot size can be readily controlled by the choice of lens and the distance from the lens to the work piece. The spot sizes can be made as small as a few thousandths of an inch ($\sim$ 0.1 mm). The beam direction can be easily controlled by the use of small, light-weight mirrors and thus the processing can be easily automated with the beam being made to move over programmed paths of any complexity. An advantage that the laser has over an electron beam is that the laser beam is not attenuated in air and can thus be sent over very long distances from the laser to the work area. This not only allows more complex paths to be followed, but also allows for the possibility of time sharing a central laser system with several work stations. Finally, the laser produces no penetrating radiation, such as x-rays, and thus protective shielding is very simple and, in fact, is usually made of plexiglass.

In order to operate successfully as an industrial processing tool, a laser must not only produce a high power density, but must also have a relatively high average power capability. The continuous carbon dioxide lasers can be operated at much higher average powers than all other laser systems and are generally simpler and less expensive. Therefore the continuous CO_2 laser will provide the basis for all of the work described in this paper. The power levels of interest to us range from about 100w up to 20kw and all of the laser systems used for this work are commercially built lasers designed for industrial application. All of these lasers operate on the same basic principles. In particular, a mixture of CO_2, N_2 and He gases flow through an optical resonant cavity and excitation is provided by an electrical discharge. The output beam has a wavelength of 10.6μm which is in the infrared and, of course, not visible.

Each of the three major areas of application will now be described in some detail and some specific examples of applications to automotive production will be discussed.

Source: Technical Paper MR74-962, SME, 1974

CUTTING AND DRILLING

Laser cutting is effective on a wide variety of materials including metals, ceramics, plastics, vinyl, rubber and composite materials. Laser cutting, of course, involves no physical contact with the part and allows greatly simplified fixturing. The cutting rates are generally quite fast and there is no limitation on the configuration of the cut. For example, inside cuts, of any shape, can be readily made. Because the beam can be focused down to a very small spot, the kerf widths can be made very narrow, as small as a few thousandths of an inch (~ 0.1 mm) in most cases. Also, the cut-edge surfaces are generally very smooth and clean due to the highly localized nature of the heating.

The cutting mechanism is somewhat different for the different materials. For the non-metals, the beam is very highly absorbed and the high power density results in an almost explosive vaporization at the region of highest intensity. This causes the surrounding molten material to be blown out, which greatly enhances the cutting rates over that which would be obtained if vaporization were the only removal mechanism. However, since this only occurs in the path of the beam, the cut-edge surfaces remain quite smooth. In general, an inert-gas jet is used to prevent oxidation and to help blow away debris. As an example, a glass-filled polyester, 0.130 in. thick (3.3 mm), which is being considered for use as a body-panel material, can be cut at 60 in/min (25 mm/sec) using a 400w laser. This material is very easily cracked during conventional cutting or drilling, but high-quality cuts result using the laser. As another example, a very hard ceramic material, silicon-nitride, which is being considered for use in an automotive turbine engine can be laser cut at a rate ten times faster than with a diamond wheel. And, of course, cuts of any shape can be made with the laser, whereas only straight lines can be made with the diamond wheel.

The cutting of steels and other reactive materials is accomplished by using an oxygen gas jet coaxial with the beam. The laser beam heats the steel sufficiently to initiate and sustain the highly exothermic oxidation reaction. This allows these materials to be cut at very fast rates and, in fact, the cutting of these materials requires substantially less laser power than does welding. The volatile oxide products are rapidly swept away by the gas jet and the cut surfaces are generally quite clean. Narrow kerf widths and small heat-affected zones are also characteristic of these cuts. Some particularly important advantages of laser cutting of steel are that the cutting rates are essentially independent of the hardness of the steel, and, of course, there is no tool wear. Typical cutting rates in steel using a 400w laser are 240 in/min (100 mm/sec) at 0.030 in. thickness (0.76 mm) and 60 in/min (25 mm/sec) for 0.075 in. thickness (1.9 mm). Increasing the laser power to 1000w results in a nearly linear increase in cutting rates.

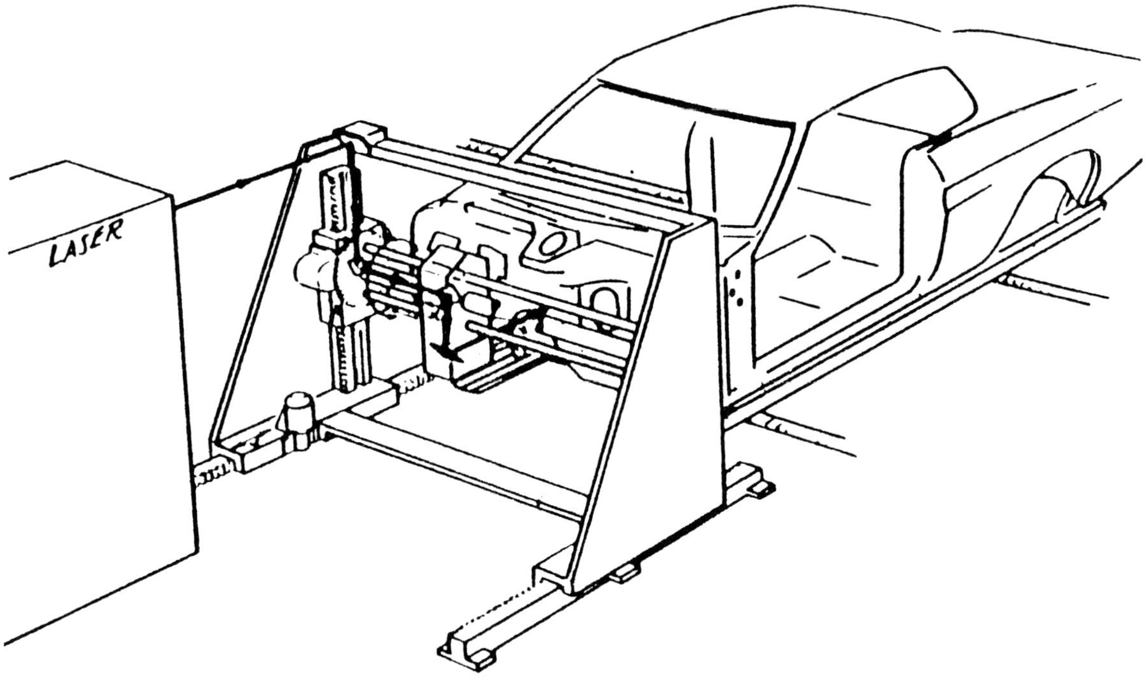

Fig. 1. Automatic Laser Hole Cutting System. The laser beam can be translated along the X, Y and Z directions and also rotated about the X-axis.

An example of a potential production application of laser cutting of steel is the hole-cutting system shown schematically in Fig. 1. This system is designed for the automatic cutting of holes of varying sizes and at various locations in the dash panel of a body shell as required by the several regular production options of any particular unit. Information concerning the size and position of each hole is read from punch cards by a computer control system which then automatically 1) moves the body into the work station, 2) registers the dash panel to the laser cutter, 3) cuts the required hole pattern and 4) moves the body out of the work station. The motion of the beam is controlled by four moving mirrors as shown in the diagram. The laser which would be used is a 400w laser. This gives sufficient cutting speeds so that the maximum number of holes can be cut well within the time required for this job. The system has the advantages of much greater precision in the location and size of the option holes and also allows for the cutting of holes of other-than-circular shapes.

The drilling of holes as opposed to the cutting of holes is accomplished with no beam motion and the mechanism of hole formation is essentially the same as for cutting. For steel an oxygen jet is used and again most of the material is removed as volatile oxides, while most of the remaining material is blown out of the hole leaving clean, smooth sides. However, the sides of the hole are not straight and tend to taper in from the top with a small flaring-out right at the bottom. The size of the holes can be readily varied over a wide range by control of the beam spot size and the length of time the beam is on. The tolerances that can be achieved depend upon the diameter and the depth of the hole and, of course, the amount of laser power available.

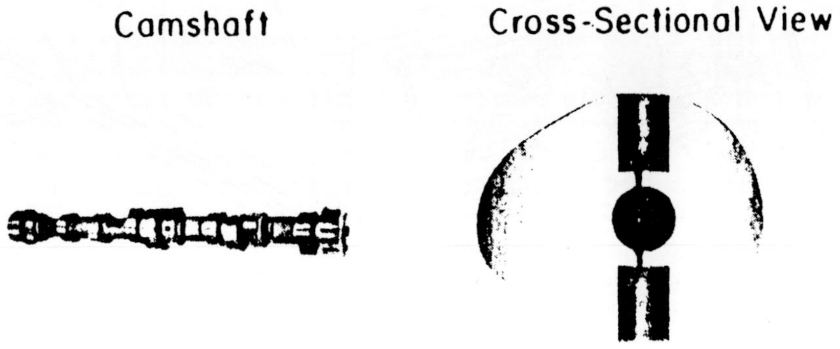

APPLICATION

Drilling the oil metering holes in the lobes for the four-cylinder, rifle-bored camshaft.

PARAMETERS

- Power-375 watts
- O_2 assist gas-80 psig
- Drilling time-0.10 sec.
- Standoff distance-0.2in. (5mm)

REMARKS

- Hole size-.030 in diameter (0.76 mm)
- Hole size variation-±.005 in. (±0.13mm)

Fig. 2. Laser Machining Application.

An example of a potential laser hole-drilling application is shown in Fig. 2. This application involves the drilling of an oil-metering hole in a rifle-bore camshaft. The problem, in essence, is the drilling of a small hole at the bottom of a big hole. The camshaft is made of a hardenable iron and that makes conventional drilling or punching very difficult. The cross-sectional view shows two holes drilled, one from each side, with a drilling time of 0.1 sec for each hole using 375w of laser power. The hole diameter is 0.030 in. (0.76 mm) and the hole depth is 0.065 in. (1.65 mm). The tolerance in diameter in this case is ± .005 in. (± 0.12 mm). The use of a higher power laser could give a higher tolerance on hole diameter if that were desired.

SURFACE HARDENING

Laser hardening is the simplest application in terms of the physical process involved since it only requires heating of the surface or, in other words, some absorption of the beam. Of course, we are concerned here only with metals and in particular with various irons and steels. The absorption of the beam in these metals depends very strongly on the surface finish and can be enhanced either by roughening the surface or by coating it with a thin layer of a highly absorbing material. In general, hardening will be accomplished by a solid-state transformation, i.e., with little or no melting involved, and careful control of the surface temperature is required. Melting would also produce a hardened area, but would be less desirable from the standpoint of potential brittleness and the effect on surface roughness. Heat transfer is determined by the conductivity of the material, and the depth and width of the hardened zone for a particular material is determined mainly by the focusing and beam shape, by the rate at which the beam moves over the surfaces, and by the microstructure of the material.

There are two major advantages of laser hardening. The first is the ability to localize the hardening at the exact spot where it is needed. This minimizes the total heat input to the material and thus eliminates the distortions which often accompany other hardening techniques. Second is the ability to apply the heat to otherwise inaccessible areas, for example, small bores or complicated shapes not amenable to induction hardening. Both of these advantages stem from the small size and ease of manipulation of the laser beam.

A particular application of laser hardening which we have been investigating is the hardening of the inner bore of the valve guide and this is shown in Fig. 3. Wear in the valve guide causes increased oil consumption and increased exhaust emissions, and is particularly severe with the use of lead-free gas. In this case, hardened rings were made in the valve guide by first positioning in the guide a copper rod with a 45° polished surface which acts as a mirror. The beam is then directed into the bore and is reflected by the copper mirror onto the surface of the guide. Rotating the copper rod then moves the beam in a ring around the inside of the valve guide. A 400w beam was spread out to a width of 0.060 in. (1.5 mm) and the speed was set to give the desired depth of hardening with no surface melting. In this case the hardened depth is 0.010 in. (0.25 mm) and the linear speed of the beam over the surface is 10 in/min (4.2 mm/sec). The micrograph in Fig. 3 shows the hardened area (martensitic) which has formed and also shows that there is no disturbance to the smoothness of the surface. Two such bands were made at the positions where the maximum stress between the valve and the guide exists. This minimal heat input to the part produces no measurable distortion, as expected.

Source: Technical Paper MR74-962, SME, 1974

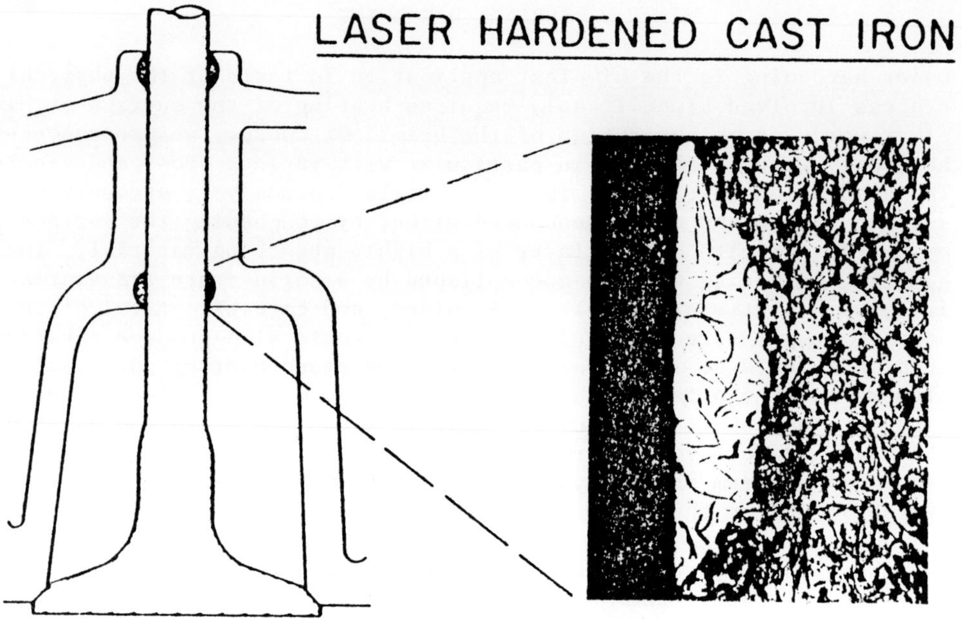

- Internal surface of valve guide can be laser hardened to minimize guide wear.
- Typical Parameters: 400 watts power, .060 in. beam spot size, 10 in./min. (1.5 mm beam spot size, 4.2 mm/sec.)

Fig. 3. Laser Hardening Application.

WELDING

The third major category of laser processing is welding and we are concerned here only with the welding of metals. In the simple case, enough power is applied to the surface so that melting occurs. The distribution of heat would then be by conduction through the material and the molten zone would look somewhat radial as in conventional arc welding. However, when the power density gets above a certain threshold level an interesting phenomenon occurs known as a keyhole or deep-penetration effect. The material at the most intense part of the beam vaporizes and the adjacent regions melt. The high vapor pressures produced then push the molten metal to the side, out of the beam path, and the beam proceeds to essentially create a deep hole. The vaporization is not intense enough to blow the molten metal out of the hole and an inert gas cover prevents any oxidation reaction. Thus a balance is reached between the pressure created in the hole and the head of molten metal being supported and this determines the depth of penetration. As the beam or sample moves, the molten metal flows in behind the hole effecting the weld. Thus quite deep and narrow welds can be produced. Welds of this deep-penetration type have been produced only by electron beams and lasers. No other method produces sufficiently high power densities. The region of 500 to 1000 watts is the threshold region for laser power to achieve deep penetration and most of our welding has been done in the 1-10 kw region.

In addition to producing deep welds, this effect also solves the problem of the reflection of the beam from shiny metal surfaces. Once the deep hole is created, it acts nearly like a black body and essentially all of the beam is absorbed as this hole moves through the metal. However, a problem is created by the evolution of metal vapor which stimulates gas breakdown and plasma generation in the region just above the metal surface. This plasma plume is very highly absorbing of the laser beam and essentially blocks the beam from the metal. In order to dissipate this plasma, an inert gas jet is directed along the surface of the metal. Properly designed gas jets have proven very effective in removing the plume and, at the same time, in protecting the surface from oxidation.

Of course, laser welding has the same advantages as the other laser processes: no contact with the work required, small heat-affected zones, easy manipulation of the beam and high processing speeds. Using very high-power lasers with 10 kw output, penetrations in steel of up to 0.75 in. (19.0 mm) have been achieved at welding speeds of 30 in/min (13 mm/sec), while at a rate of 600 in/min (254 mm/sec) penetrations up to 0.080 in. (2.0 mm) have been achieved. These penetrations drop fairly rapidly with power, but even at 1000 watts deep penetration effects still occur and penetrations of 0.060 in. (1.5 mm) have been achieved at a weld speed of 60 in/min (25 mm/sec).

Source: Technical Paper MR74-962, SME, 1974

However, achieving good welds not only depends upon obtaining sufficient penetration, but also depends upon the effect of the laser welding on the metallurgical structure of the material being welded. In general, the heating and cooling rates during laser welding are much higher than during conventional welding and the heat-affected zones are much smaller. In the welding of carbon steels, for example, as the carbon content increases, the rapid cooling rates tend to result in cracking and increased notch sensitivity. Only the addition of a low-carbon filler metal allows good welds to be made, and generally laser welds are restricted to low-carbon steels. However, for steels which depend upon the addition of other alloying elements for strength, such as the high-strength low-alloy steels, the fast cooling rate can be advantageous. By judicious selection of welding parameters, laser welds can be made in HSLA steels with essentially equivalent mechanical properties to the base metal. This is in contrast to the case of arc welding of HSLA steels which usually results in a decrease in strength due to the greater total heat input and the slower cooling rates.

In the laser welding of aluminum, the localized nature of the heating helps keep the surfaces clean and minimizes the build-up of an oxide layer which would otherwise prevent wetting. Therefore, this material, which is generally difficult to weld, can be, with care, successfully laser welded.

An example of a potential laser welding application using a 1000w laser is shown in Fig. 4. The heater motor yoke assembly is comprised of a stack of laminations joined together. The height of the stack is 1.5 in. (38 mm) and the thickness of each lamination sheet is 0.050 in. (1.3 mm) The stack shown in the photograph has been laser welded at a weld speed of 70 in/min (300 mm/sec) so that two weld passes could be made in a total time of 3 sec, well within the time required for this application. A close-up of the weld bead shows a very smooth bead with no apparent porosity. A longitudinal section and a transverse section through the weld are also shown. The depth of the weld is about 0.050 in. (1.3 mm) which is the same as the thickness of the lamination, and there is very little or no evidence of porosity in the weld bead. Also there are no cracks in the weld bead. This configuration could be welded either by moving the beam or by moving the part. The laminations do not have to be cleaned as a small amount of oil or other contaminant does not affect the weld. Also, of course, no filler metal is required.

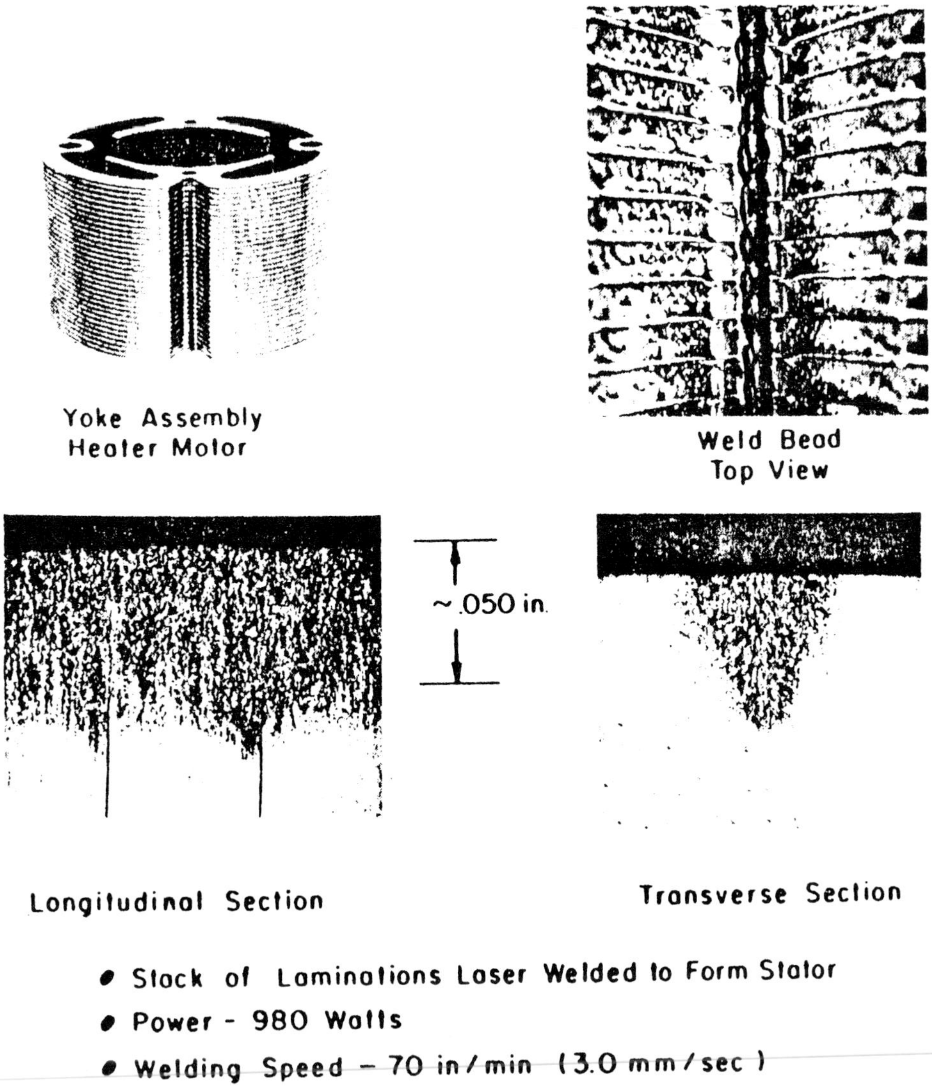

Fig. 4. Medium-Power Laser Welding Application.

An example of a higher power welding application is the 6kw underbody welding system recently built for Ford by Hamilton Standard. In this system, shown schematically in Fig. 5, the several stamped panels which are to be welded together to form the underbody are loaded in a clamping assembly fixture which is part of a transfer table. The transfer table provides one axis of weld motion, here called the x-axis. The laser itself, which has an output power of 6kw, is mounted on a platform and the beam is directed through a beam transport system which provides four axes of motion. Two of these axes are translations, designated as y and z-motions, and two are rotations, one in the horizontal plane, and one

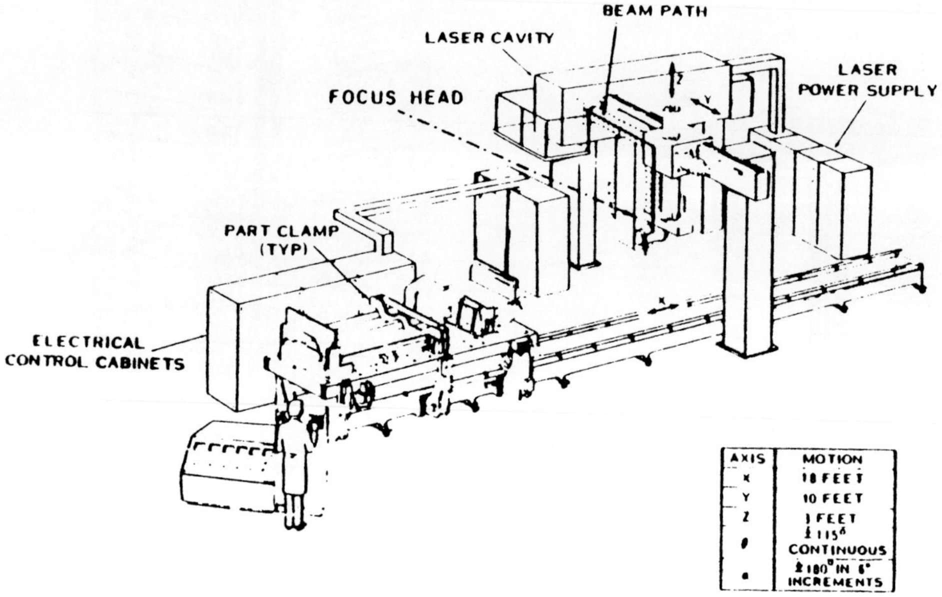

Fig. 5. 6kw Laser Underbody Welding System. (Reprinted with permission of Hamilton Standard Division, United Aircraft Corporation.)

in the vertical plane. All of the beam motions, performed by moving mirrors, as well as the transfer table motion, are directed by a computer control system which will perform welds along pre-programmed paths at selected weld speeds. The clamping and unclamping systems are also computer controlled. The beam is focused, by a suitably curved mirror in the focus head, onto the parts to be welded. Most of the welding is the burn-through type weld joining two sheets of 0.035 in. thickness each (0.9 mm). This welding is performed at 400-450 in/min (170-190 mm/sec). This laser weld is continuous which results in high structural integrity and eliminates the need for a sealing operation. The programmability of the system gives the advantage that underbodies for any car line or model can be welded simply by calling up the right program from the computer memory. This system is presently in a pilot run producing underbodies for the 1975 Torino/Montego cars.

CONCLUSION

We have seen that the laser is a very versatile tool and can be used for a large number of different types of applications. Laser processing has many advantages over conventional processing, mostly associated with the high power densities and the high degree of control of the laser beam. It is expected that these advantages should result in a significant role for laser processing in automotive production in the next few years.

ACKNOWLEDGEMENT

The development of the Automatic Laser Hole Cutting System and the Underbody Welding System is largely due to the efforts of E. E. Lohmolder of the Ford Motor Company Automotive Assembly Division. The idea for the laser welding of the heater motor yoke assembly is due to L. Mihalyfy of the General Products Division. The contributions of these and others within the Ford Motor Company are gratefully acknowledged.

MICRO-CIRCUIT FLATPACK SEALING BY LASER WELDING

H. N. KRISHNASWAMY and V. E. BOCCELLI

ABSTRACT

The development of a laser welding process for the hermetic sealing of microcircuit flatpack packages is described. Details of the manufacturing process, including weld schedules, and equipment and tooling details, are included.

1. INTRODUCTION

The flatpack is a metal case containing conformally coated microcircuits on a ceramic substrate, with leads passing through dielectric glass beads in the case wall. The case and its cover, both made of KOVAR, are welded together to form a hermetic seal. Previously, this operation was performed by resistance seam welding; however, the available equipment is size and configuration limited. Laser welding was developed as a more versatile alternate method.

2. MATERIALS AND EQUIPMENT

KOVAR, an iron base alloy containing 29% nickel, 17% cobalt and 0.3% manganese nominally, is the material for both the case and cover. This alloy was selected due to its thermal expansion compatibility with thermal shock resistant borosilicate type glass. The thermal compatibility of KOVAR with glass permits the manufacture of glass-to-metal seals with good thermal shock resistance. The cases are environmentally protected with gold plating of approximately 0.0001 inches.

A typical configuration, for which the laser sealing process was developed, is shown in Figure 1. It is approximately 3" long, 1" wide and 0.180" high, with a nominal wall-thickness of 0.040", and a cover of 0.020" thickness.

The details of the 0.025 inch diameter KOVAR leads that pass through dielectric glass beads fused into 0.056 inch diameter holes in the side-walls of the case are also shown in the sketch. The glass-to-metal seals are obtained by heating to temperatures in the range of 1800°F to 1900°F. Figure 2 shows a photograph of a flatpack with the ceramic substrate bonded to the bottom and the electronic components in place, but the cover yet to be installed. This top view of the assembly depicts the case mounted on a test fixture.

Equipment

The equipment used in this development program consists of a CO_2 laser rated at 1 KW (continuous wave) manufactured by Photon Sources, Inc. The equipment is capable of being operated in either a continuous wave (CW) mode or various pulse modes. A large amount of the work for the flatpack development was done in the continuous pulse mode, because of the minimal heat inputs and narrower weld beads possible with this mode as compared to CW mode.

A work station that was formerly part of a numerically controlled drilling machine has been converted for use with the laser. The work station is a sturdy table that can move in the X and Y directions precisely at set speeds up to 200

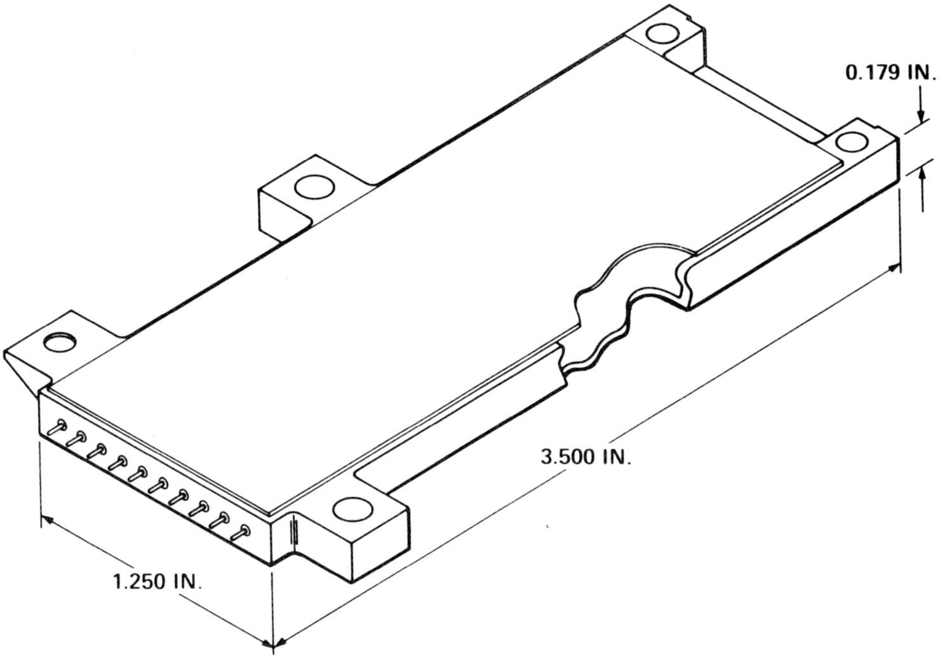

Figure 1. A Typical Flatpack Configuration.

Figure 2. A Photograph of a Flatpack, with the Cover Removed, Mounted on a Test Fixture.

Figure 3. A Photograph of the 1KW CO_2 Laser and the N/C Work Station

inches per minute, by numerical control or manual control, as required. Figure 3 shows a view of the laser and work station set-up.

A stream of argon gas is directed at the part being welded from a nozzle below the lens in order to prevent any debris from flying into the lens and to minimize oxidation of the welded metal. An arrangement is available for protecting the underside of weldments with another stream of argon. Use of other gases to achieve desirable results in specific applications is also possible.

3. Discussion

The weld joint must be designed so as to minimize power input requirements and consistently provide a good hermetic case-to-cover seal; at the same time there should be no deleterious effects on: a) the glass beads due to excessive differential heating or thermal shock, and b) the enclosed electronic components which generally should not be heated beyond 150°C.

When attempts were made to weld the 0.020" thick covers to the case walls from the top, it was realized that the heat input was high since a considerable fraction of the energy was consumed in melting the cover thickness before any useful work of sealing of the cover to case took place. With this in mind, several different joint design configurations and beam directions for welding were investigated. Figure 4 shows details of the various joint designs and beam directions, all of which can accomplish the required seal. 4A shows a weld from the side, making a butt type weld. This would require minimal energy input since the energy is directed at the joint to be sealed, and only partial penetration is necessary. However, there is some chance of metal splatter into the case, which is undesirable. 4B shows a weld with the beam coming from the top making a lap type (spike) weld. The reduced cover thickness at the weld area results in less heat input to the part than if the full cover thickness is used. 4C shows a design which simplifies fixturing and alignment, and results in low heat input as in 4A, at the same time reducing chances of splatter of molten metal into the case during welding. However, this design requires more precise machining and fit-up and hence is more expensive to manufacture than designs 4A and 4B.

Preliminary Parameter Development

Initial parameter development work was accomplished using 0.020" thick KOVAR sheet stock (same thickness as the enclosure covers). "Bead-on-plate" welds, butt welds and lap welds were made with various parameters. The parameters were refined to produce optimum weld bead contour with minimal heat input. Some tensile tests were conducted on the samples. Details of the laser parameters used, mechanical properties and appropriate remarks, are shown on Table 1. The results indicate that, with proper parameters and gas shielding, bead-on-plate and butt welds can be made to produce strengths nearly equal to that of the parent metal. In the lap shear tests, the specimens did not fail in the weld, but in the adjacent heat affected area. The increased heat input necessary in this type of joint apparently reduced the tensile strength slightly as indicated in parentheses in Table 1. However, the bending stress during the lap shear testing could be a factor, as well. The weld shear area was measured to be 0.030 square inches per inch length of weld. The calculated shear stress indicated in the table is less than the actual shear strength, since the specimens did not fail in shear.

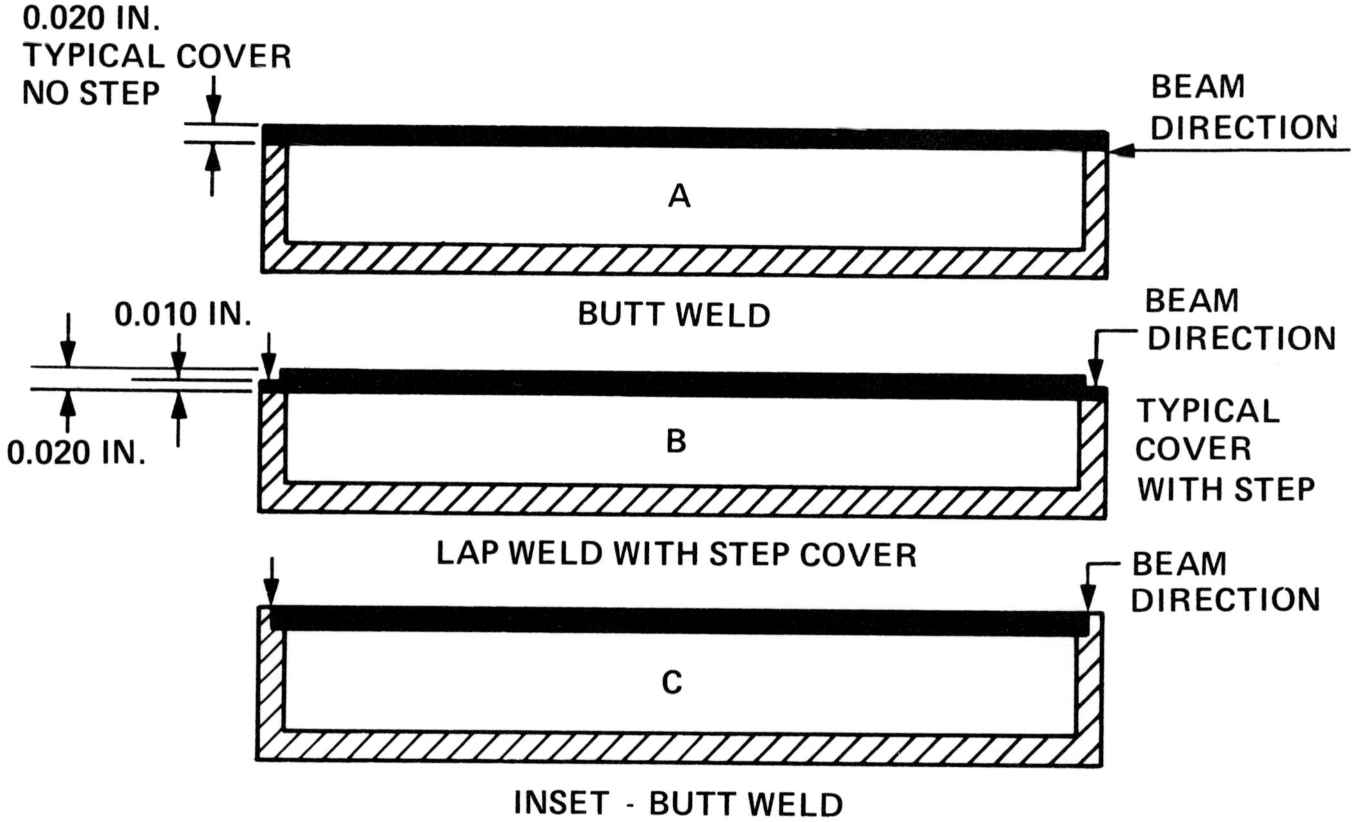

Figure 4. Details of the Joint Designs Studied.

TABLE I
PLATE SAMPLE WELD PARAMETERS & MECHANICAL PROPERTIES

NO.	TYPE	SIZE (INCHES)	WELD PARAMETERS PULSE (MS) ON	OFF	POWER (WATTS)	WELDING SPEED (IN./MIN.)	TENSILE STRENGTH (PSI)	LAP SHEAR STRENGTH (PSI)	REMARKS
1	PARENT MATERIAL	0.020 X 0.500	---	---	---	---	73,9000	---	CONTROL SAMPLE
2	BEAD-ON-PLATE	0.020 X 0.500	0.5	1.2	230	25	73,850	---	FRACTURED OUTSIDE THE WELD - IN PARENT MATERIAL
3	BUTT WELD	0.020 X 0.500	0.5	1.2	230	25	73,300	---	FRACTURED IN WELD
4	BUTT WELD	0.020 X 0.500	0.5	1.2	230	25	73,100	---	FRACTURED IN WELD
5	BUTT WELD	0.020 X 0.500	0.5	1.2	230	25	73,000	---	FRACTURED IN WELD
6	LAP WELD	0.020/0.020 X 0.500	4	5	300	25	(65,200)	>43,300	FAILED IN PARENT MATERIAL, IN HEAT AFFECTED ZONE
7	LAP WELD	0.020/0.020 X 0.500	4	5	300	25	(68,600)	>45,670	FAILED IN PARENT MATERIAL, IN HEAT AFFECTED ZONE
8	LAP WELD	0.020/0.020 X 0.500	4	5	300	25	(69,600)	>46,300	FAILED IN PARENT MATERIAL, IN HEAT AFFECTED ZONE

TABLE II
LASER WELD SCHEDULES FOR SEALING FLATPACKS

JOB TITLE: FLATPACK WELDING
MATERIAL: KOVAR TO KOVAR
THICKNESS: 0.010-0.020 LIDS TO 0.040 WALL

MODE	FUNCTION	OPTIMUM WELD PARAMETERS PULSE (MILLISECONDS) ON	OFF	COUNT	BURST COUNT	CURRENT SETTING	POWER (WATTS)	SPEED (IPM)	OPERATION	JOINT CONFIGURATION
2	T-SEQ.	2.0	--	1	1	999	800	---	LID SIDE TACK SPOT WELD	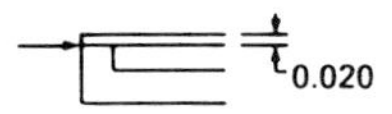
2	T-SEQ.	4.0	--	1	1	999	800	---	LID TOP TACK SPOT WELD	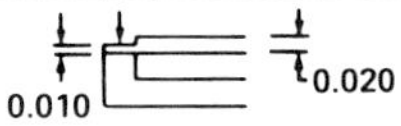
1	CP	0.5	1.9	-	-	956	200	25.7	CONTINUOUS WELD	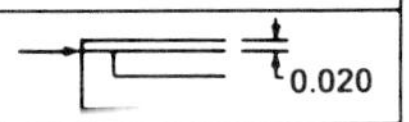
1	CP	0.5	1.3	-	-	956	250	25.7	CONTINUOUS WELD	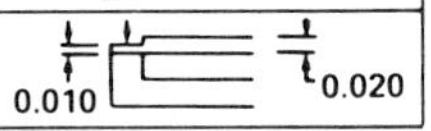

RESONATOR	
GAS MIXTURE	G-1
TYPE	FLOW (CFH)
HELIUM	4.3
CO_2	1.8
NITROGEN	3.5
GAS ASSIST (NOZZLE)	
ARGON	20

Flatpack Welding

The weld parameters developed for welding the covers to the cases from the side (butt) and from the top (spike or lap) to assure adequate strength and a good hermetic seal are summarized in Table II.

The case and cover are cleaned after the electronic components are installed by wiping with alcohol or acetone. A small countersunk hole is drilled in the cover and a drop of solder alloy deposited on it. A small vent hole is left through the solder so that when the case is completely sealed at the edges, there is no possibility of blow-outs due to heated air from inside the case building up excessive pressure. After completion of the welding, the solder is reflowed with minimal heat input by touching with a hot soldering iron, to re-seal the hole. The cover is fixtured over the case and tack welded at several places as detailed in Table 2. The fixtures for holding the part during welding are made of copper to minimize heating of the part during

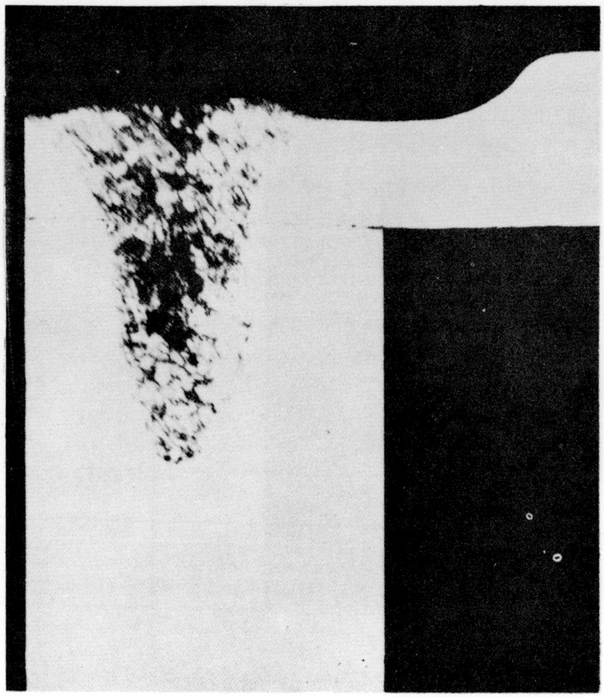

Figure 5. A Typical Weld Bead Cross-section of a Lap Weld Using Stepped Cover. Magnification - 50X Approx.

welding. For the parameters listed, typical welds from the top and from the side were examined microscopically. The photomicrograph in Figure 5 illustrates a typical cross-section of a lap (from the top) weld. Figure 6 is a photomicrograph of a typical longitudinal section of the lap weld and shows the variation of the spike depth along the weld, which in this case is from 0.023" to 0.032". A cross-section of a butt weld (from the side) is shown in Figure 7.

Hermeticity Testing

The hermeticity test requirements for this application are explained in MIL-STD-883A, Method 1014.1. The sealed cases are tested for fine and gross leaks. Fine leaks are detected by first pressurizing the sealed case in a helium pressure chamber. It is subsequently placed in a vacuum chamber equipped with a mass spectrometer type leak detector preset to read measured helium leak rates of 10^{-9} atm. cc/sec. and greater. Gross leaks are detected by placing the helium pressurized case in an observation container containing a fluorocarbon detecting fluid. Any bubbles emanating from the cases are observed through a (3X to 30X) magnifier.

Temperature Profile Measurement Studies

Early experiments in welding test units indicated that temperatures of the case walls and cover remained quite low. However, temperature profiles at various locations

Figure 6. A Longitudinal Section of a Lap Weld [Cover Stepped Down to 0.010 Inches for Welding]: Magnification - 80X Approx.

Figure 7. A Typical Weld Bead Configuration of a Butt Weld [Cover Welded to Case from the Side]: Magnification - 50X Approx.

during and immediately after welding were considered desirable. In order to accomplish this task, 3 mil-diameter chromel-alumel thermocouples were attached to the case walls and bottom at various locations and platinum-platinum 10% rhodium thermocouples were used in locations adjacent to the weld. (laser tack welded in place). Figure 8 shows a view of the assembly before the cover is put on and welded. The weld schedules used for this task are shown in Table III. The pulse off time is shorter than the number finally selected for this type of joint, and the case wall thickness in this test is 0.030" and the final design configurations is 0.040." These two factors result in a lower temperature profile in the actual production parts than the numbers arrived at in this experiment. A recording oscillograph was used to record the time-temperature curves during welding.

Figure 9 and 10 show the location of the thermocouples (inset) and the temperature profiles recorded in two tests. The temperature profiles are plotted to indicate temperature excursions at seconds prior to, at, and after crossover of the laser beam for each thermocouple location. An examination of the graphs shows the temperatures reached at the bottom of the cases are well below the maximum allowable temperature of approximately 300°F for the electronics. The temperatures on the side walls remain at high levels only for very short times before dissipation. The temperature profiles obtained should be considered approximate since minor changes in some of the factors involved could influence the profiles considerably.

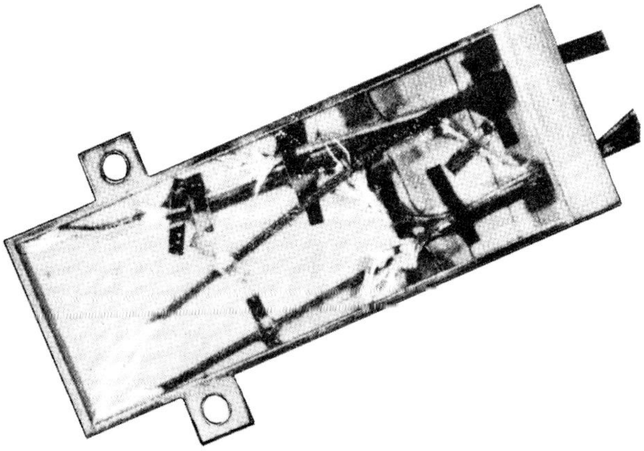

Figure 8. A View of the Thermocouple Leads Attached to Case Before Welding Cover to Case.

However, a computer model of heat flow is being developed using nodal representation of the transient heat transfer problem. Preliminary data indicate reasonably good correlation with experimental results.

A third test was made to measure temperatures reached during welding at various points on a ceramic substrate mounted in the case. Thermocouples were soldered to the substrate at various points close to the edges. The measured temperature rise during welding ranged from 117°F to 197°F.

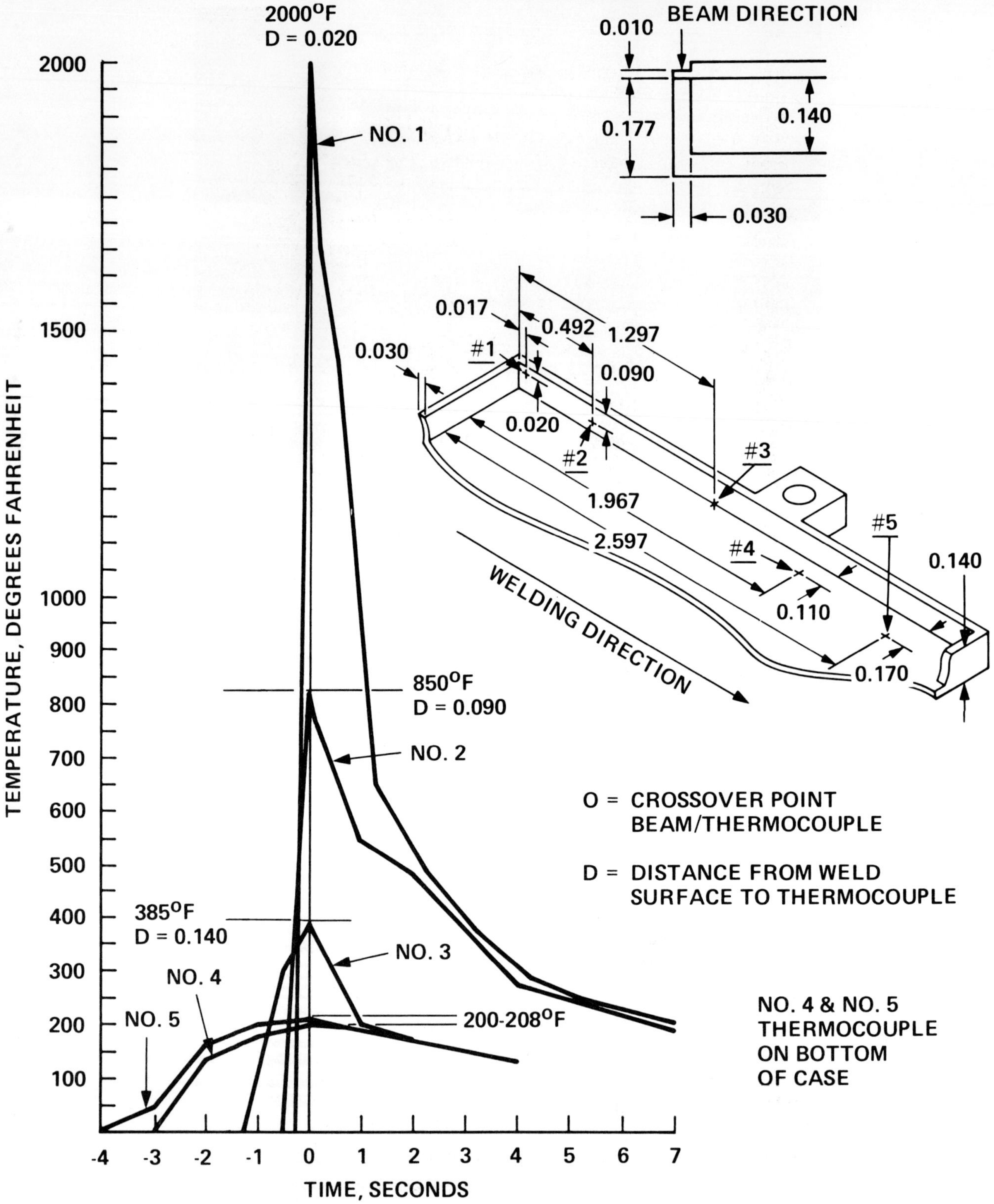

Figure 9. Temperature Profile Measurement for Lap Weld - Stepped Cover, Test 1

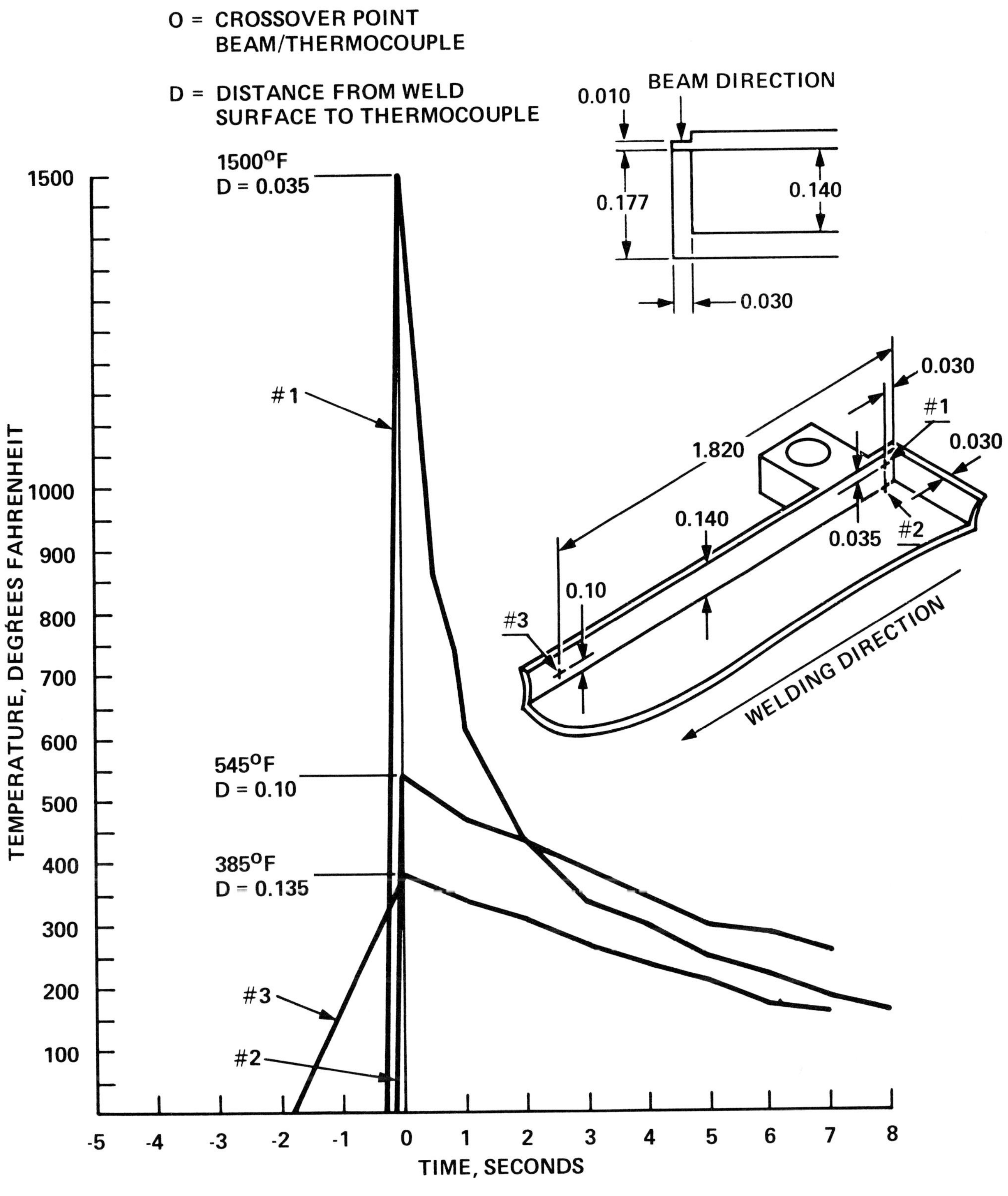

Figure 10. Temperature Profile Measurement for Lap Weld - Stepped Cover, Test 2.

Source: *SAMPE Quarterly*, Jul 1977

TABLE III
LASER WELD SCHEDULES FOR TEMPERATURE PROFILE MEASUREMENT STUDY

MATERIAL: KOVAR TO KOVAR
THICKNESS: 0.010 LIT TO 0.040 WALL

WELD PARAMETERS											JOINT CONFIGURATION
TEST #	MODE	FUNCTION	PULSE (MILLISECONDS) ON	PULSE (MILLISECONDS) OFF	COUNT	BURST COUNT	FOCUS	CURRENT SETTING	POWER (WATTS)	SPEED (IPM)	
1	1	CP	0.5	1.2	–	–	–	936	230	25.6	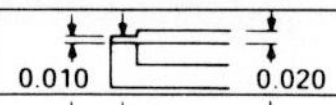
2	1	CP	0.5	1.2	–	–	–	936	230	25.6	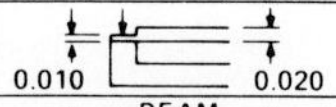
3	1	CP	0.5	1.2	–	–	–	936	230	25.6	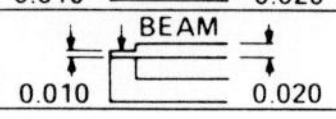

RESONATOR	
GAS MIXTURE	G-1
TYPE	FLOW
HELIUM	4.3
CO_2	1.8
NITROGEN	3.5
GAS ASSIST (NOZZLE)	
ARGON	20.

Repair Procedure

It was demonstrated that a laser weld which does not produce a hermetic seal can be repaired by making a second pass over the original weld. However, if any changes have to be made to the electronic circuitry, the cover must be removed by grinding to within 0.001 inches thickness and then peeling off. Subsequently a new cover is welded in place.

Other Problem Areas

During the development phase, the covers exhibited bowing and warpage in some cases. The covers were being plated with approximately 0.7 mils of nickel for environmental protection. During the welding, some of the nickel was alloying with the molten KOVAR, effectively changing the composition of the nugget and its thermal expansion characteristics. Elimination of the nickel plating was the solution for this problem.

Another problem sometimes encountered in this application involves cracking of the glass beads in the side-walls, causing loss of hermetic seal. In most instances, this is attributed to mishandling and bending of the wire leads. It was also found early in the program that deep weld penetrations too close to the beads also result in bead cracking.

Any soldered connections through the sidewalls should be located as far away from the welds as feasible. However, since any temperature beyond the flow temperature of solder and drop-off take place extremely fast, no significant problems are anticipated.

4. Conclusions

The results of this development study indicate that the CO_2 laser can produce excellent welds in KOVAR, and that hermetic sealing of micro-circuit flatpacks can be accomplished using the proper parameters.

In order to prevent overheating of the case and the electronic components, the weld joints should be designed so that minimal energy inputs are necessary to obtain a good seal.

H. N. Krishnaswamy is a Process Development Engineer with the Re-Entry & Environmental Systems Division of General Electric in Philadelphia. He holds a Master's Degree in Metallurgical Engineering from Syracuse University. His areas of responsibility at General Electric include Metallurgy and Metal Joining.

H. N. Krishnaswamy

Victor E. Boccelli is a Process Development Engineer at General Electric Re-Entry & Environmental Systems Division. He holds a B.S. in Mechanical Engineering and a M.S. in Environmental Engineering and is a registered Professional Engineer in the State of Pennsylvania.

Victor E. Boccelli

SECTION B-IV:

Properties of Welds

MECHANICAL PROPERTIES, FRACTURE TOUGHNESSES AND MICROSTRUCTURES OF LASER WELDS OF HIGH STRENGTH ALLOYS

E.A. Metzbower and D.W. Moon
Material Science and Technology Division
Naval Research Laboratory
Washington, DC 20375

INTRODUCTION

The laser provides us with a unique source of heat-a highly energetic, coherent, monochromatic beam of photons that can be focused. If the energy density of the laser beam is high enough, melting and vaporization of the recipient surface can occur. By utilizing a sufficiently high power laser, materials can be welded. In welding, vaporization of the surface results in the formation of a plasma which, if controlled, allows the formation of a deep penetration column of vapor extending into the metal. The walls of this vapor column are molten and as the hole is moved along the joint of the two metals, the metal on the advancing side of the hole, throughout its depth, is melted. The metal flows along the bore of the hole and solidifies along the rear side of the hole to make the weld. This hole results from a balance between the energy density of the laser beam and the welding speed. Thus it is important that the energy density of the laser beam at the workpiece and the welding speed be chosen so as to complement each other. An energy density that is too high will result in an unstable hole which can cause drop through, whereas an energy density that is too low will not permit vaporization or formation of the keyhole. A welding speed that is too fast will not permit a full penetration weld, whereas a welding speed that is too slow will cause a very wide weld and possible drop through.

The development of multi-kilowatt, continuous carbon dioxide gas laser systems has provided us with a laser energy source that has the capability for deep penetration welding. Laser beam welding has been utilized to join a variety of metals and alloys (1) including low alloy and stainless steel; aluminum alloys; titanium alloys; refractory and high temperature alloys. Porosity-free, generally ductile welds can be attained with tensile strengths equivalent to or exceeding those of the base plate. Hardness increases are obtained in alloys having microstructures related to cooling rates. The cooling rates of laser beam welds are typically very high as a consequence of the low heat input per unit length of weld. The impact strength (1,2) of laser beam welds in several high strength low alloy steels has been attributed to a lowering of the impurity inclusion content and a change in the inclusion distribution.

Deep penetration welds produced with high power CO_2 lasers are similar to electron beam welds. However, laser beam welding (2) offers several advantages:

a) a vacuum requirement is not required for the workpiece since the laser beam is easily transmitted through the air;

b) x-rays are not generated in the laser beam/workpiece interaction;

c) the laser beam may be readily focused, aligned and redirected by optical elements; and

d) due to the slightly lower energy density of the focused laser beam, the tendency for spiking, underbead spatter, incomplete fusion, and root porosity is reduced.

This study was underaken to measure mechanical properties and fracture toughnesses of laser beam welds in a variety of high strength alloys and to correlate these properties with their microstructure, hardness, and fractography. The welds were fabricated by United Technologies Research Center and represent "a best effort" type of approach rather than an extensive research program aimed at developing optimum properties.

PROCEDURE

The materials that were laser beam welded are listed in Table 1 along with their nominal compositions. The plates that were welded were about 12 mm (0.5 in) thick and were 100 mm (4.0 in) x 200 mm (8.0 in). These plates were welded to form a 200 mm x 200 mm (8 in x 8 in) weldment. All welds were butt welds. Welding parameters are given in Table 2. Helium was used in all cases to provide an inert gas shield over the reactive, hot welds and also to assist in plasma control.

The weldments were given visual and radiographic examinations. Some of the weldments especially the aluminum alloy showed excessive porosity. The weldments were then cut into blanks for mechanical property and fracture toughness specimens.

RESULTS

Mechanical Properties

Mechanical properties were measured on a standard ASTM round specimen 6.4 mm (0.252 in) in diameter. The weld was transverse to the loading direction in these tests. The yield strength, ultimate strength, percent elongation and percent reduction in area were determined for each specimen. A 25 mm (1.0 in) gauge length was used to determine percent elongation. The average values of these tests are given in Table 3. Twenty laser beam welds of HY-130 were tested and all fractured in the base plate. In the HY-80 laser beam welds sixteen specimens were tested; twelve of these fractured in the base plate and the other four failed in the weld. These four specimens whose fracture surfaces indicated a considerable amount of porosity had adequate yield and ultimate strengths but had significantly reduced values of elongation. In the titanium laser beam welds sixteen specimens were tested. Eleven of these failed in the base plate, whereas five failed in the weld. A significant decrease in yield strength corresponded to the reduction in the ductility of these specimens. Fractography revealed a considerable amount of porosity in these specimens. All of the aluminum laser beam welds failed in the weld. The ductility of these specimens was low and the amount of porosity visible on the fracture surfaces was high.

Table 1

Nominal Compositions of Base Plates

Material	Fe	C	Cr	Mo	Ta	Ni	Al	Ti	Nb	Mg	Mn
HY-80	bal	0.18	1.68	0.45	-	3.0	-	-	-	-	-
HY-130	bal	0.11	0.55	0.55	-	5.0	-	-	-	-	0.8
Aluminum	-	-	0.12	-	-	-	bal	-	-	5.10	0.8
Titanium	-	-	-	0.80	1.0	-	6.0	bal	2.0	-	-

Table 2

Laser Beam Welding Parameters

Material	Power (kw)	Speed mm/sec	Speed (ipm)	Heat Input kj/mm	Heat Input (kj/in)
HY-80	10.6	12.7	(30)	0.83	(21.2)
HY-130	11.0	12.7	(30)	0.87	(22.0)
Al alloy	8.0	27.5	(65)	0.29	(7.4)
Ti alloy	11.0	14.8	(35)	0.74	(19.0)

Table 3

Average Tensile Properties

Material	0.2% YS (MPa)	(ksi)	UTS (MPa)	(ksi)	% Elongation	% Reduction-in-Area
HY-80	627	91	752	109	23.5	75.8
HY-130	952	138	987	143	18.5	68.4
Al alloy	193	28	289	42	9.2	16.9
Ti alloy	758	110	855	124	11.5	31.7

Table 4

DT Energies of Laser Welds

Material	AU DT Energy Nm	(ft.lb)	Range of DT Energy Nm	(ft.lb)
HY-80	76	(56)	612-176	(38-130)
HY-130	707	(530)	836- 66	(459-627)
Titanium alloy	249	(184)	176-311	(130-230)
Aluminum alloy	116	(86)	101-128	(75- 95)

Source: *Applications of Lasers in Materials Processing*, ASM, 1979

Fracture Toughness

Fracture toughness was measured on a subsize, 12 mm (0.5 in) thick, dynamic tear (DT) test specimen that was notched in the weld. The standard size DT specimen is 16 mm (0.625 in). The fracture resistance of the laser beam welds were evaluated at room temperature in a double pendulum testing machine. These results are given in Table 4. In the HY-130 and titanium alloys, the fracture starts in the weld and then quickly moves into the base plate. Thus, there is a strengthening effect from the base plate. Evidence for porosity was observed on the HY-130 steel as well as both the aluminum and titanium alloys. The HY-80 steel fracture surface had what seems to be a coarse grain appearance as if fusion was incomplete. The HY-130 laser beam welds were also tested as a function of temperature and these results are shown in Fig. 1. The dramatic decrease in fracture toughness that occurs at 10°F is a real effect. Two specimens were in the same constant temperature bath. One was removed and tested. A value of 95 ft.lbs. was recorded. The second specimen was removed and tested. A value of 508 ft.lbs. was recorded. The difference in testing temperature between the two tests should not have exceeded 1°F. Obviously this temperature is at the edge of the upper shelf and also the lower shelf. A macrograph of the fracture surface of these two specimens is shown in Fig. 2.

Fractography

A field emission scanning electron microscope was used to examine the fracture surfaces of both the tensile and the fracture toughness specimens. All fractographic observations were made using the technique of stereographic pairs. In the cases where the fracture occurred in the base plate, the separation process was microvoid coalescence. Where porosity was observed on the fracture surface, the porosity was surrounded by microvoid coalescence as seen in Fig. 3. The size, number and distribution of pores varied from specimen to specimen in an alloy system as well as from alloy to alloy. The fracture toughness of specimens of the HY-80 alloy showed indications of hot cracking (see Fig. 4)

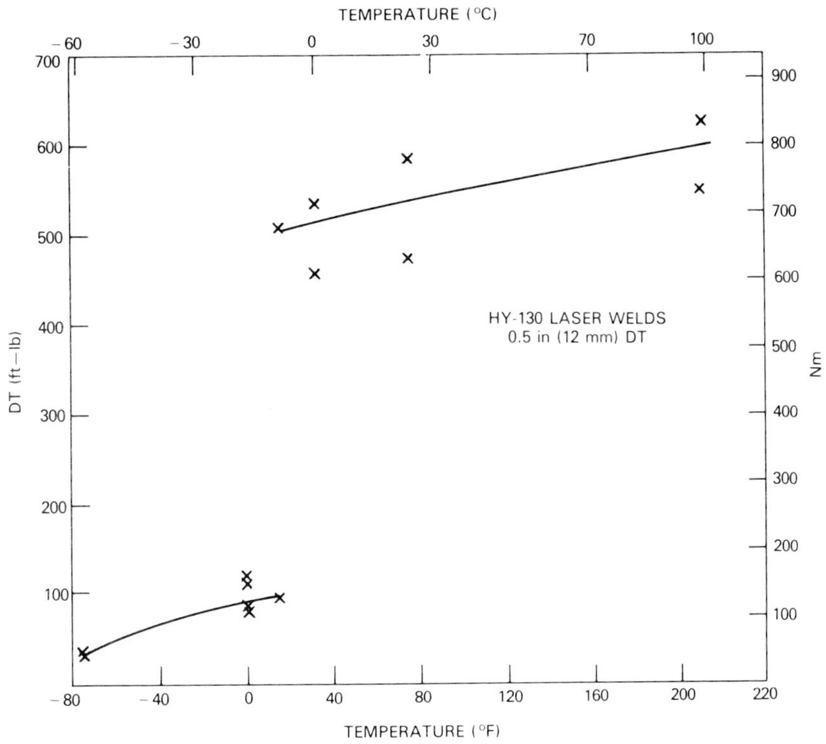

Fig. 1. Dynamic tear (DT) energy as a function of temperature for HY-130 laser beam welds.

Fig. 2. Macrograph of DT specimens tested at -18°C (10°F).

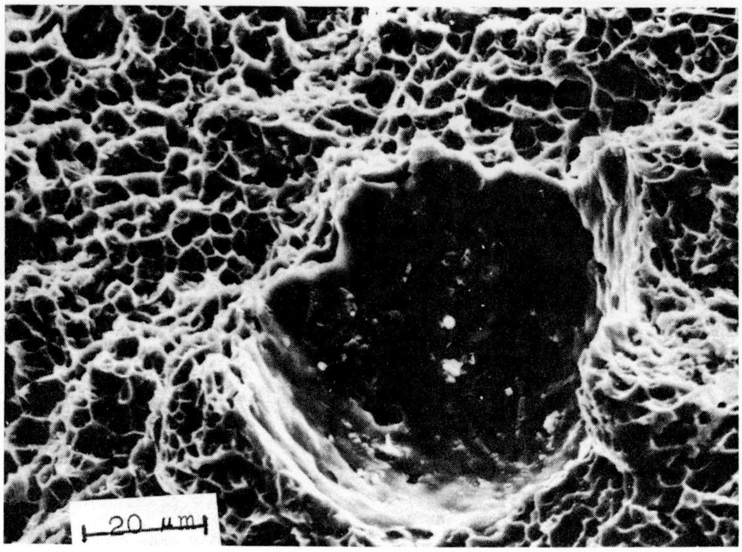

Fig. 3. Porosity seen in DT fracture surface of aluminum laser weld. Porosity is surrounded by microvoid coalescence.

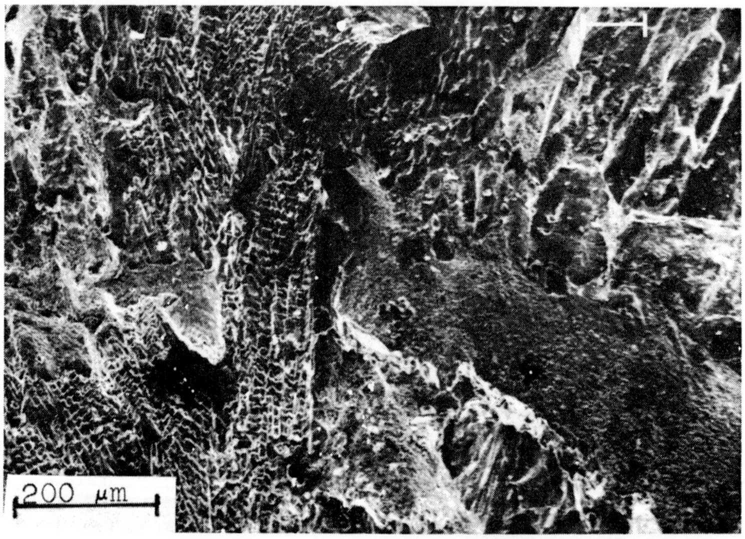

Fig. 4. Fracture surface of DT specimen of HY-80. Lack of fusion and secondary cracking are prevalent.

and extensive secondary cracking which was most likely present prior to testing. Extensive fractographic analyses were made on the HY-130 DT laser welds tested at 10°F. A comparison of the fracture mechanisms can be seen in Fig. 5 where the low toughness fracture separation process is cleavage whereas the high toughness weld is characterized by microvoid coalescence.

Metallography

Transverse cross sections of the weldments were prepared for microstructural examination and solidification structure analysis. The microstructures of each material were examined in the regions of the fusion zone, heat-affected zone (HAZ) and the base metal.

In the HY-80 laser beam welds, the microstructure of the fusion zone is composed mostly of martensite and some bainite (Fig. 6a). The HAZ contains the same microstructural components (Fig. 6b) but the overall size of these features appears to be reduced. HY-80 base plate consists of tempered martensite as is seen in Fig. 6c.

The microstructural features of HY-130 laser welds are similar except for the HAZ. The HAZ of the laser beam welds of HY-130 can be divided into two areas: near the fusion zone in a medium grain size region, where martensite and bainite are found; and near the base plate in a finer grain size region, where some ferrite is found interspersed with the martensite.

The fusion zone of the titanium alloy contains a microstructure consisting of α' (martensitic α) and β' as seen in Fig. 7c. Also evident in the micrograph are the prior β grain boundaries. The HAZ consists of primary α , Widmanstatten α and β as seen in Fig. 7b. The size and number of the primary α in the HAZ decreases and the β phase increases as the fusion zone is approached. These various structures evolved from the welding of the base plate which consists of elongated and equiaxed primary α in a matrix of β and Widmanstatten α - (Fig. 7a).

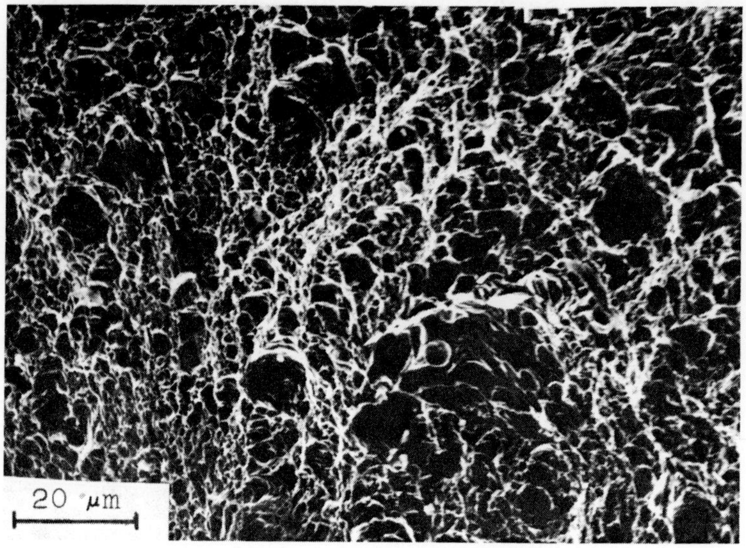

(a)

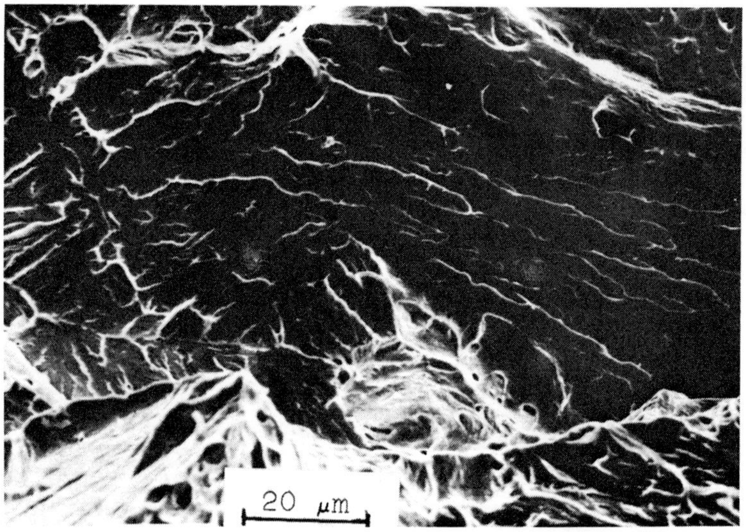

(b)

Fig. 5. Fracture surface of DT specimens tested at -18°C (10°F) (a) microvoid coalescence on higher energy specimen; (b) cleavage on lower energy specimen.

(a)

(b)

(c)

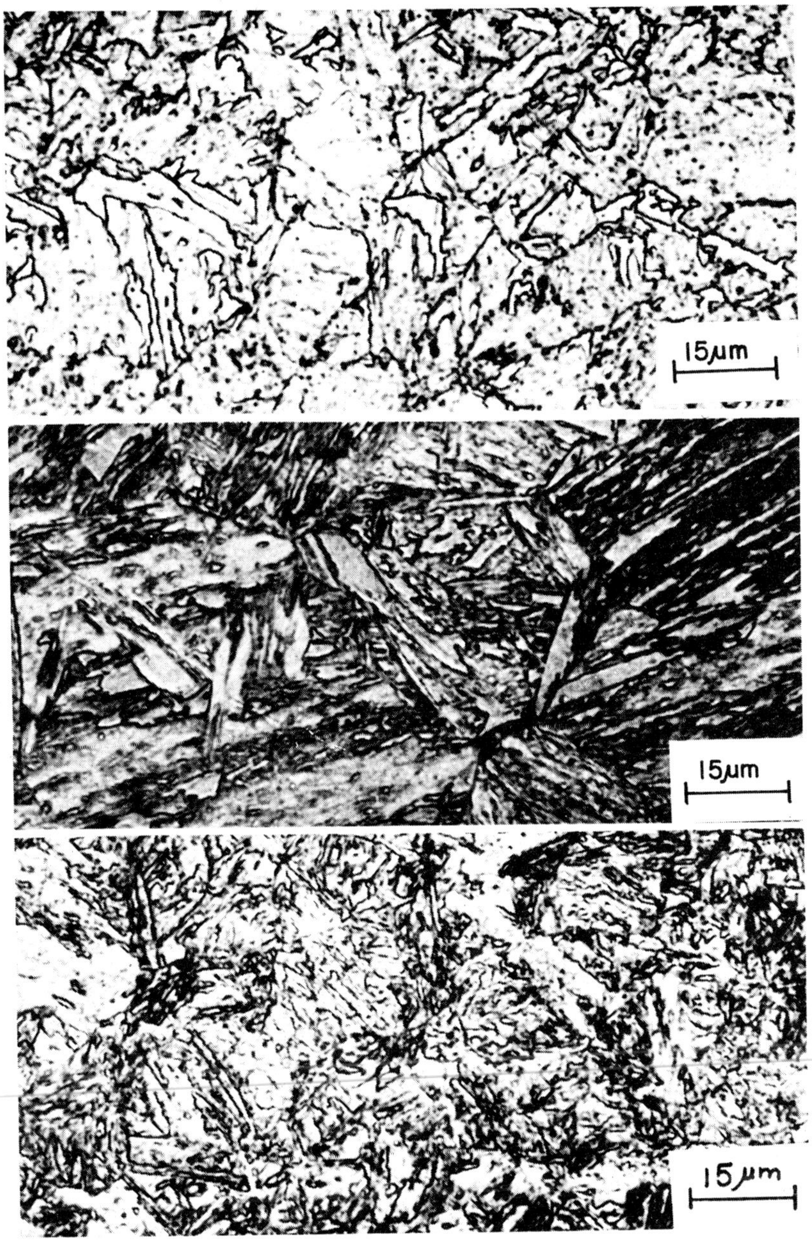

Fig. 6. Microstructures from different regions of laser beam weldment of HY-80. (a) Tempered martensite in base metal Rc 20.0 (b) Martensite and some bainite in heat-affected zone. Rc 42.0 (c) Martensite and some bainite in fusion zone. Rc 44.0. Etched in 1% nital.

(a)

(b)

(c)

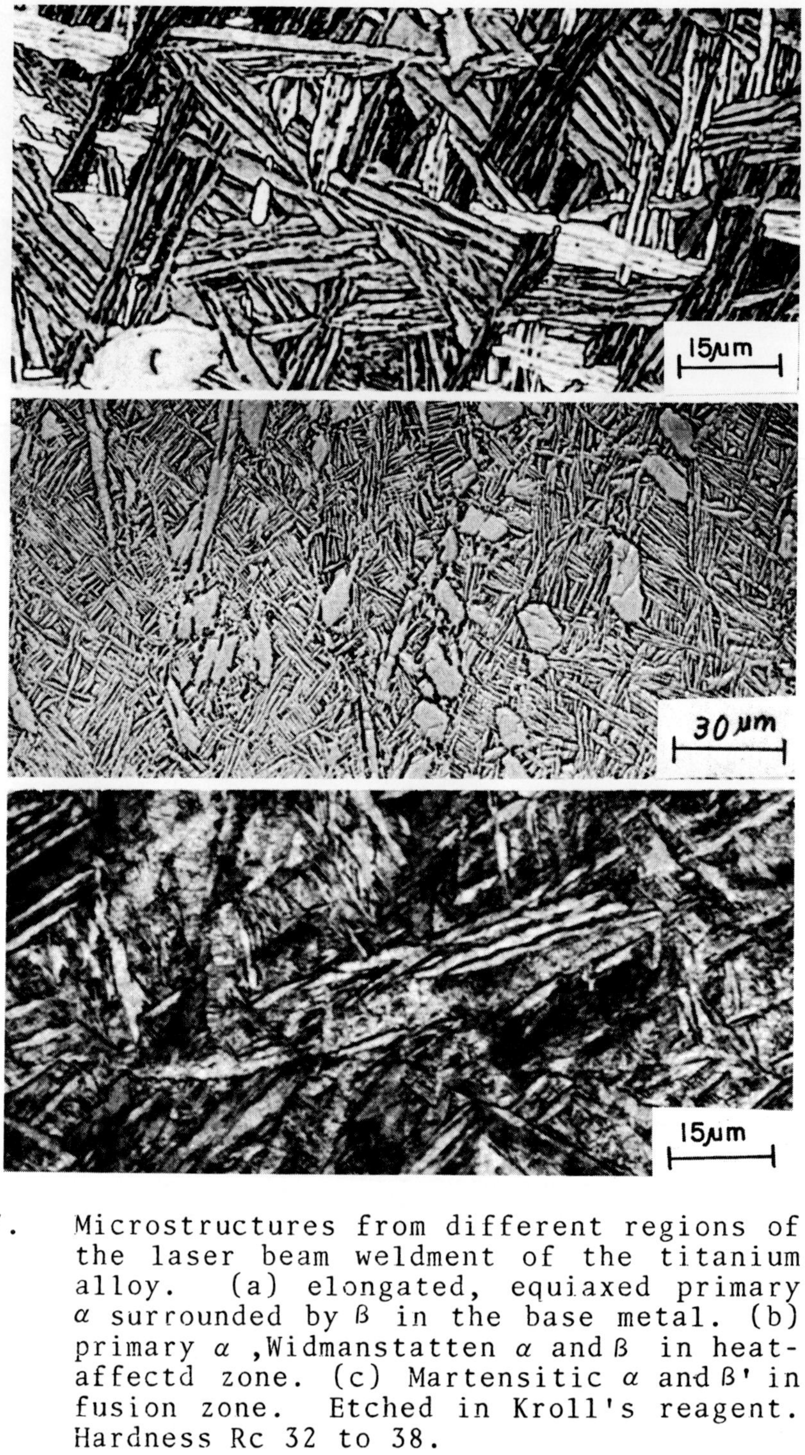

Fig. 7. Microstructures from different regions of the laser beam weldment of the titanium alloy. (a) elongated, equiaxed primary α surrounded by β in the base metal. (b) primary α ,Widmanstatten α and β in heat-affectd zone. (c) Martensitic α and β' in fusion zone. Etched in Kroll's reagent. Hardness Rc 32 to 38.

The microstructure of the Al 5456 alloy (Fig. 8a) consists of a fine distribution of the precipitate Mg_2Al_3 in a matrix of Al. Also present in the microstructure are large insoluble particles of Mg_2Si and $(Fe,Mn)Al_6$. When this alloy is laser beam welded, the fusion zone consists solely of α Al and the precipitate Mg_2Al_3, (Fig. 8b) whereas the HAZ displays a redistribution and shrinkening of the large insoluble particles (Fig. 8c).

In general the fusion zone of laser beam welds undergo a lowering of the impurity inclusion content and a change in the inclusion size distribution. In the fusion zone in the Al alloy, only four percent of the large insoluble particles are left after laser beam welding; the number of inclusions in the Ti alloy is reduced by 50 percent; in the HY-80 alloy only 21 percent of the inclusions remain, whereas in the HY-130 alloy, 28 percent of the inclusions are left.

HARDNESS

Transverse cross sections of the weldments were cut for microhardness testing. Using a diamond pyramid indenter with a 300 gm load, the microhardnesses were measured from one side of the base metal across the weld metal to the other side of the base metal.

The hardness of the weld and the HAZ in steels is much higher than that of the base metal as shown in Figs. 9 and 10. Average hardness values of the weld are 43 Rc for HY-80 and 38 Rc for HY-130 steel. The hardness of the HAZ varied from 27 Rc to 46 Rc in HY-80 and from 37 Rc to 39 Rc in HY-130 steel. The average hardness of the base metals is 18 Rc for HY-80 and 32 Rc for HY-130 steel.

The hardness traverse of the Ti alloy exhibits little change across the three different zones with average value of 36 Rc as seen in Fig. 11.

The hardness behavior of the Al alloy is presented in Fig. 12. The Vickers number varied from 87.0 in the base metal to 77.5 in the HAZ adjacent to the weld-HAZ interface. The highest Vickers number of the weld metal is 88.0.

(a)

(b)

(c)

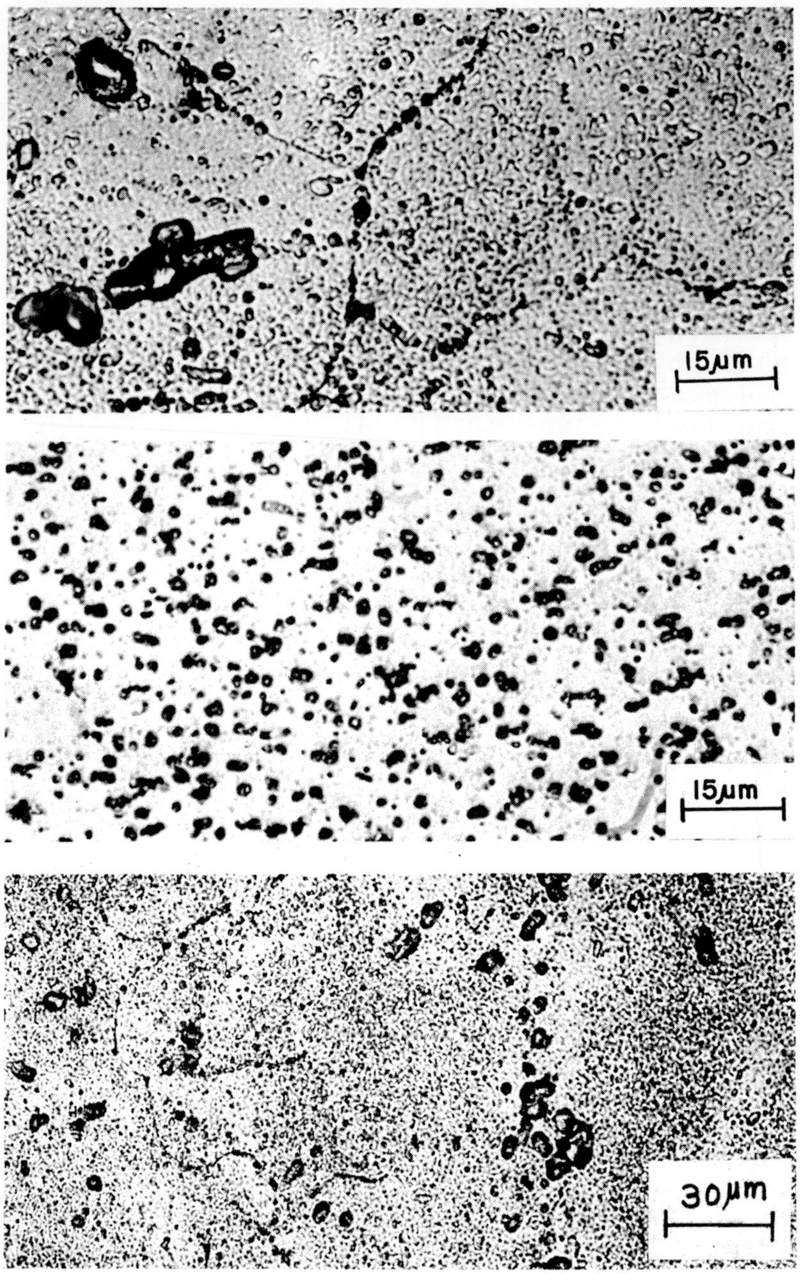

Fig. 8. Microstructures from different regions of the laser beam weldment of the aluminum alloy. (a) α Al and Mg_2Al_3 in base metal; (b)very fine grained α Al and β Al in fusion zone; (c) α Al and Mg_2Al_3 in HAZ. Etched in Keller's reagent.

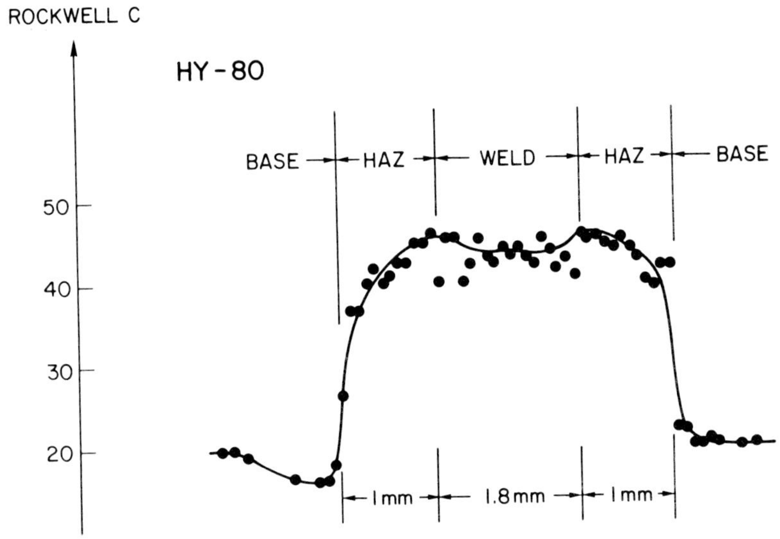

Fig. 9. Hardness traverse of HY-80.

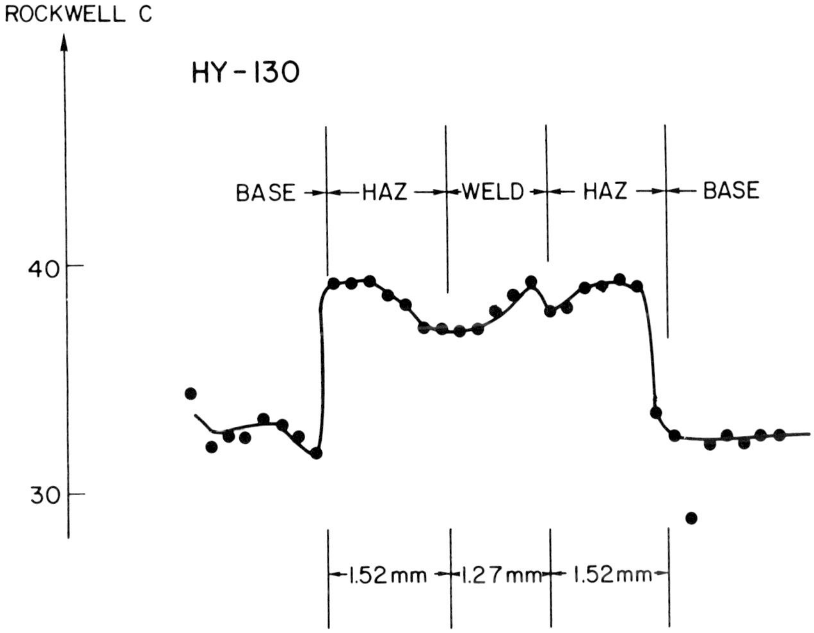

Fig. 10. Hardness traverse of HY-130.

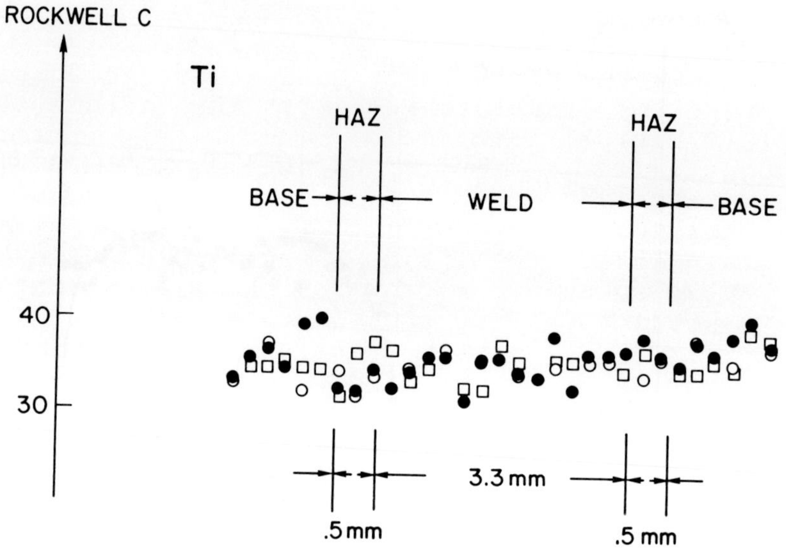

Fig. 11. Hardness traverse of titanium alloys.

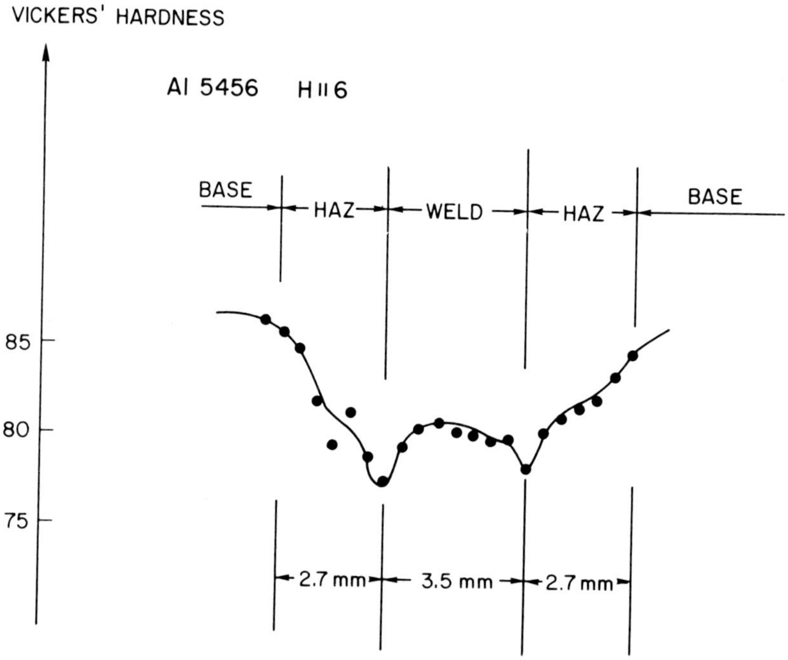

Fig. 12. Hardness traverse of aluminum alloy.

DISCUSSION

In the HY-80 laser beam welds, good mechanical propeties correlate well with the martensitic/bainitic microstructure found in the fusion area. The poor fracture toughness can be correlated through fractography to an apparent lack of fusion, hot cracking, extensive secondary cracking and large grain size. Porosity was negligible. The heat affected zones are narrow and the microstructure corresponds to a fast cooling rate. The hardness values indicate a fast transformation product and no tempering. The solidification structure indicates a small grain size. However, fractographic observations in the fracture surfaces of the toughness specimens indicate a fairly large grain size especially in the areas near lack of fusion.

In the HY-130 laser beam welds, good mechanical properties and excellent fracture toughness at room temperature were measured. These proerties correlate well with the fine solidification structure and the martensitic/bainitic microstructures. The narrow heat affected zone also contributes to these properties. The high hardness values indicate a fast cooling rate both in the fusion zone and the HAZ. These high hardness values also result in the high temperature at which the ductile to brittle transition as measured by DT energy occurs in these welds. Porosity, while evident, did not degrade either the strength or toughness of these welds.

In the titanium alloy, the mechanical properties of the laser weld compare favorably to those of the base plate and gas metal arc welds(3). The fracture toughness is excellent. Fractography revealed considerable small porosity in the fracture surfaces of the DT specimens. This is attributable to titanium high affinity to oxygen. A variety of transformed and untransformed phases (α , α' and β , β') were observed in the metallographic surfaces. The fluctuations in hardness are probably due to the different hardness values of these phases.

The aluminum laser beam weld had considerable porosity which was found both in the tensile and DT fracture surfaces. This porosity led to an obvious reduction in ductility and ultimate strength. The yield strengths and fracture toughnesses were near that to be expected in conventional welds of aluminum. Near the HAZ-weld interface the alloy has softened. In the HAZ the microstructure reveals elongated grains and an increase in density of insoluble constitutents (Mg_2Si) and a second phase (Mg_2Al_3), both in the matrix and at the grain boundaries, as compared to the weld zone.

CONCLUSIONS

Laser beam welds of high strength alloys have been shown to possess adequate mechanical propeties and good fracture toughness. Hardnesses are related to their fast cooling rate. Porosity is a definite problem which probably can be alleviated by judicious selections of shielding gases and effective shielding gas coverage. The fine grain size produced in the fusion zone and the narrow heat affected zone contribute significantly to the good mechanical proerties and fracture toughnesses found in these welds. A certain amount of second phase vaporization lowers the number of inclusions and changes the inclusion distribution especially in the lower melting point alloys.

REFERENCES

1. Welding Handbook, Volume 1, Seventh Edition, The American Welding Society, Miami, FL, 1976, pp. 21-22.

2. E.M. Breinan and C.M. Banas, "Fusion Purification During Welding with High Power CO_2 Lasers," Proceedings of the Second International Symposium of the Japan Welding Society, Osaka, Japan, Aug 24-28, 1975.

3. Welding Handbook, Section 4, Sixth Edition, The American Welding Society, Miami, FL, 1972, p. 73.36.

Fusion Zone Purification During Welding with High Power CO_2 Lasers

Edward M. Breinan* and Conrad M. Banas*

A significant metallurgical phenomenon is reported which has been observed following laser welding of several steels. Results of measurements in HY-130 alloy are presented which illustrate that the impact resistance of the laser weld zone is substantially greater than that of the as-fabricated base plate. This effect, which is linked to a measured reduction in the volume fraction of impurity inclusions within the weld zone, occurs simultaneously with increases in weld zone hardness and tensile strength. Possible mechanisms for the observed fusion zone purification are proposed. It is also noted that the observed mechanical properties may result, in part, from changes in the inclusion size distribution due to laser welding; however, direct measurements of such distributions have not yet been made. Similar results have been obtained on X-80 alloy, but are not included in this condensed version. The reported results suggest that laser welding can be effectively utilized to produce high quality, clean and tough welds in some alloy steels without the requirements for addition of filler metal of a complementary composition.

I. Introduction

Deep penetration, direct fusion welding with a high-power carbon-dioxide laser may offer an attractive means to join metal for certain applications, especially those where high volume production is involved. For example, laser welding of longitudinal or spiral seams for line pipe (1) offers the possibility of making square butt welds without filler metal.

In order to realize the potential economic advantages of welding without filler metal, it is necessary to have a process which is capable of joining the metal without requiring a region of different composition at the joint. Filler metal is normally used for joining HY-130 (2,3), for example, in order to provide good joint strength while maintaining adequate toughness. A filler metal composition which is different from the base metal composition is normally required to offset any contamination by oxygen or hydrogen during welding. The base metal HY-130 composition is quite susceptible to contamination by small amounts of oxygen and hydrogen, and such contamination seriously decreases the toughness of the material.

The laser welding process, as utilized to date, has, generally, not been used in conjunction with filler metal addition. The basic process used for forming butt welds is an autogenous square butt joint, fabricated in the deep-penetration mode. As with electron beam equipment, deep-penetration laser welds are accomplished by focusing the laser radiation to a power density in excess of 10^6 watts/cm^2 at the workpiece surface. A column of metal vapor surrounded by an equilibrium liquid pool is formed and creates the weld

* United Aircraft Research Laboratories, East Hartford, Ct. 06108 USA

Source: *Advanced Welding Technology*, The Japan Welding Society, 1975 by translation along the seam. Any top-surface or underbead reinforcements result from lateral drawing together of the plates being joined.

The fusion zone purification effect reported herein was found in laser welded HY-130 alloy, and most of the investigative work on the effect was performed on such welds. In addition, welds in X-80 pipe steel were investigated relative to the potential application discussed in reference 1, and impact strength improvement coupled with microstructural evidence of fusion zone purification was again found. The effect has also been shown qualitatively in grade B ship steel, a less-refined alloy in which impact behavior was greatly improved and in which a reduction in impurities had clearly taken place during the laser fusion process. Only HY-130 results are reported here.

II. Experimental Procedure

The composition of the HY-130 alloy is given in Table I.

Table I - Composition of HY-130 (all elements in weight percent)

Fe	Bal	Mo	0.40
C	0.09-0.11	V	0.07
Mn	0.25	Ni	5.50
Cr	0.50		

HY-130 welds were formed with a continuous-wave, co-axial flow, CO_2 laser. This laser, which is described in reference 4, was equipped with oscillator-amplifier optics to provide a maximum output beam of approximately 7 kW with a Gaussian mode beam. Welds were made at 5 kW at speeds of 0.85, 1.27, 1.48 and 1.90 cm/sec in 0.64 cm thick material. Welds were fabricated by clamping the plates into a fixture which was equipped with provisions for backup inert gas shielding with argon. A top surface trailer shield similar to that employed for GTAW was utilized with the beam passing through the aperture normally used for the electrode. A He-A leader in the vicinity of the beam served to shield and minimize plasma formation; the trailing section was supplied with argon. The leader gases were applied essentially parallel to the workpiece surface. A combined total flow of 472 cm^3/sec of argon and 157 cm^3/sec of helium was supplied to the top and bottom shields.

Following welding, welds were radiographically inspected and then sectioned to provide specimens for metallographic examination in accordance with standard techniques. Remaining areas of welded plate were used for tensile, bend, and impact specimens. Impact testing of HY-130 was performed according to ASTM E23-66 (5) using a specimen which was subsize (one-half size) in the width direction for specimens welded in the 0.64 cm thick plate. In addition to mechanical test pieces, weld beads for chemical analysis were fabricated at 0.85 and 1.27 cm/sec. The weld and base metal from adjacent areas were carefully analyzed for carbon, nitrogen, oxygen, sulfur and hydrogen contents. Small cubes of weld metal and adjacent base metal were carefully cut and then the surface layers of these cubes were ground away to remove potential sources of error from contamination. Oxygen and nitrogen contents were determined by inert gas fusion; carbon was determined by combustion to CO_2 followed by gas chromatography; sulfur was determined by combustion in oxygen and iodometric titration of sulfur oxides; and hydrogen was determined by a hot extraction technique. Results in all analyses were accurate to ± 10 percent of the amount present or ± 5 ppm, whichever was greater.

III. Results

Visually acceptable welds exhibiting uniformly spaced trailing edge isotherms as shown in Fig. 1b were produced in HY-130. Differences in weld profiles obtained at three representative speeds may also be noted in Fig. 1. Good penetration was obtained for all welds with some slumping of the top surface bead and excessive drop-through being characteristic of the bead made at 0.85 cm/sec.

(0.64 CM THICK PLATE 5.0 kW)

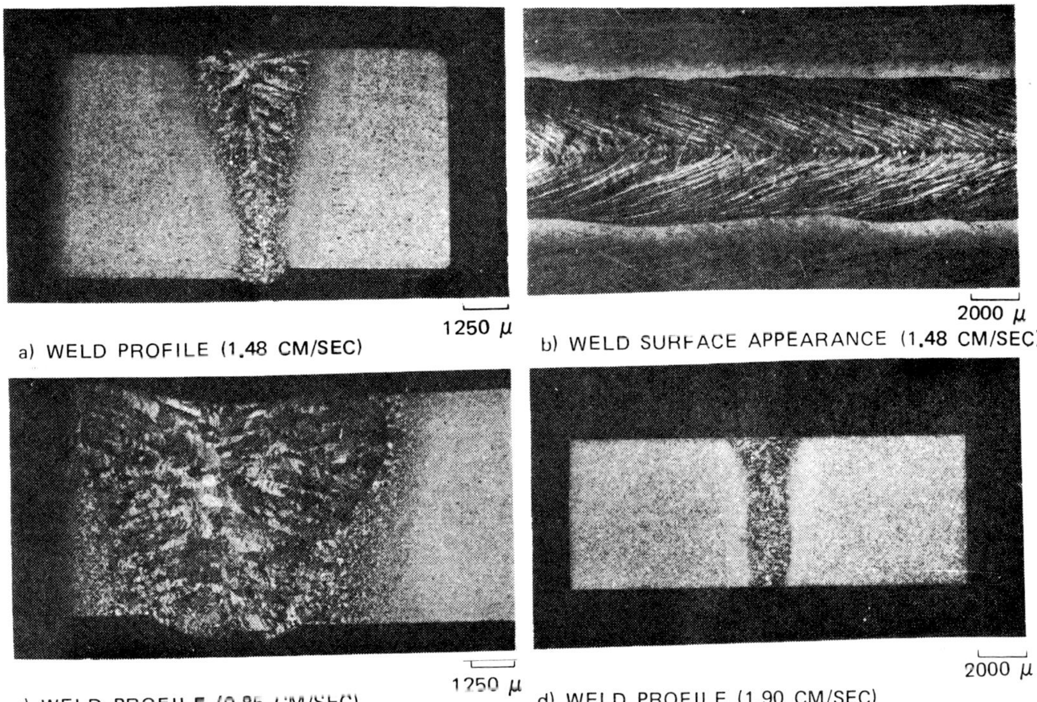

a) WELD PROFILE (1.48 CM/SEC)

b) WELD SURFACE APPEARANCE (1.48 CM/SEC)

c) WELD PROFILE (0.85 CM/SEC)

d) WELD PROFILE (1.90 CM/SEC)
(SURFACES GROUND BEFORE SECTIONING)

FIG. 1. HY–130 LASER WELDS

Following initial mechanical tests, which established that weld metal tensile strength and hardness were in excess of base metal values and which produced no failures in the weld metal or HAZ, a series of impact tests was conducted in welds fabricated at three welding speeds. The results of these impact tests are summarized in Table II. It can be seen that the impact strengths of all welds were equivalent to or greater than those of the base metal from the same plate. Significant improvement in toughness was observed in the weld formed at 1.90 cm/sec; an increase of forty-six percent above base metal was measured. Low magnification scanning electron micrographs of the impact fracture surfaces at both -1.1°C and 23.9°C are displayed in Fig. 2 for material welded at 1.90 cm/sec. It is notable in the scanning micrographs that the inclusion shape (aspect ratio) in the weld metal is considerably

Table II - Results of Impact Tests on HY-130 Laser Welds

Laser Power (kW)	Welding Speed (cm/sec)	Temp. (°C)	Weld Impact Strength (Joules)		Base Metal Impact Strength (Joules)
5.0	1.90	-1.1	52.9	half-size Charpy specimens	35.8
5.0	1.90	23.9	52.9		36.6
5.0	1.48	23.9	38.4		32.5
5.0	0.85	23.9	36.6		33.9

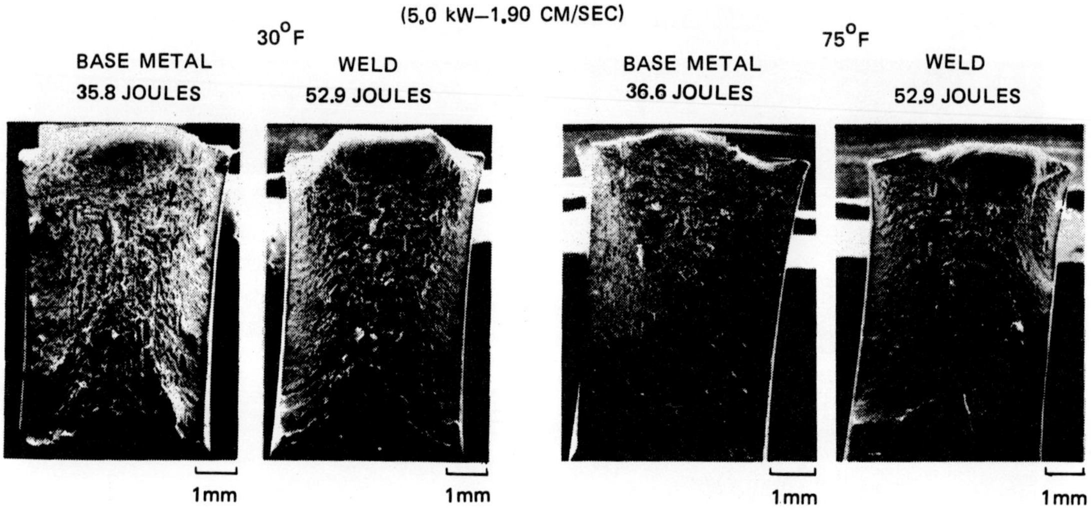

FIG. 2. HY–130 LASER WELD IMPACT FRACTURES

different from that of the base metal. The rolled out or "stringer" type inclusions have spheroidized during welding and, upon fracture, open larger voids or cracks due to their higher aspect ratio. In addition, it appeared qualitatively that there were fewer visible inclusions in the weld metal. According to the work of Hill and Passoja (6), a change in the inclusion size distribution with a decrease in the volume fraction of inclusions which participate in fracture also correlates with improved impact resistance.

To estimate the inclusion volume fraction, blocks of weld metal were formed at 0.85 and 1.27 cm/sec. Additional welds were also formed at 1.48 cm/sec for mechanical tests. Careful chemical analyses were conducted on the weld metal and immediately-adjacent base metal. The 0.85 and 1.48 cm/sec welds were metallographically sectioned and mechanically tested, and the impact test results are included in Table II above. Chemical analysis results are included in Table III which show that carbon content was essentially unchanged during welding. Nitrogen content was reduced between 12 and 18 percent. Similarly, oxygen content was significantly reduced by between 35

Table III - Results of Chemical Analysis of HY-130 Laser Welds Made at 0.85 and 1.27 cm/sec

Data are Averages of Several Determinations

	0.85 cm/sec		1.27 cm/sec	
Element	Base Metal	Weld Metal	Base Metal	Weld Metal
Carbon %	0.127	0.120	0.122	0.117
Nitrogen (ppm)	125	103	130	115
Oxygen (ppm)	44	21	78	51
Sulfur (ppm)	69	74	73	70
Hydrogen (ppm)	<10	<10	<10	<10

and 52 percent. Sulfur was found to be essentially unchanged. Finally, hydrogen was found to be less than 10 parts per million in all specimens, establishing the fact that the welds were not contaminated by excessive hydrogen. The reduction in oxygen content indicated a probable reduction in the content of manganese silicate type inclusions - the type identified by Hill and Passoja (6) as being principally responsible for the impact behavior of the alloy.

Further evidence of the reduced inclusion content of the fusion zone appears in Fig. 3. Magnified cross sections of the 0.85 and 1.48 cm/sec welds are shown extending from the fusion zone center line out to the unaffected base metal, and from the top of the weld, half way through the plate thickness. In both welds the larger inclusions in the base metal and in the heat-affected zone (which were never molten) are easily visible. In the case of the 1.48 cm/sec weld, some stringer type inclusions are found in the plane of polish. At the edge of the fusion zone, the number of large, visible inclusions suddenly becomes sharply reduced, and in fact, the fusion zones of both welds show a considerably lower visual inclusion content than the unmelted areas of the HAZ and base metal. This agrees qualitatively with the SEM observations on base metal and weld metal fracture surfaces, and with the reduced oxygen content of the weld metal as compared with the base metal. A further verification could come from a quantitative analysis of the inclusion size distribution on the impact fracture surfaces; weld specimens for such a study are being prepared at the time of this writing.

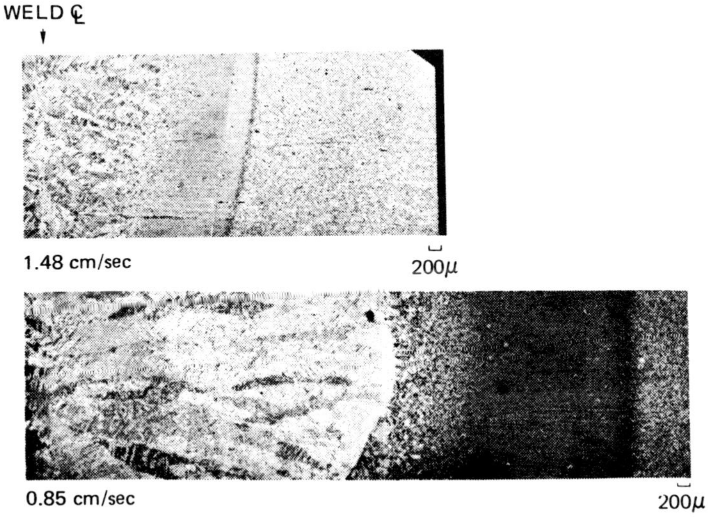

FIG. 3. MICROSTRUCTURES OF HY–130 LASER WELDS

IV. Discussion

In general, it appeared that following laser welding of HY-130 alloy under gas shielding in the atmosphere, some reduction in the inclusion content of the fusion zone below that of the base metal was found. A change in the inclusion size distribution may also have occurred, but this has not been as yet determined. The resultant impact resistance of the welds was found to be equivalent to, or better than that of the base metal, with a maximum increase of 46 percent occurring at a near optimum welding speed at 5.0 kW and 1.90 cm/sec in 0.64 cm thick plate. At lower welding speeds, smaller impact improvements (between 8 and 18 percent) were observed, indicating that the effectiveness of the mechanism by which fusion zone purification takes place is probably dependent on the local dynamics in the region of the deep penetration cavity, as influenced by the welding speed, shielding, and focusing.

One possible explanation for the fusion zone purification observed is based on the fact that fluids (such as vaporized impurities and vaporized adsorbed surface moisture) naturally tend to move away from high energy regions. In arc welding, the highest energy region is the arc itself which can be represented as an inverted cup of high energy over the weld puddle. Any vaporized impurities on or in the metal may therefore be driven into the puddle by the energy gradient; such behavior may also be augmented by the momentum of the arc gases as well as their own content of contaminants.

In contrast, for laser welding, the highest energy region is the weld puddle and deep-penetration cavity provided that formation of a high energy plasma above the workpiece surface is successfully prevented. Under these conditions the energy gradient is directed away from the puddle and penetration cavity, allowing vaporized impurities to escape from the weld puddle. In addition, shielding and plasma control gases for laser welding are applied parallel to the workpiece surface in order to help aspirate impurities from the penetration cavity. This behavior may be further enhanced due to the high absorptivity of nonmetallic inclusions (compared to metals) for carbon dioxide laser radiation. This characteristic may lead to preferential vaporization of nonmetallic inclusions during welding.

REFERENCES

(1) E. M. Breinan and C. M. Banas: WRC Bulletin #201, Welding Research Council, New York, New York, Dec. 1974.

(2) A. Enis and R. T. Telford: Welding Research Supplement to the Welding Journal, June 1968, pp 271-S - 278-S.

(3) S. T. Rolfe, R. P. Haak, and J. H. Gross: Welding Research Supplement to the Welding Journal, Jan. 1965, pp 40-S - 48-S.

(4) W. G. Burwell: Review of CW High-Power Laser Technology, United Aircraft Research Laboratories Report UAR-M132, Oct. 1973.

(5) "Standard Methods for Notched Bar Impact Testing of Metallic Materials" ASTM Designation E23-66, ASTM Standards, Part 31, May 1968, p 285.

(6) D. C. Hill and D. E. Passoja: Welding Research Supplement to the Welding Journal, Nov. 1974, p 481-S - 485-S.

INDEX

Part A: Electron Beam Welding

Part B: Laser Welding

T

W